JOURNAL OF CHROMATOGRAPHY LIBRARY — volume 31

gradient elution in column liquid chromatography

theory and practice

JOURNAL OF CHROMATOGRAPHY LIBRARY - volume 31

gradient elution in column liquid chromatography

theory and practice

P. Jandera and J. Churáček

Department of Analytical Chemistry, University of Chemical Technology, Pardubice, Czechoslovakia

ELSEVIER
Amsterdam — Oxford — New York — Tokyo 1985

ELSEVIER SCIENCE PUBLISHERS B.V.
Molenwerf 1
P.O. Box 211, 1000 AE Amsterdam, The Netherlands

Distributors for the United States and Canada:

ELSEVIER SCIENCE PUBLISHING COMPANY INC.
52, Vanderbilt Avenue
New York, NY 10017

Library of Congress Cataloging in Publication Data

Jandera, P. (Pavel), 1944-
Gradient elution in column liquid chromatography.

(Journal of chromatography library ; v. 31)
Bibliography: p.
Includes index.
1. Liquid chromatography. I. Churáček, Jaroslav.
II. Title. III. Series.
QD79.C454J36 1985 543'.0894 85-1498
ISBN 0-444-42124-6

ISBN 0-444-42124-6 (Vol. 31)
ISBN 0-444-41616-1 (Series)

Printed in The Netherlands

CONTENTS

PART III

INSTRUMENTAL AND OTHER PRACTICAL ASPECTS OF CHROMATOGRAPHY WITH PROGRAMMED COMPOSITION OF THE MOBILE PHASE

PART IV

APPLICATIONS OF GRADIENT ELUTION IN LIQUID CHROMATOGRAPHY

JOURNAL OF CHROMATOGRAPHY LIBRARY

A Series of Books Devoted to Chromatographic and Electrophoretic Techniques and their Applications

Although complementary to the *Journal of Chromatography*, each volume in the Library Series is an important and independent contribution in the field of chromatography and electrophoresis. The Library contains no material reprinted from the journal itself.

Other volumes in this series

Volume 1	**Chromatography of Antibiotics** (*see also* Volume 26) by G.H. Wagman and M.J. Weinstein
Volume 2	**Extraction Chromatography** edited by T. Braun and G. Ghersini
Volume 3	**Liquid Column Chromatography**. A Survey of Modern Techniques and Applications edited by Z. Deyl, K. Macek and J. Janák
Volume 4	**Detectors in Gas Chromatography** by J. Ševčík
Volume 5	**Instrumental Liquid Chromatography**. A Practical Manual on High-Performance Liquid Chromatographic Methods (*see also* Volume 27) by N.A. Parris
Volume 6	**Isotachophoresis**. Theory, Instrumentation and Applications by F.M. Everaerts, J.L. Beckers and Th.P.E.M. Verheggen
Volume 7	**Chemical Derivatization in Liquid Chromatography** by J.F. Lawrence and R.W. Frei
Volume 8	**Chromatography of Steroids** by E. Heftmann
Volume 9	**HPTLC — High Performance Thin-Layer Chromatography** edited by A. Zlatkis and R.E. Kaiser
Volume 10	**Gas Chromatography of Polymers** by V.G. Berezkin, V.R. Alishoyev and I.B. Nemirovskaya
Volume 11	**Liquid Chromatography Detectors** by R.P.W. Scott
Volume 12	**Affinity Chromatography** by J. Turková
Volume 13	**Instrumentation for High-Performance Liquid Chromatography** edited by J.F.K. Huber
Volume 14	**Radiochromatography**. The Chromatography and Electrophoresis of Radiolabelled Compounds by T.R. Roberts
Volume 15	**Antibiotics**. Isolation, Separation and Purification edited by M.J. Weinstein and G.H. Wagman
Volume 16	**Porous Silica**. Its Properties and Use as Support in Column Liquid Chromatography by K.K. Unger
Volume 17	**75 Years of Chromatography — A Historical Dialogue** edited by L.S. Ettre and A. Zlatkis

Volume 18A **Electrophoresis.** A Survey of Techniques and Applications.
Part A: Techniques
edited by Z. Deyl

Volume 18B **Electrophoresis.** A Survey of Techniques and Applications.
Part B: Applications
edited by Z. Deyl

Volume 19 **Chemical Derivatization in Gas Chromatography**
by J. Drozd

Volume 20 **Electron Capture.** Theory and Practice in Chromatography
edited by A. Zlatkis and C.F. Poole

Volume 21 **Environmental Problem Solving using Gas and Liquid Chromatography**
by R.L. Grob and M.A. Kaiser

Volume 22A **Chromatography.** Fundamentals and Applications of Chromatographic and Electrophoretic Methods. **Part A: Fundamentals**
edited by E. Heftmann

Volume 22B **Chromatography.** Fundamentals and Applications of Chromatographic and Electrophoretic Methods. **Part B: Applications**
edited by E. Heftmann

Volume 23A **Chromatography of Alkaloids. Part A: Thin-Layer Chromatography**
by A. Baerheim Svendsen and R. Verpoorte

Volume 23B **Chromatography of Alkaloids. Part B: Gas—Liquid Chromatography and High-Performance Liquid Chromatography**
by R. Verpoorte and A. Baerheim Svendsen

Volume 24 **Chemical Methods in Gas Chromatography**
by V.G. Berezkin

Volume 25 **Modern Liquid Chromatography of Macromolecules**
by B.G. Belenkii and L.Z. Vilenchik

Volume 26 **Chromatography of Antibiotics**
Second, Completely Revised Edition
by G.H. Wagman and M.J. Weinstein

Volume 27 **Instrumental Liquid Chromatography.** A Practical Manual on High-Performance Liquid Chromatographic Methods
Second, Completely Revised Edition
by N.A. Parris

Volume 28 **Microcolumn High-Performance Liquid Chromatography**
by P. Kucera

Volume 29 **Quantitative Column Liquid Chromatography.** A Survey of Chemometric Methods
by S.T. Balke

Volume 30 **Microcolumn Separations.** Columns, Instrumentation and Ancillary Techniques
edited by M.V. Novotny and D. Ishii

Volume 31 **Gradient Elution in Column Liquid Chromatography.** Theory and Practice
by P. Jandera and J. Churáček

Volume 32 **The Science of Chromatography.** Lectures Presented at the A.J.P. Martin Honorary Symposium, Urbino, May 27—31, 1985
edited by F. Bruner

PREFACE

In the last 15 years, high-performance liquid chromatography (HPLC) has become one of the most widely used analytical methods with a rapid increase in the number of applications to the analysis of various compounds. This increase was mainly due to the advent of sophisticated instrumentation and to the development of new working techniques. It is not surprising that many books on HPLC have been (and are still being) published. In addition to works concerning the general aspects of HPLC, several books appeared treating special fields of application, such as instrumentation or special detection techniques. It is understandable that the main concern of these books was instrumentation and the development of efficient columns, which made possible the success of HPLC.

We wanted to write a book on gradient elution in HPLC. Gradient elution means programming the composition of the mobile phase during the separation, in contrast to isocratic elution, in which the composition of the mobile phase is held constant. This is the reason why the primary concern of this book is the influence of the mobile phase and its composition on separation. The correct selection of the mobile phase for a given separation problem gives to the practising chromatographer a powerful tool for achieving successful separations of a wide variety of complex samples, and this is further increased by the possibility of changing the composition of the mobile phase during the elution. In virtually all fields of application gradient elution has been applied with the aim of improving resolution and shortening the time of separation. The utility of this technique increases with increasing complexity of samples that have to be analysed using HPLC.

Liteanu and Gocan[1] wrote a book on gradient liquid chromatography 10 years ago. However, it was necessarily concerned mainly with classical column (and thin-layer) chromatography and could not take full account either of specific problems of gradient elution in HPLC or of recent advances in the understanding of the role of the mobile phase in the chromatographic process. Therefore, we felt that there is a need for another book on this topic.

The theory of gradient elution chromatography has been developed to such an extent that relatively precise calculations of the retention data and the establishment of optimum conditions with respect to the separations required are possible in many useful gradient elution systems. Unfortunately, the benefits of the theory for the better identification of compounds and the faster selection of optimum separation conditions have not yet become widely known and accepted among practising chromatographers, in spite of the increasing number of papers devoted to the theory of gradient elution published in recent years. We wanted to write a book that might be helpful in this respect.

We have divided the book into four parts. To obtain an idea of the theoretical basis of gradient elution separations, it is necessary to become acquainted with the fundamentals of the phenomena that control the influence of the mobile phase and its composition on retention behaviour in various liquid chromatographic systems under isocratic conditions. Therefore, it was felt necessary to devote one part to this topic. In this first part, several models are discussed that allow descriptions of the interrelationships between the composition of the mobile phase and retention, as characterized by capacity factors, in chromatography on polar adsorbents, on polar bonded phases and on ion exchangers, in reversed-phase systems and in ion-pair chromatography. The quantitative description of these relationships is essential for developing any theory of gradient elution liquid chromatography.

Part II is concerned with the theory of gradient elution chromatography. We have attempted to bring together different theories developed by several groups of workers in the field rather than to give a simple survey of various approaches. The prediction of retention behaviour (retention volumes and times, bandwidths, resolution, peak capacity, etc.) by calculation is treated in detail and possibilities are outlined for the rational optimization of gradient elution conditions.

In Part III, the instrumentation for the formation of mobile phase gradients is described and possible sources of instrumental errors are discussed. Most attention is paid to commercially available gradient elution systems for HPLC, and the "classical" laboratory-made gradient formers are only briefly mentioned, because detailed descriptions of the various types and the methods for the calculation of gradient profiles formed in these devices have been presented by other workers[1-3]. Other instrumental aspects are discussed as far as they concern specific problems connected with the practical use of the gradient elution technique.

Part IV gives examples of the application of gradient elution techniques to separations of different classes of compounds in various chromatographic systems. The number of published applications clearly shows the increasing use of gradient elution in chromatography on non-polar chemically bonded phases (which is by far the most frequently used method in contemporary HPLC). However, these examples can by no means cover all of the published applications of gradient elution and are intended rather as illustrations of the potential power of this technique. Therefore, the examples of applications are presented mainly with respect to the type of mobile phase gradient used.

Hydrophobic or hydrophilic gels are occasionally used in the gradient elution technique for certain separation problems. However, these applications cannot be considered as the application of gel chromatography in the sense of steric exclusion chromatography, because the compounds are eluted later than would correspond to the column dead volume and therefore the sample compounds interact with the column packing materials. In real steric exclusion chromatography, these interactions are absent and the mobile phase cannot influence the separation other than by improving or impairing the solubility of sample compounds (or, possibly, by solvating them). Therefore, gel chromatography is not considered in this book and the applications of gels are classified among other reversed-phase or normal-phase chromatographic systems, according to the nature of the predominating interactions.

Because we did not intend just to write another book on column liquid chromatography, the reader is referred to some of the outstanding books dealing with HPLC in general for information on the general aspects of HPLC. Only those topics are treated which concern the theory, practice or applications of the gradient elution technique and the book does not cover areas such as column efficiency, general instrumentation and techniques.

We thank the many companies who supplied technical information. Dipl. Ing. Zdeněk Dušek is thanked for his assistance with the literature search, dipl. Ing. Blanka Wanková for careful drawing of the diagrams and Mrs. Marie Valášková for technical assistance during the preparation of the manuscript.

Pardubice, 1984

Pavel Jandera
Jaroslav Churáček

REFERENCES

1. C. Liteanu and S. Gocan, *Gradient Liquid Chromatography*, Wiley, New York, 1974.
2. L.R. Snyder, *Chromatogr. Rev.*, 7 (1965) 1.
3. O. Mikeš and R. Vespalec, in Z. Deyl, K. Macek and J. Janák (Editors), *Liquid Column Chromatography*, Elsevier, Amsterdam, 1975, Ch. 10.

PART I

ISOCRATIC ELUTION CHROMATOGRAPHY

Chapter 1

THE MOBILE PHASE AND CHROMATOGRAPHIC BEHAVIOUR UNDER ISOCRATIC CONDITIONS

The quantitative theory of gradient elution chromatography is based on the interrelationships between the composition of the mobile phase and retention behaviour in isocratic elution chromatography, where the eluent composition is held constant throughout the separation[1,2]. The theory of gradient elution chromatography is much more complicated than that of isocratic elution, but a number of phenomena in elution chromatography with programmed composition of the mobile phase can be readily understood from analogies with isocratic elution chromatography[3,4]. Therefore, it is necessary to become familiar at least with the basic principles of the influence of the mobile phase composition on the chromatographic behaviour of sample compounds under isocratic conditions before dealing with the theory of gradient elution chromatography. A detailed discussion of the theory of isocratic elution chromatography is beyond the scope of this book, but it is useful to consider briefly several basic quantities that are important for the description of the chromatographic process.

Separation in elution liquid column chromatography can be characterized by several important parameters, which can be evaluated directly from the chromatogram: retention volume, V_R; corrected (net) retention volume, V'_R; bandwidth, w (in volume units); and resolution of compounds 1 and 2, R_s. The retention in liquid column chromatography can be characterized either by retention time, t_R, or by retention volume, V_R, which are the time (or the volume of eluate from the column) from the injection of the sample on to the column to the elution of the maximum of the elution curve (peak) of a chromatographed compound (Fig. 1.1). The two quantities are interrelated by the equation

$$V_R = t_R F_m \tag{1.1a}$$

$$V'_R = t'_R F_m \tag{1.1b}$$

where F_m is the volume flow-rate of the mobile phase (volume of eluate per unit time) and t'_R is the corrected (net) retention time.

From the retention volume, V_R, (or retention time, t_R) the capacity factor of a sample compound, k', can be calculated, if the volume of the mobile phase in the column (column void volume), V_m, is known:

References on p. 52.

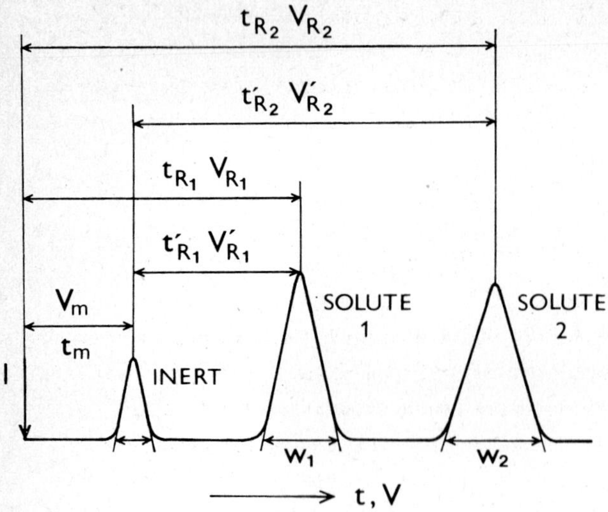

Fig. 1.1. Interpretation of a chromatogram in liquid column chromatography. Retention times (t_R, t'_R), retention volumes (V_R, V'_R), column dead time (t_m), dead volume (V_m) and bandwidths (w) in liquid column chromatography. I, Sample injection; t, time elapsed; V, volume of eluate.

$$k' = \frac{V_R - V_m}{V_m} = \frac{V'_R}{V_m} = \frac{t_R - t_m}{t_m} = \frac{t'_R}{t_m} \qquad (1.2)$$

The capacity factor is a dimensionless quantity that depends neither on the dimensions of column nor on the flow-rate of the mobile phase. Essentially, it is a quantity that gives the ratio of the total amount of sample compound in the stationary phase to its amount in the mobile phase. Thus, k' is proportional to the distribution coefficient, K_D, which gives the ratio of the concentrations of the sample compound in the two phases:

$$k' = K_D \cdot \frac{V_s}{V_m} = K_D\phi \qquad (1.3)$$

where ϕ is the phase ratio in the column, i.e., the ratio of the volume of the stationary phase (V_s) to that of the mobile phase (V_m). Of course, the mass of the stationary phase can be used instead of V_s, with a corresponding change in definition of K_D.

Chromatography is used to separate sample compounds, and consequently a quantitative description of the separation is of major interest for both the practice and the theory of chromatography. Resolution, R_s, is the quantity most frequently and obviously most appropriately used to characterize the separation of two sample compounds 1 and 2. This is the distance between the two elution maxima (the difference in the elution volumes or elution times) divided by the average bandwidth of the two compounds (in volume or time units). For the sake of simplicity, the bandwidth of the later eluted compound (2) is often used instead of the average bandwidth:

$$R_S = \frac{2(V_{R_2} - V_{R_1})}{w_1 + w_2} = \frac{2(V'_{R_2} - V'_{R_1})}{w_1 + w_2} \approx \frac{V'_{R_2} - V'_{R_1}}{w_2} \tag{1.4}$$

From eqn. 1.4 it follows that resolution increases with increasing difference in retention volumes (difference in the retention of the two sample compounds) and decreases with increasing bandwidth, which should be therefore kept as narrow as possible. The sample is usually injected on the column as a very narrow band (volumes of a few microlitres), but broadening of this band inevitably occurs during migration along the column. The width of a band eluted from the column is proportional to the retention volume of the sample compound and it depends on the efficiency of the column under given conditions, which is conveniently characterized by the number of theoretical plates, n:

$$w = \frac{4\sigma}{L} \cdot V_m(1 + k') = \frac{4V_m}{\sqrt{n}} \cdot (1 + k') \tag{1.5}$$

where σ is the standard deviation of the bandwidth at the end of the column in length units.

The total number of theoretical plates in the column is proportional to the column length, L:

$$n = \frac{L}{H} = 16\left(\frac{V_R}{w}\right)^2 \tag{1.6}$$

where H is the height equivalent to a theoretical plate, which depends on a number of parameters (temperature, mobile phase viscosity and linear velocity, sample compound diffusion coefficient, particle diameter and thickness of stationary phase layer of the packing material, etc.[5]. The linear velocity of the mobile phase, u, is a very important parameter, which influences not only bandwidth and resolution, but also the speed of analysis. It is related to the volume flow-rate of the mobile phase, F_m, by the equation

$$u = \frac{L}{t_m} = \frac{F_m}{\pi r_i^2 \varepsilon_T} \tag{1.7}$$

where r_i is the inner radius of the column and ε_T is the total porosity of the column packing:

$$\varepsilon_T = \frac{V_m}{V_c} \tag{1.8}$$

where V_c is the total inner volume of the void column.

The relationship between the height equivalent to a theoretical plate and the linear velocity of the mobile phase is described by the well known Van Deemter equation:

$$H = A + \frac{B}{u} + Cu \tag{1.9}$$

where the terms A, B and C are used to characterize the contributions of eddy diffussion (A), molecular diffusion (B) and mass-transfer resistance (C) to H. In liquid column chromatography, the Knox equation is very often used instead of the Van Deemter equation:

$$h = \frac{B}{\gamma} + A\gamma^{0.33} + C\gamma \tag{1.10}$$

This semi-empirical equation relates the reduced quantities, reduced height equivalent to a theoretical plate:

$$h = \frac{H}{d_p} \tag{1.11}$$

and reduced linear velocity of the mobile phase:

$$\gamma = \frac{ud_p}{D_m} \tag{1.12}$$

where d_p is particle diameter of the column packing material, D_m is the diffusion coefficient of the solute in the mobile phase and A, B and C are as in the Van Deemter equation.

The theory of efficiency is very important for a good understanding of the chromatographic process, but it would be beyond the scope of this book to go into more detail, and it has been discussed thoroughly in other books[5,6].

Using eqns. 1.2 and 1.6, we can rearrange eqn. 1.4 for resolution into the following form, which makes it possible to differenciate between three factors I-III influencing the resolution:

$$R_s = \frac{k'_2 - k'_1}{\frac{2(k'_1 + 1)}{\sqrt{n_1}} + \frac{2(k'_2 + 1)}{\sqrt{n_2}}} \approx \underbrace{\frac{1}{4}\sqrt{n_2}}_{\text{I}} \cdot \underbrace{\frac{\alpha - 1}{\alpha}}_{\text{II}} \cdot \underbrace{\frac{k'_2}{1 + k'_2}}_{\text{III}} \quad (1.13)$$

where the subscripts 1 and 2 relate to the corresponding compounds. Here, α is the separation factor for elution bands of compounds 1 and 2:

$$\alpha = \frac{k'_2}{k'_1} \quad (1.14)$$

The terms I-III in eqn. 1.13 are essentially independent and characterize separation efficiency (I), separation selectivity (II) and capacity (III). This means that each of these three contributions to resolution may be adjusted separately to control the resolution[5].

Column dimensions, particle diameter of the column packing material, linear velocity of the mobile phase and extra-column contributions to the band broadening influence the efficiency, but not the other two terms. In contrast, the nature and composition of the mobile phase and of the stationary phase (sorbent) can influence all three terms, but not to identical extents. The efficiency (height equivalent to a theoretical plate) depends to some extent on the capacity factors of sample compounds and (more significantly) on their diffusion coefficients in the mobile phase, which are a function of the viscosity and thus of the composition of the mobile phase[5,6]. However, the influence of the composition of the mobile phase on the height equivalent to a theoretical plate is difficult to describe because of the complexity of the effects involved and it seems to be hardly significant in most practical systems. It is reasonable to suppose that the efficiency remains approximately constant with changing composition of the mobile phase in a given chromatographic system (such as in mixtures of organic solvents of different polarities in chromatography on polar adsorbents, in aqueous-organic mobile phases used in chromatography on chemically bonded non-polar stationary phases, or in aqueous or aqueous - organic solutions of salts or buffers used in ion-exchange chromatography)[2]. The nature of the compounds chromatographed also usually does not influence the efficiency term much.

The selectivity is understood here as the ability of the stationary phase - mobile phase system to retain sample compounds (solutes) to different extents. It depends on the nature of the stationary phase (column packing material) and on the nature of the mobile phase. The nature of the stronger eluting component in the mobile phase generally influences the selectivity more significantly than

References on p. 52.

the nature of the less efficient eluting component and the concentration of the two components in solvent binary systems[3,7] (see discussion in Section 2.2).

The choice of the components of the mobile phase and of their concentration has a very significant influence on the capacity term of the resolution (III). The separation of sample compounds generally improves with increasing capacity factors, but at the cost of increased analysis time. A maximum resolution per time unit is obtained when working at $k' = 2$ (ref. 5).

The influence of the composition of the mobile phase on capacity factors is controlled by the separation mechanism in a given system. The separation mechanism can be complex and often two or more different effects are operating simultaneously. A thermodynamically rigorous treatment of such systems is not always possible or it would lead to very complex mathematical description of the operating phenomena that would have only very limited practical use for relating the capacity factors of solutes to the composition of the mobile phase. For practically useful chromatographic purposes it is not necessary to know in detail the chromatographic behaviour over the whole concentration range of the mobile phase, but it is sufficient to develop as simple a description of this relationship as possible that fits reasonably well the composition limits of the mobile phase corresponding to the useful range of the capacity factors, i.e., from $k' \approx 0.4$ to $k' \simeq 10$-20. (Solutes with lower k' are only very slightly retained and solutes with larger k' almost do not move along the column; thus a great majority of solutes in practical chromatographic systems migrate with k' within the above limits, under either isocratic or gradient elution conditions. For gradient elution chromatography, the range from $k' = 1$ to 10 is usually sufficient.) For this purpose, the separation models must be simplified to a certain extent if the theoretical treatment is to be expected to yield useful results[8].

It is impossible to give a detailed treatment of chromatographic mechanisms in different liquid chromatographic systems here. However, it is necessary to survey briefly some simplified models of liquid chromatography on polar adsorbents, chemically bonded non-polar phases, ion exchangers and several other systems involving selective interactions in the stationary or mobile phase.

1.1. CHROMATOGRAPHY ON POLAR ADSORBENTS

Three models have been developed for adsorption chromatography on polar adsorbents; a number of published works in this field either further elaborate upon or slightly modify these models. The first model was proposed by Snyder in the early 1960s[9-12]. Flat adsorption in a monomolecular layer on a homogeneous adsorbent surface is assumed in this model, so that the adsorption energies of solute (and solvent) molecules at different positions on the surface are constant. It is further assumed that the interactions between the adsorbent and the molecules of the solute and between the adsorbent and the molecules of the solvent largely predominate over the interactions in the mobile phase, which can be neglected (the interactions between the solute and the polar solvent, if significant, are assumed to be cancelled out by similar interactions in the adsorbed phase). The asorption is understood as a competition phenomenon between the molecules of the solute (X) and of the solvent (S), the stoichiometry of which can be described by the equation

$$X_m + mS_a \rightleftarrows X_a + mS_m \tag{1.15}$$

The subscripts m and a are used for the molecules in the mobile and adsorbed phases, respectively, and

$$m = \frac{A_s}{n_b} \tag{1.16}$$

gives the ratio of the molecular areas on the adsorbent surface occupied by one molecule of the solute (A_s) to that occupied by one molecule of the solvent (n_b).

Taking into account the above assumptions, Snyder derived an equation for the capacity factors, k'_{ab}, in binary mobile phases composed of solvents a and b as a function of the solvent strength, ε:

$$\log k'_{ab} = \log k'_a + \alpha' A_s(\varepsilon_a - \varepsilon_{ab}) \tag{1.17}$$

where k'_a is the capacity factor of the solute in a pure, less polar solvent a, α' is an activity parameter of the adsorbent and ε_{ab} is the solvent strength of the binary mobile phase, the dependence of which on the concentration (molar fraction), x_b, of the more polar solvent b in the mobile phase is controlled by the following equation[9]:

$$\varepsilon_{ab} = \varepsilon_a + \frac{\log \left[x_b\ 10^{\alpha' n_b(\varepsilon_b - \varepsilon_a)} + 1 - x_b\right]}{\alpha' n_b} \tag{1.18}$$

and ε_a and ε_b are the solvent strengths of the pure solvents a and b, respectively.

This model can also be applied in systems involving adsorbents with a heterogeneous surface, where the molecules of the solute and of solvent are localized preferentially on strong adsorption sites[9,13]. The so-called secondary solvent effects may result in preferential localization on the adsorbent, whereas the

References on p. 52 .

solute - solvent interactions in the mobile phase appear to be of minor importance. The localization effects can be considered in eqns. 1.17 and 1.18 by using corrected (higher) values for n_b and A_s or by introducing empirical correction terms[9]. These correction terms were studied in detail and were related to the selectivity of the solvents, which made it possible to predict quantitatively the selectivity effects of the mobile phase in chromatography on polar adsorbents[144-148].

The adsorption model developed by Soczewiński[14] is based on the assumption of localized adsorption sites on the heterogeneous surface of the adsorbent which interact with functional groups of the adsorbed molecules to form some kind of complexes, such as proton-acceptor complexes with silanol groups of silica. Adsorption in the monomolecular layer and cancellation of solute-solvent interactions in the mobile and stationary phases are also assumed in this model[13,14].

Using these postulates, Soczewiński and Golkiewicz[15,16] suggested the equation

$$R_M \equiv \log k' = \text{constant} - m \log c \tag{1.19}$$

where c is the concentration of the more polar solvent in the mobile phase. The constant m is correlated with the number of equivalent, strongly adsorbing groups in the solute molecules.

The other theory of adsorption on polar adsorbents, which differs essentially from the models of Snyder and Soczewiński, is due to Scott and Kucera[17-21], who defined the distribution coefficient of a solute between the stationary and mobile phases as the ratio of the total forces acting on the solute in the two phases. The total forces were set equal to the product of the probability of interactions and the forces between the solute and the respective phase. In this model both polar and dispersive forces in the mobile phase were assumed, and dispersive forces on the adsorbent were neglected. The activity of the adsorbent (or the number of interacting sites on the surface of the adsorbent) was assumed not to depend on the composition of the mobile phase when the content of the polar solvent is kept above about 3%. In this region, the change in retention should be due primarily to changing interactions in the mobile phase which increase with increasing content of the polar solvent in the mobile phase (c) and thus diminish the retention of the solute[19,21]. With these assumptions, Scott and Kucera derived the relationship

$$\frac{1}{V'_R} = A + Bc \tag{1.20}$$

where A and B are experimental constants.

The calculations of the capacity factors of solutes in practical chromatographic systems using eqns. 1.17 and 1.18 derived by Snyder are difficult. With a few exceptions[9], the tabulated data for the parameters of these equations are insufficient and these parameters are difficult to determine experimentally with good precision to allow for accurate predictions of the retention volume of an arbitrary compound in a given system under isocratic or gradient elution conditions. For such a prediction, an adequate and simple function is needed to describe the dependence of the capacity factors of solutes on the composition of the mobile phase and on the bulk properties of the system (adsorbent, mixed mobile phase and solute). However, the parameters of such a function should be determined empirically in a given system.

Jandera and co-workers tried to simplify Snyder's equations in the following way[7,8]. Elimination of ε_{ab} from eqns. 1.17 and 1.18 gives

$$\log k'_{ab} = \log k' = (a + bc)^{-m} \tag{1.21}$$

where

$$a = k_a'^{-(n_b/A_s)} \tag{1.22a}$$

$$b = k_a'^{-(n_b/A_s)}(10^{\alpha' n_b(\varepsilon_b - \varepsilon_a)} - 1) \tag{1.22b}$$

$$m = \frac{A_s}{n_b} \tag{1.22c}$$

are experimental values, which do not depend on the concentration of the polar solvent b in the mobile phase (c) and can be determined from the experimental $k' = f(c)$ values using non-linear regression analysis.

In a large number of practical separations on polar adsorbents, a non-polar aliphatic hydrocarbon is used as the weak solvent a. Most compounds are retained strongly in pure hydrocarbon mobile phases and consequently the term a in eqn. 1.21 is very low and can be neglected to first approximation. Then eqn. 1.21 can be simplified as follows[8]:

$$k' = (bc)^{-m} \equiv k_0' c^{-m} \tag{1.23}$$

where $k_0' = b^{-m}$ refers to the capacity factor in pure polar solvent b, $k_0' = k_b'$.

It can easily be shown that eqn. 1.23 is virtually identical with eqn. 1.19 introduced by Soczewiński and Golkiewicz[15,16]. The validity of these equations

References on p. 52.

has been verified by Jandera and co-workers[7,22,23], by Soczewiński and co-workers[15,16,24-28] and other workers[29-33] in a number of practical systems in both column and thin-layer chromatography on polar adsorbents (silica, alumina).

Eqn. 1.20 due to Scott and Kucera can be rewritten as

$$k' = (a + bc)^{-1} \tag{1.24}$$

which is essentially identical with eqn. 1.21 based on Snyder's model of adsorption for $n = 1$, i.e., with the assumption that one molecule of the polar solvent b replaces exactly one molecule of the solute on the adsorbent surface[7]. This assumption, however, was implied in the way Scott and Kucera defined the distribution coefficient in their model.

From comparison of eqns. 1.21, 1.23 and 1.24, if follows that it is difficult to distinguish between different adsorption models on the basis of liquid chromatographic retention data only. A more detailed discussion of the adsorption models may be found in Snyder's book[9] and elsewhere[7,8,13-16]. A critical comparison of the models of Snyder and of Soczewiński and Golkiewicz with the model of Scott and Kucera was given by Snyder[13], Soczewiński[25] and, very thoroughly, Snyder and Poppe[34]. For experimental evidence in favour of one of the two models, an independent method for the determination of solute activity in the mobile phase would be necessary, such as headspace gas chromatography[35].

For the purpose of the calculation of retention data in chromatography under isocratic or gradient elution conditions, it is not essential to discriminate between different adsorption models, because eqn. 1.21 derived from Snyder's model formally describes the other models discussed, which represent certain limiting cases[4,7]. Eqn. 1.23 describes the systems in which the solutes are strongly retained in the pure, weaker solvent a and eqn. 1.24 is suitable for the systems in which the adsorption stoichiometry between the solute and the polar solvent b is close to unity[7].

At present, more experimental systems have been described in which the simplified eqn. 1.23 is obeyed[7,15,16,22-33] than systems described by eqn. 1.24. Scott and Kucera verified the validity of eqn. 1.24 in certain systems using polar adsorbents and found significant deviations in mobile phases containing less than 3-5% of the polar solvent in the mobile phase[18-20], which agreed with our results[7].

The validity of eqns. 1.23 and 1.24 may be illustrated by Figs. 1.2 and 1.3.

Deviations from eqn. 1.23, which have been observed for certain solutes in several practical systems, can be attributed to the omission of activity coefficients in the derivation of the theoretical models discussed above, to the effects of selective interactions between certain solutes and solvents and to

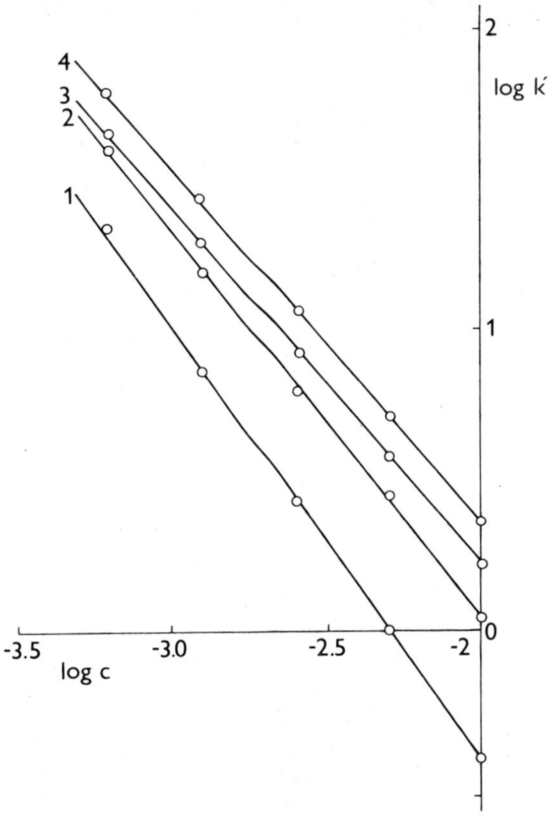

Fig. 1.2. Logarithmic plots of capacity factors, k', of lumisterol (1), tachysterol (2), calciferol (3) and ergosterol (4) versus concentration of n-propanol, c [% (v/v) · 10^{-2}], in n-heptane as the mobile phase in chromatography on LiChrosorb ALOX T, 30 μm, 596 x 2.3 mm I.D. column; V_m = 2.30 ml. (Reproduced from ref. 7.)

the combination of adsorption phenomena and distribution between two liquid phases - the mobile phase and the polar solvent adsorbed preferentially on the surface of the adsorbent[35-37] (see, for instance, Fig. 1.4).

It seems reasonable, however, to expect that a great majority of adsorption chromatographic systems can be described adequately by eqn. 1.23 or 1.21, at least in the range of practical k' values, and that these equations can be used for calculations of retention volumes under isocratic and gradient elution conditions to a first approximation[2,4,22].

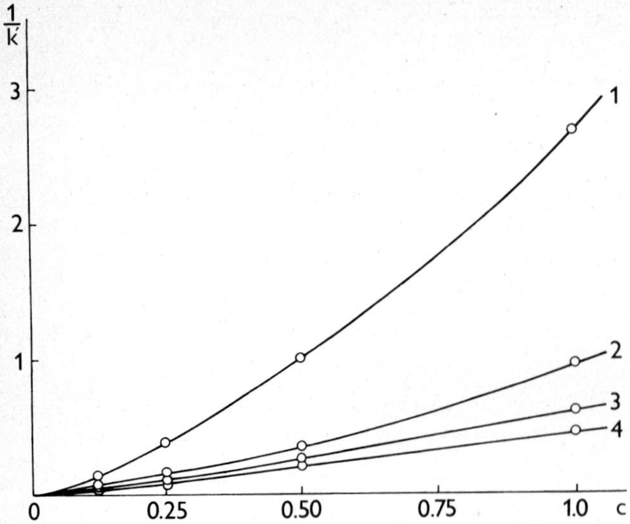

Fig. 1.3. Reciprocal plots of capacity factors, k', of lumisterol (1), tachysterol (2), calciferol (3) and ergosterol (4) versus concentration of n-propanol, c[% (v/v) · 10^{-2}], in n-heptane as the mobile phase in chromatography on LiChrosorb ALOX T, 30 μm, 596 x 2.3 mm I.D. column, V_m = 2.30 ml. (Reproduced from ref. 7.)

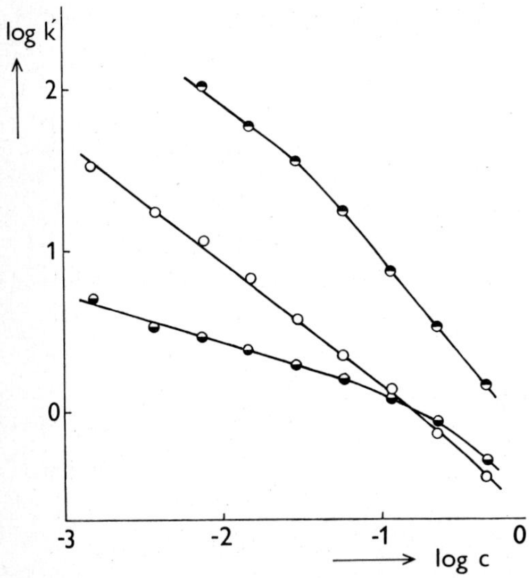

Fig. 1.4. Logarithmic plots of capacity factors, k', of nitrobenzene (1), benzyl acetate (2) and 1-phenylethanol (3) versus concentration of ethyl acetate, c, in n-heptane as the mobile phase in chromatography on Partisil 10. (Adapted from ref. 35.)

1.2. POLAR AND NON-POLAR BONDED PHASE CHROMATOGRAPHY

A continuous change in the composition of the mobile phase during gradient elution would lead to stripping of the stationary phase from the column in liquid-liquid chromatography. Therefore, columns with anchored liquid phases cannot be used for gradient elution liquid chromatography.

This problem has been solved by the introduction of chemically bonded stationary phases, which have become the most useful materials for liquid column chromatography during the last few years. Various bonded phases have been developed, differing in their functional groups, types of bonds with the surface of the support (an inorganic material, usually silica), and the nature (polymeric, monomeric or free-standing organic chains, the so-called brushes)[6,38,39]. The most popular column packing materials of this class are non-polar phases with aliphatic alkyl chains chemically bonded to the surface of silica, which can be used for separations of compound classes with very large differences in polarities (from hydrocarbons to sulphonic acids).

Further, the problems with deactivation of the adsorbent surface and with the so-called solvent-demixing effect (see Chapter 10.1) are largely eliminated with non-polar chemically bonded phases[3].

In spite of the frequent use of these packing materials, the theory of liquid chromatography in reversed-phase systems is not yet fully understood. The main difficulty consists in an appropriate description of the behaviour of chemically bonded hydrocarbon chains in contact with the mobile phase, which obviously controls the nature of the retention process. Various descriptions of the separation mechanism have been presented:

(1) Liquid-liquid partition is assumed to occur, either between the mobile phase and the stationary bonded phase which behaves in much the same way as a liquid alkane[40,41], or between the mobile phase and the mixed stationary liquid phase formed by preferential uptake of the less polar component (organic solvent) from the mobile phase[42-45] (thus, the stationary phase should have an intermediate composition and a polarity between that of a pure alkane and the pure organic solvent).

(2) Adsorption is assumed to determine the retention[46-49].

The solute molecules either adhere directly to the individual alkyl chains or are adsorbed at the interface between the bulk mobile phase and the region containing solvated alkyl chains.

In reality, it is very difficult to provide clear evidence for a partition or adsorption mechanism[50] and it is probable that the real process is neither one nor the other[51]. The bonded alkyl chains cannot be expected to behave as a bulk liquid, as they are only a monolayer thick and have fewer translational and

References on p. 52.

rotational degrees of freedom than the molecules in an unbonded liquid phase[52]. On the other hand, various experimental approaches have demonstrated the predominating role of the mobile phase interactions in the retention mechanism (the so-called solvophobic effect) and, consequently, the stationary phase has the role of a passive acceptor rather than of an active attractor of solute molecules and the retention is determined primarily by the competition between the interaction of the solute with the mobile phase molecules and the mutual interactions of the molecules of the mobile phase, which are stronger and repulse the solute molecules from the mobile phase and force them to enter the stationary phase[50-54]. The attractive forces between the solute molecules and the stationary phase can contribute to retention but, in most instances, this effect is far weaker than the solvophobic effect in the mobile phase.

The organic solvent is preferentially sorbed from the mobile phase to the hydrocarbonaceous non-polar surface of the bonded phase and probably causes some swelling of alkane chains. It is not certain if the total surface of the alkane chains is in direct contact with the sorbed solvent, but it seems that the contact region is proportional to the length of the bonded alkane chains. Thus, coming from the surface of silica, we can find a layer of less or more solvated alkane chains in contact with pure organic solvent, followed by further layers in which the concentration of organic solvent gradually diminishes until the composition of the mobile phase is reached. We have no clear idea about the thickness of this liquid interface layer, nor do we know how deeply the solute molecules penetrate into this layer or into the layer of swollen bonded alkane chains and whether any solvent molecules are displaced in this process. It is probable that the depth of penetration of solute molecules into the stationary phase is proportional to the length of the bonded alkane chains and to the size of the non-polar hydrocarbonaceous part of the solute molecule[51] which can be exposed to the stationary phase. It was found experimentally that the retention increases with increasing bonded alkyl chain length only until a critical limit is reached, which depends on the structure of solute and on the composition of the mobile phase[45,51]. This phenomenon has been interpreted as the evidence of partition mechanism[45] and, by other authors as the contrary evidence of adsorption behaviour[51].

However, the exact discrimination between an adsorption and a partition mechanism in reversed-phase chromatography, if possible at all, would have little value for the understanding of the influence of the mobile phase composition on retention.

Several workers have presented quantitative descriptions of reversed-phase systems via various models in which the retention is correlated to different thermodynamic properties of the systems. Thus, Karger et al.[55] attempted to predict retention on the basis of a topological index-molecular connectivity. Horváth et al.[52-54] elaborated a model based on the solvophobic theory, developed by Sinanoglu and Abdulnur[56,57]. Here, the driving force of retention is related to the energy required to form the cavity in the highly associated aqueous mobile phase necessary to accommodate a relatively non-polar solute molecule. This solvation energy is proportional to the area of this cavity, which is determined by the hydrocarbonaceous surface area of the solute and is decreased when the non-polar molecule complexes with (adheres to) the alkyl chains of the bonded phase[52-54]. The mathematical description of these models yields complex mathematical equations and requires a knowledge of a number of physico-chemical constants, which are often difficult to find in the literature or to determine experimentally. Hence this relatively sophisticated model cannot be used easily in practice, where a simpler model would be more useful, even if not as rigorous and precise as the model of Horváth et al.

Jandera and Churáček[8] used the Hildebrand solubility parameter theory[58] to express the relationship between the capacity factor of a solute, k, and the concentration, c, of the stronger eluting solvent in a binary mobile phase for liquid-liquid chromatography as follows[8]:

$$\log k' = \left(\log \frac{V_s}{V_m} + \frac{V_x}{2.31\, RT}\right)\left[(\delta_x - \delta_a)^2 - (\delta_s - \delta_x)^2\right] -$$

$$\frac{2V_x}{2.3\, RT}(\delta_x - \delta_a)(\delta_b - \delta_a)c + \frac{V_x}{2.3\, RT}(\delta_b - \delta_a)^2 c^2 =$$

$$= a - mc + pc^2 \tag{1.25}$$

or

$$k' = k'_0 \cdot 10^{(-mc+pc^2)} \tag{1.26}$$

where δ_x, δ_s, δ_a and δ_b are the solubility parameters of the sample component (δ_x), stationary (δ_s) and the weaker (δ_a) and stronger (δ_b) solvents used in the binary mobile phase, V_x is the molal volume of the sample solute, V_s and V_m are the total volumes of the stationary and mobile phases, respectively, in the column, R is the gas constant and T is absolute temperature. Ideal behaviour in dilute solutions is assumed and consequently the activity coefficients are neglected. The quadratic term can be neglected[8] in many instances without affecting significantly the precision in calculations of k'.

If a reversed-phase system with a chemically bonded stationary phase is understood as a liquid-liquid system, eqn. 1.26 should be applicable here also.

References on p. 52.

Schoenmakers and co-workers[59,60] applied the regular solution theory[58] to chromatography on alkane bonded stationary phases in aqueous - organic mobile phases and later derived an equation virtually identical with eqn. 1.25 or 1.26. They also showed that the consideration of possible changes in properties of the stationary phase with variations in the composition of the binary mobile phase (preferential adsorption of the organic solvent from the mobile phase) leads to a quadratic equation formally identical with eqn. 1.25, but with a changed meaning and values of the constants a, m and p.

A qualitative agreement between the influence of various parameters on capacity factors and on constants a, m and p was observed for a number of solutes in practically used reversed-phase separation systems[60,61], but the attempt by Schoenmakers et al. to calculate these constants from published solubility parameters was unsuccessful; the discrepancies between the calculated values and the experimental data were too large[60].

Jandera and co-workers[62,63] presented another simplified model for calculations of capacity factors in reversed-phase systems. Here, the role of the stationary phase is neglected to a first approximation, as it is assumed that dispersive interactions only are effective in this phase, which are approximately cancelled out by analogous interactions in the mobile phase. Consequently, the difference in the mutual interactions between the molecules of the mobile phase and in the interactions between the molecules of the solute and those of the mobile phase are the driving force of retention (sovophobic effect). The retention is proportional to the size of the molecule and to the interaction forces per unit volume, which are characterized by so-called "interaction indices" for individual solutes and solvents. The empirical scale of interaction indices is based on a set of arbitrarily chosen standard compounds. As the stationary phase is acting as a passive acceptor of solute molecules only, no exact assumption concerning the partition or adsorption mechanism is necessary.

The mathematical solution of the interaction indices model[62,63] leads to an equation for capacity factors formally identical with eqn. 1.25, but with different meanings of the constants a, m and p:

$$a = \log\left(\frac{V_{\mathrm{s}}}{V_{\mathrm{m}}}\right) + \frac{V_{\mathrm{x}} I_{\mathrm{H_2O}} c_{\mathrm{m}}}{2.31\,RT} \cdot (I_{\mathrm{H_2O}} c_{\mathrm{m}} - I_{\mathrm{x}} c_{\mathrm{x}}) \tag{1.27a}$$

$$m = \frac{V_{\mathrm{x}} c_{\mathrm{m}}}{2.31\,RT} \cdot (2 I_{\mathrm{H_2O}} c_{\mathrm{m}} - I_{\mathrm{x}} c_{\mathrm{x}})(I_{\mathrm{H_2O}} - I_{\mathrm{org}}) \tag{1.27b}$$

$$p = \frac{V_{\mathrm{x}} c_{\mathrm{m}}^2}{2.31\,RT} \cdot (I_{\mathrm{H_2O}} - I_{\mathrm{org}})^2 \tag{1.27c}$$

where I_x, I_{H_2O} and I_{org} are interaction indices of the solute (I_x), water (I_{H_2O}) and the organic solvent used in the binary mobile phase (I_{org}), c_m and c_x are experimental constants depending on the reference solutes chosen and on the mobile phase (c_m) or on the solute (c_x) and the meaning of the other constants is as in eqn. 1.25.

Using the interaction indices model[62,63], approximate estimates are possible of capacity factors on different reversed-phase columns in mobile phases containing various organic solvents in varying concentrations. Thus, the neglect of the role of the stationary phase seems justified in many systems using non-polar chemically bonded stationary phases. For very precise predictions of chromatographic behaviour, the interactions in the stationary phase will probably have to be taken into account, as the interactions either with the bonded moieties or the sorbed organic solvent[64] and with unreacted silanol groups[65] can influence the retention to a certain extent but, as far as we know, no mathematical correction approach for these effects has been presented.

Good validity of the equation

$$k' = k'_0 \cdot 10^{-mc} \qquad (1.28)$$

has been found experimentally for many solutes in reversed-phase systems using chemically bonded hydrocarbon phases and various organic solvents in water as the mobile phase (e.g. refs. 3, 49, 53-55, 60, 61, 66-77). An example is given in Fig. 1.5.

Eqn. 1.28 is obtained from the more general eqn. 1.26 by neglecting the quadratic term in the exponent, which is obviously justified for many practical separations. For certain combinations of solutes and systems, however, curvatures of the experimental plots of log k' versus c were found. These curvatures can be respected by using eqn. 1.26 with a quadratic term[60,62,67]. There are certain indications that these curvatures increase with decreasing polarity of the organic solvent used in the mobile phase and/or with increasing polarity of the solute[62,73] (see, e.g., Fig. 1.6).

Eqn. 1.21 can sometimes fit the experimental data better than eqn. 1.28, e.g., in the reversed-phase chromatography of certain polar compounds such as nucleic acid bases. However, eqn. 1.21 can be rewritten in a polynomic form, resembling formally eqn. 1.26, if only the first three terms of the Taylor series are used[78].

Thus, eqn. 1.28 and (less frequently) eqn. 1.26 seem to be appropriate for a description of the relationship between the capacity factors of solutes and the concentration of the organic solvent in an aqueous mobile phase. By ex-

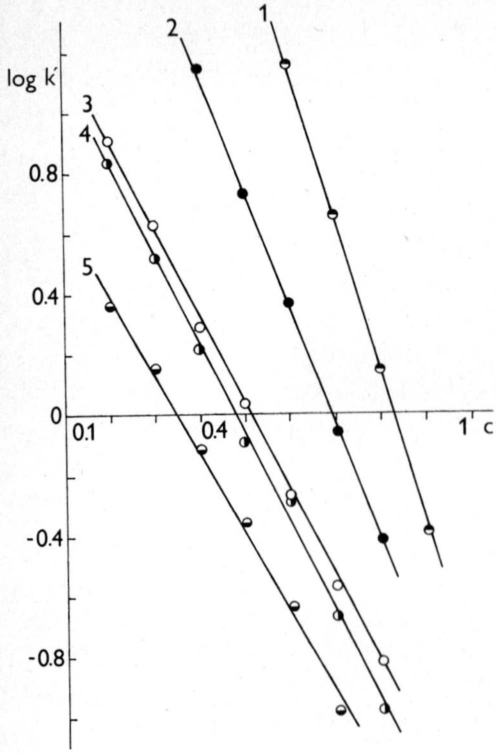

Fig. 1.5. Semi-logarithmic plots of capacity factors, k', of phenolic compounds versus concentration of methanol, c [% (v/v) · 10^{-2}], in water as the mobile phase in chromatography on octadecylsilica (LiChrosorb Si-100 as support), 10 μm, 300 x 4.2 mm I.D. column, V_m = 3.05 ml. 1, 4-*tert.*-Octylpyrocatechol; 2, 4-*tert.*-butylpyrocatechol; 3, 3-methylpyrocatechol; 4, 4-methylpyrocatechol; 5, pyrocatechol.

tending both the model of interaction indices and the solubility parameter theory, similar simple equations were derived for capacity factors in ternary and more complex solvent mixtures[62,63,128,137,138].

Chromatography on chemically bonded stationary phases containing more or less polar functional groups at the end of the bonded alkyl chain (such as nitrilo, ether, nitro or hydroxy groups) has become popular for some special applications in liquid column chromatography. Here, the influence of the mobile phase on retention has not been investigated in a systematic manner so far. However, these phases may act either as polar stationary phases in non-polar mobile phases or as non-polar stationary phases in aqueous mobile phases. It is likely that the separation mechanism would be more complex and that more effects can influ-

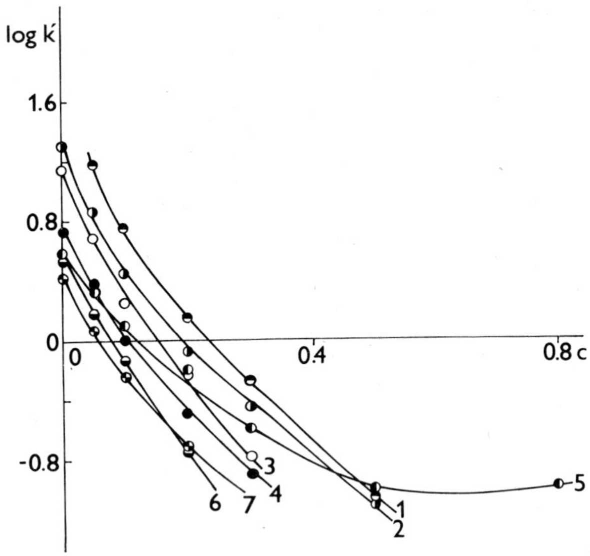

Fig. 1.6. Semi-logarithmic plots of capacity factors, k', of nucleosides versus concentration of methanol, c [% (v/v) · 10^{-2}], in water as the mobile phase in chromatography on octadecylsilica (LiChrosorb Si-100 as support), 10 μm, 300 x 4.2 mm I.D. column, V_m = 3.20 ml. 1 = Adenosine; 2 = thymidine; 3 = guanosine; 4 = 2'-deoxycytidine; 5 = 5-methylcytidine; 6 = uridine; 7 = cytidine.

ence the retention simultaneously. For example, in experiments with certain compounds, such as nucleic acid bases, nucleosides, barbiturates, sulphonamides and phenols on unmodified and modified Spheron gels (with an organic matrix carrying hydroxy or other functional groups) we found a reversal of the retention mechanism at certain compositions of aqueous-methanolic mobile phases[75-77].

The system behaved as a reversed-phase system in mobile phases with lower contents of methanol, while normal-phase properties were found in mobile phases rich in methanol. Correspondingly, a parabolic relationship between the capacity factors of solutes and the concentration of methanol was found for many compounds, which often showed a distinct minimum of retention at certain compositions of the mobile phase. This is illustrated in Fig. 1.7.

1.3. CHROMATOGRAPHY ON ION-EXCHANGE MATERIALS

In the chromatography of organic compounds on ion-exchange columns, several effects can influence the retention simultaneously[79-82]:

(a) Ionized solutes are subject to ion-exchange reactions with functional groups of the exchanger (sulphonic or carboxylic groups in strong or weak

References on p. 52.

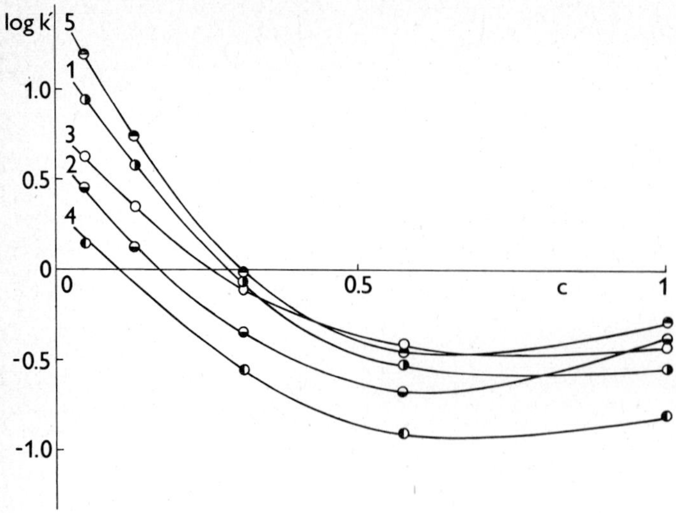

Fig. 1.7. Semi-logarithmic plots of capacity factors, k', of sulphonamides versus concentration of methanol, c [% (v/v) · 10^{-2}], in water as the mobile phase in chromatography on glycol methacrylate gel (Spheron P 1000), 25 μm, 300 x 4.2 mm I.D. column, V_m = 3.55 ml. 1 = Acetylsulphanilamide; 2 = sulphaguanidine; 3 = sulphanilamide; 4 = sulphamethazine; 5 = sulphathiazole. (Reproduced from ref. 76 with permission.)

cation exchangers and quaternary, tertiary or secondary amino groups in strong or weak anion exchangers).

(b) Ionized or non-ionized solutes enter into attractive and repulsive interactions with the mobile phase and the matrix of the ion exchangers (which resemble the interactions in reversed-phase chromatography on non-polar bonded phases), and interactions with the sorbed stagnant phase in the pores of an ion exchanger may contribute to the retention (as in liquid-liquid chromatography systems).

(c) If the solute has the character of a partially dissociated weak acid or base, or of an ampholyte, it can enter into ionization or protonization equilibria either in the mobile phase or in contact with the exchange groups of the ion exchanger.

(d) Further chemical reactions, such as complex formation, are possible in the presence of different complex-forming ions or molecules present in the mobile or exchanger phases.

Hence the mobile phase may influence the retention in chromatography on ion exchangers in various ways and the retention can be controlled by adjusting the ionic strength, pH or content of the organic solvent or of the complex-forming compounds.

The potential usefulness of these various parameters of the mobile phase depends on the nature of the exchanger and of the chromatographed solutes.

1.3.1. *Chromatography of ionized solutes in the absence of complex-forming equilibria*

Ion exchange of completely ionized compounds can be described by a stoichiometric equation for the exchange of a solute ion B carrying a positive or negative charge of value $q+$ ($q-$) with an ion A^{s+} (A^{s-}) of the exchanger:

$$q(\mathrm{A}) + s\mathrm{B} \rightleftarrows q\mathrm{A} + s(\mathrm{B}) \tag{1.29}$$

This exchange reaction can be characterized by means of a conventional selectivity constant:

$$K_{\mathrm{A}}^{\mathrm{B}} = \frac{[\mathrm{A}]^{q}(\mathrm{B})^{s}}{(\mathrm{A})^{q}[\mathrm{B}]^{s}} \tag{1.30}$$

where the parentheses refer to the concentrations in the "inner" exchanger phase and the square brackets to these in the outer solution. The activity coefficients are assumed not to depend significantly on the change in concentration of the ions A and B and are included in the value of the selectivity constant, to a first approximation.

Provided that the mobile phase containing ion A at a concentration $[\mathrm{A}]$ is used in the chromatography of trace amounts of the ion B, the relationship between the capacity factor of the chromatographed ion B and the concentration of the eluting agent in the mobile phase can be derived as[8]

$$k' = \frac{V_{\mathrm{s}}}{V_{\mathrm{m}}}\frac{(\mathrm{B})}{[\mathrm{B}]} \approx \frac{V_{\mathrm{s}}}{V_{\mathrm{m}}}(K_{\mathrm{A}}^{\mathrm{B}})^{\frac{1}{s}}\,\mathrm{Q}^{\frac{q}{s}}\,[\mathrm{A}]^{-\frac{q}{s}} = k'c^{-m} \tag{1.31}$$

This equation is formally identical with eqn. 1.23, derived for adsorption chromatography (Section 1.1), with

$$c = [\mathrm{A}] \tag{1.32a}$$

$$m = q/s \tag{1.32b}$$

and

$$k'_0 = \frac{V_s}{V_m}(K_A^B)^{\frac{1}{s}}\, Q^{\frac{q}{s}} \tag{1.32c}$$

where Q is the total exchange capacity of the exchanger for the ion A.

Fig. 1.8 demonstrates the validity of eqn. 1.31 in the anion-exchange chromatography of nucleotides, where the slopes of the linear plots of log k' of sample compounds versus the logarithms of molarity of the potassium dihydrogen phosphate ions in the mobile phases are very close to the charges of the nucleotide anions[7].

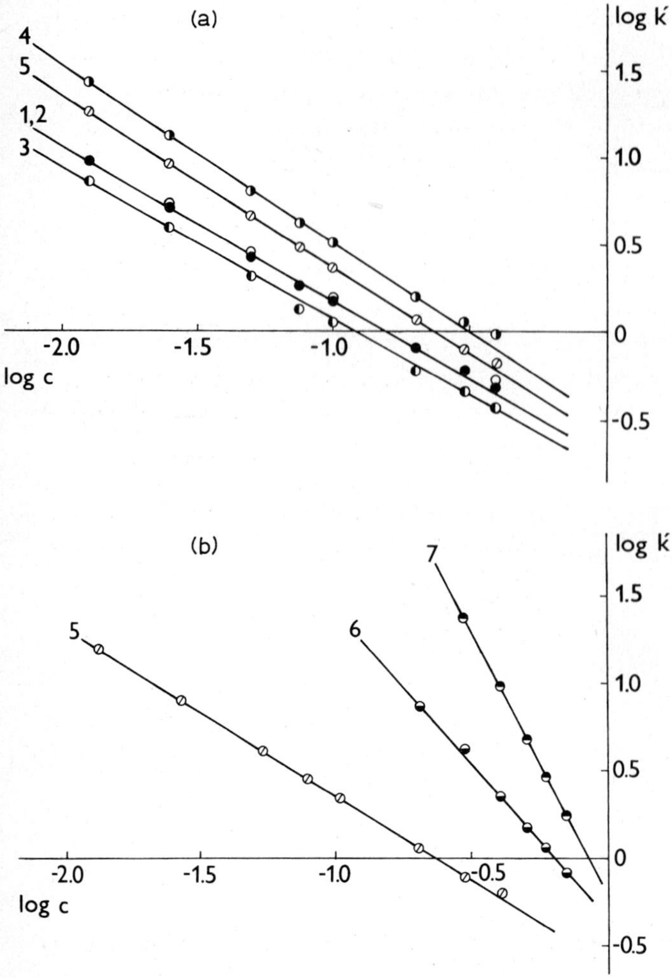

Fig. 1.8. Logarithmic plots of capacity factors, k', of nucleotides versus concentration of potassium dihydrogen phosphate, c (mol/l), in water as the mobile phase (pH = 3.15) in anion-exchange chromatography on Perisorb AN, 30-40 μm, 905 x 2.3 mm I.D. column, V_m = 2.05 ml. 1 = Thymidine 5'-monophosphate; 2 = ribothymidine 5'-monophosphate; 3 = deoxyuridine 5'-monophosphate; 4 = deoxyguanosine 5'-monophosphate; 5 = guanosine 5'-monophosphate; 6 = guanosine 5'-diphosphate; 7 = guanosine 5'-triphosphate. (Reproduced from ref. 7.)

Two types of ion-exchange column packing materials are used in modern liquid chromatography: microparticulate resinous ion exchangers with an organic matrix and microparticulate or pellicular ion exchangers based on an inorganic support material (silica) with chemically bonded alkyl chains carrying ending ion-exchange functional groups, or with an organic polymer layer containing exchange groups. Hence the matrix of the exchanger contributes in a different manner to the total retention of solutes, which is usually larger on ion exchangers with an organic skeleton than on inorganic exchangers. Usually, the contribution of the matrix to the retention does not change very significantly with changing concentration of the salt and is included in the value of the selectivity constant, and eqn. 1.31 can be applied satisfactorily to describe the influence of the mobile phase composition on capacity factors.

In the chromatography of weak acids and bases, only the ionized form is subject to ion exchange, while the non-ionized form may be sorbed due either to the solvophobic effect or to the attractive forces of the matrix or functional groups of the ion exchanger. Consequently, in deriving the relationships for the capacity factors, acid-base equilibria must be considered. This can be demonstrated on the example of a weak n-protic acid HB.

Let $D_{\mathrm{H}_n\mathrm{B}}$ be the distribution coefficient of the undissociated acid:

$$D_{\mathrm{H}_n\mathrm{B}} = \frac{(\mathrm{H}_n\mathrm{B})}{[\mathrm{H}_n\mathrm{B}]} \tag{1.33}$$

K_i is the dissociation (acidity) constant of the acid for the ith step in the mobile phase:

$$K_i = \frac{[\mathrm{H}^+]^i [\mathrm{H}_{n-i}\mathrm{B}^{i-}]}{[\mathrm{H}_n\mathrm{B}]} \tag{1.34}$$

and $K_{\mathrm{A}}^{\mathrm{H}_i\mathrm{B}}$ is the selectivity constant of the ion $\mathrm{H}_{n-i}\mathrm{B}^{i-}$:

$$K_{\mathrm{A}}^{\mathrm{H}_i\mathrm{B}} = \frac{[\mathrm{A}^{s-}]^i (\mathrm{H}_{n-i}\mathrm{B}^{i-})^s}{(\mathrm{A}^{s-})^i [\mathrm{H}_{n-i}\mathrm{B}^{i-}]^s} \tag{1.35}$$

Then, by analogy with eqn. 1.31, the following equation for k' can be derived:

$$k' = \frac{V_{\mathrm{s}}}{V_{\mathrm{m}}} \cdot \frac{D_{\mathrm{H}_n\mathrm{B}} + \sum_{i=1}^{n} \left(K_{\mathrm{A}}^{\mathrm{H}_i\mathrm{B}}\right)^{1/s} \frac{Q^{i/s}}{[\mathrm{H}^+]^i} \cdot K_i [\mathrm{A}]^{-\frac{i}{s}}}{1 + \sum_{i=1}^{n} \left(\frac{K_i}{[\mathrm{H}^+]^i}\right)} \tag{1.36}$$

If the pH of the mobile phase is kept constant during the elution at a value where one dissociated form of the acid predominates in the solution (and $D_{\mathrm{H}_n\mathrm{B}}$ has a sufficiently low value), then eqn. 1.36 can be simplified considerably and the influence of the concentration of a salt in the mobile phase (counter ions A) on k' can be described by a simple equation, formally identical with eqn. 1.31 or 1.23:

$$k' \approx \frac{V_\mathrm{s}}{V_\mathrm{m}} \cdot \frac{(K_\mathrm{A}^{\mathrm{H}_i\mathrm{B}})^{\frac{1}{s}} \, Q^{i/s} \, K_i}{K_i + [\mathrm{H}^+]^i} \cdot [\mathrm{A}]^{-\frac{i}{s}} = k'_0 c^{-m} \tag{1.37}$$

An analogous approach can be applied to the cation-exchange chromatography of weak bases.

In agreement with these considerations, linear experimental plots of log k' of carboxylic acids versus the logarithms of the molarities of acetate and tetraborate ions were found in chromatography on resinous anion exchangers[83,84], and analogous relationships were found for the cation-exchange chromatography of bases[85].

In the chromatography of acidic or basic compounds, solutions of acids and bases can be used as the eluting agents, where H^+ or OH^- ions, respectively, act as the competing ions (counter ions) and simultaneously determine the degree of dissociation of sample compounds. In these instances, ion exchangers are used in their free forms, i.e. cation exchangers in the H^+ form and anion exchangers in the OH^- form. If the pH of the eluent is changed within the range where one ionic form of the chromatographed acid or base predominates in solution and the concentration of H^+ or OH^- ions, respectively, is large enough in comparison with K_i, a linear relationship between log k' and pH can be expected (eqn. 1.37). This has been found experimentally, e.g., in the cation-exchange chromatography of protonated, weakly basic compounds in 0.2-1.6 *M* hydrochloric acid in aqueous ethanolic solutions. In this range of H^+ concentration, the fraction of the monoprotonated form of the bases varied from about 75% to about 98%[86,87] (see Fig. 1.9).

In many practical chromatographic ion-exchange separations of weak acids or bases, the exchangers in ionic forms other than H^+ (or OH^-) are used and the pH is appropriately adjusted by using buffered mobile phases. Here, linear plots of log k' against the counter ion concentration are frequently found if the pH is far enough from the pK of the acid or base, otherwise curvatures of these plots may occur.

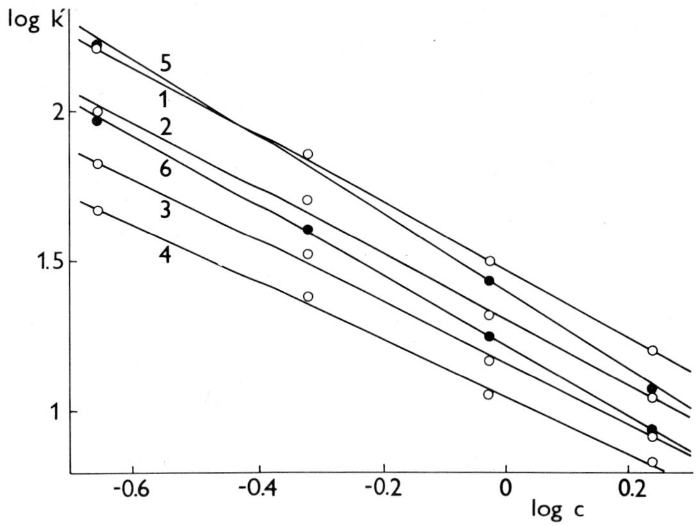

Fig. 1.9. Logarithmic plots of capacity factors, k', of p-N,N-dimethyl-p'-amino-benzeneazobenzoyl amides versus concentration of hydrochloric acid, c (mol/l), in 76.5% aqueous ethanol as the mobile phase in chromatography on cation exchanger Dowex 50W-X2 (H^+), 200-400 mesh, 240 x 2.7 mm I.D. column, V_m = 0.43 ml. 1 = Dimethylamide; 2 = diethylamide; 3 = di-n-propylamide; 4 = di-n-butylamide; 5 = methylamide; 6 = n-butylamide. (Reproduced from ref. 87.)

If the change in the composition of the mobile phase, such as buffer concentration, significantly affects the equilibrium between the individual ionic forms of the solute, the influence of the salt concentration on k' usually has the form of curved plots and is often difficult to predict theoretically.

The simplest example of these chromatographic systems is the anion-exchange chromatography of weak monoprotic acids (or bases), where the pH is changed while the counter ion concentration is approximately constant, in the range where both dissociated and undissociated forms of the acids are present in the mobile phase. Here, i = 1 in eqn. 1.37, which then acquires the form of eqn. 1.24, where

$$a = \frac{V_m}{V_s}\left\{\frac{[A]}{(K_A^{HB})Q}\right\}^{\frac{1}{s}} \qquad (1.37a)$$

$$b = \frac{V_m}{V_s}\cdot\frac{1}{K}\left\{\frac{[A]}{(K_A^{HB})Q}\right\}^{\frac{1}{s}} \qquad (1.37b)$$

In certain, rather rare instances, curvatures are observed of the logarithmic plots of k' of totally ionized solutes versus the concentration of salt in the mobile phase, owing to changing properties of the exchanger phase, or of activity coefficients, with changing salt concentration.

Svoboda[88] suggested the use of an empirical three-parameter parabolic equation to describe the $k' = f(c)$ plots in the anion-exchange chromatography of nucleotides:

$$\log k' = \log k'_0 - m \log c + p\,(\log c)^2 \tag{1.38}$$

where k'_0, m and p are experimental constants which depend on the solute and the chromatographic system used. He used this empirical equation, which has nothing in common with a real retention mechanism, to calculate the retention in gradient elution chromatography using numerical methods. However, most of the experimental $k'_0 = f(c)$ plots published by Svoboda[88] can be described well by eqn. 1.31 to a first approximation, which was in agreement with our experimental results.

The ion-exchange chromatography of high-molecular-weight biologically important compounds, such as nucleic acids, peptides and proteins, differs in certain respects from the ion-exchange chromatography of low-molecular-weight solutes. The biological macromolecules possess an extremely large number of ionizable groups that can interact strongly with the functional groups of ion exchangers. Consequently, such compounds are strongly retained and could be eluted only if most of these ion-exchange bonds are eliminated by an appropriate adjustment of the pH near to their isoelectric points. A change in pH is far more efficient for this purpose than a change in the concentration of the electrolyte in the mobile phase.

1.3.2. The use of complexes in ion-exchange chromatography

Two different methods of utilization of complex-forming equilibria in the mobile phase have been applied in ion-exchange chromatography.

(a) A chromatographed ion B^{q+} (B^{q-}) is able to form a non-sorbed complex M_qB with a complex-forming ion M^- (M^+) (for the sake of simplicity, a univalent ion M is considered), whereas the competing ion A^{s+} (A^{s-}) does not form complexes with the ion M. Then, the following equation for k' of the solute B on a cation- or anion-exchange column, respectively, can be derived, provided that the stability of the complex M_qB is sufficiently high:

$$k' = \frac{V_\mathrm{s}}{V_\mathrm{m}} \cdot \frac{(K_\mathrm{A}^\mathrm{B})^{1/s}\; Q^{q/s}}{K_{\mathrm{M}_q\mathrm{B}}\; \boxed{\mathrm{A}}^{q/s}\; \boxed{\mathrm{M}}^{q}} \tag{1.39}$$

where K_A^B is the selectivity constant for the ion exchange of the ions A and B, as defined by eqn. 1.30, Q is the capacity of the exchanger, $[A]$ and $[M]$ are the concentrations of the respective ions in the mobile phase and K_{M_qB} is the complex formation (stability) constant in the mobile phase:

$$K_{M_qB} = \frac{[M_qB]}{[M]^q\ [B]} \tag{1.40}$$

Obviously, eqn. 1.39 is formally identical with eqn. 1.31 or 1.23.

Hence the capacity factors can be influenced by changing the concentration of the complex-forming ion M in the mobile phase while the concentration of the ion A is kept constant (or changed simultaneously with changing concentration of ion A), such as in the anion-exchange chromatography of carboxylic acids with eluents containing acetates of metal cations (Mg^{2+}, Zn^{2+}) that form complexes with the acids[89,90]. Here, a linear decrease in log k' of the acids with the logarithm of the molarity of the metal acetate in the mobile phase was found, in agreement with eqn. 1.39, as is illustrated in Fig. 1.10.

(b) A non-ionized solute B is chromatographed in form of an ionized complex $B_mM_b^{mb+}$ ($B_mM_b^{mb-}$) with an appropriate complex-forming ion M^{m+} (M^{m-}), which is used as the competing ion in the mobile phase (the exchanger is in the form of the ion M). Provided that the non-ionized solute is not retained significantly and that only one, sufficiently stable complex is formed, the capacity factor of the complex ion B_mM_b is given by

$$k' = \frac{V_s}{V_m}\left(K_M^{B_mM_b}\right) Q^b\ [M]^{-b} = k_0'c^{-m} \tag{1.41}$$

where $K_M^{B_mM_b}$ is the selectivity constant for the exchange of the complex ion B_mM_b and the ion M as defined by eqn. 1.30, Q is the exchange capacity of the ion exchanger and $[M]$ is the concentration of the ion M in the mobile phase. Here again, eqn. 1.41 is formally identical with eqn. 1.31 or 1.23.

The anion-exchange chromatography of alcohols and sugars in borate eluents, of carbonyl compounds in bisulphite medium and of rare earth ions in ethylenediaminetetraacetate solutions represent practical examples of this approach to the use of complex-forming equilibria in ion-exchange chromatography. Linear relationships between log k' of solutes and the logarithms of the molarity of the complex-forming anion in agreement with eqn. 1.41 were found experimentally in a number of anion-exchange systems (e.g., refs. 91-94).

If more than one complex is formed in the mobile phase containing complex-forming ions, approximate validity of eqn. 1.39 or 1.41 may be expected if one

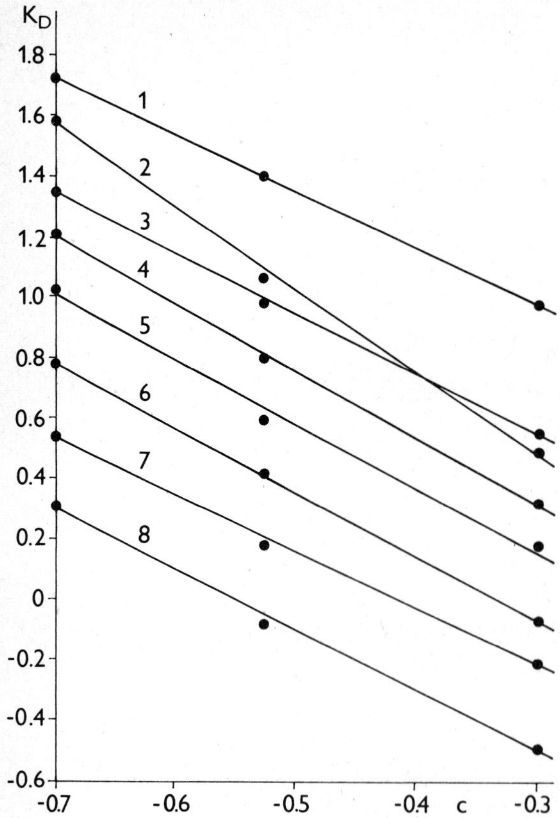

Fig. 1.10. Logarithmic plots of distribution coefficients, K_D, of carboxylic acids versus concentration of magnesium acetate, *c* (mol/l), in water as the mobile phase in chromatography on anion exchanger Dowex 1-X8, 17-20 μm, 1350 x 3.2 mm I.D. column, 70°C. 1 = Fumaric acid; 2 = *cis*-aconitic acid; 3 = itaconic acid; 4 = L-tartaric acid; 5 = β-hydroxy-β-methylglutaric acid; 6 = methylmalonic acid; 7 = malonic acid; 8 = oxalic acid. (Adapted from ref. 90.)

complex form predominates in solution. A more detailed discussion of ion-exchange equilibria in the presence of complex-forming and/or acid-base equilibria was presented by Inczédy[95,96].

Complex-forming equilibria can be utilized not only in the mobile but also in the exchanger phase, if a cation exchanger is used in the form of a complexing metal ion (Cu^{2+}, Ag^{+}, Ni^{2+}, Fe^{3+}, etc.). In this chromatographic technique (ligand-exchange chromatography), the solutes (potential ligands) are sorbed from solutions and form complexes with metal ions in the exchanger, and can displace other ligands that have previously complexed the metal ion. A solution containing other, competing, ligands is used as the mobile phase. In contrast to conventional ion exchange, in which counter ions are exchanged, ligand exchange represents the exchange of ligands for other ligands or for solvent molecules in the ionic solvation shells[97].

Provided that the metal ion M forms a single complex MB with the solute B and a single complex ML_m with the competing ligand L and assuming that neither the ligand L nor the metal M are subject to any other competitive reaction, a simple equation can be written[98] for the k' value of the solute B as a function on the concentration of L in the mobile phase, $[L]$:

$$k' = \frac{V_S}{V_m} \cdot K_L^B Q^m [L]^{-m} = k'_0 c^{-m} \tag{1.42}$$

where Q is the exchange capacity of the ion exchanger, m = moles of the ligand L bonded per millimole of the metal ion and K_L^B is the selectivity constant, defined as

$$K_L^B = \frac{(B)\,[L]^m}{[B]\,(L)^m} \tag{1.43}$$

This equation, again, is formally identical with eqn. 1.31 or 1.23. The details of the calculations for systems where other complexes are formed can be found elsewhere[98,99].

Ligand-exchange chromatography has been most frequently used for the chromatography of amines and amino acids on cation exchangers in the Cu^{2+} or Fe^{3+} form with ammonia or amine solutions as the mobile phases[81,97-99].

1.3.3. Chromatography of non-electrolytes on ion exchangers

Non-ionized organic compounds can be retained on ion-exchange columns because of the solvophobic effect and/or the attractive forces between the solutes and the matrix (or functional groups) of the exchanger. Hence these forces are essentially of the same nature as the interactions that are operative in reversed-phase chromatography. The retention depends mainly on the character of the exchanger and can be influenced by the composition of the mobile phase.

Generally, the retention increases with increasing ionic strength of the mobile phase as the repulsive solvophobic forces acting on non-ionized solutes are enhanced. This phenomenon is termed the "salting-out effect" and has been utilized by Sargent and Rieman[100-102] and other workers[103-105] for the so-called salting-out chromatography of various organic non-ionized compounds, such as alcohols[100], esters[104], ethers[101], aldehydes and ketones[103], very weak organic

References on p. 52.

acids[101,104], amines[102] and nitro compounds[105] on both cation- and anion-exchange resins. In these systems, an approximate validity of the empirical equation

$$k' = k'_0\ 10^{mc} \tag{1.44}$$

has been reported[100-104], where c is the concentration (molarity) of salt in the mobile phase and k'_0 and m are experimental constants. This equation is formally identical with eqn. 1.28, with the opposite sign of the exponent, however.

As in chromatography on non-polar chemically bonded phases, the retention of non-ionized solutes on ion exchangers decreases with increasing content of the organic solvent in the mobile phase. This has been utilized in the so-called "solubilization chromatography" of various organic compounds on ion-exchange columns[106,107]. In experimental systems, a linear decrease in log k' with increasing concentration of the organic solvent in the mobile phase was found for ketones, aldehydes and other compounds, in agreement with eqn. 1.28.

If very polar compounds, such as polyols and sugars, which are more polar than the organic solvent used in the mobile phase, are chromatographed on ion exchangers with an organic matrix, their retention usually increases with increasing content of the organic solvent in the mobile phase and, as can be found from published experimental data[108,109], the plots of log k' versus c are approximately linear.

1.4. ION-PAIR CHROMATOGRAPHY AND SALTING-OUT CHROMATOGRAPHY OF IONIZED ORGANIC COMPOUNDS ON CHEMICALLY BONDED NON-POLAR PHASES

The use of ion-pair forming reagents has become very popular in the chromatography of ionized and ionizable organic compounds[110]. The formation of ion pairs of the solutes with large organic ions is utilized, such as of anions of acids with quaternary ammonium cations and of cations of bases with alkanesulphonic acid anions. The technique has been called ion-pair chromatography, soap chromatography or paired-ion chromatography. It is possible to use polar liquid stationary phases on an appropriate support (silanized silica) in combination with low-polarity mobile phases. However, the gradient elution technique cannot be applied in these systems, as in other liquid-liquid chromatography systems, owing to stripping of the stationary phase from the column on changing the composition of the mobile phase. Ion-pair chromatography in reversed-phase systems with chemically bonded non-polar phases is much simpler to use in practice and makes possible the easy application of gradient elution. Here, an ion-pair forming reagent is added to aqueous solutions of organic solvents used as the mobile phases[111-117].

The mechanism of ion-pair reversed-phase chromatography has not yet been fully elucidated. According to one theory, ion pairs of solutes with the reagent are

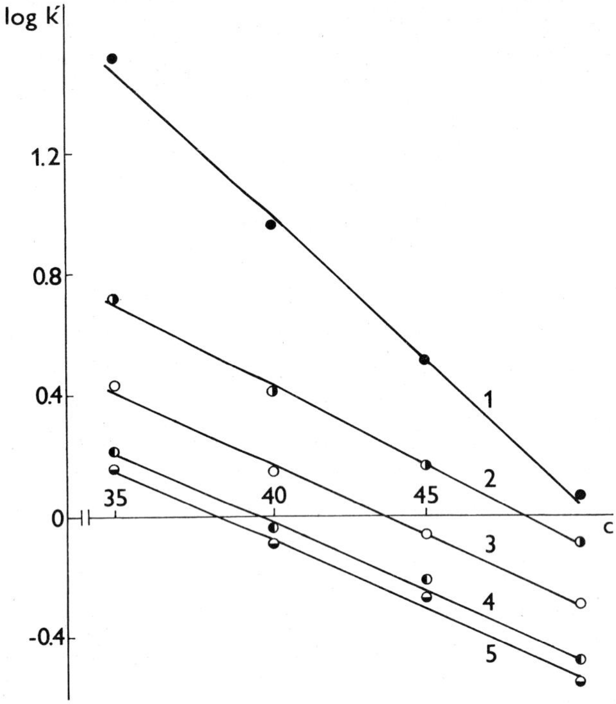

Fig. 1.11. Semi-logarithmic plots of capacity factors, k', of sulphonic acids versus concentration of methanol, c [% (v/v)], in aqueous-methanolic solutions containing 0.005 mol/l of tetrabutylammonium sulphate as the mobile phase in ion-pair chromatography on octylsilica (Silasorb C8), 10 μm, 300 x 4.2 mm I.D. column, V_m = 3.21 ml. 1 = 8-Aminonaphthalene-1-sulphonic acid; 2 = 7-aminonaphthalene-1-sulphonic acid; 3 = 5-aminonaphthalene-2-sulphonic acid; 4 = 4-aminonaphthalene-1-sulphonic acid; 5 = 5-aminonaphthalene-1-sulphonic acid.

formed in the mobile phase and the ion pair formed, which has a neutral total charge, is subject of a solvophobic effect like non-ionized solutes[111-114]. The other theory assumes sorption of the ion-pair forming ion on the surface of the bonded phase, where a layer of a liquid ion exchanger is formed, which is in equilibrium with the ion-pair containing mobile phase. Consequently, the ionized solutes are subject to ion-exchange interactions with the stationary phase[115]. Combination of the two effects may occur.

The retention of solutes in ion-pair reversed-phase chromatography is influenced both by the organic solvent and by the ion-pair forming reagent in the mobile phase[111,116,117]. The logarithms of the capacity factors of solutes decrease linearly with increasing concentration of the organic solvent in the mobile phase[112,116,117] (an example is illustrated on Fig. 1.11), as in the reversed-phase chromatography of non-ionized solutes without the addition of an ion-pair forming reagent. The retention of solutes increases with increasing length

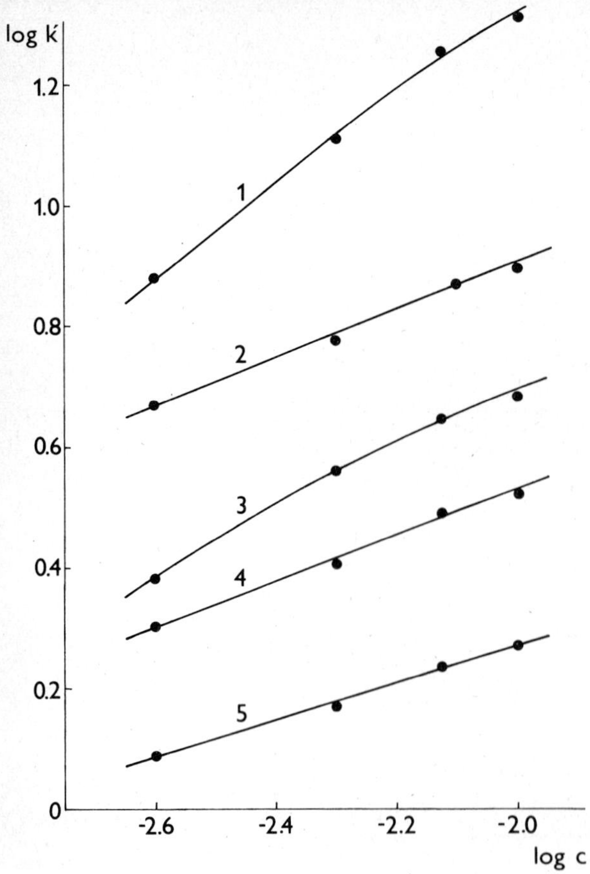

Fig. 1.12. Logarithmic plots of capacity factors, k', of sulphonic acids versus concentration of tetrabutylammonium sulphate, c (mol/l), in 35% (v/v) aqueous methanol as the mobile phase in ion-pair chromatography on octadecylsilica (Silasorb C18), 10 μm, 300 x 4.2 mm I.D. column, V_m = 2.92 ml. 1 = 1,3,7-Naphthalenetrisulphonic acid; 2 = 5-amino-3-naphthalenesulphonic acid; 3 = 1,5-naphthalenedisulphonic acid; 4 = 5-aminonaphthalene-1-sulphonic acid; 5 = 4-aminonaphthalene-1-sulphonic acid.

and number of alkyl chains in ion-pair forming ions[111] and depends to a lesser extent on the nature of the inorganic counter ion[116] (e.g., tetrabutylammonium sulphate, phosphate or perchlorate).

Very similar equations to describe the influence of the concentration of ion-pair forming ions in the mobile phase on retention to those for ion exchange can be derived for ion-pair formation equilibria in the mobile phase[118]. In practice, the log k' values of solutes increase approximately linearly with increasing logarithm of the concentration of the ion-pair forming agent in the mobile phase (log c), at least for ions with relatively shorter alkyl chains[116,118] (methyl to n-butyl), as it is shown in Fig. 1.12. This means that eqn. 1.23 can be for-

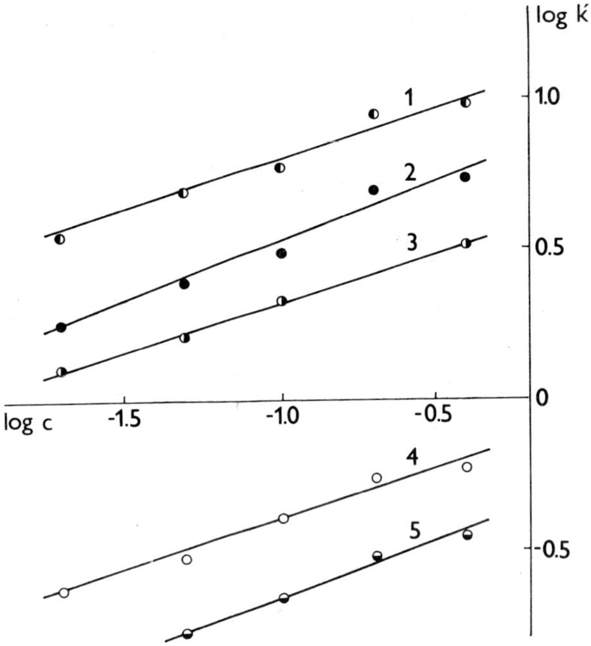

Fig. 1.13. Logarithmic plots of capacity factors, k', of sulphonic acids versus concentration of sodium sulphate, c (mol/l), in 10% (v/v) aqueous methanol as the mobile phase in chromatography on octadecylsilica (LiChrosorb Si-100 as support), 10 μm, 300 x 4.2 mm I.D. column, V_m = 3.05 ml. 1 = *p*-Nitrotoluene-*o*-sulphonic acid; 2 = *p*-toluenesulphonic acid; 3 = *m*-nitrobenzenesulphonic acid; 4 = 3-hydroxynaphthalene-2,7-disulphonic acid; 5 = metanilic acid.

mally used to describe this relationship. If ion-pair forming reagents with longer aliphatic chains are used, such as cetrimide (cetyltrimethylammonium bromide), the influence of the ion-pair forming reagent concentration on retention is more complex; the relationships between log k' and log c are usually curved and maxima on these plots may occur.

A number of ionized organic compounds may be chromatographed on chemically bonded non-polar phases if simple electrolytes (salts) are added to the aqueous-organic mobile phase in a relatively high concentration (0.1-0.4 *M* solutions)[117,119,120]. The nature of the salt (sodium sulphate, potassium hydrogen phosphate, lithium nitrate, etc.) has only a minor influence on retention[119], which is increased mainly owing to an increase in the solvophobic effect (an increase in the ionic strength of the mobile phase means an increase in its polarity). This phenomenon is largely analogous to the salting-out effect reported by Sargent and Rieman[100-102] in the chromatography of non-ionized organic substances on ion exchangers. It may be noted that the retention of both ionized and non-ionized solutes on chemically bonded alkane phases is increased by the addition of a

References on p. 52.

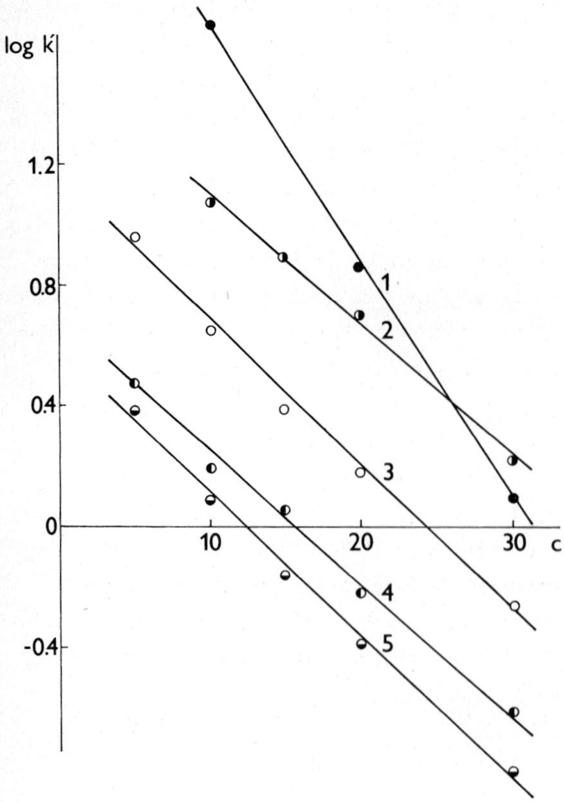

Fig. 1.14. Semi-logarithmic plots of capacity factors, k', of sulphonic acids versus concentration of methanol, c [% (v/v)], in 0.4 M sodium sulphate solution as the mobile phase in chromatography on octadecylsilica (Silasorb C18), 10 μm, 300 x 4.2 mm I.D. column, V_m = 2.92 ml. 1 = 8-Aminonaphthalene-1-sulphonic acid; 2 = 7-aminonaphthalene-1-sulphonic acid; 3 = 5-aminonaphthalene-2-sulphonic acid; 4 = 4-aminonaphthalene-1-sulphonic acid; 5 = 5-aminonaphthalene-1-sulphonic acid.

salt to the mobile phase. However, certain totally ionized compounds, such as benzene- and naphthalenesulphonic acids, are eluted from the columns packed with chemically bonded phases with an elution volume lower than the column void volume (are excluded) if mobile phases without the addition of a salt (even pure water) are used[116,117,119]. In these instances, good retention and separation could be achieved in salt solutions[119,120].

According to Horváth et al.[54], the capacity factors of ionized substances should increase with increasing ionic strength, I, of the mobile phase (in the absence of an ion-pair forming substance), in agreement with the following approximate equation (slightly modified symbols are used):

$$\log k' = \log k'_0 + AI + BI^{1/3} \tag{1.45}$$

where k'_0 is the capacity factor of the sample compound in the absence of the salt and A and B are constants for a given solute, solvent and column. In our experiments, log k' were found to increase approximately linearly with increase in the logarithm of the concentration of the salt in the mobile phase (c_s) (Fig. 1.13), whereas the log k' versus c_s plots were significantly curved[119]. This means that the simple eqn. 1.23 can be used to describe the influence of the salt concentration on k', to first approximation. Of course, neither eqn. 1.23 nor eqn. 1.45 can be used to describe the retention of ionic solutes in solutions with low contents of salts if these solutes are excluded in mobile phases without the addition of a salt, as $k' < 0$ in this region.

Changing the content of the organic solvent (c) in salt-containing mobile phases has an analogous effect on retention to that in the reversed-phase chromatography of non-ionized compounds without the addition of a salt[119] and to that in ion-pair reversed-phase chromatography, i.e., eqn. 1.28 can be used to describe log k' versus c plots (Fig. 1.14).

Of course, variations in pH can be used to influence the distribution between the non-ionized and ionized forms of weak acids or bases, respectively, in reversed-phase chromatography on non-polar chemically bonded phases. The dissociation can be either suppressed if pure aqueous organic mobile phases are used or promoted if ion-pair chromatography is applied.

References on p. 52.

Chapter 2

RETENTION CHARACTERISTICS IN ISOCRATIC ELUTION CHROMATOGRAPHY

As will be shown later, there are numerous similarities between the influence of the composition of the mobile phase on the separation process in isocratic elution chromatography and the influence of the profile of the concentration gradient in the mobile phase on separation under gradient elution conditions. Therefore, we shall discuss briefly some general relationships between the mobile phase composition and certain important characteristics of the chromatographic process (retention volume, bandwidth, retention ratio and resolution). A good understanding of these relationships will be useful later, in discussing these characteristics in gradient elution chromatography.

2.1. CALCULATION OF RETENTION CHARACTERISTICS

Most workers in practical liquid chromatography usually use retention times for the identification of peaks of solutes. The use of the capacity factors to characterize the retention has an obvious advantage, because k' values depend neither on the flow-rate of the mobile phase nor on the dimensions of the column used and is characteristic of the chromatographic system. Any general calculations of retention characteristics in either isocratic or gradient elution chromatography are possible on the basis of the capacity factors only[1-4]. Further, the retention of solutes measured with different columns and flow-rates of the mobile phases may be compared easily using capacity factors.

For a quantitative description of the influence of the mobile phase composition on retention characteristics, the influence of this composition on capacity factors in a given chromatographic system must be known. As has been discussed in detail in Sections 1.1-1.4, a few relatively simple two- or three-parameter equations can describe the relationship between the capacity factors of solutes and the concentration of the more efficient eluting component in various binary mobile phases for most of the systems used in liquid column chromatography (with respect to those systems, where gradient elution is or may be applied). Different $k' = f(c)$ functions, either derived theoretically or found experimentally in practical LC systems, are listed in Table 2.1.

If the form of the $k' = f(c)$ equation and numerical values of the constants of this equation are known, the influence of the composition of the mobile phase

References on p. 52.

TABLE 2.1

LITERATURE SURVEY OF USEFUL FUNCTIONS OF CAPACITY FACTORS, k', DEPENDING ON THE CONCENTRATION OF THE STRONGER ELUTION COMPONENT IN THE MOBILE PHASE, c

$k' = f(c)$*	Eqn. No.	Applies in liquid chromatographic system	Ref.	Notes
$k' = (a + bc)^{-m}$	1.21	Adsorption chromatography, ion-exchange chromatography, ion-pair chromatography (conditionally in some partition chromatography systems)	7-9	General three-parameter equation derived from Snyder's theory
$k' = k'_0 c^{-m}$	1.23	Adsorption chromatography, ion-exchange chromatography, ion-pair chromatography (c = concentration of ion-pair forming ion in mobile phase), salting-out chromatography on non-polar phases (c = concentration of salt in mobile phase)	7,8,15,16, 22-33,83-87, 89-94,98,116, 118,119	Simplified eqn. 1.21, two-parameter equation: $k' \to \infty$ in less polar solvent (adsorption chromatography); one ionic form predominates in ion-exchange chromatography
$k' = (a + bc)^{-1}$	1.24	Adsorption chromatography; ion-exchange chromatography	17-21	Simplified eqn. 1.21, two-parameter equation; one molecule of stronger eluting agent is assumed to replace one molecule of solute; $c >$ by about 3%

References on p. 52.

$k' = k'_0 \cdot 10^{(-mc+pc^2)}$	1.26	Partition chromatography, including reversed-phase chromatography; c for organic solvent or electrolyte in chromatography of non-ionized organic compounds on ion exchangers; c for organic solvent in ion-pair chromatography and salting-out chromatography on non-polar phases	8,59-63	General three-parameter equation derived from solubility parameter theory
$k' = k'_0 \cdot 10^{-mc}$	1.28	Partition chromatography, including reversed-phase chromatography; c for organic solvent or electrolyte in chromatography of non-ionized organic compounds on ion exchangers; c for organic solvent in ion-pair chromatography and salting-out chromatography on non-polar phases	49,53-55, 60,61,66-77,100-109, 112,116,117, 119	Simplified eqn. 1.26, two-parameter equation; limited range of c; not too large differences in solubility parameters of the components of the mobile phase assumed
$\log k' = \log k'_0 - m \cdot \log c + p \cdot (\log c)^2$	1.38	Ion-exchange chromatography	88	Empirical equation; no relation to separation mechanism
$\log k' = \log k'_0 + mc + pc^{\frac{1}{3}}$	1.45	c = concentration of salt in chromatography on non-polar phases	54	Simplified equation from the model of Horváth et al.

*a, b, d, k'_0, m and p are experimental constants in the given liquid chromatographic system.

on the retention characteristics (retention volumes, bandwidths, retention ratios and resolution) can be derived simply after introducing the $k' = f(c)$ function into eqns. 1.2, 1.4, 1.13 and 1.14[4,7,8]. The resulting equations are surveyed in Appendix 1.

This operation is very simple in principle, but a few notes concerning practical measurements may be useful.

To verify the function $k' = f(c)$ and to find the constants of this equation, it is necessary to measure the k' values of solutes in several mobile phases with different concentrations of the more efficient eluting component (c). The measurements should be performed at at least three different compositions of the mobile phase; four or more different mobile phase compositions are to be preferred, however. If $k' = f(c)$ is a two-parameter function, its constants can be obtained easily from the experimental data by linear regression analysis. For the determination of the constants of three-parameter equations, non-linear regression analysis is required. The calculation procedure involves approximate methods that can be performed only with a computer and usually require a large number of experimental data.

The capacity factors of solutes under practical chromatographic conditions only rarely exceed about 10, otherwise the analysis time would be very long. On the other hand, even relatively large errors in k' of early eluted compounds (solutes with $k' \leqslant 0.5$) do not have a very significant influence on the calculated retention volumes. Consequently, only the range between $k' \simeq 0.5$ and $k' \simeq 10$ of $k' = f(c)$ functions is of practical importance for the calculation of retention characteristics. Within these limits, the $k' = f(c)$ plots can usually be approximated by a two-parameter equation, which can be easily linearized for regression analysis[7,8].

Practical measurements of capacity factors are subject to certain problems relating to the determination of the volume of the mobile phase in the column, V_m.

If the sample of solutes injected on to the column contains an appropriately chosen inert compound, which is not retained under the conditions used, the peak maximum of this compound can be used as the marker of V_m. Here, the capacity ratios of solutes can be measured easily from the distances of peak maxima of the solute and of the inert compound from the start of the chromatogram (see Fig. 1.1). (It is advisable to check frequently the flow-rate declared on the pump using the chronometer and a small calibrated volumetric flask (or cylinder) and to calculate the elution volumes of retained compounds and of the inert compound from the known chart speed of the recorder and volume flow-rate of the mobile phase.) Unfortunately, it is not often possible to find an appropriate non-retained standard compound that yields a signal with the detector used, or such a compound might interfere with certain solute peaks. Sometimes, characteristic disturbances of the detector signal are observed on the baseline in the vicinity of V_m, which could be used for determining V_m. This method suffers from insufficient accuracy, but it is very useful for rough and rapid estimates of k' directly from the chromatogram.

Usually, V_m of the column used is determined in an independent run with an appropriate detector (usually a universal one, such as a refractive index detector). The main problem consists in an appropriate choice of the inert standard for measurements of V_m, because the standard solute should not be retained under the experimental conditions. In adsorption chromatography on polar adsorbents, tetrachloroethylene or benzene (UV detection) can be used in most mobile phases, with the exception of pure alkanes or pure alkanes containing only 1-2% or less of the polar solvent.

The measurement of V_m in reversed-phase chromatography on chemically bonded alkylsilica phases is more difficult than in other systems and has recently attracted considerable attention[51,64,70,125-127,129-132]. The main problem consists in solvation of the bonded alkyl chains that may form a solvated layer of a thickness dependent on the length of bonded alkyl chains and on the mobile phase composition[51,125,133]. The mobile phase-stationary phase boundary involves a gradual, rather than a sharp change in the composition of the interfacial layer and the volumes of the stationary and of the mobile phases are not shraply discriminated. Depending on the experimental conditions, V_m may theoretically vary in the limits between the interparticulate column volume and the sum of the interparticulate volume and the total volume of the pores in the particles. There are some indications that various compounds can penetrate into the pores of the particles to different extents, because of steric or ionic exclusion effects, and, possibly, each compound would require "its own" column dead volume[130].

For these reasons, it is difficult to find a standard inert solute that is neither retained nor excluded under the experimental conditions. Components of the mobile phase (water, methanol, acetonitrile)[51,64,70,125,127,132], either deuterated or non-deuterated, uracil[55], cytosine, acetone[134], N,N-dimethylformamide[135] and various inorganic and organic salts[51,125,129,136] have been proposed as dead volume markers. However, salts may be excluded from the particles by ionic exclusion (this can be compensated for by using other inorganic salts as components of the mobile phase in this measurement[51,125], but the solvation of the bonded phase is then probably changed) and other compounds differ in their elution volumes, which also may depend on the mobile phase composition[51,125] (for illustration see Fig. 2.1).

The static method for the determination of V_m consists in weighing the column filled and equilibrated with a solvent of a low density (such as methanol) and then with another solvent of a high density (such as tetrachloromethane). V_m,

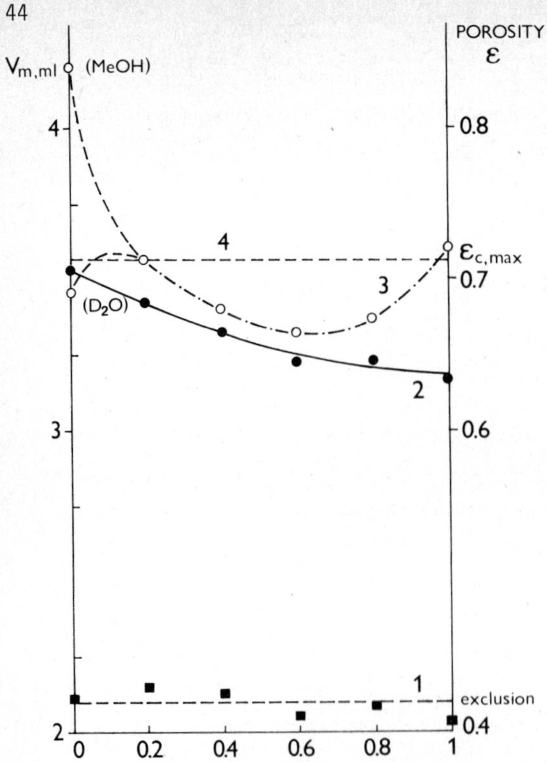

Fig. 2.1. Estimates of column void volume (volume of the mobile phase in column), V_m, and column porosity, ε, of an octadecylsilica column as a function of the composition of mobile phase methanol-water, c $[\%(v/v)\cdot 10^{-2}]$. 1, Peak start of completely excluded compound, 10^{-4} *M* KBr, corresponding to interparticle porosity ε_0; 2, V_m corresponding to linearization of retention volumes in homologous series of *n*-alkanols; 3, peak maximum of mobile phase components (CH_3OH, H_2O, D_2O); 4, maximum V_m determined by weighing the columns containing methanol and tetrachloromethane. Column: octadecylsilica on LiChrosorb Si-100, 10 μm, 300 x 4.6 mm I.D. (Adapted from ref. 51.)

calculated from the difference in weights, corresponds to the maximum column porosity[51,75,125].

The most time-consuming, but apparently the most precise, method for the calculation of V_m is the linearization of the relationship between the logarithms of net retention volumes and the carbon number for the members of homologous series[51,125,126,131]. However, the members of the homologous series should be carefully selected in order to obtain accurate results[131].

For practical purposes, the method of Glajch et al.[121] seems to be adequate. In this method, a small volume of the solution containing a slightly smaller amount of the organic solvent than the mobile phase (about 5%) is injected and the first peak detected by a refractometric detector is used to determine V_m.

The errors in V_m influence more seriously lower capacity factors ($k' < 1$) than higher capacity factors. To minimize the error, it is important to measure

V_m in the same, clearly defined way. A number of possible misunderstandings may be avoided and the possibility of reproducing published data improved when the method for the determination of V_m is always given together with k' values.

2.2. PREDICTION OF SEPARATION CONDITIONS

From the equations in Appendix 1, it is clear that both the capacity factors and bandwidths of sample solutes dcrease with increasing concentration of the more efficient eluting component in the binary mobile phase. Using these equations, it is possible to calculate these characteristics at any composition of the mobile phase, where the respective $k' = f(c)$ function applies.

From the practical point of view, the influence of the mobile phase composition on separation factors and on resolution is much more intersting. In systems where the solutes of a similar structure interact in an approximately equal way with the stationary phase on one hand and with the components of the mobile phase on the other, the influence of the mobile phase composition on the separation factors of these compounds is generally negligible and $\alpha \simeq \alpha_0$ is constant[8].

In this instance, the parameters of the equations describing $k' = f(c)$ functions must be related in a certain manner, as follows from the analysis of the equations in Appendix 1 and as summarized in Table 2.2.

If solutes differing in structure are subject to less or more different interactions with the stationary and mobile phases, the separation factor would be more influenced by the concentration of the more efficient eluting component in the mobile phase (c). In certain instances, two solutes 1 and 2 have equal retention volumes and $\alpha = 1$ at a certain concentration $c_{\alpha=1}$ of the more efficient eluting component, while a change in c, either an increase or a decrease, leads to a change in α and often makes possible the separation of the two solutes (see Figs. 1.6, 1.9, 1.10 and 1.11).

It is possible to calculate $c_{\alpha=1}$ from the equations for α in dependence on c in Appendix 1 after introducing $\alpha=1$ into the respective equation[7]. For most practical separation examples, such values are outside the practically available concentration range ($0 < c < 1$, where $c = 1$ for pure more efficient eluting component).

In practical separation systems, the differences in the dependence of capacity factors on the composition of the mobile phase may lead to maximum selectivity for certain pairs of compounds at certain compositions of the mobile phase, such as in the chromatography of sterols on alumina (Fig. 2.2)[7].

To understand better the influence of the composition of the mobile phase on resolution, we shall recall the three terms that contribute to R_s (eqn. 1.13). The efficiency term is essentially independent of the composition of the mobile

TABLE 2.2

RELATIONSHIPS BETWEEN THE PARAMETERS OF THE $k' = f(c)$ EQUATIONS WHICH MUST BE FULFILLED WHEN THE SEPARATION FACTORS OF TWO COMPOUNDS 1 AND 2 DO NOT DEPEND ON THE COMPOSITION OF THE MOBILE PHASE

Equation No.	Parameters*
1.21	(1) $m_1 = m_2$, $a_1 \gg b_1c$ and $a_2 \gg b_2c$ (2) $m_1 = m_2$, $b_1c \gg a_1$ and $b_2c \gg a_2$
1.23	$m_1 = m_2$
1.24	(1) $a_1 \gg b_1c$ and $a_2 \gg b_2c$ (2) $b_1c \gg a_1$ and $b_2c \gg a_2$
1.26	$m_1 = m_2$ and $p_1 = p_2$
1.28	$m_1 = m_2$
1.38	$m_1 = m_2$ and $p_1 = p_2$
1.45	$m_1 = m_2$ and $p_1 = p_2$

*Subscripts 1 and 2 relate to compounds 1 and 2, respectively.

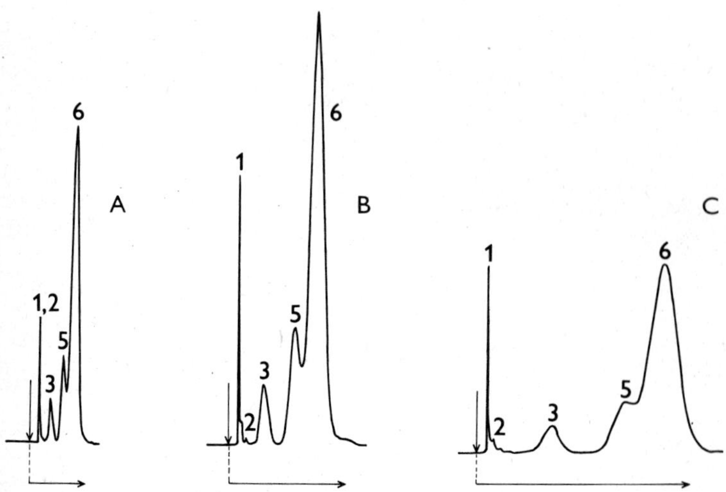

Fig. 2.2. Chromatographic separation of a mixture containing lumisterol (3), tachysterol (5) and calciferol (6) on LiChrosorb ALOX T, 30 μm, 596 x 2.3 mm I.D. column; V_m = 2.30 ml, in a mobile phase containing (A) 0.5%, (B) 0.25% and (C) 0.125% (v/v) of *n*-heptane in *n*-propanol. Flow-rate of mobile phase, 1.93 ml/min; detection, UV (254 nm). Compounds 1 and 2 are impurities. (Reproduced from ref. 7.)

phase, to a first approximation. As has been discussed above, the capacity term generally decreases with increasing concentration of the more efficient eluting component in the mobile phase (with a few exceptions, such as those shown in Fig. 1.7). This means that the resolution generally improves as the concentration of the more efficient component in the mobile phase decreases (the time of analysis increases, however). The selectivity term (separation factor) may complicate this simple rule in certain instances.

If the separation factor is approximately constant over the whole composition range of the mobile phase, the capacity term controls the influence of the mobile

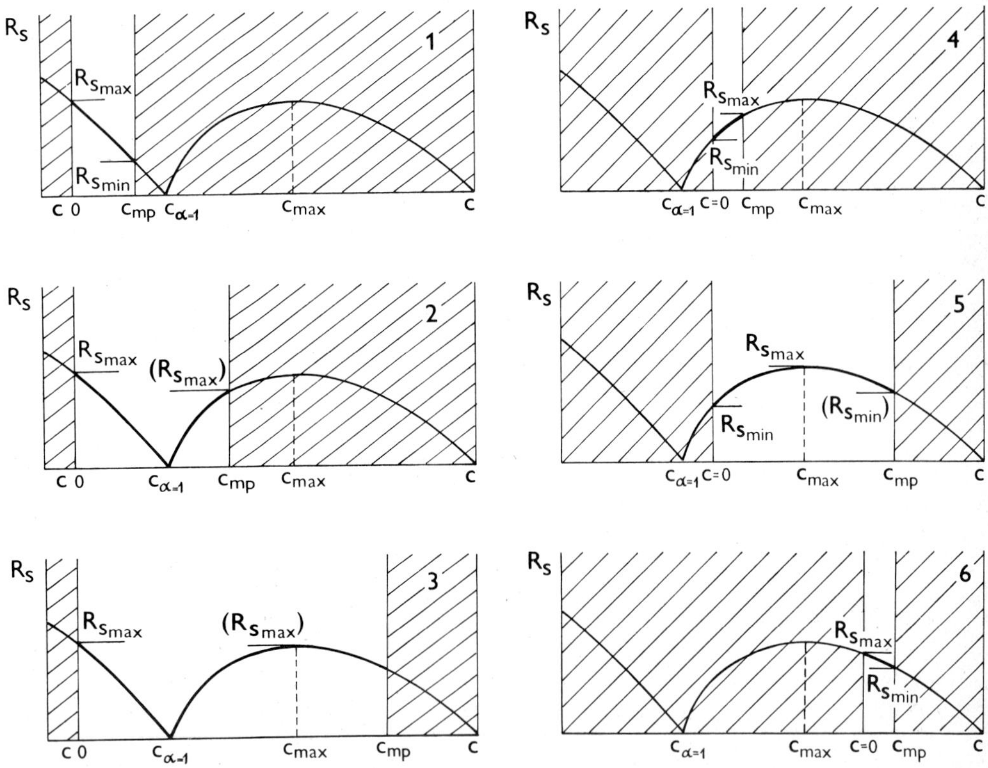

Fig. 2.3. Representative relationships between resolution, R_S, and concentration, c, of the more efficient eluting agent in the mobile phase. 1-6 demonstrate different profiles of the function $R_S = f(c)$ in the practically possible range of concentration c, represented as white areas. $R_{S\,min}$ and $R_{S\,max}$ are the minimum and maximum resolutions, respectively, which can be achieved within real limits of c ($c = 0$ and $c = c_{mp}$); $c_{\alpha} = 1 = c$, at which $\alpha = 1$; $c_{max} = c$, at which maximum on $R_S = f(c)$ plot occurs. (Reproduced from ref. 7.)

phase composition on resolution and R_s decreases with increasing concentration of the more efficient eluting component in the mobile phase over the whole composition range. The resolution of those solute pairs whose separation factors change with composition of the mobile phase may depend on the concentration of the more efficient eluting component in the mobile phase in a complex manner. The situation is most complex for pairs of solutes for which $\alpha = 1$ at a certain real concentration of the more efficient eluting component in the mobile phase, $c_{\alpha=1}$, as is illustrated in Fig. 2.3. Here, both the capacity and the selectivity terms decrease as the concentration of the more efficient component in the mobile phase, c, increases from zero and consequently the resolution also decreases, until $R_s = 0$ is reached at $c_{\alpha=1}$, where $\alpha = 1$. If we further increase c, the influence of the increasing selectivity term predominates over the decreasing capacity term and the resolution increases. At a certain concentration, $c = c_{max}$, the influence of the two terms will compensate one another and R_s will reach a maximum value in this part of the $R_s = f(c)$ plot. A further increase in c is connected with the predominating influence of the decreasing capacity term on resolution, which again decreases[7].

In real chromatographic systems we can only rarely observe the whole development of the $R_s = f(c)$ plot as discussed above. Usually only part of this plot can be observed within the real composition range of the mobile phase. Thus, the plots of $R_s = f(c)$ may have different forms, as indicated in Fig. 2.3 (1-6), where the real concentration range is represented by blank areas between $c = 0$ and c_{mp} (maximum possible c). It is obvious that, at a constant efficiency, the resolution may vary only within certain limits ($R_{s\ min}$ and $R_{s\ max}$ in Fig. 2.3) with changing composition of the mobile phase[7].

The limiting values of resolution may be calculated from the functions $R_s = f(c)$ in Appendix 1 after introducing $c = 0$, $c = c_{mp}$, $c = c_{\alpha=1}$ and $c = c_{max}$. The concentration c_{max}, at which the maximum on the $R_s = f(c)$ plot occurs, may be calculated after introducing the appropriate $R_s = f(c)$ function into the condition

$$dR_s/dc = 0 \tag{2.1}$$

Fig. 2.4 shows a practical example of $R_s = f(c)$ plots for three pairs of sterols chromatographed on alumina in mobile phases containing n-propanol in n-heptane (the chromatograms are shown in Fig. 2.2). Here, the different slopes of the logarithmic $k' = f(c)$ plots (Fig. 1.2) mean that the selectivity of separation depends on the composition of the mobile phase. Both the separation factors of the pairs tachysterol-lumisterol and calciferol-tachysterol increase with increasing concentration of n-propanol (c) in the mobile phase. However, the contribution of the selectivity term to the resolution of tachysterol and calciferol predominates over the contribution of the capacity term, while the

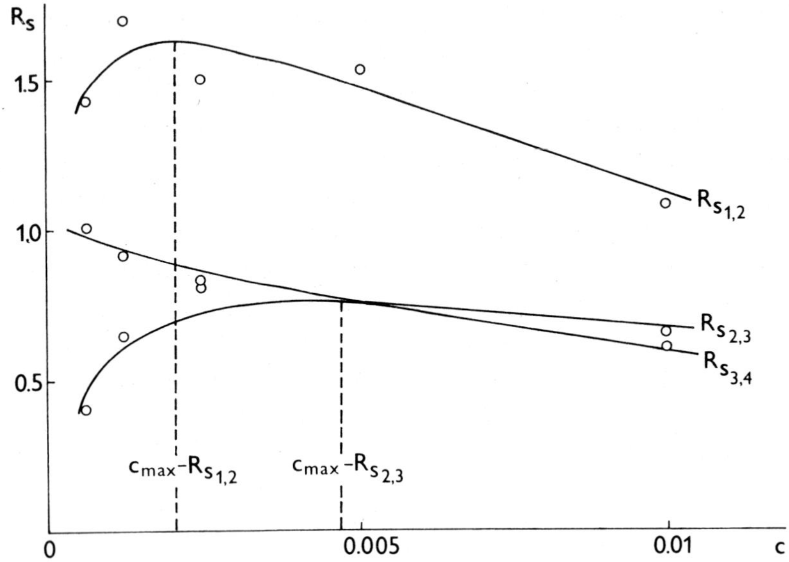

Fig. 2.4. Relationships between resolution, R_S, of steroids and concentration, c [%(v/v)·10^{-2}], of n-propanol in n-heptane as the mobile phase on LiChrosorb ALOX T, 30 μm, 596 x 2.3 mm I.D. column, V_m = 2.30 ml. $R_{S\ 1,2}$ = resolution of lumisterol and tachysterol; $R_{S\ 2,3}$ = resolution of tachysterol and calciferol; $R_{S\ 3,4}$ = resolution of calciferol and ergosterol. The points represent experimental values; the curves were calculated from the parameters m and k'_0 in eqn. 1.23 for the individual compounds and the efficiency of the column. (Reproduced from ref. 7.)

opposite holds for the resolution of lumisterol and tachysterol. Consequently, the best separation of the mixture of these three sterols can be achieved in the mobile phase containing 0.5% (v/v) of n-propanol in n-heptane, as is obvious from Fig. 2.4. The resolution is impaired at both lower and higher concentrations of n-propanol in the mobile phase (see Fig. 2.2). A complete separation of the above mixture of sterols was achieved using a more efficient column packed with alumina in 0.5% n-propanol in n-heptane[7].

Using this approach, conditions can be calculated for the best separation of solute mixtures. The precision of such calculations is not critical, as the maxima on R_S = f(c) plots in real systems are usually rather flat (Fig. 2.4) and a minor error in the calculated composition of the mobile phase would not influence the results seriously. An apparent drawback of this method consists in the need for previous experiments in three or four mobile phases with different contents of the more efficient eluting component, which are time consuming. Hence the calculation approach seems to offer certain advantages over trial-and-error experiments mainly for developing a routine method for the analysis of a large number of samples. This approach has been extended to separations that require ternary or more complex mobile phases[137,138], where it can obviate much of the time and effort connected with trial-and-error experiments.

An alternative approach for the optimization of mobile phase composition in reversed-phase liquid chromatography was suggested by Watson and Carr[123]. This method makes use of an optimization function which gives a single-value criterion for the quality of separation of the multi-component sample mixture. This criterion was originally used in the form of the so-called chromatographic resolution function (see Section 7.5 for a more detailed discussion). The optimization approach was later applied to both reversed-phase[139] and normal-phase[140] liquid chromatography. Glajch et al.[121] reported an amendment consisting in definition of the so-called chromatographic optimization function (COF) in the following form:

$$\mathrm{COF} = \sum_{i=1}^{k} A_i \ln\left(\frac{R_i}{R_{i\mathrm{d}}}\right) + B(t_\mathrm{M}-t_\mathrm{L}) \tag{2.4}$$

where R_i is the resolution of the ith pair of sample compounds and $R_{i\mathrm{d}}$ is the desired resolution for that pair, t_L is the experimental time, t_M is the maximum acceptable analysis time and B and A_i are the arbitrary weighting factors. A_i, included for each pair of peaks of interest, emphasizes less or more separations of pairs of peaks that are more important than the others. The optimized separation should provide a COF value as near to zero as possible. As neither the CRF nor the COF offers the best description of the separation, the product of resolutions between all adjacent peaks[141] and, further refined, the so-called relative resolution product (resolution product related to the maximum possible value for resolution product within a given chromatogram)[142] were suggested as better optimization criteria. The use of these functions is similar to the optimization method using the CRF function as described by Watson and Carr[123] and discussed Section 7.5.

A disadvantage of this method is that it does not take into account a possible reversal of the elution order for certain pairs of compounds with changing mobile phase composition. Further, equal numerical values of COF (or CRF), as a single number, may characterize separations that are not equally satisfactory from the chromatographic point of view. However, this statistical optimization approach may be useful in designing routine chromatographic separation methods.

The "overlapping resolution mapping" (ORM) method suggested by Glajch et al.[121] consists in a statistical analysis of data obtained in several mobile phases such as, for reversed-phase systems, those composed of methanol, acetonitrile, tetrahydrofuran and water in different concentrations. This data analysis makes use of the solvent selectivity triangle, as introduced by Snyder[124] (Fig. 2.5).

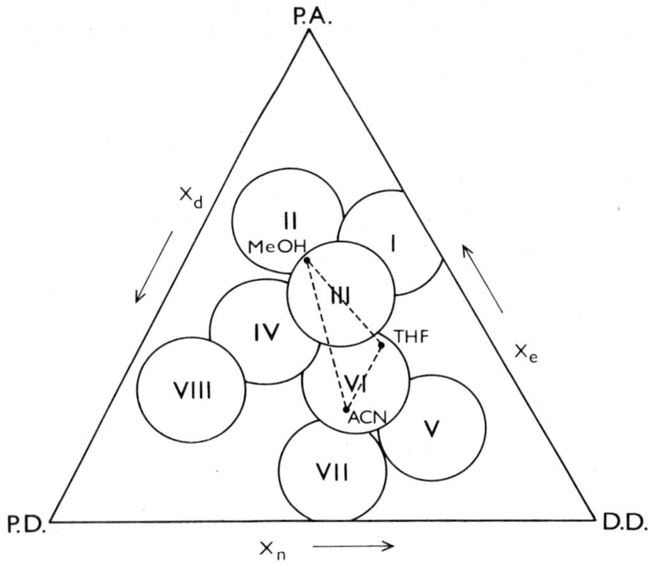

Fig. 2.5. Selectivity triangle for different contributions to polar interactions of common solvents used in liquid chromatography. x_n = contribution of dipole-dipole interactions (D.D.); x_d = contribution of proton-donor interactions (P.D.); x_e = contribution of proton-acceptor interactions (P.A.). The apices of the selectivity triangle correspond to 100% interaction of a certain type. The circles with Roman figures correspond to the approximate region of different solvent selectivity groups, as defined by Snyder. The inner dashed triangle corresponds to the selectivity range in reversed-phase chromatography, limited by the most common solvents methanol (MeOH), acetonitrile (ACN) and tetrahydrofuran (THF) used in optimization of separation conditions by the ORM method. (Adapted from refs. 121 and 124.)

Each of the corners of this triangle is represented by the mixture of water and the pure organic solvent (methanol, acetonitrile and tetrahydrofuran). The three mixtures have approximately equal elution strengths, but are differentiated by different relative roles of various contributions of the organic solvents to the total interaction force (dipole-dipole, proton-donor, proton-acceptor). The ORM optimization method searches for an optimum composition of a mixture of these three solvents for a given separation. This may be performed in a series of ten experiments with mobile phase compositions selected as the points, located in regular distances on the selectivity triangle. The computer selects the important pairs of compounds in the sample mixture, the separation of which may be difficult in a given system. The resolution is calculated for each such pair of compounds and then a resolution contour map is constructed for calculating the resolution for all important pairs of compounds in all solvent compositions within a selected solvent triangle. Here, the regions are traced in which the actual resolution is lower than the required resolution level. Consequently, the

References on p. 52.

remaining "blank" area outside these overlapping regions of the solvent selectivity triangle gives the range from which the composition of the mobile phase should be selected to yield a sufficient selectivity to obtain the resolution required.

In addition to the optimization of reversed-phase chromatographic separations[121], this method was also applied to chromatography on polar adsorbents[143-146] and can be performed automatically using the so-called "SENTINEL" (DuPont) system.

This optimization method is superior to the approaches using a single optimization criterion such as the ORF or COF function, because it gives a detailed picture of the resolution of all sample components. However, it assumes a previous determination of the elution strength of the mobile phase and does not take into account possible changes in the elution order with changing content of water in the mobile phase, in contrast to the calculation approach from $R_s = f(c)$ and $k' = f(c)$ functions.

This discussion of the relationships between the composition of the mobile phase and the retention, selectivity and resolution of chromatographed solutes was necessary to allow a better understanding of the influence of the gradient profile on separation in gradient elution chromatography. Numerous similarities exist between the isocratic and gradient elution techniques, as we shall see in Part II.

REFERENCES TO PART I

1 L.R. Snyder, *Chromatogr. Rev.*, 7 (1965) 1.
2 P. Jandera and J. Churáček, *J. Chromatogr.*, 91 (1974) 223.
3 L.R. Snyder, in Cs. Horváth (Editor), *High Performance Liquid Chromatography, Advances and Perspectives*, Academic Press, New York, 1980, p. 207.
4 P. Jandera and J. Churáček, *Advan. Chromatogr.*, 19 (1981) 125.
5 L.R. Snyder and J.J. Kirkland, *Introduction to Modern Liquid Chromatography*, Wiley, New York, 1974.
6 H. Engelhardt, *Hochdruck-Flüssigkeits-Chromatographie*, Springer, New York, 1977.
7 P. Jandera, M. Janderová and J. Churáček, *J. Chromatogr.*, 148 (1978) 79.
8 P. Jandera and J. Churáček, *J. Chromatogr.*, 91 (1974) 207.
9 L.R. Snyder, *Principles of Adsorption Chromatography*, Marcel Dekker, New York, 1968.
10 L.R. Snyder, *J. Chromatogr.*, 16 (1964) 55.
11 L.R. Snyder, *Advan. Anal. Chem. Instrum.*, 3 (1964) 251.
12 L.R. Snyder, *J. Chromatogr.*, 8 (1962) 178.
13 L.R. Snyder, *Anal. Chem.*, 46 (1974) 1384.
14 E. Soczewiński, *Anal. Chem.*, 41 (1969) 179.
15 E. Soczewiński and W. Golkiewicz, *Chromatographia*, 4 (1971) 501.
16 E. Soczewiński and W. Golkiewicz, *Chromatographia*, 6 (1973) 269.
17 R.P.W. Scott and P. Kucera, *Anal. Chem.*, 45 (1973) 749.
18 R.P.W. Scott and P. Kucera, *J. Chromatogr.*, 112 (1975) 425.
19 R.P.W. Scott, *J. Chromatogr. Sci.*, 8 (1970) 619.
20 R.P.W. Scott, *J. Chromatogr.*, 122 (1976) 35.
21 R.P.W. Scott and P. Kucera, *J. Chromatogr.*, 149 (1978) 93.

22 P. Jandera and J. Churáček, *J. Chromatogr.*, 93 (1974) 17.
23 P. Jandera, M. Janderová and J. Churáček, *J. Chromatogr.*, 115 (1975) 9.
24 E. Soczewiński, W. Golkiewicz and W. Markowski, *Chromatographia*, 8 (1975) 13.
25 E. Soczewiński, *J. Chromatogr.*, 130 (1977) 23.
26 E. Soczewiński and W. Golkiewicz, *J. Chromatogr.*, 118 (1976) 91.
27 E. Soczewiński, T. Dzido, W. Golkiewicz and K. Gazda, *J. Chromatogr.*, 131 (1977) 408.
28 E. Soczewiński and J. Kucmierczyk, *J. Chromatogr.*, 150 (1978) 53.
29 W. Golkiewicz, *Chromatographia*, 9 (1976) 113.
30 T. Wawrzynowicz, *Chem. Anal. (Warsaw)*, 22 (1977) 17.
31 T. Wawrzynowicz and T. Dzido, *Talanta*, 24 (1977) 669.
32 S. Hara, Y. Fuji, M. Hirasawa and S. Miyamoto, *J. Chromatogr.*, 149 (1978) 143.
33 S. Hara, N. Yamauchi, C. Nakae and S. Sakai, *Anal. Chem.*, 52 (1980) 33.
34 R.L. Snyder and H. Poppe, *J. Chromatogr.*, 184 (1980) 364.
35 E.H. Slaats, J.C. Kraak, W.T.J. Brugman and H. Poppe, *J. Chromatogr.*, 149 (1978) 255.
36 J.E. Paanakker, J.C. Kraak and H. Poppe, *J. Chromatogr.*, 143 (1978) 111.
37 M. Borówko and M. Jaroniec, *Chromatographia*, 12 (1979) 672.
38 A. Pryde, *J. Chromatogr. Sci.*, 12 (1974) 486.
39 V. Řehák and E. Smolková, *Chromatographia*, 9 (1976) 219.
40 J.J. Kirkland and J.J. DeStefano, *J. Chromatogr. Sci.*, 8 (1970) 39.
41 D.C. Locke, *J. Chromatogr. Sci.*, 11 (1973) 120.
42 J.H. Knox and A. Pryde, *J. Chromatogr.*, 112 (1975) 171.
43 E.J. Kikta and E. Grushka, *Anal. Chem.*, 48 (1976) 1098.
44 R.P.W. Scott and P. Kucera, *J. Chromatogr.*, 142 (1977) 213.
45 C.H. Lochmüller and D.R. Wilder, *J. Chromatogr. Sci.*, 17 (1979) 17.
46 H. Colin and G. Guiochon, *J. Chromatogr.*, 158 (1978) 185.
47 M.J. Telepchak, *Chromatographia*, 6 (1973) 234.
48 H. Colin and G. Guiochon, *J. Chromatogr. Sci.*, 18 (1980) 54.
49 K. Karch, I. Sebestian, I. Halász and H. Engelhardt, *J. Chromatogr.*, 122 (1976) 171.
50 E.H. Slaats, S. Heemstra and H. Poppe, cited in E.H. Slaats, *Thesis*, University of Amsterdam, Amsterdam, 1980.
51 G.E. Berendsen, *Thesis*, Technische Hogeschool, Delft, Delft, 1980.
52 Cs. Horváth and W. Melander, *J. Chromatogr. Sci.*, 15 (1977) 393.
53 Cs. Horváth, W. Melander and I. Molnár, *J. Chromatogr.*, 125 (1976) 129.
54 Cs. Horváth, W. Melander and I. Molnár, *Anal. Chem.*, 49 (1977) 142.
55 B.L. Karger, J.R. Gant, A. Hartkopf and P.H. Weiner, *J. Chromatogr.*, 128 (1976) 65.
56 O. Sinanoglu, in R. Pullman (Editor), *Molecular Associations in Biology*, Academic Press, New York, 1968, p. 427.
57 O. Sinanoglu and S. Abdulnur, *Fed. Proc., Fed. Amer. Soc. Exp. Biol.*, 24 (1965) 12.
58 J.H. Hildebrand and L.R. Scott, *The Solubility of Non-electrolytes*, Dover, New York, 3rd ed., 1964.
59 R. Tijssen, H.A.H. Billiet and P.J. Schoenmakers, *J. Chromatogr.*, 128 (1976) 65.
60 P.J. Schoenmakers, H.A.H. Billiet, R. Tijssen and L. de Galan, *J. Chromatogr.*, 149 (1978) 519.
61 P. Jandera, J. Churáček and L. Svoboda, *J. Chromatogr.*, 174 (1979) 35.
62 P. Jandera and H. Colin, *13th International Symposium on Chromatography, Cannes, France, 1980.*
63 P. Jandera, H. Colin and G. Guiochon, *Anal. Chem.*, 54 (1982) 435.
64 R. McCormic and B.L. Karger, *Anal. Chem.*, 52 (1980) 2249.
65 W. Melander, J. Stoveken and Cs. Horváth, *J. Chromatogr.*, 199 (1980) 35.
66 R.A. Hartwick, C.M. Grill and P.R. Brown, *Anal. Chem.*, 51 (1979) 34.
67 P.J. Schoenmakers, H.A.H. Billiet and L. de Galan, *J. Chromatogr.*, 185 (1979) 179.
68 L.R. Snyder, J.W. Dolan and J.R. Gant, *J. Chromatogr.*, 165 (1979) 3.

69 J.W. Dolan, J.R. Gant and L.R. Snyder, *J. Chromatogr.*, 165 (1979) 31.
70 H. Elgass, *Thesis*, University of Saarbrücken, Saarbrücken, 1978.
71 R.B. Sleight, *J. Chromatogr.*, 83 (1973) 31.
72 S.R. Abbott, J.R. Berg, P. Achener and R.L. Stevenson, *J. Chromatogr.*, 126 (1976) 421.
73 J.A. Schmit, R.A. Henry, R.C. Williams and J.F. Dieckman, *J. Chromatogr. Sci.*, 9 (1971) 645.
74 H. Colin, N. Ward and G. Guiochon, *J. Chromatogr.*, 149 (1978) 169.
75 P. Jandera, J. Churáček, J. Čáslavský and M. Vojáčková, *Chromatographia*, 13 (1980) 734.
76 P. Jandera, J. Churáček and D. Szabó, *Chromatographia*, 14 (1981) 7.
77 P. Jandera, J. Churáček, J. Čáslavský and D. Szabó, *Chromatographia*, 14 (1981) 100.
78 P. Jandera, unpublished results.
79 P. Jandera and J. Churáček, *Chromatogr. Rev.*, 86 (1973) 351.
80 P. Jandera and J. Churáček, *Chromatogr. Rev.*, 86 (1973) 423.
81 P. Jandera and J. Churáček, *Chromatogr. Rev.*, 98 (1974) 1.
82 P. Jandera and J. Churáček, *Chromatogr. Rev.*, 98 (1974) 55.
83 B. Alfredsson, S. Bergdahl and O. Samuelson, *Anal. Chim. Acta*, 28 (1963) 371.
84 U.B. Larsson and O. Samuelson, *J. Chromatogr.*, 19 (1965) 404.
85 F. Murakami, *Anal. Chem.*, 47 (1975) 390.
86 P. Jandera and J. Churáček, *J. Chromatogr.*, 104 (1975) 23.
87 P. Jandera and J. Churáček, *J. Chromatogr.*, 104 (1975) 257.
88 V. Svoboda, *Radioisotopy*, 18 (1977) 775.
89 K.S. Lee and O. Samuelson, *Anal. Chim. Acta*, 37 (1967) 359.
90 L. Bengtsson and O. Samuelson, *Anal. Chim. Acta*, 44 (1969) 217.
91 N.C. Li and J.M. White, *J. Inorg. Nucl. Chem.*, 16 (1960) 131.
92 M. Walo, *Monatsh. Chem.*, 92 (1961) 605.
93 B. Alfredsson, L. Gedda and O. Samuelson, *Anal. Chim. Acta*, 27 (1962) 63.
94 O. Samuelson and R. Simonson, *Svensk Papperstidn.*, 65 (1962) 363.
95 J. Inczédy, *Kunstharz-Ionenaustauscher Symposium, Leipzig, 1968*, Akademie Verlag, Berlin, 1970.
96 J. Inczédy, *Analytical Applications of Complex Equilibria*, Akadémiai Kiadó, Budapest, 1976.
97 F. Helfferich, *Nature (London)*, 189 (1961) 1001.
98 J.J. Latterell and H.F. Walton, *Anal. Chim. Acta*, 32 (1965) 101.
99 F. Helfferich, *J. Amer. Chem. Soc.*, 84 (1962) 3237 and 3242.
100 R. Sargent and W. Rieman, *J. Phys. Chem.*, 61 (1957) 354.
101 R. Sargent and W. Rieman, *Anal. Chim. Acta*, 18 (1958) 197.
102 R. Sargent and W. Rieman, *Anal. Chim. Acta*, 17 (1957) 408.
103 A. Breyer and W. Rieman, *Anal. Chim. Acta*, 18 (1958) 204.
104 F. Jakob, K.C. Park, J. Ciric and W. Rieman, *Talanta*, 8 (1961) 431.
105 W. Kemula and S. Brzozowski, *Roczn. Chem.*, 35 (1961) 711.
106 J. Sherma and W. Rieman, *Anal. Chim. Acta*, 18 (1958) 214.
107 J. Sherma and W. Rieman, *Anal. Chim. Acta*, 19 (1958) 134.
108 O. Samuelson and H. Strömberg, *Carbohydr. Res.*, 3 (1966) 89.
109 E. Martinsson and O. Samuelson, *J. Chromatogr.*, 50 (1970) 429.
110 S. Eksborg and G. Schill, *Anal. Chem.*, 45 (1973) 2092.
111 D.P. Wittmer, N.O. Nuessle and W.G. Haney, *Anal. Chem.*, 47 (1975) 1422.
112 D. Westerlund and A. Theodorsen, *J. Chromatogr.*, 144 (1977) 27.
113 R. Gloor and E.L. Johnson, *J. Chromatogr. Sci.*, 15 (1977) 413.
114 E. Tomlinson, T.M. Jefferies and C.M. Riley, *J. Chromatogr.*, 159 (1978) 315.
115 B.A. Bidlingmeyer, S.N. Deming, W.P. Price, B. Sachok and M. Petrusek, *J. Chromatogr.*, 186 (1979) 419.
116 H.V. Ehmcke, H. Kelker, K.H. König and H. Ullner, *Z. Anal. Chem.*, 294 (1979) 251.
117 P. Jandera and H. Engelhardt, *Chromatographia*, 13 (1980) 18.
118 K.G. Wahlund, *J. Chromatogr.*, 115 (1975) 411.
119 P. Jandera, J. Churáček and J. Bartošová, *Chromatographia*, 13 (1980) 485.

120 P. Jandera and J. Churáček, *J. Chromatogr.*, 197 (1980) 181.
121 J.L. Glajch, J.J. Kirkland, K.M. Squire and J.M. Minor, *J. Chromatogr.*, 199 (1980) 57.
122 W.K. Al-Thamir, J.H. Purnell, C.A. Wellington and R.J. Laub, *J. Chromatogr.*, 173 (1979) 388.
123 M.W. Watson and P.W. Carr, *Anal. Chem.*, 51 (1979) 1835.
124 L.R. Snyder, *J. Chromatogr. Sci.*, 16 (1978) 223.
125 G.E. Berendsen, P.J. Schoenmakers, L. de Galan, G. Vigh and Z. Varga-Puchony, *J. Liquid Chromatogr.*, 3 (1980) 1669.
126 G.Y. Vigh and Z. Varga-Puchony, *J. Chromatogr.*, 196 (1980) 1.
127 J.H. Knox, R. Kawszan and G.J. Kennedy, *Faraday Symp. Chem. Soc. (London)*, No. 15 (1980) 113.
128 P.J. Schoenmakers, H.A.H. Billiet and L. de Galan, *J. Chromatogr.*, 218 (1981) 261.
129 M.J.M. Wells and C.R. Clark, *Anal. Chem.*, 53 (1981) 1341.
130 E. Kováts, to be published.
131 A.M. Krstulović, H. Colin and G. Guiochon, *Anal. Chem.*, 54 (1982) 2438.
132 E.H. Slaats, W. Markowski, J. Fekete and H. Poppe, *J. Chromatogr.*, 207 (1981) 299.
133 Cs. Horváth and H.-J. Lin, *J. Chromatogr.*, 126 (1976) 401.
134 J.K. Johnson, S.F. Cernosek and R.M. Gutierez-Cernosek, *J. Chromatogr.*, 177 (1979) 277.
135 S.H. Unger and T.F. Feuerman, *J. Chromatogr.*, 176 (1979) 426.
136 O.A.G.J. van der Houwen, J.A.A. van der Linden and A.W.M. Indemans, presented at 14th Int. Symp. on Chromatography, London, 1982.
137 P. Jandera, J. Churáček and H. Colin, *J. Chromatogr.*, 214 (1981) 35.
138 P. Jandera, H. Colin and G. Guiochon, *Chromatographia*, 16 (1982) 132.
139 J.C. Berridge, *J. Chromatogr.*, 244 (1982) 1.
140 J.C. Berridge, presented at 14th Int. Symp. on Chromatography, London, 1982. *Chromatographia*, 16 (1982) 172.
141 P.J. Schoenmakers, A.C.J.H. Drouen, H.A.H. Billiet and L. de Galan, *Chromatographia*, 15 (1982) 688.
142 A.C.J.H. Drouen, H.A.H. Billiet, P.J. Schoenmakers and L. de Galan, *Chromatographia*, 16 (1982) 48.
143 H.J. Issaq, J.R. Klose, K.L. McNitt, J.E. Haky and G.M. Muschik, *J. Liquid Chromatogr.*, 4 (1981) 2091.
144 L.R. Snyder, J.L. Glajch and J.J. Kirkland, *J. Chromatogr.*, 218 (1981) 299.
145 J.L. Glajch, J.J. Kirkland and L.R. Snyder, *J. Chromatogr.*, 238 (1982) 269.
146 L.R. Snyder, *J. Chromatogr.*, 255 (1983) 3.
147 L.R. Snyder and J.L. Glajch, *J. Chromatogr.*, 214 (1981) 1.
148 J.L. Glajch and L.R. Snyder, *J. Chromatogr.*, 214 (1981) 21.

PART II

THEORY OF CHROMATOGRAPHY WITH PROGRAMMED COMPOSITION OF THE MOBILE PHASE

Chapter 3

CLASSIFICATION OF MOBILE PHASE GRADIENTS

In chromatography with programmed composition of the mobile phase, the composition of the mobile phase is changed during the elution with time according to a pre-set programme. In this way, it is possible to change the elution strength (or to change the elution strength and selectivity at the same time) for elution of individual sample compounds. This technique makes it possible to elute in a single chromatographic run compounds that differ widely in retention on a given column and to overcome so-called "general elution problem"[1]. If such samples are chromatographed under isocratic conditions, a relatively strong elution strength of the mobile phase is required to elute all sample components in a reasonable time. However, weakly retained compounds are eluted with retention volumes close to V_m and are very poorly separated, if at all (Fig. 3.1A). With the elution strength of the mobile phase adjusted so as to achieve the separation of weakly retained compounds, the elution of strongly retained sample components requires a very long time, the peaks are very broad and the compounds may be so diluted in the eluate that their detection becomes very difficult or impossible (Fig. 3.1B).

In gradient elution chromatography, the elution strength of the mobile phase is increased with time, the capacity factors of sample compounds decrease with time during the elution and, consequently, weakly retained sample components are eluted in the mobile phase of lower elution strength as well resolved peaks whereas the elution of strongly retained compounds is accelerated in the mobile phase of relatively high elution strength in the last part of the gradient run (Fig. 3.1C). Thus, all sample components may be eluted in a reasonable time as relatively narrow and well resolved peaks (in the ideal case). The number of sample components that can be analysed in a single chromatographic run is increased in gradient elution compared with isocratic operation[1].

The tailing of the eluted sample bands is generally much decreased in gradient elution chromatography compared with isocratic elution chromatography, where this effect may occasionally occur by virtue of non-linearity of the isotherm or column overloading. Thus, a relatively accelerated migration of the tail of the band in comparison with the front of the peak leads to an automatic correction effect for band tailing in gradient elution chromatography. Further, the column may be loaded with a larger total amount of sample than under iso-

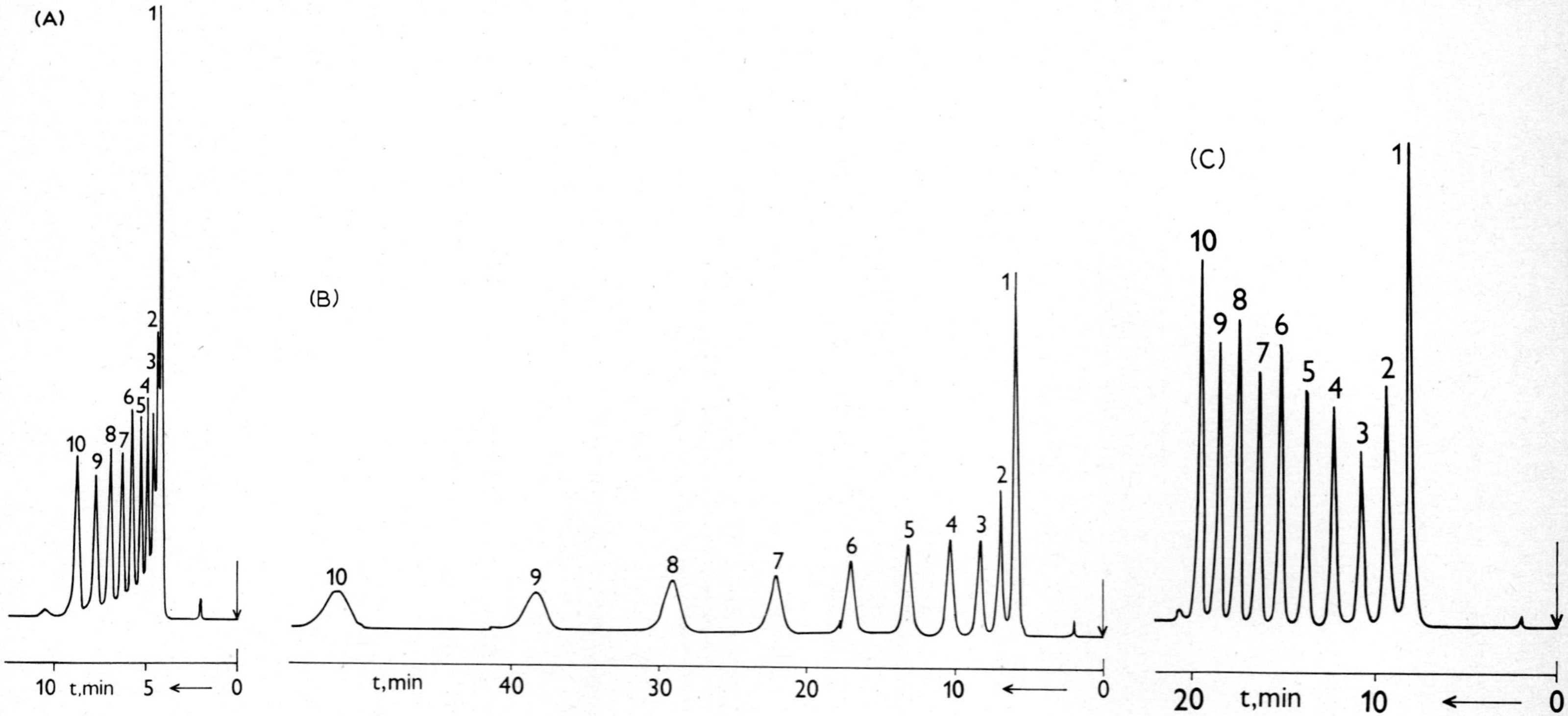

Fig. 3.1. Reversed-phase chromatographic separation of 1,2-naphthoylenebenzimidazolesulphonamide derivatives of a homologous series of C_1-C_{10} *n*-alkylamines. Column, C18 on LiChrosorb Si 100, 10 µm; 300x4.2 mm I.D.; V_m=3.1 ml; flow-rate, 1 ml/min; detection, fluorimetric, λ_{ex}=365 nm; λ_{em} > 410 nm. Mobile phase: (A) methanol-water (95:5, v/v), (B) methanol-water (80:20, v/v), (C) linear concentration gradient from methanol-water (70:30) to 100% methanol in 20 min. Numbers on peaks agree with the carbon numbers in the *n*-alkylamines.

cratic conditions. Large volumes of sample may be loaded on the column under the initial conditions, where the capacity factors of sample compounds are relatively high and the sample compounds are sorbed on the top of the column. This does not lead to deterioration of resolution under gradient elution conditions. This effect, together with band compression, leads to an increased sensitivity in gradient elution compared with isocratic elution chromatography.

Mobile phase gradients may be classified according to the number of components of the mobile phase or according to the form of the mobile phase concentration change with time (i.e., mobile phase programme or mobile phase gradient)[1-4]. Binary gradients are formed from two components of the mobile phase, *a* (less efficient eluting agent) and *b* (more efficient eluting component). The two solutions A and B used for preparing a binary gradient are either pure component *a* or *b*, or contain their mixtures in different proportions. Further compounds can be added at same concentration to the two solutions. If the concentrations of three components of the mobile phase are changed simultaneously during gradient elution, such gradients are termed ternary gradients. Recently, the use of multi-solvent gradients has been proposed, in which the concentrations of four or more components in the mobile phase are changed at the same time. Such gradients may prove potentially useful for programming the separation selectivity and retention either simultaneously or independently of one another (to a certain extent) during the chromatographic run. This aim can be approached theoretically using gradients formed from solvents having the greatest difference in potential chemical interaction with the solutes to be separated. Thus, in reversed-phase liquid chromatography, water, methanol (proton acceptor), acetonitrile (proton donor) and tetrahydrofuran (dipole-dipole interactions) can be used to form four-component mobile phase gradients, while a non-polar hydrocarbon (such as *n*-hexane or *n*-heptane), dichloromethane (non-localized dipole interactions), acetonitrile (localized dipole interactions) and di-*tert*.-butyl ether (localized proton acceptor interactions) are potentially suitable for this purpose in liquid-solid chromatography[84].

Instruments where three, four or more liquids (components of the multi-solvent gradient) are mixed in pre-programmed proportions changing with time have recently become commercially available[5-7]. However, a number of ternary gradients (those where the concentration is changed in a linear manner with time) may be formed by mixing only two solutions if each solution contains the three components of the mobile phase in different proportions.

Binary mobile phase gradients are at present used far more frequently than ternary gradients, because they are much simpler to understand and control. However, the following classification of different forms of binary mobile phase gradients may also be used for ternary or even more complex gradients.

Mobile phase gradients are usually characterized as a time function of the concentration of the more efficient eluting component in the mobile phase at the outlet from the gradient-generating device (or, more precisely, at the top of the column):

$$c = f(t) \tag{3.1}$$

It should be noted that this time function is not always identical with the gradient programme that is used to control the mixing of the two solutions A and B contained in reservoirs of the gradient elution equipment. This gradient programme represents the change in the content of solvent B in the mixed mobile phase with time, which can be expressed as

$$\phi = f(t) \tag{3.2}$$

where ϕ is the percentage by volume of B x 10^{-2}. The solutions A and B are not necessarilly the pure components a and b of the mobile phase, but each solution may contain different concentrations of the more efficient eluting agent b in the less efficient component of the mobile phase a. Let c_A be concentration of b (c) in solution A and c_B concentration of b in solution B. In this instance, the time function of the concentration c may be expressed as

$$c = c_A + (c_B - c_A)f(t) \tag{3.3}$$

which may differ from the function described by eqn. 3.2. The same function $c = f(t)$ can be produced using different gradient programs $\phi = f(t)$ for various combinations of the concentrations of the more efficient eluting component in the two solvents, as is demonstrated by the example in Fig. 3.2, where five different methods that can be employed to achieve the required concentration gradient of the more efficient eluting component in the mobile phase are indicated.

The time function $\phi = f(t)$ may be directly pre-programmed in the electronic part of most commercially available modern instrumentation. With certain instruments, any form of the function may be used; other equipment simulates the curved gradients as a sequence of short linear steps; finally, some gradient elution chromatographs make possible the use of only a few forms of gradients (linear or curved). This will be discussed in more detail in Part III.

Simple gradients are described by a continuous $c = f(t)$ function (Fig. 3.3A,B). Simple gradients may differ from one another in three respects, which it is always necessary to define clearly[8]:

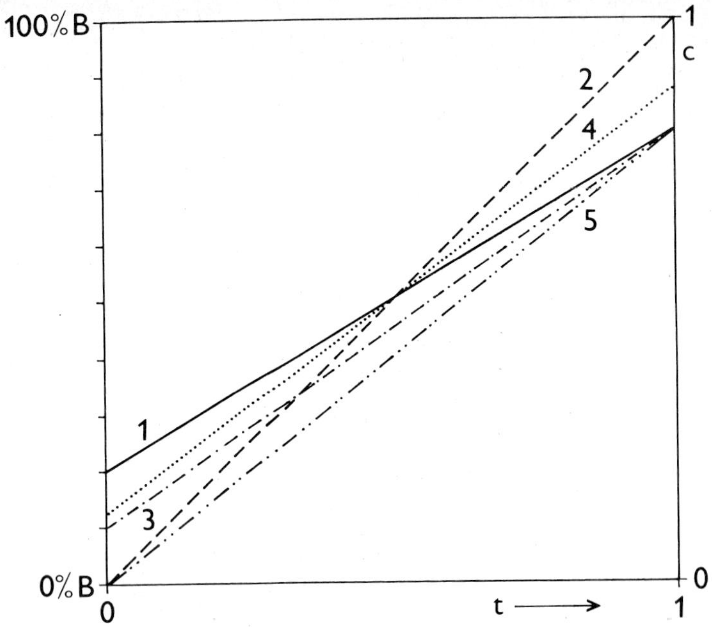

Fig. 3.2. Five examples of different binary gradient programmes (percentage of solvent B versus time) that can be used to form the same concentration gradient of the more efficient eluting agent b in the mobile phase: $c = 0.2 + 0.6t$ [$c \equiv$ % (v/v) of $b \cdot 10^{-2}$]. Curves: 1, %B = 20 + 60t; $c_A = 0$, $c_B = 1$; 2, %B = 100t, $c_A = 0.2$, $c_B = 0.8$; 3, %B = 10 + 70t, $c_A = 0.114$, $c_B = 0.971$; 4, %B = 12.5 + 75t, $c_A = 0.1$, $c_B = 0.9$; 5, %B = 80t, $c_A = 0.2$, $c_B = 0.95$. c_A, c_B are concentrations (c) of b in solvents A and B, respectively, which are used to form the binary gradient; t = time of gradient (arbitrary units).

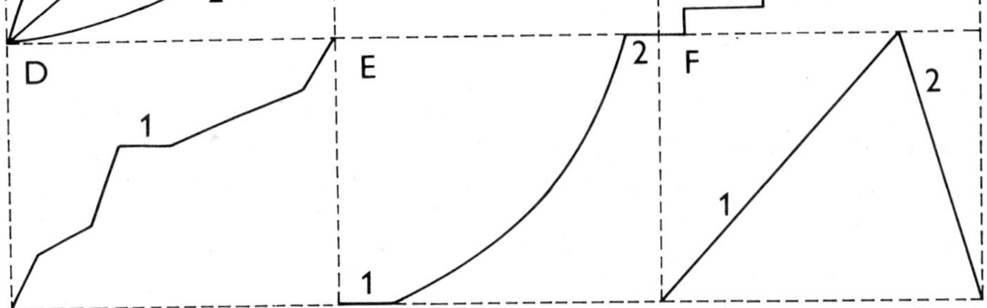

Fig. 3.3. Examples of various types of mobile phase gradients: A and B, simple continuous gradients - linear (1), concave (2) and convex (3); C, stepwise gradient; D, composed multilinear gradient with an isocratic hold-up step (1); E, concave gradient with initial delay (1) and a hold-up step after the end of the gradient (2); F, linear gradient (1) followed by a steep reverse gradient to speed-up the re-equilibration of the initial conditions.

References on p. 180.

(a) The shape (curvature) of the gradient. So-called linear gradients are controlled by a linear $c = f(t)$ function (curve 1 in Fig. 3.3A and B). Curved gradients may be either concave (curve 2 in Fig. 3.3A and B) or convex (curve 3 in Fig. 3.3A and B).

(b) The slope (steepness) of the gradient. This may be understood as an average change in c per unit time in the course of the gradient run. The instantaneous slope is constant for linear gradients, increases with time for concave gradients and decreases with time for convex gradients (the gradients in Fig. 3.3A are steeper than those in Fig. 3.3B).

(c) The initial concentration of the more efficient component in the mobile phase at the beginning of the gradient run, c_0, which may be either zero (Fig. 3.3A) or greater than zero (Fig. 3.3B).

In addition to simple gradients, composed gradients are frequently used, which are divided into several sequences (steps), differing in the character of the $c = f(t)$ function in each step. If the composition of the mobile phase is kept constant during each step and is changed to another constant value between the subsequent steps (Fig. 3.3C), such a gradient composed of isocratic steps is termed a stepwise gradient.

The most complex gradients are composed of several subsequent steps, where the concentration of the more efficient eluting component in the mobile phase is changed according to different continuous $c = f(t)$ functions in each step, as is shown in Fig. 3.3D, where linear gradients are used in each step. An isocratic step may be inserted into such a composed gradient (1 in Fig. 3.3D) or such a step may be formed by stopping the simple continuous gradient for a certain time period. Such an isocratic step inserted into the gradient run is called gradient hold-up.

In practice, an isocratic step, with either $c_0 = 0$ or $c_0 > 0$, often precedes the simple or composed gradient. This initial hold-up of the gradient is known as a gradient delay (1 in Fig. 3.3E). In practical operation with most instruments for gradient elution chromatography one should be aware of a certain minimum gradient delay, t_z, which is determined by the inner volume of the instrumentation between the point of mixing of the mobile phase components and the top of the column (V_z), which usually amounts to 1-5 ml. This gradient delay may be compensated for by delaying the sample injection onto the column at time t_z. If the mobile phase contained in the space between the point of mixing and the column at the beginning of the gradient has a very low elution strength for all sample compounds, gradient delay does not change the differences in their elution volumes, but only increases the retention times by t_z.

Often certain sample compounds are retained so strongly that they are not eluted in course of the gradient and a further, isocratic step using pure component *b* of the mobile phase is necessary to achieve their elution (2 in Fig. 3.3E). This is a special instance of two-step elution with the first, gradient step followed by second, isocratic step; a three-step elution may occur, if t_z cannot be neglected (Fig. 3.3E). This situation occurs mainly if the solvent *b* used to form the mobile phase gradient is not a sufficiently stronger eluent than solvent *a*.

Several workers have preferred to use two solvents with relatively small differences in polarities (elution strengths) to form binary gradients in chromatography on polar adsorbents to suppress the so-called solvent demixing effect (see discussion in Part III), which may adversely affect the separation under certain circumstances. The relative influence of this effect increases with increasing difference in the polarities of the mobile phase components used to produce the gradient. To avoid large differences in solvent polarities and to obtain a large polarity change during the gradient run, which is necessary for eluting compounds that differ widely in retention, Scott and Kucera[9-13] introduced the so-called "incremental gradient elution", where the gradient is composed of several subsequent steps. In the first step beginning with pure least polar solvent *a*, the composition of the mobile phase is changed to a more polar solvent *b*. In the following step, *b* is the weaker solvent and the mobile phase composition is changed from pure solvent *b* to pure solvent *c*, more polar than solvent *b*, and so on until the last step is finished and the mobile phase is represented by the pure most polar solvent of the solvent series. Originally, twelve solvents of increasing polarity were suggested (from *n*-heptane to water)[12], but later six solvents were proposed as sufficient[14].

However, special instrumentation is required for incremental gradient elution, which is not available commercially, and the operation is complicated from the practical point of view (mass transport detectors must be used). Therefore, this technique has not been widely adopted by practicing chromatographers. Moreover, the solvent demixing effect does not seem to be as detrimental to chromatographic runs, if some precautions are taken, as had been supposed originally.

So-called reverse gradients (2 in Fig. 3.3F), i.e., gradients where the elution strength of the mobile phase decreases with time, are not used in the chromatographic run, but they are advantageous for restoring quickly the initial conditions after the completion of the gradient run. These gradients are usually much steeper than those used for elution of sample compounds and the time of re-equilibration of the column before the start of a further experiment is usually shorter than when the composition of the mobile phase is abruptly changed to the initial conditions after the end of the gradient elution experiment.

References on p. 180.

Chapter 4

SIMPLE (CONTINUOUS) GRADIENT ELUTION CHROMATOGRAPHY

4.1. BASIC CONCEPTS OF THEORETICAL SOLUTION

Calculations of retention characteristics in gradient elution chromatography are based on the fundamental equation describing the distribution of a solute between the stationary and mobile phases[1,3,4]. In principle, it is possible to calculate the whole elution curve of a solute using numerical methods[15], as in isocratic elution chromatography, but the calculation approach is complex and does not yield an immediate picture of the influence of various operating conditions on the chromatographic behaviour of sample compounds. In practice, it is sufficient to calculate retention volumes and bandwidths, which can be further used for calculation of separation factors and resolution.

Several workers have published equations that make possible calculations of retention volumes in gradient elution chromatography for certain specific separation problems and for certain types of gradients[16-36]. The symbols and the treatments of the problem used differ, and it therefore does not seem very useful to reproduce the treatments of the individual workers in detail here (this has already been done to a certain extent in the book by Liteanu and Gocan[16]). Rather, a general method of calculation will be briefly described. The resulting equations for important retention characteristics can be found in Appendix 2 for various types of gradients and chromatographic systems, including most of previously published contributions, both the older ones and the more recent works of Snyder and co-workers[3,36-38], Schoenmakers and co-workers[39-41], Jandera and co-workers[4,8,42-57] and other groups[58-63]. The symbols in these equations have been unified so that it is possible to use easily the results for direct calculations in various chromatographic systems using the gradient elution technique[4].

The calculation of retention volumes in gradient elution chromatography is based on the following basic idea. A differential increase in the volume of the eluate from the column (the volume of the mobile phase that has passed through the column), dV, causes a migration of the solute band maximum by a distance corresponding to a differential fraction of the column void volume, dV_m. During this differential migration, the capacity factor, k', of the solute does not change and consequently

References on p. 180.

$$dV = k'dV_m \tag{4.1}$$

This equation may be written in another form, suitable for integration between $V = 0$ and $V = V'_g$ (net retention volume in gradient elution chromatography):

$$\int_0^{V_m} dV_m = V_m = \int_0^{V'_g} \frac{1}{k'} \cdot dV \tag{4.2}$$

This fundamental equation, derived by Drake[17] and Freiling[20,21], has been used in various modifications.

To make possible the solution of eqn. 4.2 for retention volume, it is necessary to know the relationship between the solute capacity factor and the volume of the eluate, V, from the start of the gradient. Of course, k' versus time relationships may also be used, as eqn. 4.2 may be written in another form[1]:

$$\int_0^{t_m} dt_m = t_m = \int_0^{t'_R} \frac{1}{k'} \cdot dt \tag{4.3}$$

Eqns. 4.2 and 4.3 are equivalent and eqn. 4.3 is obtained when eqn. 4.2 is divided by the volume flow-rate of the mobile phase. Thus, it is possible to obtain a solution either for retention time or for retention volume in gradient elution chromatography. The latter quantity is to be preferred, as the retention volumes do not depend on the mobile phase flow-rate for a given column.

4.2. LINEAR SOLVENT STRENGTH GRADIENTS

To provide a simple solution to eqn. 4.3, Snyder and co-workers[1,3,34-38] introduced the concept of "linear solvent strength gradients". It is assumed that the mobile phase composition gradients are designed for each chromatographic system in such a way that the logarithms of the capacity factors of all of the chromatographed compounds decrease linearly with increasing volume of the eluate (or with increasing time from the start of the gradient run):

$$\log k' = \log k'_a - \beta\left(\frac{t}{t_m}\right) = \log k'_a - \beta\left(\frac{V}{V_m}\right) \tag{4.4}$$

where k'_a is the capacity factor of a given solute at the beginning of the gradient elution and β is the parameter defining the steepness (slope) of the "linear solvent strength" gradient. In such a definition, β involves a contribution depending on the chromatographic behaviour of solutes in a given chromatographic system. This contribution, however, usually differs for individual sample components and consequently it is not possible to have an exactly equal value of the parameter β for all solutes chromatographed in a single gradient run. Thus, the steepness parameter β in eqn. 4.4 should be understood as an approximate parameter, being an average value for various sample components.

In reversed-phase chromatography on chemically bonded non-polar stationary phases, where the dependence of capacity factors on the concentration of the organic solvent in the mobile phase may be conveniently described by eqn. 1.28, "linear solvent strength" gradients are virtually identical with the gradients where the content of the organic solvent in the mobile phase (c) is changed in a linear manner (linear concentration gradients):

$$c = A + BV \tag{4.5}$$

After combination with eqn. 1.28 we obtain

$$\log k' = \log k'_0 - mA - mBV \tag{4.6}$$

where k'_0 and m are the parameters of the $k' = f(c)$ relationship (eqn. 1.28). This equation has a form equivalent to eqn. 4.4, where

$$\log k'_a = \log k'_0 - mA \tag{4.7a}$$

and

$$\beta = mBt_mF_m = mBV_m \tag{4.7b}$$

where F_m is the volume flow-rate of the mobile phase and V_m is the column dead volume. Thus, the steepness parameter β of eqn. 4.4 includes both the slope (steepness) of the linear concentration gradient and the parameter m, which depends on the nature of solute and of the chromatographic system used (namely on the nature of the chemically bonded phase and of the organic solvent in the mobile phase).

In linear solvent strength gradients, β and the parameter m in eqn. 1.28 are assumed not to vary very significantly with the nature of the sample solutes. For such gradients, the capacity factors of all sample compounds change in approximately the same manner with time (or with the volume of the eluate). The bandwidths are approximately equal for all sample compounds eluted in course of a "linear solvent strength" gradient, and should yield equal resolution and regular spacing in different parts of the chromatogram for pairs of compounds with equal separation factors, α.

Linear gradients of the concentration of the organic solvent in reversed-phase chromatography yield chromatograms that deviate more or less from the

References on p. 180.

"ideal" chromatogram for "linear solvent strength" gradients, because the idealized assumptions do not apply fully in practical chromatographic systems. First, regular spacing of the peaks of chromatographed compounds in a chromatogram is possible, even under idealized conditions, only for samples for which the separation factors of all neighbouring pairs of compounds in the chromatogram are approximately equal and do not depend significantly on the composition of the mobile phase. Such mixtures are found relatively rarely in practice. However, homologous series, benzologous series, oligomeric series and samples containing solutes that differ regularly in the number of functional groups obey closely this requirement.

Further, the solutes in practical samples differ more or less in the values of the gradient steepness, β, because the parameter m in eqn. 1.28 depends, more or less, on the structure of the sample compounds[40,41,50]. An approximately linear increase in m was reported for increasing parameters k'_0 in this equation[40,41]. This means that the parameter m increases with increasing retention of solutes and a regular increase in m can be observed in homologous, benzologous and oligomeric series. To compensate for this effect, the steepness of the concentration gradient should decrease slightly with time (or with increasing volume of the eluate) so as to keep an approximately equal steepness parameter β for all sample solutes. This means that, theoretically, slightly convex concentration gradients provide a better approximation to "linear solvent strength" gradients than the linear concentration gradients in reversed-phase chromatography[37,38,40,41]. In practice, linear concentration gradients are often preferred for the sake of simplicity and they can be considered as a good approximation to "linear solvent strength" gradients with respect to bandwidth and peak spacing in the chromatogram[3,37,38]. However, we shall take into account that such gradients cannot be understood as exactly "linear solvent strength" gradients for individual sample compounds. Fig. 4.1 shows schematically examples of concentration gradients required to achieve an approximately "linear solvent strength" gradient. The gradient of concentration of the organic solvent represented by curve 1 assumes approximately equal values of m for all sample compounds, and the gradient shown by curve 2 assumes a regular increase in m with increasing retention of solutes.

Curve 3 in Fig. 4.1 shows an approximately "linear solvent strength" gradient in a normal-phase system (chromatography on a polar adsorbent, such as silica or alumina). For this purpose, a concave gradient of concentration of a more polar solvent (c) in a less polar solvent should be used[1-5]. If we assume an approximate validity of a two-parameter relationship between k' and c (eqn. 1.23), it can be easily shown that an exponential concentration gradient yields the best approximation to the "linear solvent strength" gradient[4]:

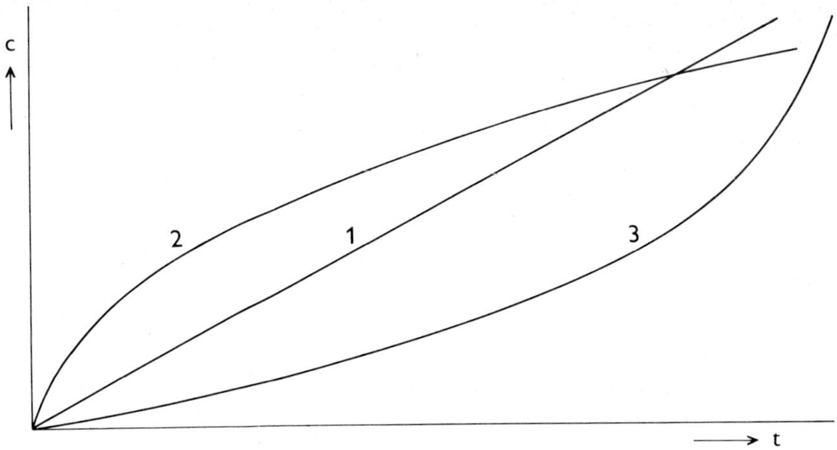

Fig. 4.1. Examples of "linear solvent strength" gradients in various chromatographic systems. 1, Reversed-phase chromatography, parallel log $k' = f(c)$ lines; 2, reversed-phase chromatography, the slopes of the log $k' = f(c)$ lines increase with increasing retention; 3, normal-phase chromatography on polar adsorbents, ion-exchange chromatography. t = time from the beginning of gradient; c = concentration of the more efficient eluting component in the mobile phase.

$$c = A'e^{B't} = 10^{A} \cdot 10^{BV} \tag{4.8}$$

After introducing eqn. 1.23, we obtain

$$\log k' = \log k'_0 - mA - mBV \equiv \log k'_a - \beta(\frac{V}{V_m}) \tag{4.9}$$

where eqns. 4.7a and 4.7b apply for k'_a and β, respectively. An exponential change in the concentration of the stronger (more polar) solvent in the mobile phase corresponds to a linear change in the solvent strength, ε, of the mobile phase (see Part I). The term "linear solvent strength" gradient was introduced originally for gradients employing a linear change in ε with time in normal-phase chromatography on polar adsorbents[1,34-36].

Analogous arguments may be applied to other liquid chromatographic systems discussed in Part I.

"Linear solvent strength" gradients may be very useful for providing adequate separations of unknown mixtures and are to be preferred in a first attempt to separate unknown complex mixtures[2,3]. However, these gradients do not yield optimum separations of certain solute mixtures and some commercial and laboratory equipment is not capable of reproducing certain shapes of concentration gradients required to form "linear solvent strength" gradients. Further, real optimization of a given separation problem under gradient conditions requires a "tailor-made" gradient, where certain more general forms of gradients often provide a more flexible tool for the selection of optimum separation conditions.

It appears to us advantageous to treat gradient elution in a more general way, whereas "linear solvent strength" gradients may be described as useful special instances of general concentration gradients[4].

4.3. THE PROFILE OF THE GRADIENT AND THE GRADIENT FUNCTION

A general approach to the solution of eqn. 4.2 or 4.3 requires the expression of the dependence of solute capacity factors on the volume of eluate (or time from the start of the gradient) in the form of two partial functions[42]. The first function describes the relationship between the solute capacity factor and the concentration of the more efficient eluting component in the binary mobile phase, $k' = \mathrm{f}(c)$. This function is discussed in detail in Part I for different liquid chromatographic systems and various useful mathematical forms of this function for different systems are listed in Table 2.1.

The second partial function is used to express the concentration, c, as a function of time or, better, of the volume of eluate; this means that it describes mathematically the concentration gradient. We suggested[8] denoting this partial function as a "gradient function", $c = \phi(V)$.

Using the two partial functions, we can rearrange eqn. 4.2 into the following general form:

$$\int_0^{V_\mathrm{m}} \mathrm{d}V_\mathrm{m} = V_\mathrm{m} = \int_0^{V'_\mathrm{g}} \frac{\mathrm{d}V}{\mathrm{f}[\phi(V)]} \tag{4.10}$$

An advantage of this approach is the clear separation of the contributions of a solute and of the given chromatographic system, which cannot be influenced by the experimenter without changing the nature of the system from the profile of the concentration gradient described by the gradient function, which can be selected, corrected and changed arbitrarily according to the experimenter's need and will. Of course, certain experimental and theoretical aspects should be respected when selecting an adequate gradient function.

First, the available forms of the gradient function may be limited by the instrumentation used, because a number of gradient liquid chromatographs are able to provide only certain types of gradient functions[1,3,4,64]. Further, certain instruments may not reproduce exactly the pre-programmed gradient function and the gradient formed may deviate more or less from the required profile (see Part III for a more detailed discussion).

If the instrumentation used makes it possible, it is advantageous for a single mathematical form of the gradient function to describe a variety of concentration gradient profiles, differing in the steepness (slope), the shape (curvature) and the initial mobile phase composition. It would be useful to chose the gradient function in such a way that these three characteristics of the gradient profile could be controlled independently of one another[8]. An ideal gradient function should have a minimum number of adjustable parameters to allow for easy and straightforward control. Finally, the mathematical form of the gradient function should allow an easy and explicit solution of eqn. 4.10 for retention volume, to allow theoretical calculations of the important retention characteristics under gradient conditions and of the optimum separation conditions for a given problem.

Table 4.1 gives a survey of potentially useful general gradient functions together with various forms of gradient functions produced by certain types of instrumentation for gradient elution, which are limited with respect to the gradient profile (eqns. 4.11-4.20). A more detailed discussion of the equipment for gradient elution will be given in Part III; a formal description of various gradient profiles by means of gradient functions in Table 4.1 is sufficient for our present purpose.

The older types of equipment for gradient elution of the exponential or proportional type[1,16] make it possible to form gradient profiles controlled by gradient functions described by eqns. 4.11-4.14 in Table 4.1. In the Varigrad system[65], a number of different gradient profiles may be created; very frequent are gradients described by gradient functions according to eqn. 4.18 in Table 4.1. For further details, see Part III.

Gradient elution devices used in modern high-pressure chromatographs are based on the direct mixing of two or more components of the mobile phase and their exact delivery, either continuously by means of two or more individual pumps or by means of time-proportioning valves[3,64]. These instruments are equipped with an electronic programmable device that controls the profile of the gradient. The gradient can be programmed either as a certain, flexible form of gradient function, such as eqn. 4.16 in Table 4.1 is used in the Waters M660 gradient programmer, or any desired gradient profile (gradient function) can be drawn and introduced into a programmable electronic device where it is read and reproduced precisely in the profile of the gradient formed. Finally, an arbitrary continuous gradient function may be approximated by a number of short linear sequences composing the gradient profile.

These devices make it possible to select an adequate gradient function that is best suited to the above requirements. Eqns. 4.14-4.16 define gradient functions sufficiently flexible to describe a wide variety of convex, concave and linear gradients. In the gradient functions given by eqn. 4.15, suggested by Jándera and Churáček[8], and eqn. 4.16, used in the Waters M660 gradient programmer[64], the three parameters A, B and κ characterize conveniently the

References on p. 180.

TABLE 4.1

SURVEY OF GRADIENT FUNCTIONS $c = \phi(V)$ COMMONLY USED OR POTENTIALLY USEFUL IN CONTINUOUS GRADIENT ELUTION CHROMATOGRAPHY

Symbols used; c_0 = initial concentration of the efficient eluting component b at the beginning of gradient elution; c_g = concentration of b at the outlet from gradient-generating device corresponding to the total volume of the eluate, V_G. For devices based on one or two mixing chambers and eventually an additional reservoir: c_0 = initial concentration of b; p_1 = initial density of the solution in mixing chamber 1 (close to the outlet from the device) of volume V_1 containing an initial volume V_{01} of the solution; c_R = initial concentration of b; p_2 = initial density of the solution in mixing chamber 2 (eventually the reservoir) of volume V_2 containing an initial volume V_{02} of solution; c_{RA} = concentration of b in an additional reservoir; Q_1, Q_2 = volume flow-rates between the two mixing chambers (Q_1) and at the outlet from mixing chamber 1 to the column (Q_2). For Varigrad system with N equal-volume reservoirs containing the same volume of solutions, with initial concentration of $b = c_R$ in the Nth chamber from the outlet; in other chambers $c = 0$: V_t = total initial volume of liquid in all chambers.

$c = \phi(V)$	Eqn. No.	Type	Meaning of constants	Available device
(1) $c = A + (DV - B)e^{-V\kappa}$	4.11	Exponential (convex)	$\kappa = \frac{1}{V_c}$; $B = c_{RA} - c_0$; $A = c_{RA}$; $D = \frac{c_{RA} - c_R}{V_c}$; $V_c = V_1 = V_2 = V_{01} = V_{02}$; $Q_1 = Q_2$	Exponential device; two constant-volume mixing chambers with an additional reservoir
(1a) $c = A - Be^{-V\kappa}$	4.12	Exponential (convex)	As in (1); $c_{RA} = c_R$; $D = 0$	Exponential device; one constant-volume mixing chamber and one reservoir

References on p. 180.

(1b)	$c = Be^{-V\kappa}$	4.13	Exponential (convex)	As in (1); $c_{RA} = c_R = 0$; $D = 0$; $A = 0$; $B = c_0$	Exponential device; continually decreased concentration of b
(2)	$c = D - (A^{1/\kappa} - BV)^{\kappa}$	4.14	Concave ($\kappa > 1$) or convex ($\kappa < 1$)	(I) $\kappa = \frac{V_2}{V_1} = \frac{V_{02}}{V_{01}}$; $\left(\kappa = \frac{V_{02}p_1}{V_{01}p_2} \text{ for } p_1 \neq p_2\right)$ $A = c_R - c_0$; $D = c_R$; $B = \frac{(c_R - c_0)^{1/\kappa}}{V_{02} + V_{01}}$	Proportional volume device; gravity flow
				(II) $\kappa = \frac{Q_1}{Q_2 - Q_1}$ $A = c_R - c_0$; $D = c_R$; $B = \frac{Q_2 - Q_1}{Q_2 V_{01}}(c_R - c_0)^{1/\kappa}$	Proportional flow-rate device; pump forced flow
				(III) $A = D = c_g$ $B = c_g^{1/\kappa}/V_G$	Programmable device

(Continued on p. 76)

TABLE 4.1 (continued)

$c = \phi(V)$	Eqn. No.	Type	Meaning of constants	Available device
(3) $c = (A^{1/\kappa} + BV)^{\kappa}$	4.15	Concave ($\kappa > 1$) or convex ($\kappa < 1$)	$A = c_0$; $B = \frac{c_g^{1/\kappa} - c_0^{1/\kappa}}{V_G}$ κ arbitrary, > 0	Programmable device
(4) $c = A + BV^{\kappa}$	4.16	Concave or convex	$A = c_0$; $B = \frac{(c_g - c_0)}{V_G^{\kappa}}$; κ arbitrary, > 0	Preprogrammed function-type electronic gradient programmer, such as Waters M660
(3a) $c = A + BV$	4.17	Linear ($\kappa = 1$)	$A = c_0$; $B = \frac{c_g - c_0}{V_G}$ $\kappa = 1$	Proportional volume device with $V_2/V_1 = V_{02}/V_{01} = 1$; proportional flow-rate device with $Q_2 = 2Q_1$
(3b) $c = BV^{\kappa}$	4.18	Concave ($\kappa > 1$) or convex ($\kappa < 0$); zero initial concentration, c_0	$A = c_0 = 0$; $B = \frac{c_g}{V_G^{\kappa}}$ κ arbitrary, > 0	Varigrad N-chamber device; $c = c_R$ in chamber N; $c = 0$ in chambers $< N$

				Varigrad system with N chambers: $B = \frac{c_R}{V_t^{(N-1)}}$; $\kappa = N - 1$	
(3c)	$c = BV$	4.19	Linear (κ = 1); zero initial concentration	$A = c_0 = 0$; $B = \frac{c_g}{V_G}$ $\kappa = 1$	As (3a)
(5)	$c = \kappa \log (A^{1/\kappa} + BV)$	4.20	Logarithmic	$A = 10^{c_0}$; $B = \frac{10^{c_g/\kappa} - 10^{c_0/\kappa}}{V_G}$; κ arbitrary, > 0	Programmable device

three different properties of the gradient profile. The parameter κ is a measure of the shape (curvature) of the gradient, which is concave at $\kappa > 1$, linear at $\kappa = 1$ and convex at $\kappa < 1$. The parameter A denotes the initial concentration of the stronger eluting agent b in the mobile phase, $A = c_0$, and B describes the steepness (slope) of the gradient, as expressed by the following equations:

$$B = \frac{c_g^{1/\kappa} - c_0^{1/\kappa}}{V_G} \tag{4.20a}$$

which applies with eqn. 4.15, and

$$B = \frac{c_g - c_0}{V_G^{\kappa}} \tag{4.20b}$$

which applies with eqn. 4.16. In eqns. 4.20a and 4.20b, c_g denotes the concentration of b at the outlet of the gradient device (or on the top of the column), corresponding to the volume V_G of the eluate. If $c_g = c_{fin}$, which is the final concentration of the component b at the end of the gradient run, then V_G denotes the volume of the gradient[61,62]. If the concentration of the more efficient eluting agent b in the mobile phase is changed linearly from 0% at the beginning to 100% at the end of the gradient ($\kappa = 1$; $c_0 = 0$ and $c_g = 1$), then

$$B = \frac{1}{V_G} \tag{4.21}$$

Fig. 4.2 shows sixteen examples of various linear, convex and concave gradient profiles controlled by a gradient function described by eqn. 4.15 with different values of the parameter κ (curvature) and B (steepness). $A = 0$ in all of these examples and consequently eqn. 4.15 becomes simplified to the form of eqn. 4.18 in Table 4.1.

The meaning of the constants κ and B in eqn. 4.14 is the same as that in eqns. 4.15 and 4.16. The initial concentration, c_0, however, is given as the combination of two other constants[4]:

$$c_0 = D - A \tag{4.22}$$

Because the three general exponential functions expressed by eqns. 4.14-4.16 can describe the same variety of gradient profiles, it is more advantageous to use the gradient function defined by eqn. 4.15 or 4.16 when a programmable gradient device is available, because the latter two functions include only three adjustable parameters, each of which describes another characteristic of

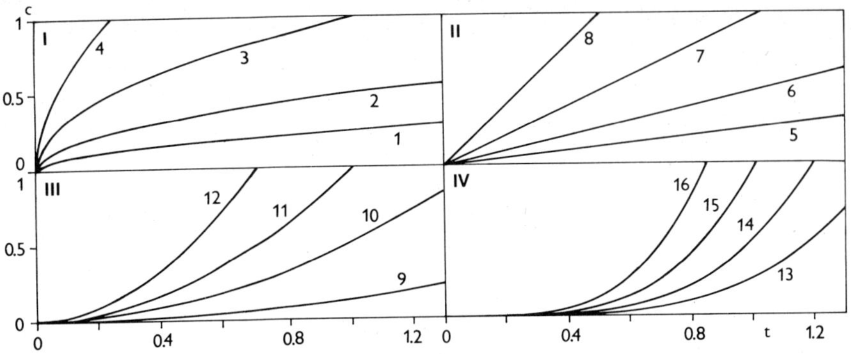

Fig. 4.2. Examples of linear, convex and concave gradient profiles controlled by gradient function described by eqn. 4.18 (A = 0 in eqn. 4.15 or 4.16). I, κ = 0.5, B = 0.0016 (1), 0.0065 (2), 0.0260 (3), 0.1039 (4); II, κ = 1.0, B = 0.0065 (5), 0.0130 (6), 0.0260 (7), 0.0519 (8); III, κ = 2.0, B = 0.0092 (9), 0.0184 (10), 0.0260 (11), 0.0367 (12); IV, κ = 4.0, B = 0.0184 (13), 0.0218 (14), 0.0260 (15), 0.0309 (16).

the gradient profile. These two equations are equivalent in versatility and flexibility, but eqn. 4.15 can yield an algebraic solution for retention volumes with a larger variety of $k' = f(c)$ functions than eqn. 4.16[4,55,56], as will be demonstrated later. For linear gradients, where κ = 1, all three gradient functions become identical and yield eqn. 4.17 in Table 4.1. If the gradient starts from zero initial concentration c_0 (A = 0), the same eqn. 4.18 results from eqn. 4.15 or 4.16, but the parameter B in eqn. 4.18 is then defined in different ways for each of the two original gradient functions[4].

Sometimes, the change in the concentration of the stronger eluting agent in the mobile phase is used only for adjusting another property of the mobile phase, such as the pH by the addition of a weak acid to the solution containing the salt of this acid with a strong base, or by addition of a weak base to the solution containing the salt of this base with a strong acid. If the concentration of the salt, c_s, is kept constant during the gradient elution, the concentration of H^+ ions in a pH gradient is proportional to the concentration of the weak acid, c_a[1,23]:

$$[H^+] = \frac{K_a}{c_s} \cdot c_a \qquad (4.23)$$

where K_a is the acidity constant of the weak acid. In this instance, gradient functions as expressed by eqns. 4.11-4.20 in Table 4.1 can be used, with the appropriate introduction of the proportionality constant K_a/c_s. This also holds for the combination of a weak base and its salt with a strong acid.

4.4. CALCULATION OF RETENTION VOLUMES IN GRADIENT ELUTION CHROMATOGRAPHY

After introduction of the appropriate $k' = \mathrm{f}(c)$ function (Table 2.1), which applies in given chromatographic system and of the gradient function used, $c = \phi(V)$ (Table 4.1), into the general equation 4.10, this may be integrated and the formulae for calculation of retention volumes under gradient conditions may be derived[4,56,57]. However, this approach does not yield algebraic solutions for all combinations of gradient functions and $k' = \mathrm{f}(c)$ relationships. The resulting equations for retention volume in gradient elution chromatography that can be derived for various combinations of $k' = \mathrm{f}(c)$ functions from Table 2.1 and of gradient functions from Table 4.1 are surveyed in Appendix 2. The list of equations in Appendix 2 involves most of the relationships published previously by various workers and also other possible solutions. To make possible a comparison of the resulting relationships for gradient functions in different chromatographic systems, the symbols used had to be unified and the parameters of the $k' = \mathrm{f}(c)$ and $c = \phi(V)$ functions from Tables 2.1 and 4.1 were used consistently. The equations for retention volumes (or retention times) published previously by different workers were checked and found to agree with the equations presented in Appendix 2, when the necessary transformation of symbols was carried out[4,57]. The mathematical solution in Appendix 2 is consistently presented for retention volumes rather than for retention times, and thus it does not depend on the flow-rate of the mobile phase.

To illustrate better the approach of the solution for retention volumes, we shall give two examples, which are perhaps the most useful from the practical point of view.

(A) Chromatography on polar adsorbents, ion-exchange chromatography and other systems, where the equation

$$k' = k'_0 c^{-m} \tag{1.23}$$

applies, in connection with the general gradient function

$$c = (A^{\frac{1}{\kappa}} + BV)^{\kappa} \tag{4.15}$$

After introducing these equations into eqn. 4.10, we obtain the following equation:

$$V_{\mathrm{m}} = \int_0^{V'_{\mathrm{g}}} \frac{1}{k'_0} \cdot (A^{\frac{1}{\kappa}} + BV)^{\kappa m} \mathrm{d}V \tag{4.24}$$

which may be easily integrated to yield the resulting relationship for retention volume[8]:

$$V'_g = \frac{1}{B}\left[(\kappa m + 1)Bk'_0 V_m + A^{\frac{\kappa m+1}{\kappa}}\right]^{\frac{1}{\kappa m+1}} - \frac{A^{\frac{1}{\kappa}}}{B} \tag{4.25}$$

(B) Reversed-phase chromatography on non-polar chemically bonded stationary phases, where the chromatographic behaviour may be described by the equation

$$k' = k'_0 \cdot 10^{-mc} \tag{1.28}$$

Here, the general gradient functions defined by eqns. 4.14 and 4.16 do not yield algebraic solutions for V'_R. If we use a linear gradient function:

$$c = A + BV \tag{4.17}$$

we can combine eqns. 1.28 and 4.17 with eqn. 4.10 to yield

$$V_m = \int_0^{V'_g} \frac{1}{k'_0} \cdot 10^{m(A+BV)}\,dV \tag{4.26}$$

After integration of this equation we obtain the following relationship for retention volume in reversed-phase chromatography where linear concentration gradients are applied[8,37-39,50,61]:

$$V'_g = \frac{1}{mB} \cdot \log\,(2.31 mBk'_0 V_m \cdot 10^{-mA} + 1) \tag{4.27}$$

Linear concentration gradients are most frequently used in reversed-phase chromatography and they represent "linear solvent strength" gradients for this system, provided that the constants m for individual sample solutes are approximately equal (see discussion in Section 4.2). An equation formally identical with eqn. 4.27 is obtained for a "linear solvent strength" gradient in any chromatographic system, as can be easily demonstrated on "linear solvent strength" gradients in chromatography on polar adsorbents. Here, according to Snyder[1-3], the solvent strength of the mobile phase increases linearly with time or with increasing volume of the eluate:

$$\varepsilon = A + BV \tag{4.28}$$

The capacity factor of a solute is related to the solvent strength of the mobile phase by the equation

$$k' = k'_0 \cdot 10^{-m\varepsilon} \tag{4.29}$$

and the solution for V'_R yields eqn. 4.27[1-3,34-38].

We shall now review briefly previously published equations for retention volumes in certain liquid chromatographic systems using gradient elution, which are listed in Appendix 2 among other solutions. Different workers have most frequently used the two-parameter $k' = f(c)$ function given by eqn. 1.23 in connection with gradients produced by either exponential or proportional gradient-forming devices (gradient functions according to eqns. 4.11-4.14, 4.17 and 4.19 in Table 4.1). Schwab et al.[23] and Ohashi and co-workers[28,29] presented the equations for V'_g in anion-exchange chromatography using exponential gradients of a salt concentration in systems where eqn. 1.23 applies. An algebraic solution for V'_g is possible only in the implicit form for an integer parameter m in this equation and it is shown in Appendix 2 for $m = 1$, $m = 2$ and $m = 3$ (eqns. A49-A51). The resulting equations were verified for the chromatography of oxyphosphorus acids[23] and of simple inorganic ions[28] using exponential concentration gradients of sodium nitrate or ammonium acetate. Ohashi and co-workers[29,30] later published an approximate approach for the calculation of V'_g in these systems for any constant m in eqn. 1.23 and verified their method for the anion-exchange chromatography of polyphosphate anions[30] using an exponential concentration gradient of potassium chloride in water. The calculations require the use of tables and nomograms and are too complex for practical work.

Molnár et al.[26] derived an explicit equation for V'_g in anion-exchange chromatography of the nitrate complexes of rare earth cations with use of an exponential gradient device. In this system, eqn. 1.23 applies with $m = -1$. A more general form of this equation for an arbitrary but negative value of m was derived by Inczédy[66] for linear pH gradients applied in the cation-exchange chromatography of metal complexes.

Inczédy[33,66] also published an equation for retention volumes in the cation-exchange chromatography of bivalent metal ions eluted with a linear concentration gradient of a monovalent cation (with zero initial concentration c_0), where eqn. 1.23 applies ($m = 2$). Schwab et al.[23] derived an analogous solution for $m = 1$, $m = 2$ and $m = 3$ in eqn. 1.23 and verified the resulting equations in the anion-exchange chromatography of simple anions using a linear concentration gradient of sodium nitrate in water.

Massart and Bossaert[27] employed eqn. 4.16 to define a general gradient function suitable for describing various linear and curved gradient profiles in systems where eqn. 1.23 applies. In this instance, an algebraic solution for retention volumes is not possible and they used a numerical integration of eqn. 4.10 by application of the six-point Newton-Cotes formula and verified their method in the anion-exchange chromatography of rare earth complexes using linear concentration gradients. In this special instance, however, they could have used an algebraic solution as given by eqn. 4.25 with $\kappa = 1$.

A gradient function according to eqn. 4.15 in Table 4.1 is equivalent to eqn. 4.16 with respect to the possibility of describing a variety of gradients of almost any linear, convex and concave profile, but it yields a general algebraic solution for V'_R in systems, where eqn. 1.23 can be used to describe the $k' = f(c)$ relationship. This solution, given by eqn. 4.25, was derived by Jandera and Churáček[8] and verified in the adsorption chromatography of azo compounds on silica using different binary mobile phases. Table 4.2 illustrates the accuracy of calculations according to eqn. 4.25 achieved in this system.

In systems where eqn. 1.28 applies as the $k' = f(c)$ function, retention volumes in gradient elution chromatography using linear concentration gradients of the more efficient eluting agent in the mobile phase can be calculated from eqn. 4.27. An equation for V_g which can be rearranged into the form of eqn. 4.27 was first derived by Drake[17], who intended originally to use it for calculations in gradient elution chromatography on polar adsorbents (in normal-phase systems). However, eqn. 1.28 does not fit well with normal-phase systems, but it can be applied very well in reversed-phase systems (see Part I). Eqn. 4.27 for the calculation of retention volumes in reversed-phase systems, where linear gradients of concentration of the organic solvent are applied, was introduced independently by Schoenmakers et al.[39], Jandera and co-workers[8,53], Snyder and co-workers[37,38] and Elgass[61]. Its use was earlier suggested and verified for "linear solvent strength" gradients[1,34,35] and for linear concentration gradients in partition systems[8]. The validity of eqn. 4.27 in reversed-phase systems on chemically bonded non-polar stationary phases was verified experimentally by Jandera et al.[50], Hartwick et al.[63], Snyder and co-workers[37,38] and Elgass[61]. Examples of calculations of V'_g in reversed-phase systems are shown in Table 4.3.

As has been shown in Section 4.2, the linear solvent strength gradients in reversed-phase systems are, for certain types of sample compounds, better approximated by slightly convex gradients[3,37,38]. Various convex gradients may be described by a general logarithmic gradient function[53] (eqn. 4.20 in Table 4.1), which makes possible an explicit algebraic calculation of retention volumes in systems where eqn. 1.28 applies. The solution leads here to eqn. 4.25. An example of the verification of calculations is shown in Table 4.4[53].

TABLE 4.2

EXPERIMENTAL AND CALCULATED VALUES OF NET RETENTION VOLUMES, V_g', AND PEAK WIDTHS, w_g, OF THE N,N-DIMETHYL-*p*-AMINOBENZENEAZOBENZOYLAMIDES IN ADSORPTION CHROMATOGRAPHY ON PORASIL A

Column dimensions, 400 x 3 mm I.D.; V_m = 2.00 ml; adsorbent, Porasil A(60), 37-75 μm; flow-rate of mobile phase, 0.64 ml/min. Chromatographed compounds: 1 = dimethylamide; 2 = diethylamide; 3 = di-*n*-propylamide; 4 = di-*n*-butylamide derivative. Gradient function: $c = (BV)^{\kappa}$. The numbers of the experiments agree with the gradient profiles shown in Fig. 4.2.

No.	κ	B	Compound	V_g' (ml)		ω_g (ml)	
				Calc.	Exptl.	Calc.	Exptl.
(A) Gradient of ethyl acetate in cyclohexane:							
1	0.5	0.0016234	1	64.61		12.12	
			2	32.74	32.88	6.48	5.77
			3	22.33	22.26	4.55	4.13
			4	17.56	17.83	3.79	3.46
2		0.0064935	1	33.13	31.69	6.55	5.13
			2	16.79	16.52	3.66	3.17
			3	11.19	11.06	2.66	2.27
			4	9.00	8.88	2.28	2.11
3		0.025974	1	16.99	16.39	3.69	3.17
			2	8.61	8.46	2.21	1.98
			3	5.74	5.64	1.70	1.66
			4	4.62	4.52	1.50	1.37
4		0.103896	1	8.71	8.40	2.23	2.34
			2	4.41	4.23	1.46	1.47
			3	2.94	2.85	1.20	
			4	2.37	2.40	1.10	
5	1.0	0.0064935	1	62.65	56.68	8.17	5.22
			2	39.59	38.85	5.42	4.97
			3	30.12	30.05	4.28	3.97
			4	26.01	26.04	3.79	3.33
6		0.012987	1	39.92	35.57	5.45	3.43
			2	25.23	23.96	3.70	3.14
			3	19.19	18.50	2.98	2.63
			4	16.57	16.00	2.66	2.30
7		0.025974	1	25.43	24.34	3.72	2.95
			2	16.07	16.19	2.60	2.40
			3	12.22	12.47	2.14	1.92
			4	10.56	10.64	1.95	1.76
8		0.051948	1	16.20	16.42	2.62	2.50
			2	10.24	10.42	1.91	1.92
			3	7.79	7.98	1.61	1.86
			4	6.73	6.86	1.49	1.60

TABLE 4.2 (continued)

No.	κ	B	Compound	V'_g (ml)		w_g (ml)	
				Calc.	Exptl.	Calc.	Exptl.
9	2.0	0.0091831	1	75.56	74.17	6.16	4.17
			2	57.22	58.61	4.83	4.10
			3	48.48	50.62	4.19	3.46
			4	44.36	46.74	3.90	3.17
10		0.018366	1	43.76	42.60	3.85	2.95
			2	33.14	33.39	3.08	2.59
			3	28.07	28.71	2.72	2.27
			4	25.69	26.37	2.54	2.08
11		0.025974	1	33.30	32.56	3.09	2.59
			2	25.22	25.27	2.51	2.24
			3	21.36	21.55	2.23	1.95
			4	19.55	19.79	2.10	1.79
12		0.036732	1	25.34	24.66	2.52	2.30
			2	19.19	18.89	2.07	1.92
			3	16.26	16.16	1.86	1.63
			4	14.88	14.82	1.76	1.53
13	4.0	0.018366	1	51.62	50.91	2.77	2.66
			2	44.18	43.57	2.47	2.40
			3	40.27	39.94	2.31	2.24
			4	38.32	37.98	2.23	2.02
		0.021842	1	44.30	43.82	2.48	2.34
			2	37.93	37.56	2.22	2.08
			3	34.57	34.35	2.08	1.95
			4	32.89	32.62	2.01	1.89
15		0.025974	1	38.03	37.53	2.22	2.21
			2	32.56	32.30	2.00	1.95
			3	29.67	29.48	1.89	1.86
			4	28.23	27.87	1.83	1.73
16		0.030888	1	32.64	32.40	2.01	2.08
			2	27.94	27.65	1.81	1.76
			3	25.47	25.31	1.71	1.63
			4	24.23	24.06	1.66	1.53
(B) Gradient of *n*-propanol in *n*-heptane:							
5	1	0.0064935	1	6.75	6.22	1.87	-
			2	8.36	7.59	2.12	-
			3	11.79	10.97	2.75	2.62
			4	20.62	19.42	4.08	3.39
6		0.012987	1	4.77	4.16	1.53	-
			2	5.88	5.09	1.71	-
			3	8.36	7.44	2.16	2.15
			4	14.35	12.90	3.06	2.86
7		0.025974	1	3.36	-	1.29	-
			2	4.14	3.85	1.41	-
			3	5.94	5.26	1.74	1.97
			4	9.98	9.31	2.35	2.26
8		0.051948	1	2.37	-	1.12	-
			2	2.91	3.07	1.21	-
			3	4.21	3.96	1.45	1.57
			4	6.95	6.64	1.85	1.84

References on p. 180.

(Continued on p. 86)

TABLE 4.2 (continued)

No.	κ	B	Compound	V'_g (ml)		w_g (ml)	
				Calc.	Exptl.	Calc.	Exptl.
10	2	0.018366	1	10.97	11.03	1.97	-
			2	12.66	12.65	2.14	-
			3	15.89	15.89	2.56	2.47
			4	22.97	22.41	3.19	2.72
11		0.025974	1	8.70	8.72	1.71	-
			2	10.02	9.85	1.85	-
			3	12.63	12.39	2.19	2.10
			4	18.11	17.47	2.67	2.38
12		0.036732	1	6.90	6.55	1.51	-
			2	7.94	7.65	1.62	-
			3	10.04	9.84	1.89	1.80
			4	14.27	13.76	2.26	1.87

Abbott et al.[67] suggested an approach based on numerical integration for calculations of retention volumes in reversed-phase systems with application of exponential or linear concentration gradients and verified this method for the reversed-phase chromatography of polynuclear aromatic hydrocarbons using a linear concentration gradient of acetonitrile in water. Inczédy[68] also used numerical calculations of retention volumes of metal ions in cation-exchange chromatography using elution with pH gradients.

Schwab et al.[23] published an implicit relationship for retention volumes in the anion-exchange chromatography of weak monobasic acids, where linear pH gradients were used for elution. These gradients are equivalent to exponential gradients of the concentration of H^+ ions. In these systems, the $k' = f(c)$ relationship is controlled by eqn. 1.24, as has been shown in Chapter 1. The solution for V'_g yields eqn. A24 in Appendix 2.

Schoenmakers et al.[39] presented an approach for the calculation of retention volumes in reversed-phase chromatographic systems, where a parabolic three-parameter equation must be used to describe the $k' = f(c)$ relationships (eqn. 1.26) and linear concentration gradients of the organic solvent are used for elution. An algebraic solution for retention volumes is not possible here, and these workers overcame this difficulty by transformation of the integral in eqn. 4.10 to a Gaussian type, the so-called error function, the values of which are tabulated. No verification of this approach has been published so far. Fortunately, few practical chromatographic systems seem to require the three-parameter eqn. 1.26 for the description instead of the simple two-parameter eqn. 1.28.

TABLE 4.3

EXPERIMENTAL AND CALCULATED VALUES OF NET RETENTION VOLUMES, V'_g, AND PEAK WIDTHS, w_g, OF XANTHINE ALKALOIDS, SUBSTITUTED URACILS AND BARBITURATES IN REVERSED-PHASE GRADIENT ELUTION CHROMATOGRAPHY

Column, octadecylsilica on LiChrosorb Si 100 (10 µm); 300 x 4.2 mm I.D.; V_m = 3.2 ml; mobile phase, water-methanol, 0.97 ml/min. Gradient function: $c = A + BV$. Chromatographed compounds: 3 = theobromine; 4 = theophylline; 5 = caffeine; 6 = 3,6-dimethyluracil; 7 = 3-ethyl-6-methyluracil; 8 = 3-*n*-propyl-6-methyluracil; 10 = 3-*sec*.-butyl-6-methyluracil; 12 = 3-*n*-butyl-6-methyluracil; 13 = 3-*tert*.-butyl-6-methyluracil; 14 = barbital; 15 = heptobarbital; 16 = allobarbital; 17 = aprobarbital; 20 = butobarbital; 21 = hexobarbital; 22 = pentobarbital.

Parameters of gradient function		No.	V'_g (ml)		w_g (ml)	
A	B		Calc.	Exptl.	Calc.	Exptl.
0.0	0.06872	3	4.71	5.14	0.45	0.40
		4	5.78	6.32	0.49	0.47
		5	6.92	7.26	0.48	0.44
		6	4.45	4.79	0.41	0.33
		7	5.84	6.21	0.43	0.40
		8	7.52	7.93	0.45	0.39
		10	8.85	9.20	0.46	0.33
		12	9.18	9.58	0.46	0.40
		13	9.52	9.85	0.45	0.48
		14	7.09	7.25	0.46	0.36
		15	8.02	8.14	0.44	0.36
		16	8.61	8.72	0.44	0.37
		17	9.14	9.28	0.44	0.37
		20	9.85	10.01	0.44	0.37
		21	10.37	10.58	0.44	0.40
		22	10.62		0.42	
0.1	0.06872	3	3.33	3.54	0.44	0.42
		4	4.42	4.74	0.49	0.42
		5	5.50	5.64	0.48	0.41
		6	3.11	3.38	0.40	0.34
		7	4.46	4.75	0.43	0.39
		8	6.11	6.40	0.45	0.40
		10	7.42	7.68	0.46	0.44
		12	7.74	8.02	0.46	0.47
		13	8.08	8.33	0.45	0.47
		14	5.69	5.84	0.45	0.38
		15	6.58	6.68	0.44	0.36
		16	7.17	7.31	0.44	0.31
		17	7.70	7.90	0.44	0.31
		20	8.40	8.63	0.43	0.32
		21	8.91	9.13	0.44	0.33
		22	9.16	9.41	0.42	

(Continued on p. 88)

TABLE 4.3 (continued)

Parameters of gradient function		No.	V_g' (ml)		w_g (ml)	
A	*B*		Calc.	Exptl.	Calc.	Exptl.
0.0	0.03436	3	7.61	7.78	0.58	0.46
		4	9.17	9.60	0.66	0.54
		5	11.57	11.39	0.64	0.66
		6	7.04	6.99	0.52	0.44
		7	9.41	9.44	0.57	0.52
		8	12.47	12.60	0.61	0.53
		10	15.02	15.23	0.63	0.54
		12	15.69	15.97	0.63	0.58
		13	16.48	16.56	0.62	0.58
		14	11.55	11.56	0.62	0.51
		15	13.71	13.55	0.59	0.45
		16	14.78	14.68	0.60	0.47
		17	15.91	15.88	0.59	0.47
		20	17.39	17.39	0.58	0.47
		21	18.42	18.35	0.59	0.50
		22	19.19	19.23	0.55	
0.1	0.03436	3	5.00	5.07	0.55	0.44
		4	6.61	6.96	0.63	0.46
		5	8.80	8.72	0.63	0.49
		6	4.53	4.54	0.49	0.41
		7	6.78	6.85	0.55	0.51
		8	9.72	9.91	0.60	0.47
		10	12.21	12.49	0.62	0.47
		12	12.86	13.23	0.62	0.47
		13	13.62	13.81	0.61	0.50
		14	8.87	8.90	0.60	0.47
		15	10.88	10.81	0.58	0.45
		16	11.94	11.91	0.59	0.48
		17	13.04	13.03	0.59	0.46
		20	14.50	14.57	0.58	0.44
		21	15.53	15.54	0.58	0.44
		22	16.29	16.39	0.55	0.47
0.0	0.01718	3	11.78	12.17	0.80	0.42
		4	13.84	15.56	0.94	0.50
		5	18.71	19.21	0.95	0.55
		6	10.57	10.14	0.56	0.43
		7	14.50	14.35	0.64	0.51
		8	19.92	20.10	0.71	0.57
		10	24.76	25.23	0.74	0.58
		12	26.12	26.63	0.74	0.57
		13	27.89	27.89	0.72	0.56
		14	18.05	17.92	0.85	0.65
		15	22.81	22.27	0.81	0.55
		16	24.74	24.34	0.83	0.59
		17	27.10	26.75	0.82	0.55
		20	30.17	29.98	0.81	0.57
		21	32.21	31.90	0.81	0.60
		22	34.30	33.97	0.75	0.55

TABLE 4.3 (continued)

Parameters of gradient function		No.	V'_g (ml)		w_g (ml)	
A	*B*		Calc.	Exptl.	Calc.	Exptl.
0.1	0.01718	3	7.04	6.93	0.71	0.53
		4	9.26	9.85	0.85	0.49
		5	13.43	13.11	0.90	0.47
		6	6.16	5.51	0.48	0.41
		7	9.70	9.06	0.58	0.49
		8	14.72	14.37	0.67	0.55
		10	19.30	19.25	0.72	0.53
		12	20.60	20.65	0.72	0.56
		13	22.27	21.85	0.71	0.53
		14	13.07	12.36	0.79	
		15	17.30	16.23	0.78	
		16	19.19	18.37	0.81	
		17	21.46	20.79	0.81	
		20	24.45	23.94	0.80	0.87
		21	26.46	25.86	0.81	0.76
		22	28.51	27.91	0.75	0.73

To summarize the possibilities of calculations of retention volumes in gradient elution chromatography for various combinations of gradient functions and $k' = f(c)$ relationships as listed in Appendix 2, it follows that an explicit algebraic calculation is possible for linear concentration gradients applied to all the systems where the $k' = f(c)$ functions from Table 2.1 fit, with the exception of those described by eqn. 1.26, where the above approach of Schoenmakers et al.[39] must be applied, and those where the empirical $k' = f(c)$ function suggested by Svoboda[69] (eqn. 1.38) should be used. This does not present a serious problem, as such systems are rarely likely to be found in practice. The explicit solution for retention volumes in gradient elution liquid chromatography is also possible with a $k' = f(c)$ function controlled by eqn. 1.23 (most chromatographic systems using polar adsorbents and ion exchangers) for exponential gradients with a gradient function as given by eqn. 4.13 and with a general gradient function for linear, convex and concave gradients according to eqn. 4.15 in Table 4.1[8]. For systems where eqn. 1.28 applies (most reversed-phase and partition chromatographic systems), retention volumes can be calculated explicitly for linear concentration gradients and for convex gradients described by a general logarithmic gradient function (eqn. 4.20 in Table 4.1)[50]. With all other combinations of chromatographic systems and gradient functions, explicit equations for retention volumes cannot be derived and only an algebraic solution can be found in the implicit form. Iteration methods of calculation should be

TABLE 4.4

EXPERIMENTAL AND CALCULATED VALUES OF NET RETENTION VOLUMES, V'_g, AND PEAK WIDTHS, w_g, OF XANTHINE ALKALOIDS, SUBSTITUTED URACILS AND BARBITURATES IN REVERSED-PHASE GRADIENT ELUTION CHROMATOGRAPHY

Numbers of compounds and experimental conditions as in Table 4.3. Gradient function: $c = \log(A^{1/\kappa} + BV)^{\kappa}$; c_0 = initial concentration of methanol in the mobile phase; c_0 = 0.1.

Parameters of gradient function			No.	V'_g (ml)		w_g (ml)	
κ	A	B		Calc.	Exptl.	Calc.	Exptl.
1	1.2589	0.3004	3	2.96	3.32	0.44	0.41
			4	4.09	4.51	0.50	0.46
			5	5.32	5.59	0.51	0.41
			6	2.75	3.17	0.30	0.44
			7	4.13	4.53	0.34	0.48
			8	6.06	6.46	0.38	0.46
			10	7.84	8.29	0.42	0.47
			12	8.33	8.77	0.42	0.50
			13	8.89	9.26	0.43	
			14	5.54	5.73	0.45	0.50
			15	6.71	6.79	0.45	0.46
			16	7.53	7.70	0.47	0.41
			17	8.34	8.62	0.48	0.43
			20	9.52	9.83	0.50	0.45
			21	10.44	10.70	0.52	0.48
			22	11.08	11.42	0.50	0.48
2	1.2589	0.0701	3	4.06	4.26	0.50	0.40
			4	5.46	5.80	0.56	0.46
			5	7.20	7.30	0.58	0.41
			6	3.72	3.88	0.35	0.39
			7	5.57	5.75	0.38	0.41
			8	8.06	8.26	0.43	0.38
			10	10.28	10.52	0.45	0.40
			12	10.88	11.24	0.46	0.47
			13	11.58	11.82	0.46	0.49
			14	7.35	7.48	0.51	0.37
			15	9.00	8.98	0.50	0.36
			16	9.99	10.06	0.52	0.38
			17	11.00	11.17	0.53	0.37
			20	12.41	12.62	0.54	0.37
			21	13.45	13.59	0.55	0.41
			22	14.23	14.46	0.52	0.39

used here and often this is possible only for certain values of the parameters m or κ. With three-parameter $k' = f(c)$ relationships controlled by eqns. 1.26, 1.38 and 1.45, an algebraic solution for retention volumes in gradient elution chromatography is not possible[4,57].

Because the solution of the implicit equations by means of an approximate method is practical only with the use of a computer, it would perhaps be more convenient to use a numerical integration to calculate the retention volumes in gradient elution chromatography. Numerical integration is also the only calculation method possible with other $k' = f(c)$ functions, more complex than the functions in Table 4.1 or for the systems where it is not possible to determine these functions in the form of an algebraic equation. Graphical integration[21] could also be used, but this approach is tedious and hardly practical in comparison with the use of a computer.

For practical calculations of retention volumes in gradient elution chromatography, it appears to be sufficient to characterize the $k' = f(c)$ relationships by means of two-parameter equations even in systems where more complex equations would be necessary for an accurate description. It is only necessary that a two-parameter curve can be fitted to the $k' = f(c)$ plot in the range $k' = 1 - 10$. This can usually be achieved precisely using eqns. 1.23 and 1.28 (Table 2.1). This can be justified as follows.

Under gradient elution conditions, the sample compounds do not move significantly from the top of the column after injection until their capacity factors decrease below $k' = 10$ and consequently the region of $k' = f(c)$ plots where $k' > 10$ has no influence on the elution process. On the other hand, the sample compounds are usually eluted before the actual capacity factors in the mobile phase decrease below $k' = 1\text{-}2$. For this aspect, see Figs. 4.3 and 4.4 and the discussion in Section 4.5 and in refs. 3, 4, 37, 38 and 50.

After the introduction of modern gradient elution liquid chromatographs equipped with electronic control units that make possible the application of various linear and curved gradient profiles, exponential and proportional gradient elution devices became less important and so did the calculations of retention volumes for gradient profiles controlled by gradient functions according to eqns. 4.11-4.14 in Table 4.1. With the modern instruments, gradient functions given by eqns. 4.15 and 4.17-4.20 may be used advantageously, because they permit direct algebraic calculations of the retention volumes from the isocratic data[4,8,50]. This can facilitate the identification of compounds and the investigation of the influence of different adjustable gradient parameters on the chromatographic behaviour of sample compounds to establish optimum gradient conditions.

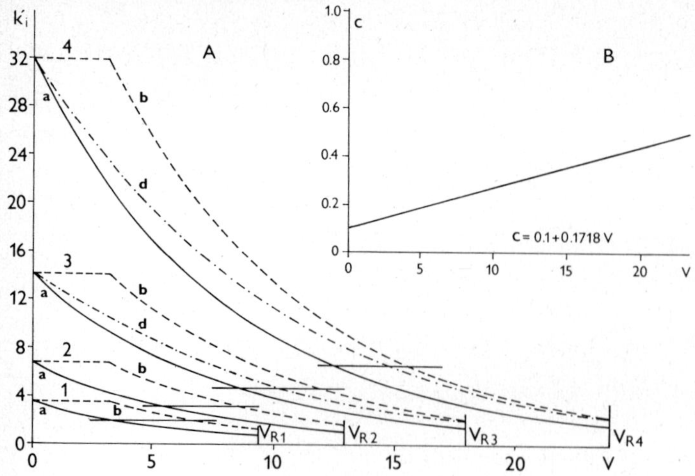

Fig. 4.3. Change in actual capacity factors in gradient elution chromatography. Column, octadecylsilica on LiChrosorb Si 100, 10 µm; 300 x 4.2 mm I.D.; V_m = 3.20 ml. Linear concentration gradient of methanol in water as shown in B applied to separation of a homologous series of 3-alkyl-6-methyluracils: 1 = methyl, 2 = ethyl, 3 = n-propyl and 4 = n-butyl homologues. Eqn. 1.28 applies in this system with constants m = 4.40, k'_0 = 9.51 for 1; m = 3.69, k'_0 = 15.63 for 2; m = 3.31, k'_0 = 29.98 for 3; and m = 3.25 and k'_0 = 67.20 for 4. Curves a correspond to changes in instantaneous capacity factors in the mobile phase at the top and curves b at the outlet of the column. Curves d (for compounds 3 and 4) show the actual change in k'_i at the place corresponding to the peak maximum as it moves along the column. Flow-rate of the mobile phase, 1 ml/min. V = volume of the eluate (ml) ≡ time from the start of the gradient (min); c = concentration of methanol in the mobile phase, [% (v/v) · 10^{-2}]; V_{R1}-V_{R4} = retention volumes of compounds 1-4. Horizontal lines show the apparent capacity factors in gradient elution, k'_g, for the four compounds.

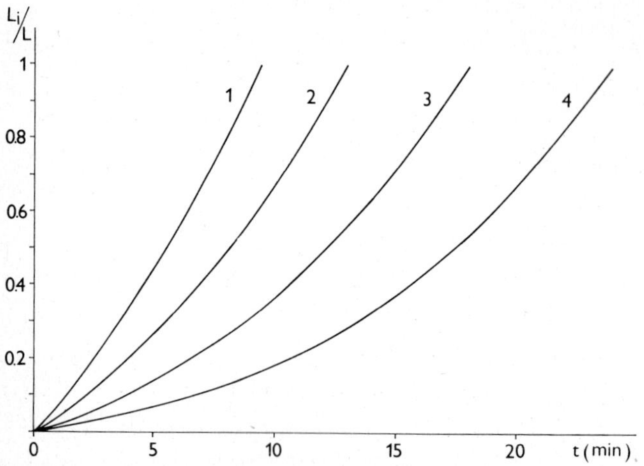

Fig. 4.4. Column length fraction, L_i, migrated by the solute band maximum as a function of time t in gradient elution chromatography. Conditions and compounds as in Fig. 4.3. L = total length of the column = 300 mm.

4.5. CAPACITY FACTORS IN GRADIENT ELUTION CHROMATOGRAPHY AND CALCULATIONS OF RETENTION VOLUMES FOR CHANGING COLUMN DIMENSIONS AND FLOW-RATE OF THE MOBILE PHASE

The capacity factors of chromatographed solutes remain constant during elution under isocratic conditions. However, in gradient elution chromatography they change with time (decrease), in agreement with increasing eluting strength of the mobile phase. Changes in k' during gradient elution are illustrated by a practical example in Fig. 4.3A for four homologous 3-alkyl-6-methyluracils chromatographed on a reversed-phase column (C18) using a linear concentration gradient of methanol in water, shown in Fig. 4.6B. This figure shows the changes in k' at the top of the column (full curves a) and at the outlet of the column (dashed curves b) and actual change in k' at the maximum of a solute band as it moves along the column (dashed-dotted curves d). The capacity factors change most dramatically at the beginning of the elution. Further, the decrease in capacity factors is much steeper for more retained solutes than for slightly retained compounds. Consequently, the compounds the capacity factors of which varied from 3.5 to 32 at the beginning of the elution have very close capacity factors at the outlet from the column (ranging from 1.2 to 2.2). This levelling of the differences in k' for different compounds at the moment of their elution from the column is a very important and characteristic feature of gradient elution.

As shown in Fig. 4.4, the migration of solute bands is relatively slow at the beginning of the gradient elution (large k') and gradually accelerates as the elution advances, whereas a constant speed of band migration is characteristic of isocratic elution. Relatively small k' values at the time of elution mean narrow bands and increased sensitivity in gradient elution chromatography compared with elution under isocratic conditions. (Using eqn. 4.27, the time t_i corresponding to the migration of the solute band maximum from the top of the column to the distance corresponding to the column length fraction, L_i, can be calculated as:

$$t_i = \frac{1}{mBFm} \cdot \log \left(2.31mBk'_0V_m \cdot \frac{L_i}{L \cdot 10^{mA}} + 1\right) + \frac{V_m L_i}{F_m L} \qquad (4.30)$$

where L is the length of the column and F_m is the flow-rate of the mobile phase.)

To calculate instantaneous capacity factors k'_i in gradient elution chromatography for compounds before their elution from the column, we have to combine the $k' = f(c)$ function with the gradient function used, $c = \phi(V)$ and to use V'_g

instead of V, using the appropriate expression for V'_g in Appendix 2, where $V_m r$ is inserted instead of V_m (r is the fractional length of the column, $r = L_i/L$, corresponding to the position of the band maximum).

The instantaneous capacity factor k'_f at the instant of the elution of the band maximum from the column is of greatest interest for theoretical considerations. Here, $r = 1$. For example, using this approach for the combination of the $k' = f(c)$ function given by eqn. 1.28 and a linear gradient function according to eqn. 4.17 (Table 4.1) - reversed-phase chromatography - we obtain

$$k'_i = k'_0(2.31mBk'_0V_m r + 10^{mA})^{-1} \tag{4.30a}$$

For the compounds not eluted very early in the gradient run, eqn. 4.30a can be simplified by neglecting the second term in parentheses, as follows:

$$k'_i = \frac{1}{2.31mBV_m r} = \frac{1}{2.31\beta r} \tag{4.30b}$$

and, for k'_f:

$$k'_f = \frac{1}{2.31mBV_m} = \frac{1}{2.31\beta} \tag{4.30c}$$

where β is the steepness of a "linear solvent strength" gradient as defined by Snyder and co-workers (eqn. 4.4).

Eqns. 4.30b and 4.30c can be used in all forms of "linear solvent strength" gradients.

Similarly, the combination of the $k' = f(c)$ function given by eqn. 1.23 and a general gradient function described by eqn. 4.15 (Table 4.1) - chromatography on polar adsorbents and ion-exchange chromatography - yields the following equations:

$$k'_i = k'_0 \left[(\kappa m + 1)Bk'_0V_m r + A^{\frac{\kappa m+1}{\kappa}}\right]^{-\frac{\kappa m}{\kappa m+1}} \tag{4.30d}$$

and

$$k'_f = k'_0 \left[(\kappa m + 1)Bk'_0V_m + A^{\frac{\kappa m+1}{\kappa}}\right]^{-\frac{\kappa m}{\kappa m+1}} \tag{4.30e}$$

As the capacity factor of a solute changes during gradient elution, we cannot use the retention volumes measured under gradient conditions to calculate k' in a straightforward manner, in contrast to isocratic elution experiments. If we divide the net retention volumes in gradient elution chromatography by the column dead volume, V_m, we obtain an "apparent capacity factor in gradient elution chromatography", k'_g[37]:

$$k'_g = V'_g/V_m \tag{4.31}$$

The values of k'_g are indicated by horizontal lines for the four compounds in the practical example shown in Fig. 4.3. It should be kept in mind that this quantity does not have any real physico-chemical meaning and cannot be directly related to the solute distribution constant in a given chromatographic system. k'_g in various systems and for different gradient functions may be calculated easily using equations for V'_g in Appendix 2. If we do so for any combination of a two-parameter $k' = f(c)$ relationship (Table 2.1) and a linear or general gradient fucntion as described by eqns. 4.15-4.20 (Table 4.1), we find that k'_g does not depend on V_m, provided that the product BV_m remains constant. In other words, k'_g does not depend on neither the column dimensions or on the flow-rate of the mobile phase under these conditions and can be used as a general retention characteristic in gradient elution chromatography, which is easy to use for the calculation of retention volumes for a column with given dimensions from the retention volumes measured on (or calculated for) another column with different geometry and for changing flow-rate of the mobile phase. Because

$$k'_g BV_m = k'_g \cdot \frac{V_m}{V_G} = \frac{V'_g}{V_G} \tag{4.31a}$$

applies to gradients run from 0 to 100% of solvent b, the ratio V'_g/V_G (the relative retention related to the volume of the gradient) is also constant, if the product BV_m is kept constant.

We can see that the condition

$$BV_m = \frac{V_m}{V_G} = \frac{V_m}{t_G F_m} = \frac{L}{t_G u} = \text{constant} \tag{4.32}$$

is analogous to the condition c = constant in isocratic elution chromatography, where also the k' value of a solute remains constant regardless the column dimensions and flow-rate of the mobile phase provided that the composition of the mobile phase is not changed. V_G and t_G in eqn. 4.32 are the volume and the time, respectively, of the gradient, F_m and u are the volume and linear flow-rate of the mobile phase, respectively, and L is the length of the column. Hence the product BV_m in gradient elution chromatography has much the same role as the concentration of the more efficient eluting component in the mobile phase under isocratic conditions.

The condition in eqn. 4.32 can be understood as a more general, theoretically based formulation of an empirical postulate verified experimentally by Elgass[61] and Engelhardt and Elgass[62] for linear concentration gradients applied in a reversed-phase chromatographic system. According to this postulate, the k'_g values on a given column do not depend on the flow-rate of the mobile phase if the volume of the gradient, V_G, remains constant. As we can see from eqns. 4.20a,b and 4.21, V_G is inversely proportional to B and, as V_m is constant for a given column, this postulate becomes equivalent to the condition in eqn. 4.32, but the latter condition applies more generally for columns with different geometries.

Here, we can again note the advantage of working with retention volumes instead of retention times in gradient elution chromatography. From the equations for V'_g in Appendix 2 this conclusion is obvious at first glance, but it is not as easy to draw directly from the equations for retention times.

The condition in eqn. 4.32 also gives the rule for how we can compensate for changed column dimensions (different lengths, L, or/and diameters, d, of columns with different V_m values) by changing the steepness parameter B when we want to keep constant values of k'_g from one column to another. This can be done very easily and may obviate much work on the identification of eluted compounds. Further, we can transfer the experimental conditions optimized for given column dimensions easily to other columns without time-consuming re-calculations or experiments. The retention volumes on a column II of given geometry can be calculated from the retention volumes measured or calculated for a column I of different dimensions using the equation

$$V'_{gII} = V'_{gI} \cdot \frac{V_{mII}}{V_{mI}} = V'_{gI} \cdot \frac{L_{II}}{L_I} \left(\frac{d_{II}}{d_I}\right)^2 \tag{4.33a}$$

if BV_m is constant, i.e., $B_I V_{mI} = B_{II} V_{mII}$.

Similarly, if the flow-rate of the mobile phase is changed from F_{mI} to F_{mII} when working with the same column and BV_m is constant, it follows that

$$V'_{gII} = V'_{gI} \cdot \frac{B_I}{B_{II}} = V'_{gI} \cdot \frac{t_{GII} F_{mII}}{t_{GI} F_{mI}} = V'_{gI} \cdot \frac{t_{GII} u_{II}}{t_{GI} u_I} \tag{4.33b}$$

From eqns. 4.33a and 4.33b, several practical rules may be formulated for work with gradient elution:

(1) We increase the volume flow-rate by a certain factor when working with a given column. Consequently, we also decrease the time of the gradient (increase the steepness of the concentration change with time) by the same factor to keep the retention volumes constant.

(2) We increase the length of the column by a certain factor, and at the same time either increase the volume flow-rate of the mobile phase (at a constant time of the gradient) or increase the time of gradient (at a constant F_m) by the same factor. Then, the retention volumes increase by the same factor.

(3) We increase the diameter of the column by a certain factor and decrease the length of the column by the square of the same factor. The retention volumes do not change.

(4) We increase the diameter of the column by a certain factor, keep the length of the column constant and at the same time either keep the linear velocity of the mobile phase constant (by increasing the volume flow-rate by the square of the same factor) at a constant time of the gradient or increase the time of the gradient by the square of the same factor (at constant F_m). The retention volumes increase by the square of this factor.

These rules are analogous to those in isocratic elution chromatography.

4.6. INFLUENCE OF THE GRADIENT PROFILE ON THE RETENTION VOLUMES OF CHROMATOGRAPHED COMPOUNDS

To understand better the influence of the gradient profile on retention in gradient elution, we shall differentiate the influence of three different characteristics of the gradient profile: (1) the curvature (shape) of the gradient, (2) the steepness (slope) of the gradient and (3) the initial concentration of the more efficient eluting component in the mobile phase at the beginning of the gradient. The exponential gradient functions given by eqns. 4.11-4.13 in Table 4.1 do not allow easy differentiation of the three characteristics of the gradient profile from the constants of these equations. This is much easier with general curved or linear gradient functions as described by eqns. 4.14-4.20 in Table 4.1. In these equations, the shape of the gradient is defined by the parameter κ, the steepness by the parameter B and the initial concentration of the more efficient component in the mobile phase by the parameter A ($D - A$ in eqn. 4.14 and log A in eqn. 4.20).

From inspection of the equations for V'_g in gradient elution controlled by eqns. 4.14-4.20 in Appendix 2, it is evident that the retention volumes of solutes decrease both with increased steepness of the gradient (B) and with

References on p. 180.

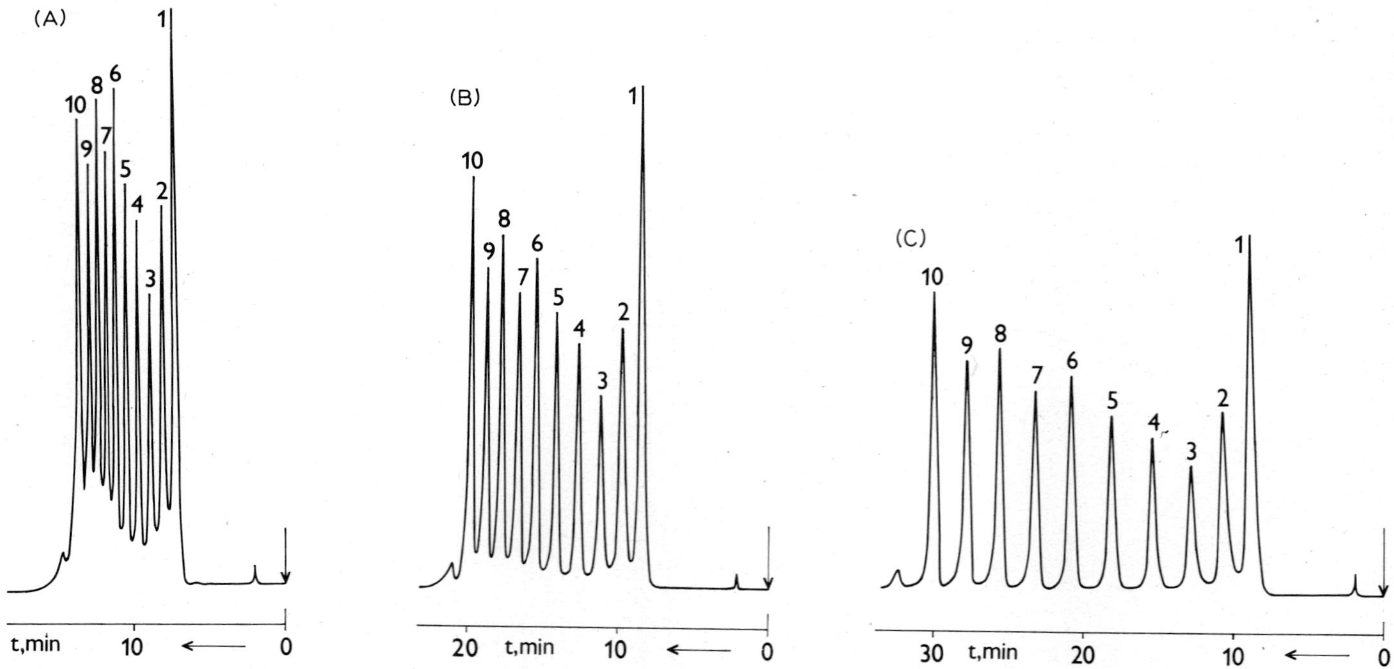

Fig. 4.5. Influence of the gradient steepness on reversed-phase chromatographic separation of fluorescent derivatives of homologous *n*-alkylamines. Compounds and operating conditions as in Fig. 3.1. Linear gradients of methanol in water starting with methanol-water (70:30). (A) 70% to 100% methanol in 10 min; (B) 70% to 100% methanol in 20 min; (C) 70% to 100% methanol in 40 min. Flow-rate: 1 ml/min.

increased initial concentration of the more efficient eluting agent in the mobile phase (A). This theoretical conclusion is in good agreement with experimental observations in a number of systems using gradient elution[2-4,43,50,51,61].

If we change the steepness of the gradient B in work with a given column (V_m is constant), then a change in B is analogous to a change in the concentration of the more efficient eluting component (c) in the mobile phase under isocratic conditions. An increase in B leads to a corresponding decrease in solute retention volumes, as increasing c decreases the retention volumes in isocratic elution chromatography. This can be seen from the example of the separation of a homologous series of fluorescent amine derivatives in reversed-phase chromatography[70] in Fig. 4.5A-C.

In normal-phase chromatography with zero initial concentration of the more efficient eluting component in the mobile phase (A = 0), eqn. 4.25 may be rewritten in the form[43]

$$\log V_g' = \bar{a} - \bar{b} \log B \tag{4.34}$$

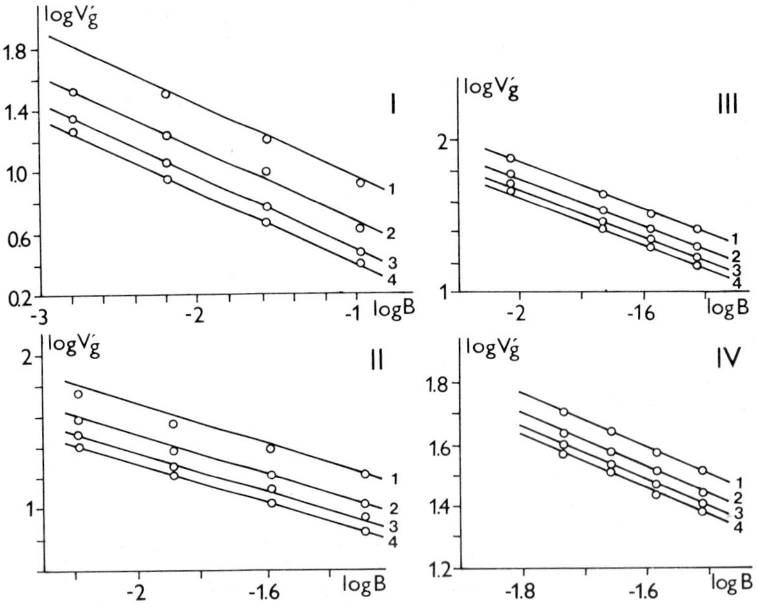

Fig. 4.6. Relationship between the retention volumes, V_g' (ml), of N,N-dimethyl-p-aminobenzeneazobenzoylamides and the steepness of gradient in chromatography on a silica column, 400 x 3.0 mm I.D., Porasil A, 37-75 μm; V_m = 2.0 ml; gradient of ethyl acetate in cyclohexane. Gradient elution according to gradient function given by eqn. 4.18, using gradients of different shapes (curvatures): I, κ = 0.5; II, κ = 1; III, κ = 2; IV, κ = 4. Flow-rate: 0.64 ml/min. Full lines, calculated relationships (eqn. 4.25, A = 0); points, experimental data. Compounds: 1 = dimethylamide; 2 = diethylamide; 3 = di-n-propylamide; 4 = di-n-butylamide. (From ref. 43.)

Here, for a given shape of the gradient (κ constant), $\bar{a}$ and $\bar{b}$ are constant and the logarithms of the retention volumes of chromatographed compounds decrease linearly with increasing steepness of the gradient (see Fig. 4.6). From the parameters $\bar{a}$ and $\bar{b}$ found by regression analysis of the experimental log V'_g versus log B plots, it is possible to calculate the parameters k'_0 and m of the equation $k' = f(c)$ (eqn. 1.23) with a precision approximately equal to that from the regression of the isocratic log k' versus log c plots[43]. This may be useful, as k'_0 and m of a number of compounds differing widely in retention may be found in a single experiment.

Linear relationships between log V'_g and log B were also found experimentally in gradient elution applied to reversed-phase chromatography on octadecylsilica[51] (Fig. 4.7).

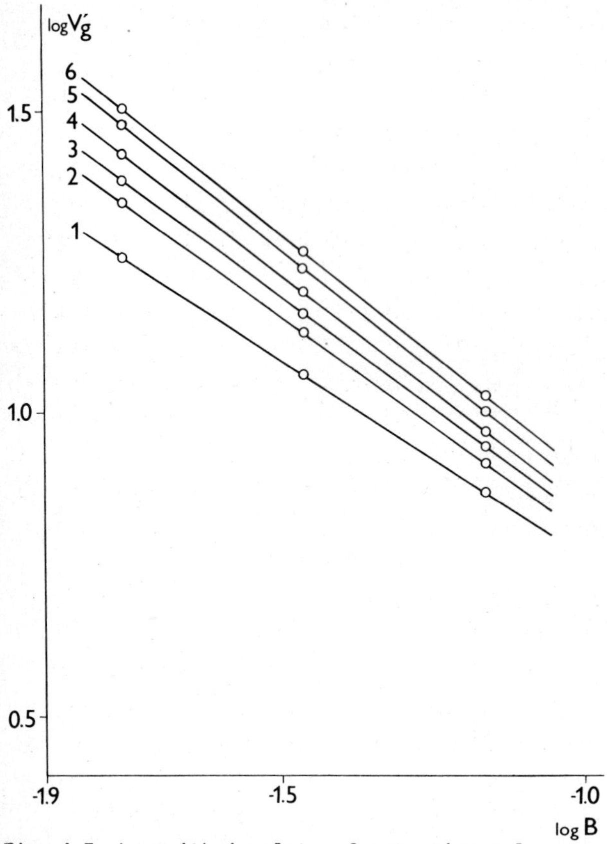

Fig. 4.7. Logarithmic plots of retention volumes, V'_g, of barbiturates versus the slope of the gradient, B, in reversed-phase gradient elution chromatography. Column: Reversed-phase C18/LiChrosorb Si 100, 10 μm; 300 x 4.2 mm I.D.; V_m = 3.2 ml. Linear concentration gradient of methanol in water; $A = 0$ (eqn. 4.19); 0.98 ml/min. V'_g in ml; B in % methanol per ml of the mobile phase x 10^{-2}. Compounds: 1 = barbital; 2 = heptobarbital; 3 = allobarbital; 4 = aprobarbital; 5 = butobarbital; 6 = hexobarbital. (From ref. 51.)

An increasing concentration of the more efficient eluting component in the mobile phase at the beginning of the gradient leads to a decrease in retention volumes, as illustrated by an example of reversed-phase separation of fluorescent derivatives of aliphatic amines[70] in Fig. 4.8.

The parameter κ in eqn. 4.15 in Table 4.1 characterizes the shape (curvature) of the gradient in such a way that, when κ increases from 0 to 1, the convexity of the gradient diminishes, the gradient is linear for $\kappa = 1$ and becomes more and more concave as κ increases further.

If we keep constant the parameter B of the gradient function, eqn. 4.15 (and $A = 0$), the volume of the gradient, V_G (eqn. 4.21), remains constant and the concentration of the more efficient eluting component in the mobile phase in a given time decreases with increasing κ. This means that the solute retention volumes increase with increasing κ, i.e., from a convex to a concave gradient shape. These conclusions may also be applied to the gradients with $A \neq 0$.

However, the relative influence of the gradient shape on retention volumes differs to a certain extent in different parts of the gradient. In convex gradients, the slope of the gradient curve decreases with time and, consequently, the elution of early eluted compounds is more accelerated than that of the later eluted compounds. Consequently, the peaks at the beginning of the chromatogram are more crowded together than those at the end of the chromatogram. The opposite holds true for concave gradients. The main practical importance of the shape of the gradient is the possibility of controlling to different extents the resolution in different parts of the separation. An increase in the concavity of the gradient leads to a better resolution of the early eluted peaks and to a certain "squeezing" of the peaks in the later part of the gradient, whereas an increase in convexity has the opposite effect. The influence of the shape of the gradient on separation is illustrated by an example in Fig. 4.9.

4.7. BANDWIDTHS IN GRADIENT ELUTION CHROMATOGRAPHY

In gradient elution chromatography, unlike under isocratic conditions, it is not possible to use the experimental peak widths to characterize band broadening in the column and column efficiency. The resulting width of the eluted band is a result of several effects which are not only properties of the column, but also of the gradient (see below), and the use of eqns. 1.5 and 1.6, which are valid for isocratic elution chromatography, would yield completely false results for the number of theoretical plates, n, or the height equivalent to a theoretical plate, H (n would increase with decreasing A or B, with increasing retention, and would depend on the shape of the gradient). Consequently, we should

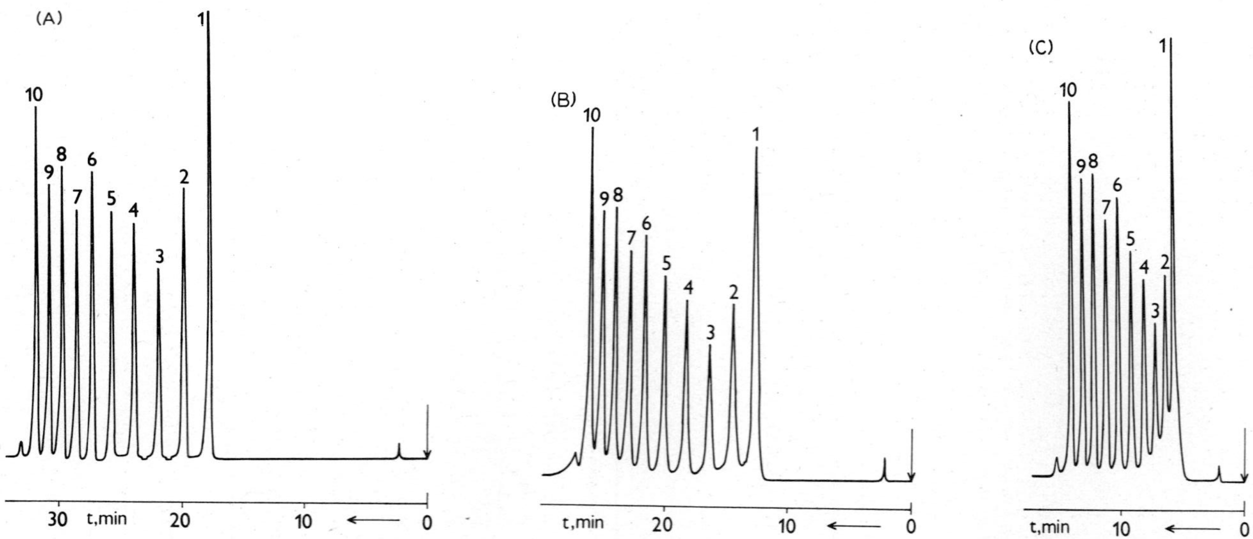

Fig. 4.8. Influence of the initial concentration of methanol in the mobile phase at the start of the gradient of the reversed-phase chromatographic separation of fluorescent derivatives of homologous *n*-alkylamines. Compounds and operating conditions as in Fig. 3.1. Linear gradients of methanol in water of an equal steepness (10% of methanol increase in 6 ml of the eluate): (A) elution started in methanol-water (50:50); (B) elution started in methanol-water (60:40); (C) elution started in methanol-water (80:20). Flow-rate: 1 ml/min.

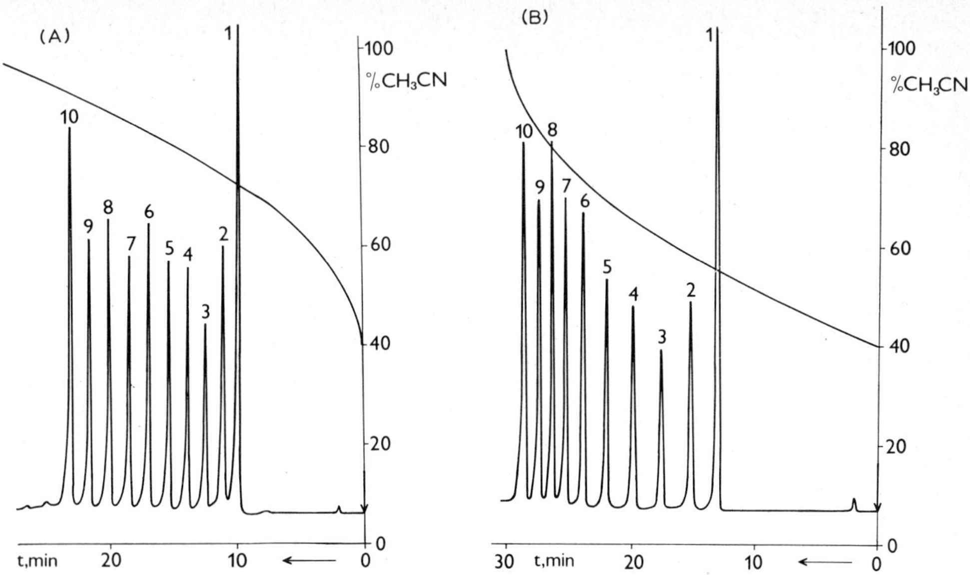

Fig. 4.9. Influence of the shape (curvature) of the gradient on the reversed-phase chromatographic separation of fluorescent derivatives of homologous *n*-alkylamides. Compounds and operating conditions as in Fig. 3.1. Curved gradients of acetonitrile in water starting at 40% (v/v) of acetonitrile in water. Flow-rate: 1 ml/min. (A) Convex gradient; (B) concave gradient.

be aware that the bandwidths in gradient elution chromatography have a meaning only as a measure of the volume in which the compound peaks are eluted.

4.7.1. *Calculation of bandwidths*

In principle, bandwidths in gradient elution chromatography may be determined exactly after calculation of all points of the elution curve[15]. This calculation is possible using numerical methods with the aid of a computer. This exact procedure seems to be too complex for most practical separation problems. If we bear in mind the relative errors in the experimental evaluation of bandwidths, which are significantly larger than those in the evaluation of retention volumes, a simplified calculation with a relative error in bandwidths of 10-30% seems to be acceptable in practice.

Snyder and co-workers[1-3,36] showed that all of the solutes chromatographed using "linear solvent strength" gradients (see Section 4.2) should be eluted with approximately equal bandwidths. In practical work, this conclusion cannot be considered as strictly exact, because it is difficult to design a gradient with an exactly "linear solvent strength" with respect to all sample compounds, but if the gradients used are not very steep (or, in contrast, if they do not have a very small steepness) and if efficient columns are used, the bandwidths of the solutes eluted in the course of a "linear solvent strength" gradient are really approximately equal.

Moreover, the differences in the bandwidths of the individual solutes under gradient conditions, even if "non-linear solvent strength" gradients are used, and if the solutes are eluted with gradients of different profiles, are considerably lower than the differences in the bandwidths of different solutes under isocratic conditions.

Consequently, the bandwidths in gradient elution chromatography can be roughly estimated (to first approximation) as equal to the bandwidths of the compounds eluted with k' = 1-2 under isocratic conditions on the same column and with the same flow-rate of the mobile phase containing the same components as used in gradient elution chromatography[52]. Of course, such a rough estimation can be used only to obtain an approximate picture of the separation and the error in the bandwidths may sometimes become larger than 100% relative.

A more accurate calculation method for bandwidths in gradient elution chromatography is based on the following arguments. The width of a solute band eluted under gradient conditions can be understood as a result of three phenomena[1-3,8,36]:

(a) broadening of the solute band as it moves down along the column (just as in isocratic elution chromatography);

(b) solute band compression caused by a steady decrease in the capacity factor of the solute as it moves along the column, so that the instantaneous capacity factor of the solute is minimal at the time of elution of peak maximum;

(c) an additional compression of the band resulting from the fact that the front of the band moves in the mobile phase with a lower eluting strength than the end of the band.

Neglecting the last effect, which is of relatively minor importance especially if efficient columns are used and the solute peaks are narrow, the width of a solute band in gradient elution chromatography can be considered as approximately equal to that under isocratic conditions employing a mobile phase of composition corresponding to the instantaneous composition in gradient elution chromatography at the moment of elution of the maximum of the band[8]. The instantaneous capacity factor at the moment of the elution of the band maximum is k'_f. Then, using the equation for bandwidth (peak width at the baseline in volume units) under isocratic conditions which results from the combination of eqns. 1.2 and 1.5, we obtain an analogous equation for peak width in gradient elution chromatography:

$$w_g = \frac{4V_m}{\sqrt{n}} \cdot (1 + k'_f) \qquad (4.35)$$

Eqn. 4.35 is a general equation for peak widths in gradient elution chromatography. In "linear solvent strength" gradients, k'_f has approximately equal values for different sample compounds eluted during the gradient run. The k'_f values of other than very early eluted compounds depend on the gradient steepness β (cf., eqn. 4.4) according to the simple relationship in eqn. 4.30c and consequently[3,37,38]

$$w_g \approx \frac{V_m\,(1 + 2.31\beta)}{0.58\beta\sqrt{n}} \qquad (4.36)$$

For calculations with other gradients, it is necessary to introduce k'_f from the appropriate $k' = f(c)$ function and gradient function $c = \phi(V)$, for $V = V'_g$ (or $V = V'_g - V_Z$, if the volume V_Z between the place of the mixing of the mobile phase components and the top of the column cannot be neglected). Then,

$$w_g \; \frac{4V_m}{\sqrt{n}} \cdot \{1 + f\,[\phi(V'_g - V_Z)]\} \qquad (4.36a)$$

where n is the column plate number under isocratic conditions.

If the $k' = f(c)$ functions from Table 2.1 and gradient functions from Table 4.1 can be used to describe the chromatographic system and the gradient used, we can introduce a corresponding equation for V'_g in Appendix 2 to eqn. 4.36a to obtain the final mathematical solution. Of course, this is possible only for gradients that yield an explicit solution for retention volumes.

For example, for the gradient function given by eqn. 4.15 (Table 4.1) and $k' = f(c)$ defined by eqn. 1.23, we obtain (if $V_Z \approx 0$)[8]

$$w_g = \frac{4V_m}{\sqrt{n}} \cdot \left\{1 + k'_0 \left[A^{\frac{\kappa m+1}{\kappa}} + (\kappa m + 1)Bk'_0V_m\right]^{-\frac{\kappa m}{\kappa m+1}}\right\} \qquad (4.37)$$

while for the combination of eqn. 1.28 with a linear gradient function (eqn. 4.17 in Table 4.1) it follows that[8]

$$w_g = \frac{4V_m}{\sqrt{n}} \cdot \{1 + k'_0\,(2.31mBk'_0V_m + 10^{mA})^{-1}\} \qquad (4.38)$$

Analogous equations can be derived using this approach for other combinations of gradient functions (Table 4.1) and $k' = f(c)$ relationships (Table 2.1) which yield explicit solutions for V'_g (Appendix 2).

Eqn. 4.37 should apply in normal-phase chromatography on polar adsorbents and in simple ion-exchange systems. Its validity was verified in adsorption chromatography on silica[43]. Eqn. 4.38, which should be useful in reversed-phase chromatography, was verified in the chromatography of various compounds on octadecylsilica using gradients of methanol in water[50]. Examples of this verification are given in Tables 4.2-4.4.

In a number of systems, the widths of peaks calculated from the above and analogous equations are on average approximately 10-20% higher than the experimental data. These discrepancies can probably be attributed mainly to the neglect of the additional band compression due to the change in k' during the elution of the sample band under gradient conditions. The neglect of the dependence of the number of plates (column efficiency) on the changing mobile phase composition seems to be less significant. Snyder and Saunders[36] used the so-called band compression factor, G, to describe the influence of the additional band compression.

For the compression due to the change from a weaker solvent (I) to a stronger solvent (II), G can be expressed by

$$G = \frac{k'_{II}(1 + k'_{I})}{k'_{I}(1 + k'_{II})} \tag{4.39}$$

where the subscripts I and II relate to the capacity ratios of the solute in mobile phases I and II, respectively. After substitution of the appropriate $k' = f(c)$ and $c = \phi(V)$ functions into the differential form of eqn. 4.39, G can be calculated (generally using numerical methods). In "linear solvent strength" gradients, G depends only on the steepness (β) (eqn. 4.4) of these gradients. Fig. 4.10 shows the plot of G versus β, as calculated by Snyder and Saunders[36].

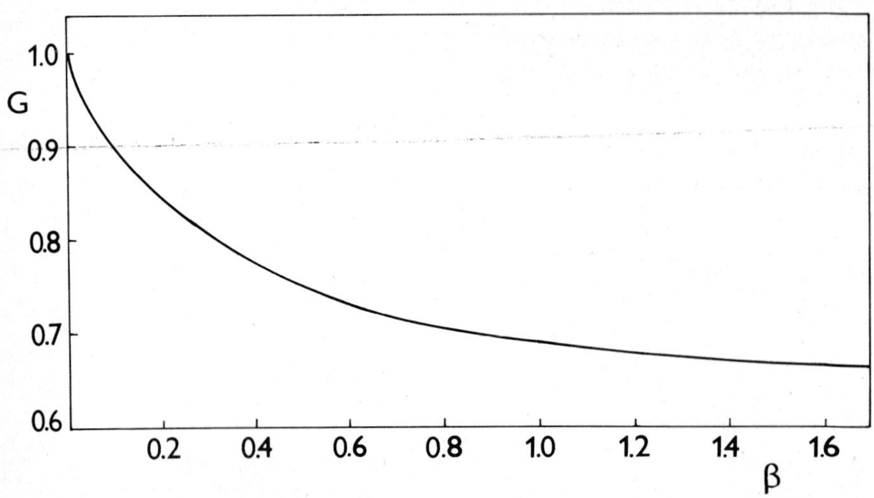

Fig. 4.10. Band compression factor, G, in "linear solvent strength" gradients of different steepnesses β. (Adapted from ref. 37.)

This relationship may be expressed as follows:

$$G^2 = (1 + p^* + \frac{p^{*2}}{3})/(1 + p^*)^2 \tag{4.39a}$$

where p^* is given by

$$p^* = \frac{2.31k'_a\beta}{k'_a + 1} \tag{4.39b}$$

and, for large k'_a [gradient started at zero initial concentration of the more efficient eluting agent ($A = 0$) in the mobile phase and compounds are relatively strongly retained at $A = 0$], this becomes[3]

$$p^* \approx 2.31\beta \tag{4.39c}$$

Thus, the "corrected bandwidths" can be calculated as

$$w_{\mathrm{corr.}} = w_g G \tag{4.40}$$

The results of the experimental verification of eqn. 4.40 involving the band compression factor G, as given by Snyder and co-workers[37,38], show that the calculated bandwidths are lower than the experimental values for $\beta > 0.2$, where $G \leqq 0.8$, whereas the agreement between the calculated and experimental values is good for $\beta < 0.2$. On the other hand, the complete neglect of G, as in Tables 4.2-4.4, leads generally to bandwidths approximately 10-20% relative larger than the experimental values. Therefore, considering the lower precision in evaluating of the experimental bandwidths, it seems to be justified to use a semi-empirical value of $G \approx 0.85$ in eqn. 4.40 for multiplying the bandwidths calculated from eqns. 4.37 and 4.38 and similar equations.

Poppe et al.[81] used the first and second moments to derive a description of peak broadening under gradient elution conditions from the mass transport equation in the dispersion model. According to these authors, it is possible to calculate the bandwidths, w_g, using the following two equations:

$$w_g^2 = y\left(\frac{k'_i}{1 + k'_i}\right)^2 \tag{4.40a}$$

and

$$\frac{dy}{dr} = H\left(\frac{k'_i + 1}{k'_i}\right)^2 \tag{4.40b}$$

References on p. 180.

where H is the height equivalent to a theoretical plate, r is the relative column distance migrated by the centre of the band from the start of the elution and y is an auxiliary quantity. The dependence of the instantaneous capacity factor at the centre of the band k'_i on time must be known to make possible the integration of the right-hand side of eqn. 4.40b and the equation for the capacity ratio, k'_f, at the time of the elution of the peak maximum should also be known, to allow the analytical solution. For linear solvent strength gradients, the solution leads to eqns. 4.39 and 4.40 as in the derivation by Snyder[3].

4.7.2. *Influence of the gradient profile on bandwidths in gradient elution chromatography*

The peak widths of different solutes eluted in one gradient run differ far less than the peak widths in an isocratic run. From eqn. 4.35 it follows that if all solutes are eluted with approximately equal instantaneous capacity factors k'_f, the bandwidths of all compounds should be equal. According to Snyder's theory[1-3], this is the case with "linear solvent strength" gradients. An approximate validity of these conclusions has been varified experimentally both for chromatography on polar adsorbents[35] and in reversed-phase systems[37,38,50,61] (for linear concentration gradients of an organic solvent in water). Approximately constant bandwidths were also found in experiments with logarithmic concentration (convex) gradients applied to reversed-phase systems[50]. As shown above, it is debatable whether the "linear solvent strength" gradients in reversed-phase systems are better approximated by linear or by slightly convex gradients of the concentration of the organic solvent in the mobile phase[3,37-40].

Larger differences are usually found between the bandwidths of solutes eluted with different gradients than between the widths of the peaks of solutes eluted in one gradient run. Gradients controlled by the gradient function given by eqn. 4.15 (Table 4.1) are especially suitable for studying the influence of various aspects of gradient profile on the widths of peaks, because they allow only one of the three basic parameters (steepness and curvature of the gradient and the initial concentration of the more efficient component in the mobile phase) to be changed, while the other two are kept constant.

Like the retention volumes, the peak widths in gradient elution chromatography decrease to a certain extent with increasing steepness of the gradient and with increasing initial concentration of the more efficient eluting component in the mobile phase (B and A in eqns. 4.37 and 4.38). However, the influence of these parameters on peak widths is smaller than their influence on

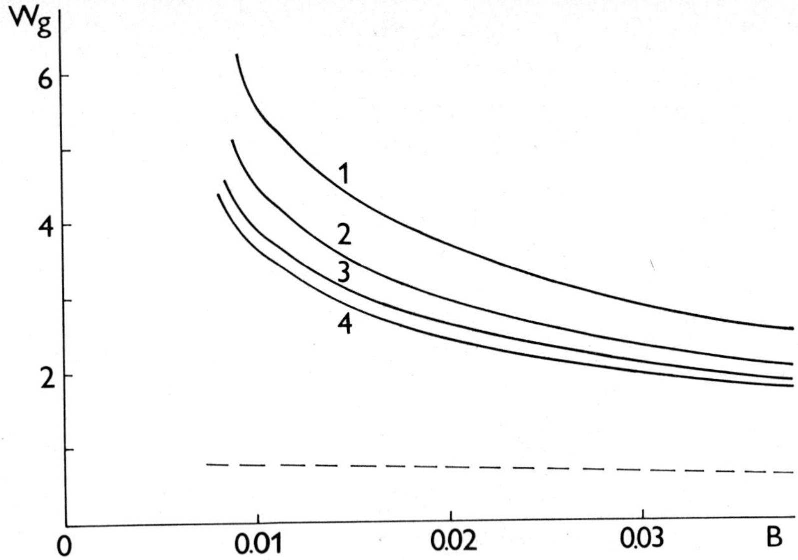

Fig. 4.11. Influence of the steepness of the gradient, B, on the peak widths, w_g, of (——) N,N-dimethyl-p-aminobenzeneazobenzoylamides in chromatography on a silica column [400 x 3.0 mm I.D., Porasil A (60), 37-75 μm]; (---) xanthine alkaloids, bases and barbiturates in chromatography on an octadecylsilica column (300 x 4.2 mm I.D., C18/LiChrosorb Si 100, 10 μm). (——) Gradient of ethyl acetate in cyclohexane, concave, according to the gradient function eqn. 4.18 ($\kappa = 2$). Conditions and compounds as in Fig. 4.6. (---) Gradient of methanol in water, linear, gradient function (eqn. 4.19). Flow-rate: 1 ml/min. Curve 5 (average data for 20 compounds); w_g in ml; B in % of solvent B x 10^{-2} per ml of the eluate. (Adapted from refs. 44 and 50.)

retention volumes. The changes in the steepness of the gradient (B) in chromatography on polar adsorbents influence peak widths more significantly if less efficient columns are used and the relative decrease of peak widths with increasing B depends on the shape of the gradient (the curvature expressed by the parameter κ in eqn. 4.37)[44,51]. The bandwidths in these systems tend to acquire constant values at a high steepness of the gradient (see the examples of the chromatography of azo compounds on a silica column of low efficiency and the reversed-phase chromatography of barbiturates and bases on a standard octadecylsilica column[50] in Fig. 4.11).

The influence of the shape (curvature) of the gradient on peak widths in chromatography on polar adsorbents is complex and depends on the character of the compounds and of the mobile phase (parameters k'_0 and m in eqn. 1.23) and on the other parameters of the gradient (A, B). All of these parameters play a certain role in the determination of the actual composition of the eluent at the time of elution of the peak maximum. An example of experimental plots of peak widths as a function of the curvature of the gradient (κ) in chromatography on silica is shown Fig. 4.12[51].

References on p. 180.

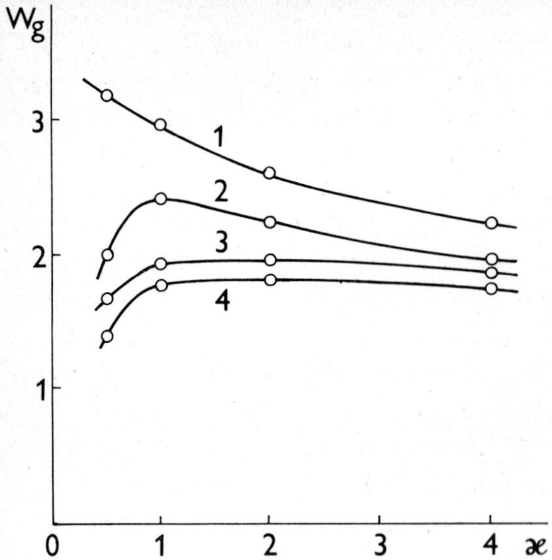

Fig. 4.12. Plots of the experimental peak widths, ω_g, of four azo compounds versus κ in gradient elution adsorption chromatography on Porasil A. Gradient function, conditions and compounds as in Fig. 4.11; B = 0.026 (% of solvent B x 10^{-2} per ml of the eluate). (From ref. 51.)

In reversed-phase chromatographic systems, the shape and the steepness of the gradient and the initial concentration of the more efficient eluting component in the mobile phase influence the peak width less than in normal-phase chromatography. Slight modifications of the curvature of the gradient and small changes in the initial composition of the mobile phase seem to have almost negligible effects on peak widths. Generally, in reversed-phase chromatography, the widths of peaks eluted in one common gradient usually do not differ from each other by more than 20-40% relative, whereas a 10-fold increase in the steepness of the gradient decreases the bandwidths approximately 4-5 fold. This holds true for standard efficiency columns (ca. 10,000-30,000 plates per metre) and the relative influence of various gradient parameters on bandwidths seems to decrease with increasing column efficiency (for very efficient columns, extra-column contributions to band broadening may become effective).

As far as other experimental effects on bandwidths are concerned, all effects that influence bandwidths under isocratic conditions have an analogous result in gradient elution chromatography (e.g., flow-rate, temperature and viscosity of the mobile phase, column dimensions and particle diameters) (see Chapter 1 and other books on liquid chromatography).

No influence of changing the mobile phase viscosity during certain gradients in reversed-phase chromatography (water-methanol, water-acetonitrile) on bandwidths is apparent in practice, in spite of relatively large changes in viscosity. This can be seen from Tables 4.3 and 4.4, where the differences between the calculated and experimental bandwidths do not depend on the composition of the mobile phase at the time of elution of the solute peak[53].

4.8. SELECTIVITY AND RELATIVE RETENTION IN GRADIENT ELUTION CHROMATOGRAPHY

The selectivity of the separation of a pair of compounds under isocratic conditions is conveniently characterized as the retention ratio, α (the ratio of the capacity factors or net retention volumes), relative retention. As shown in Part I, α does not depend on the content of the more efficient eluting component in the binary mobile phase if the parameters m_1, m_2 of the appropriate two-parameter $k' = f(c)$ function describing a given system (eqn. 1.23 or 1.28) are approximately equal for the two compounds to be separated. Then,

$$\alpha = \frac{k'_2}{k'_1} = \frac{k'_{02}}{k'_{01}} \qquad (4.41)$$

Thus, constant selectivity under isocratic conditions means that the ratio of net retention volumes of sample compounds does not depend on the composition of the mobile phase.

What happens if we apply a "linear solvent strength" gradient (eqn. 4.4) to the separation of such compounds? The difference of retention volumes for a pair of compounds is here given by

$$\Delta V'_g = V'_{g2} - V'_{g1} = \frac{1}{\beta} \cdot \log\,(2.31\beta k'_{a2} V_m + 1) - \frac{1}{\beta} \cdot \log\,(2.31\beta k'_{a1} V_m + 1) \qquad (4.42)$$

If we start a "linear solvent strength" gradient at zero initial concentration of the more efficient eluting component in the mobile phase, $k'_a = k'_0$. As k'_0 are sufficiently large for all but very early eluted compounds, we can write

$$\Delta V'_g = \frac{1}{\beta} \cdot \log \frac{2.31\beta k'_{02} V_m + 1}{2.31\beta k'_{01} V_m + 1} \approx \frac{1}{\beta} \cdot \log \frac{k'_{02}}{k'_{01}} = \frac{1}{\beta} \cdot \log \alpha \qquad (4.43)$$

Thus, if "linear solvent strength" gradients are applied to separation with a constant selectivity in a given system, the pairs of compounds with equal α values (isocratic) are eluted with approximately equal differences in retention volumes, regardless of their absolute retention volumes. In other words, "linear solvent strength" gradients should yield regular spacing of the peaks on the

References on p. 180.

chromatogram for pairs of compounds with equal (isocratic) retention ratios, as is indicated in Fig. 3.1C[1,3] (recall that bandwidths are approximately constant under these conditions).

However, to obtain a constant selectivity in gradient elution chromatography (i.e., a constant value of the parameter β in eqn. 4.4 for different sample solutes), it is necessary that the selectivity under isocratic conditions be constant also, and does not depend on the concentration of the more efficient eluting agent in the mobile phase. In this instance, the parameters m of eqn. 1.23 or 1.28 [the $k' = f(c)$ equation that applies in a given system] for different sample solutes should be approximately equal.

These assumptions are reasonably valid in the practical liquid chromatography of a number of compounds with a similar structure on columns packed with chemically bonded non-polar stationary phases when a linear concentration gradient of the organic solvent in water is used. However, for many compounds there are significant differences in the parameters m of $k' = f(c)$ functions and, consequently, the selectivity changes with time during the gradient elution of such compounds. For certain types of compounds, such as in homologous series, there is a regular increase with increasing absolute retention of the solutes[37-41,70]. Here, slightly convex concentration gradients approximate better "linear solvent strength" gradients than linear concentration gradients, and for these gradients, the above conclusions hold approximately true[3]. However, for other compounds and gradient types, significant changes in selectivity during gradient elution may occasionally be found[4,50,51]. Hence the rule of constant selectivity in gradient elution using "linear solvent strength" gradients should be understood as an approximate one, which applies generally for most compounds, but from which exceptions are often found for certain solutes.

As we have seen already, the relative retention related to the volume of the gradient, V'_g/V_G, is constant, if the product BV_m is kept constant. We can define the relative retention in gradient elution chromatography, α_g, in another way, by analogy with isocratic elution, as the ratio of the net retention volumes of two compounds 1 and 2, i.e., the "apparent" retention ratio in gradient elution chromatography, α_g[8]:

$$\alpha_g = \frac{V'_{g2}}{V'_{g1}} = \frac{k'_{g2}}{k'_{g1}} \tag{4.44}$$

As we shall see later, α_g enters into the relationship for resolution in gradient elution chromatography, analogous to the retention ratio, α, in the expression for resolution under isocratic conditions. Generally, α_g is not a

TABLE 4.5

INFLUENCE OF THE SHAPE OF THE GRADIENT FUNCTION (PARAMETER κ IN EQN. 4.15) ON THE RETENTION RATIOS OF AZO COMPOUNDS IN GRADIENT ELUTION ADSORPTION CHROMATOGRAPHY ON SILICA

Column, Porasil A (37-75 µm), 400 x 3.0 mm I.D.; V_m = 2.0 ml. Gradient of ethyl acetate in cyclohexane according to the gradient function given by eqn. 4.18 with different parameters B and κ (A = 0). α_g 2/1, α_g 3/1 and α_g 4/1 are the retention ratios for compounds 2-4 relative to compound 1. The experimental values are compared with those calculated using eqn. 4.45. Compounds: amides of p-N,N-dimethyl-p'-aminobenzeneazobenzoic acid: 1, di(n-butyl)amide, k_0' = 0.242, m = 1.68; 2, di(n-propyl)amide, k_0' = 0.330, m = 1.68; 3, diethylamide, k_0' = 0.65, m = 1.76; 4, dimethylamide, k_0' = 1.94, m = 1.93. k_0', m are the experimental parameters of eqn. 1.23 evaluated by linear regression from experimental log k'= f(log c) plots.

κ	Retention ratio	α_g (calc.)	α_g (exptl.)			
			B = 0.00162	B = 0.00649	B = 0.02597	B = 0.1039
0.5	α_g 2/1	1.24	1.25	1.24	1.25	1.19
	α_g 3/1	1.86	1.84	1.86	1.87	1.76
	α_g 4/1	3.68	-	3.57	3.62	3.50
1.0	α_g 2/1	1.16	1.15	1.16	1.17	1.16
	α_g 3/1	1.52	1.49	1.50	1.52	1.52
	α_g 4/1	2.41	2.18	2.22	2.29	2.39
2.0	α_g 2/1	1.09	1.08	1.09	1.09	1.09
	α_g 3/1	1.29	1.25	1.27	1.28	1.28
	α_g 4/1	1.70	1.59	1.62	1.64	1.66
4.0	α_g 2/1	1.05	1.05	1.05	1.06	1.05
	α_g 3/1	1.15	1.15	1.15	1.16	1.15
	α_g 4/1	1.35	1.34	1.34	1.35	1.35

true measure of the selectivity in a given system, as it depends on the profile of the gradient and can, to a certain extent, be adjusted experimentally, even if the selectivity is constant under isocratic conditions.

For example in the systems where the k' = f(c) function approximately follows eqn. 1.23 and the parameters m of this equation are approximately constant for different sample solutes (constant selectivity under isocratic conditions) and where curved gradients described by eqn. 4.18 (zero initial concentration A of the more efficient eluting component in the mobile phase) are applied, α_g for different pairs of compounds do not depend on the steepness of the gradient, but they depend on its shape (curvature) described by the parameter κ (Table 4.5)[8]:

$$\alpha_g = \left(\frac{k'_{02}}{k'_{01}}\right)^{\frac{1}{\kappa m+1}} \qquad (4.45)$$

References on p. 180.

It is obvious that a decrease in α_g with increasing κ is by no means caused by a change in the real selectivity in the system (as k'_0 and m of all compounds are constant), but is only a consequence of the shift of all retention volumes to higher values when κ is increased and the gradient becomes more concave.

In many other gradients (including "linear solvent strength" gradients) applied in practical chromatographic systems, a decrease of the steepness of the gradient (parameter B) leads to a more or less significant increase in α_g. On the other hand, the initial concentration of the more efficient eluting component in the mobile phase generally shows only a negligible influence on the differences in retention volumes $\Delta V'_g$, namely for later eluted compounds. Consequently,

$$\alpha_g = 1 + \frac{\Delta V'_g}{V'_{g1}} \approx 1 + \frac{\text{constant}}{V'_{g1}} \qquad (4.46)$$

and, because an increase in A leads to a decrease in retention volumes (V'_{g1} decreases), α_g increases with increasing A. Practical example of these effects in reversed-phase systems is shown in Table 4.6[51].

As discussed in Part I, the parameters m of the $k' = f(c)$ functions differ significantly from one another for certain pairs of compounds and the selectivity of their separation depends on the concentration of the more efficient eluting component in the mobile phase. Sometimes, such pairs of compounds are eluted with equal retention volumes at a certain composition of the mobile phase, where these compounds cannot be separated[48]. In gradient elution chromatography, the situation is analogous. As the profile of the gradients has an analogous influence on retention in gradient elution chromatography to the concentration of the more efficient eluting component in the mobile phase under isocratic conditions, certain combinations of the parameters A, B and κ of the gradient function (initial composition of the mobile phase, steepness and shape of the gradient) may be found, where $\alpha_g = 1$ and these compounds cannot be separated[51].

A knowledge of the conditions for which $\alpha_g = 1$ may be useful for appropriate selection of gradients for certain separation problems. In a similar manner as the composition of the mobile phase for $\alpha = 1$ in isocratic elution chromatography can be calculated, we can also calculate the combinations of the parameters A, B and κ for $\alpha_g = 1$ in gradient elution chromatography, after substitution of the appropriate relationships from Appendix 2 into eqn. 4.44 while setting $\alpha_g = 1$. The resulting equations usually must be solved by using approximate methods and the relationships for the required parameters can be expressed in the explicit form in special instances only[51]. If eqn. 1.23

TABLE 4.6

INFLUENCE OF THE INITIAL CONCENTRATION OF THE MORE EFFICIENT ELUTING COMPONENT IN THE MOBILE PHASE AND OF THE SLOPE OF THE GRADIENT FUNCTION (PARAMETERS A AND B OF EQN. 4.17) ON THE RETENTION RATIOS OF BARBITURATES IN REVERSED-PHASE GRADIENT ELUTION CHROMATOGRAPHY

Column, mobile phase gradient and experimental conditions as in Table 4.3. All retention ratios, α_g, are related to barbital (compound 1). The experimental values are compared with those calculated using eqns. 4.44 and 4.27. Compounds: 1, barbital, $k_0' = 21.81$, $m = 3.2$; 2, heptobarbital, $k_0' = 58.44$, $m = 3.71$; 3, allobarbital, $k_0' = 69.44$, $m = 3.55$; 4, aprobarbital, $k_0' = 106.96$, $m = 3.66$, 5, butobarbital, $k_0' = 187.41$, $m = 3.78$; 6, hexobarbital, $k_0' = 252.29$, $m = 3.77$; 7, pentobarbital, $k_0' = 470.65$, $m = 4.07$ (k_0', m are the experimental parameters of eqn. 1.28 evaluated by linear regression from experimental log $k' = f(c)$ plots, where c = concentration of methanol in the mobile phase in isocratic elution experiments).

A	Retention ratio	B					
		0.06872		0.03436		0.01718	
		α_g (calc.)	α_g (exptl.)	α_g (calc.)	α_g (exptl.)	α_g (calc.)	α_g (exptl.)
0	α_g 2/1	1.13	1.12	1.19	1.17	1.26	1.24
	α_g 3/1	1.21	1.20	1.28	1.27	1.37	1.36
	α_g 4/1	1.29	1.28	1.38	1.37	1.50	1.49
	α_g 5/1	1.39	1.38	1.51	1.50	1.67	1.67
	α_g 6/1	1.46	1.46	1.60	1.59	1.78	1.78
	α_g 7/1	1.50	-	1.66	1.66	1.90	1.90
0.1	α_g 2/1	1.16	1.14	1.23	1.22	1.32	1.31
	α_g 3/1	1.26	1.25	1.35	1.34	1.47	1.49
	α_g 4/1	1.35	1.35	1.47	1.46	1.64	1.68
	α_g 5/1	1.48	1.48	1.64	1.64	1.87	1.94
	α_g 6/1	1.57	1.56	1.75	1.75	2.02	2.09
	α_g 7/1	1.61	1.61	1.84	1.84	2.18	2.26

applies in a given system and the gradient is started with pure solvent A, we obtain the value of B for which $\alpha_g = 1$ (gradient function according to eqn. 4.18):

$$B_{(1)} = \frac{1}{V_m} \cdot \left\{ \frac{\left[k_{01}'(\kappa m_1 + 1)^{(\kappa m_2 + 1)}\right]}{\left[k_{02}'(\kappa m_2 + 1)^{(\kappa m_1 + 1)}\right]} \right\}^{\frac{1}{\kappa(m_1 - m_2)}} \qquad (4.47)$$

Thus, at a given curvature of the gradient (κ), the order of elution of compounds 1 and 2 is reversed for $B > B_{(1)}$ with respect to the order of elution for $B < B_{(1)}$. For linear concentration gradients starting at zero concentration of the more efficient eluting agent in the mobile phase applied to reversed-phase and other systems where eqn. 1.28 can describe $k' = f(c)$ functions, we can write under analogous conditions an approximate equation, assuming large k'_0 values[51]:

$$B_{(1)} = \frac{1}{2.31V_m} \cdot \frac{(k'_{02}m_2)^{\frac{m_1}{m_2-m_1}}}{(k'_{01}m_1)^{\frac{m_2}{m_2-m_1}}} \tag{4.48}$$

The reversal of the elution order with changing gradient profile is not often found in practical systems, but it could be observed for certain pairs of compounds and sometimes it may complicate the selection of the optimum gradient for a given separation problem. An example is illustrated in Table 4.7 for the reversed-phase separation of barbiturates and bases using linear concentration gradients of methanol in water[51]. The same situation was observed, e.g., for the separation of acetophenone from *o*-nitroaniline in normal-phase chromatography on silica using a linear concentration gradient of methylene chloride in *n*-heptane[61].

4.9. RESOLUTION IN GRADIENT ELUTION CHROMATOGRAPHY

The resolution of compounds 1 and 2 in gradient elution chromatography, $R_{g1,2}$, may be defined by analogy with isocratic elution chromatography (for convenience, R_g will be used in the following instead of $R_{g1,2}$):

$$R_g = \frac{V'_{g2} - V'_{g1}}{w_{g2}} \tag{4.49}$$

Snyder and Saunders[36] rearranged this equation into the following form, which applies to "linear solvent strength" gradients, where the selectivity is approximately constant and does not depend on changes in the composition of the mobile phase during the gradient elution:

$$R_g = \frac{\sqrt{n}}{4} \cdot \frac{\alpha - 1}{\alpha} \cdot Q' \tag{4.50}$$

where n is the plate number of the column under the conditions used and $\alpha = k'_2/k'_1$ is the retention ratio of the compounds 1 and 2 under isocratic conditions. Both n and α are assumed not to depend significantly on the concentration of the more

TABLE 4.7

INFLUENCE OF THE PARAMETERS *A* AND *B* OF THE GRADIENT FUNCTION (EQN. 4.17) ON THE ELUTION SEQUENCE IN REVERSED-PHASE GRADIENT ELUTION CHROMATOGRAPHY

Column as in Table 4.3. Isocratic elution: mobile phase composed of methanol (different concentrations, c) in water. k'_1-k'_4, experimental capacity factors for compounds 1-4; $\alpha_{2/1}$, $\alpha_{4/3}$ are experimental retention ratios for compounds 2,1 and 4,3, respectively. Gradient elution: gradient of methanol in water according to the gradient function given by eqn. 4.17 with different parameters A and B. V'_{Rg1}-V'_{Rg4} are experimental net retention volumes for compounds 1-4; $\alpha_{g\ 2/1}$, $\alpha_{g\ 4/3}$ are retention ratios for compounds 2,1 and 4,3, respectively, found experimentally and calculated from eqns. 4.44 and 4.27. Compounds: 1, caffeine, k'_0 = 32.28, m = 3.78; 2, barbital, k'_0 = 21.81, m = 3.20; 3, 3-n-butyl-6-methyluracil, k'_0 = 67.20, m = 3.25; 4, aprobarbital, k'_0 = 106.96, m = 3.66. k'_0, m are the experimental parameters of eqn. 1.28 evaluated by linear regression from experimental log k' = f(c) plots.

Isocratic elution

c (%)	k'_1	k'_2	$\alpha_{2/1}$	k'_3	k'_4	$\alpha_{4/3}$
20	5.82	5.12	0.88	21.89	21.06	0.96
30	1.78	2.29	1.29	7.49	8.42	1.12
40	0.90	1.16	1.29	3.24	3.47	1.07

Gradient elution

A	B	V'_{Rg} 1 (ml)	V'_{Rg} 2 (ml)	α_{g} 2/1 Exptl.	α_{g} 2/1 Calc.	V'_{Rg} 3 (ml)	V'_{Rg} 4 (ml)	α_{g} 4/3 Exptl.	α_{g} 4/3 Calc.
0	0.06872	7.26	7.25	1.00	1.02	9.58	9.28	0.97	1.00
	0.03436	11.39	11.56	1.01	1.00	15.97	15.88	0.99	1.01
	0.01718	19.21	17.92	0.93	0.96	26.63	26.75	1.00	1.04
0.1	0.06872	5.64	5.84	1.04	1.03	8.02	7.90	0.98	0.99
	0.03436	8.72	8.90	1.02	1.01	13.23	13.03	0.98	1.01
	0.01718	13.11	12.36	0.94	0.97	20.65	20.79	1.01	1.04

efficient eluting component in the mobile phase. Eqn. 4.50 is analogous to the expression for resolution under isocratic conditions (eqn. 1.13), and the contribution of the efficiency, selectivity and capacity terms to resolution in eqn. 4.50 can be clearly differentiated. Hence the factor Q' expresses the contribution of capacity to resolution in gradient elution chromatography and can be calculated as follows.

The capacity term Q' is approximately constant for different sample compounds eluted in different times and consequently the resolution is also constant during the gradient elution for all pairs of compounds with equal selectivities. By analogy with the capacity term of resolution in isocratic elution chromatography (eqn. 1.13), Q' is defined as[36]

References on p. 180.

$$Q' = \frac{\bar{k}}{1 + \bar{k}} \tag{4.51}$$

where $\bar{k}$ is a median capacity factor during the gradient elution chromatography, which is equal to $2k'_f$ (cf., eqn. 4.30c) and consequently[37,38]

$$\bar{k} \approx \frac{1}{1.15\beta} \tag{4.52}$$

The combination of eqns. 4.51 and 4.52 yields for Q'

$$Q' = \frac{1}{1 + 1.15\beta} \tag{4.53}$$

Fig. 4.13 shows the plot of Q' versus β according to eqn. 4.53 (dashed line) and the plot of $Q'G$ versus β, which may be useful for more exact calculations of R_g. It may be seen that the neglect of G has only a minor effect on the precision of the calculations[3].

A further similarity between gradient and isocratic elution may be found in the occurrence of maximum resolution per unit time at a certain steepness of the gradient ($\beta \approx 0.4$ and $Q' \approx 0.7$) as at a certain capacity factor ($k' \approx 2$) under isocratic conditions. The optimum steepness β for practical work with "linear solvent strength" gradients is in the range $0.1 \leqslant \beta \leqslant 0.3$, which roughly corresponds to $2 \leqslant k' \leqslant 5$ for operation under isocratic conditions[3].

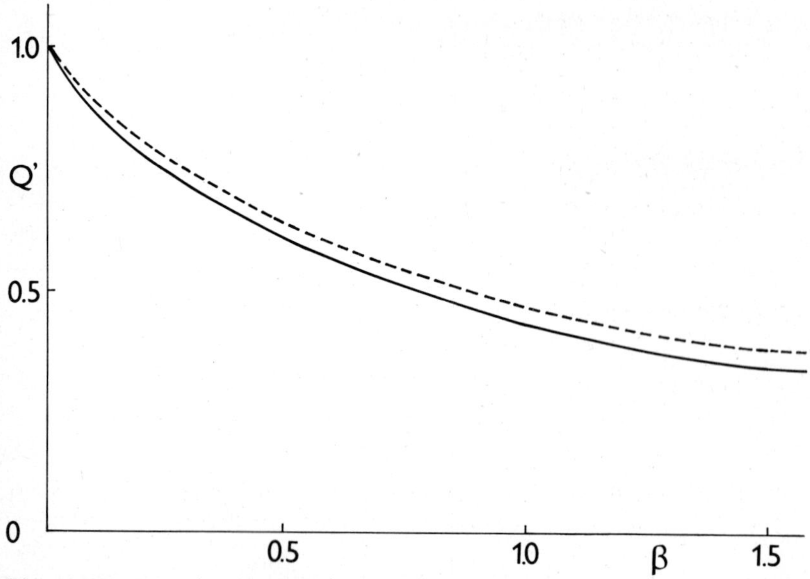

Fig. 4.13. Capacity term, Q', of the resolution equation as a function of gradient steepness β in gradient elution chromatography using "linear solvent strength" gradients. Dashed curve, function Q' versus β given by eqn. 4.53; solid curve, exact relationship of $Q'G$ versus β. (Reproduced with permission from ref. 3.)

If general, both "linear solvent strength" and "non-linear solvent strength", gradients are considered, resolution in gradient elution chromatography can be calculated after the introduction of the appropriate equations for retention volumes (Appendix 2) and bandwidths into eqn. 4.49. Thus, for example, for gradients controlled by a general gradient function (eqn. 4.15) applied to systems where the two-parameter $k' = f(c)$ function according to eqn. 1.23 is valid (normal adsorption, ion exchange), the following relationship for resolution under gradient conditions results[51]:

$$R_g = \underbrace{\frac{\sqrt{n}}{4}}_{\text{I}} \cdot \underbrace{\frac{\alpha_g - 1}{\alpha_g}}_{\text{II}} \cdot \underbrace{\frac{(\kappa m_2 + 1)Bk'_{02}V_m + A^{\frac{\kappa m_2+1}{\kappa}} - A^{\frac{1}{\kappa}}\left[(\kappa m_2 + 1)Bk'_{02}V_m + A^{\frac{\kappa m_2+1}{\kappa}}\right]^{\frac{\kappa m_2}{\kappa m_2+1}}}{V_m B\left\{\left[(\kappa m_2 + 1)Bk'_{02}V_m + A^{\frac{\kappa m_2+1}{\kappa}}\right]^{-\frac{\kappa m_2}{\kappa m_2+1}} + k'_{02}\right\}}}_{\text{III}} \tag{4.54}$$

A similar equation can be derived for resolution in reversed-phase chromatography with linear concentration gradients of the organic solvent in the mobile phase (gradient function according to eqn. 4.17), assuming the validity of eqn. 1.28 for the $k' = f(c)$ dependence[51]:

$$R_g = \underbrace{\frac{\sqrt{n}}{4}}_{\text{I}} \cdot \underbrace{\frac{\alpha_g - 1}{\alpha_g}}_{\text{II}} \cdot \underbrace{\frac{\log\,(2.31m_2BV_mk'_{02} \cdot 10^{-m_2A} + 1) - 1}{\frac{B}{A} \cdot V_m\left[1 + k'_{02}(2.31m_2BV_mk'_{02} + 10^{m_2A})^{-1}\right]}}_{\text{III}} \tag{4.55}$$

An additional compression of bandwidth due to accelerated migration of the end of the peak with respect to its front is neglected here but, if necessary, it can be taken into account by introducing the compression factor G, either calculated or an empirical value of 0.85, into the denominator of eqns. 4.54 and 4.55.

As in the analogous expression for resolution under isocratic conditions (eqn. 1.13) and as in eqn. 4.50, three different terms contribute to resolution in eqns. 4.54 and 4.55 and in other equations for R_g: I, efficiency - n is the plate number of the column used (measured under isocratic conditions), which is approximately equal for different solutes and does not depend significantly on the composition of the mobile phase; II, selectivity, where α_g is the "apparent" retention ratio in gradient elution chromatography (see eqn. 4.44), which depends on the profile of the gradient (in contrast to eqn. 4.50, where the isocratic retention ratio α is used); and III, the capacity term, which is a function of the retention of the later eluted compound 2 under the gradient conditions used.

Resolution in gradient elution chromatography depends not only on the profile of the gradient, but also on the character of the $k' = f(c)$ functions of compounds 1 and 2 [on the combination of the parameters k'_{01}, k'_{02}, m_1 and m_2 of the two-parameter equations describing adequately these functions, or on a larger number of parameters if three-parameter $k' = f(c)$ functions (Table 2.1) must be considered in a given chromatographic system]. The expressions describing directly the resolution as a function of these parameters and of the parameters of the gradient function used may be obtained easily after combination of eqns. 4.54 and 4.55 or analogous equations with eqn. 4.44 and introduction of the appropriate equations for V'_{g1} and V'_{g2} from Appendix 2[51].

As discussed in Section 4.8, if the exponents m_1 and m_2 of the $k' = f(c)$ functions of the two compounds differ, the apparent retention ratio $\alpha_g = 1$ may eventually be found for certain combinations of the parameters of the gradient function. In this instance, if the steepness of the gradient (parameter B of the gradient function) is gradually increased while the shape (curvature) of the gradient and the initial composition of the mobile phase are kept constant, both the capacity (III) and the selectivity (II) contributions to resolution decrease (eqns. 4.54 and 4.55 and analogous equations), until $\alpha_g = 1$ is reached where the resolution $R_g = 0$. A further increase in parameter B leads to reversal of the order of elution of compounds 1 and 2 (compound 2 is eluted first), but the selectivity contribution (III) increases whereas the capacity contribution still decreases. Consequently, R_g also increases and eventually a maximum of R_g may be achieved for a certain value of the parameter B, where the contribution of the capacity term equals the contribution of selectivity. If B is increased further, the influence of the capacity contribution to resolution predominates over the selectivity term and resolution again decreases[51].

This behaviour is completely analogous to the influence of increasing concentration of the more efficient component in the mobile phase on resolution, R_s, under isocratic conditions, as indicated on Fig. 2.3[48]. As in isocratic elution chromatography, in gradient elution chromatography it is also very unlikely that one will find the whole relationship of R_g versus B as indicated above, and usually only part of it can be observed within practically possible limits of B.

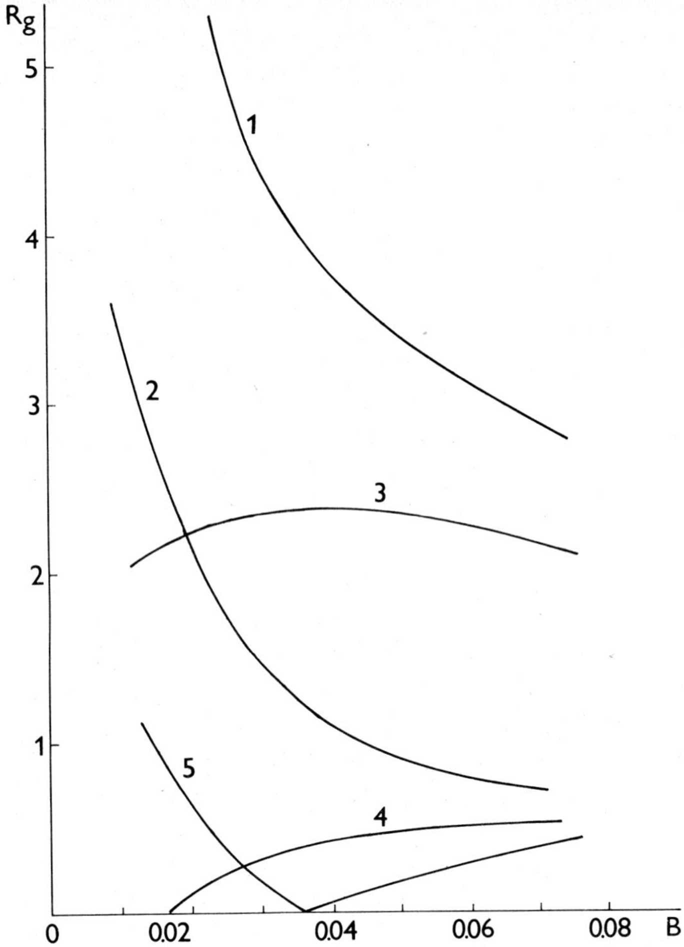

Fig. 4.14. Examples of R_g = f(B) plots in reversed-phase chromatography on an octadecylsilica column (C18/LiChrosorb Si 100, 10 μm; 300 x 4.2 mm I.D.; flow-rate 1.0 ml/min), where linear concentration gradients of methanol in water (with zero initial concentration of methanol) are applied; gradient functions described by eqn. 4.19 in Table 4.1. R_g is plotted for the following pairs of compounds: (1) 3-*n*-propyl-6-methyluracil (m = 3.3)/3-*sec*.-butyl-6-methyluracil (m = 3.2); (2) 3-*n*-butyl-6-methyluracil (m = 3.3)/3-*tert*.-butyl-6-methyluracil (m = 3.4); (3) theobromine (m = 4.6)/theophylline (m = 3.4); (4) allobarbital (m = 3.6)/3-*sec*.-butyl-6-methyluracil (m = 3.2); (5) caffeine (m = 3.8)/barbital (m = 3.2). m are the exponents in the k' = f(c) function for individual compounds in the system used (eqn. 1.28).

Examples of R_g = f(B) plots measured in linear concentration gradients of methanol in water (with zero initial concentration of methanol) applied to the chromatography of bases and barbiturates on an octadecylsilica column[50] are shown in Fig. 4.14. The R_g = f(B) plots for individual pairs of compounds represent parts of a "general" dependence of R_g on B as discussed above. The plots 1 and 2 represent typical dependences of R_g on B for most pairs of chromatographed compounds [the parameters m_1 and m_2 of $k' = f(c)$ functions are approximately equal]. Occasionally, an increase in R_g with increasing B (plot 4, $m_1 = 3.6$, $m_2 = 3.2$), maximum resolution (plot 3, $m_1 = 4.6$, $m_2 = 3.4$) or zero resolution (plot 5, $m_1 = 3.8$, $m_2 = 3.2$) (here, reversal of the elution order occurs) at a certain value of B are observed, as was found also by Elgass[61] in chromatography on silica using a gradient of methylene chloride in n-heptane and on octadecylsilica using linear gradients of methanol or acetonitrile in water. As can be seen from Fig. 4.14, the probability of the occurrence of "irregular" R_g versus B plots increases with increasing differences between the exponents m_1 and m_2 of the $k' = f(c)$ functions of the individual compounds, which is in agreement with the above theoretical discussion of the problem and is completely analogous to the behaviour under isocratic conditions (Part I, Section 2.2).

The maximum on the R_g versus B plots and the corresponding B may be calculated at constant A and κ by introducing R_g from eqns. 4.54 and 4.55 or from similar appropriate equations into the following relationship:

$$\left(\frac{dR_g}{dB}\right)_{A,\kappa} = 0 \tag{4.56}$$

and then by derivation of this equation. Similarly, the value of B corresponding to $R_g = 0$ can be found after setting $\alpha_g = 1$ and calculated from eqns. 4.47 and 4.48 or analogous equations (see Section 4.8). Of course, such calculations would yield real values of B only for such systems where the separated compounds can behave in the above indicated manner, otherwise no meaningful results are obtained.

In most practical separation problems where gradient elution is applied, the resolution does not decrease to zero, nor is maximum resolution found at certain B, but R_g decreases with increasing B over the whole really possible range of the gradient profiles.

The steepness of the gradient usually affects the resolution far more than the initial concentration of the more efficient eluting component in the mobile phase (parameter A). If we keep the shape and the steepness of the gradient (κ and B) constant and change A, a similar dependence of R_g on A to the dependence of R_g on B at constant A and κ could be expected. In practical systems, however, the initial composition of the mobile phase has relatively little influence on both the differences in retention volumes and the bandwidths of the eluted compounds (see previous paragraphs), and consequently changing A does not have much influence on resolution in the later part of the gradient elution (within certain limits). Here, an increase in selectivity, α_g, seems to compensate approximately a decrease in the capacity contribution to resolution when

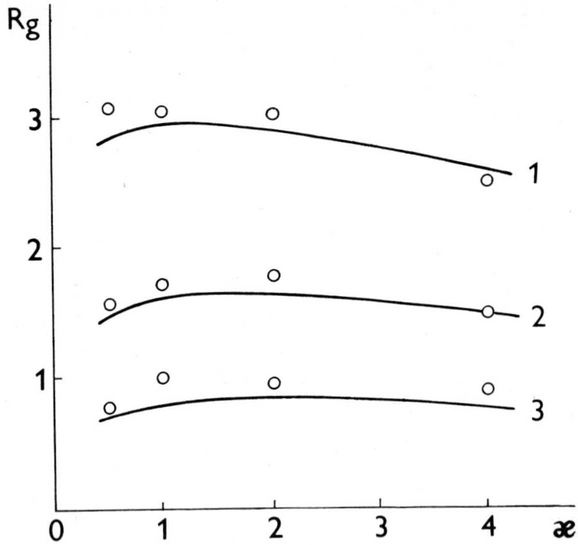

Fig. 4.15. Plots of resolution, R_g, versus the shape of gradient in adsorption chromatography on a silica column (400 x 3.0 mm I.D., Porasil A, 37-75 μm). The shape of the gradient is characterized by the parameter κ in eqn. 4.18. Gradient of ethyl acetate in cyclohexane; conditions and numbers of compounds as in Fig. 4.11a). Points, experimental values; full curves, calculated plots. Curves: 1 = R_g for compounds 1 and 2; 2 = R_g for compounds 2 and 3; 3 = R_g for compounds 3 and 4. (From ref. 51.)

the parameter A is increased[51]. However, on increasing A, the time of separation can be significantly shortened and the resolution in the initial part of the chromatogram usually decreases considerably.

The shape of the concentration gradient characterized by the parameter κ in eqns. 4.14-4.20 has a more complex influence on resolution. It can be shown theoretically that maximum resolution can be found for certain values of κ even in the simplest instances, where $m_1 = m_2$ in eqn. 1.23 or 1.28 [two-parameter $k' = f(c)$ functions] and $A = 0$ (gradient started at zero initial concentration of the more efficient eluting component in the mobile phase). The resolution approaches zero for very large κ and becomes identical with the resolution in isocratic elution using the pure, more efficient eluent for $\kappa = 0$. The resolution obtained in gradient elution chromatography is always larger than in the above two extreme situations and a maximum on the $R_g = f(\kappa)$ curve must occur. In practice, however, these maxima are likely to be rather flat and even large changes in κ would not influence the resolution very significantly, as shown by the example in Fig. 4.15[51].

References on p. 180.

4.10. CHARACTERISTICS OF GRADIENT ELUTION SEPARATION OF MULTI-COMPONENT MIXTURES

It is difficult to characterize adequately in a quantitative manner the chromatographic behaviour of a sample mixture containing more than two components.

We can relatively easily arrive at a few empirical rules concerning the influence of the gradient profile on the spacing of peaks on the chromatogram and on the resolution in different parts of the chromatogram. An increase in the steepness of the gradient (parameter B of the gradient function) leads to earlier elution of all chromatographed compounds. All peaks in different parts of the chromatogram are shifted to lower elution volumes, the bandwidths become narrower and the peaks become higher. Consequently, the sensitivity of detection is improved and the time of analysis is shortened, but the resolution is impaired in all parts of the chromatogram. This is illustrated by Fig. 4.5 for a practical example of the reversed-phase separation of fluorescent derivatives of homologous n-alkylamines. A change in the initial composition of the mobile phase has a relatively minor influence on bandwidths, peak heights, resolution and differences in retention volumes of the later eluted compounds (usually in the second part of the chromatogram).

An increase in the initial concentration of the more efficient eluting component in the mobile phase (A) leads to a decrease in the retention volumes of all compounds, to a decrease in bandwidths and in differences in retention volumes and to an increase in peak heights of early eluted compounds (Fig. 4.8). Consequently, the resolution is impaired and the sensitivity is improved in the initial part of the chromatogram, and the time of separation is decreased.

If the shape of the gradient is changed in such a way that it becomes more convex, the slope of the gradient profile becomes steeper at the beginning of the gradient relative to the later part of the gradient. Thus, the initial peaks are relatively narrower, higher and more "bunched together" and the resolution is lower in the initial part of the gradient, whereas the resolution is better, but the peaks are broader and lower in the later part of the chromatogram. In contrast, if the gradient becomes more concave, the increase in the concentration of the more efficient eluting component in the mobile phase is relatively steeper in the later part of the gradient than in the initial part. The early eluted peaks are relatively broader, lower and better separated from one another than the peaks in the later part of the chromatogram (Fig. 4.9). Here, the resolution is relatively decreased towards the end of the gradient. Generally, concave gradients are more useful than convex gradients, as there is usually a need to bunch together more the later eluted peaks than the earlier ones, especially in adsorption chromatography on polar adsorbents and in ion-exchange chromatography. Slightly convex gradients may be useful in the reversed-phase separation of homologous and benzologous series, for which they may yield improved separations compared with linear concentration gradients[2,3] (for more details, see Section 4.2).

The most important quantity characterizing the chromatographic separation of a multi-component mixture is the so-called peak capacity, P_c. The peak capacity gives the number of peaks that can be resolved in one chromatographic run provided that all compounds are separated from one another with equal resolution, $R_s = 1$. According to Giddings[71], the peak capacity under isocratic conditions is given by

$$P_c = \frac{\sqrt{n}}{4} \cdot \ln \left(\frac{V_{R_Z}}{V_{R_A}} \right) \tag{4.57}$$

where V_{R_A} and V_{R_Z} are the retention volumes of the first and the last compound eluted, respectively, and n is the plate number of the column.

Horváth and Lipsky[72] derived an analogous equation for the peak capacity in gradient elution chromatography, P_g:

$$P_g = \frac{\sqrt{n}}{4} \cdot \left(\frac{V_{g_Z}}{V_{g_A}} - 1 \right) \tag{4.58}$$

where V_{g_Z} and V_{g_A} have the same meaning as V_{R_Z} and V_{R_A} under isocratic conditions in eqn. 4.57. Equal bandwidths of all solutes are assumed as the bandwidth of A.

If the first compound is eluted with the column dead volume and the last compound is eluted in the instant when the gradient is finished, then $V_{g_A} = V_m$ and $V_{g_Z} = V_G$, and the peak capacity, P_g, gives the total number of peaks that can be resolved with $R_g = 1$ from the beginning to the end of the gradient:

$$P_g = \frac{V_G - V_m}{w} \approx \frac{\sqrt{n}}{4} \cdot \left(\frac{V_G}{V_m} - 1 \right) \cdot \frac{1}{1 + k'_f} \tag{4.58a}$$

This equation is more useful in practice than eqn. 4.58. For "linear solvent strength" gradients, eqn. 4.58a can be rewritten into the following, approximate form[3]:

$$P_g \approx \frac{2.31\sqrt{n}}{4} \cdot \frac{\Delta S}{1 + 2.31\beta} \tag{4.58b}$$

where $\Delta S = \log k'_a - \log k'_b$ is the difference in the logarithms of the capacity factors in the pure solvents a and b, i.e., ΔS is a measure of the difference in the elution strengths of these solvents.

TABLE 4.8

PEAK CAPACITIES IN ISOCRATIC (P_c) AND GRADIENT (P_g) ELUTION CHROMATOGRAPHY FOR DIFFERENT COLUMN EFFICIENCIES AND LENGTHS OF THE CHROMATOGRAM

n = Number of theoretical plates of the column; $f = P_g/P_c$ (eqn. 4.59) and length of the chromatogram is expressed by the ratio V_{g_Z}/V_{g_A}, where V_{g_A}, V_{g_Z} are the retention volumes of the first and last compounds eluted, respectively. Resolution = 1 is assumed for all sample solutes.

V_{g_Z}/V_{g_A}	f	n = 100		n = 1000		n = 10000	
		P_c	P_g	P_c	P_g	P_c	P_g
5	2.5	4	10	13	32	40	100
10	3.9	6	22	18	71	57	225
15	5.2	7	35	21	111	68	350
20	6.3	7	47	24	150	75	475
25	7.5	8	60	25	190	80	600
30	8.5	8	72	27	229	85	725

The peak capacity in gradient elution chromatography is considerably higher than that under isocratic conditions, where the bandwidths increase considerably with increasing capacity factors. This can be easily demonstrated using the factor f, expressed as the ratio between P_g and P_c provided that $V_{g_A} = V_{R_A}$ and $V_{g_Z} = V_{R_Z}$:

$$f = \frac{P_g}{P_c} = \frac{\dfrac{V_{g_Z}}{V_{g_A}} - 1}{\ln\left(\dfrac{V_{R_Z}}{V_{R_A}}\right)} \tag{4.59}$$

Table 4.8 shows the values of f, P_g and P_c calculated from eqns. 4.57-4.59 for various column efficiencies and lengths of the chromatogram expressed by the ratio $V_{g_Z}/V_{g_A} = V_{R_Z}/V_{R_A}$. The factor f increases with increasing ratio V_{g_Z}/V_{g_A}, i.e., with increasing length of the chromatogram. This factor gives the increase in the number of compounds eluted in gradient elution chromatography with respect to the number of compounds eluted in the same time under isocratic conditions. For example, if we consider a column 300 mm x 4 mm I.D., packed with material of 5 μm particle size, where 10,000 plates can be achieved, $V_m \approx 3$ ml, and the first compound is eluted with $V_R = V_m$, the flow-rate of the mobile phase is 2.0 ml/min, then, theoretically, 100 compounds can be separated and eluted in 8 min (compared with 40 under isocratic conditions) and over 700 compounds in ca. 45 min (compared with 85 under isocratic conditions). This may be understood as a rather extreme example, because most columns under practical conditions yield a lower efficiency, bandwidths are not ideally equal for different solutes eluted at different times from the beginning of gradient elution, are usually larger than the bandwidth of the first compound eluted and increase to a certain extent with decreasing steepness of the gradient as the ratio V_{g_Z}/V_{g_A} increases. Large differences in the solvent strengths of the solvents a and b would be necessary for large V_{g_Z}/V_{g_A} ratios.

The time due to the instrumental gradient delay and the time necessary for re-equilibration of the column are not considered here, but the increase in the number of compounds separated in a given time using the gradient elution technique is obvious. For a given number of chromatographed compounds, it is possible to decrease the time of separation in gradient elution chromatography also by a factor f with respect to isocratic conditions. It can also be seen from Table 4.8 that this advantage of gradient elution is far more significant for the separation of mixtures containing a large number of compounds than for relatively simple mixtures containing less than ten solutes. The great power of the gradient elution technique for the separation of complex multi-component mixtures is illustrated by an example in Fig. 4.16.

For standard analytical columns packed with 5-10 μm particles, peak capacities between 100 and 200 may be reasonably expected in practice for a 30-60 min gradient run at a flow-rate of the mobile phase of 1-2 ml/min[37].

The peak capacity in gradient elution chromatography increases with decreasing steepness of the gradient (parameter B of the gradient function), because the volume of the gradient, V_G, increases and V_G is proportional to V_{g_Z}, the volume of the last compound that may be eluted under gradient conditions. At a constant steepness of the gradient, B, the peak capacity decreases with increasing flow-rate of the mobile phase, because the efficiency (number of plates) of the column decreases. This has been demonstrated on practical examples by Elgass[61]. Further, as can be seen from eqn. 4.58b, the peak capacity increases with increasing differences in the elution strengths of the two solvents used to form the gradient.

Somewhat similar to peak capacity is the so-called "compression criterion", E, which can be used to characterize the "squeezing" of the chromatogram of a given sample mixture depending on the gradient profile[51]. This is defined as the ratio of the difference of the retention volumes of the last eluted compound (V_{g_Z}) and the first eluted compound (V_{g_A}) to the difference of the retention volumes of the second eluted compound (V_{g_B}) and the first eluted compound:

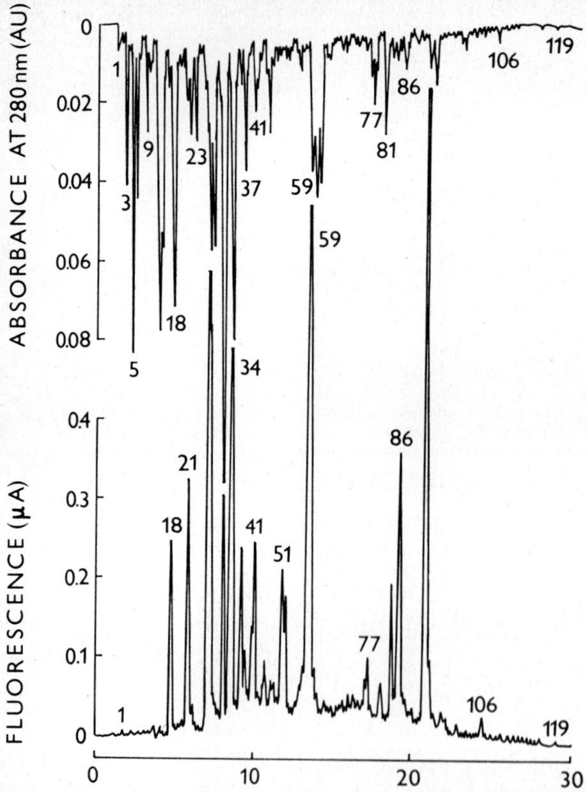

Fig. 4.16. Reversed-phase chromatographic separation of urinary aromatic acids on a LiChrosorb RP-18 (5 μm) column, 250 x 4.6 mm I.D., using a linear concentration gradient from 0.1 *M* phosphate buffer to 100% acetonitrile. Flow-rate, 2 ml/min; temperature, 70°C; detection, UV at 280 nm (upper chromatogram) and fluorimetric (lower chromatogram). (Reproduced with permission from ref. 86.)

$$E = \frac{V_{g_Z} - V_{g_A}}{V_{g_B} - V_{g_A}} = \frac{V'_{g_Z} - V'_{g_A}}{V'_{g_B} - V'_{g_A}} = \frac{\alpha_{g_{Z/A}} - 1}{\alpha_{g_{B/A}} - 1} \qquad (4.60)$$

E is the ratio of the length of the gradient program to the length of the part of this program utilized for the separation of the two first eluted sample solutes. If the bandwidths are constant in the chromatogram and if the two first peaks are eluted with resolution $R_g = 1$, E becomes identical with the peak capacity, P_g, as defined by Horváth and Lipsky[72], but E is essentially independent of these two conditions and can provide certain information concerning the "bunching together" of peaks in the initial part relative to the later part of the chromatogram and the means by which we can influence this effect experimentally.

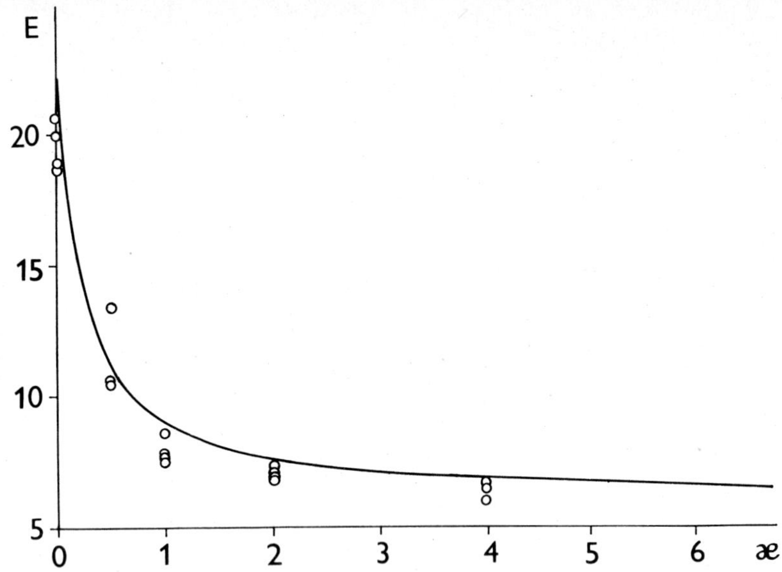

Fig. 4.17. Plot of the compression criterion, E, versus the shape of the gradient (κ) in gradient elution on silica. Column, compounds and conditions as in Fig. 4.11a. Gradient function, eqn. 4.18. Points, experimental data measured at various B; full curve, theoretical relationship. (From ref. 51.)

E depends generally on the steepness and shape of the gradient and on the initial composition of the mobile phase. However, in chromatography on polar adsorbents and in ion-exchange chromatography [if eqn. 1.23 fits the $k' = f(c)$ relationship], where gradients according to eqn. 4.18 in Table 4.1 are used ($A = 0$), the compression criterion does not depend on the steepness, but only on the shape (curvature) of the gradient as expressed by the parameter κ (see Fig. 4.17)[51]:

$$E = \frac{(\alpha_{0_{Z/A}})^{\frac{1}{\kappa m+1}} - 1}{(\alpha_{0_{B/A}})^{\frac{1}{\kappa m+1}} - 1} \qquad (4.61)$$

Here, E decreases with increasing concavity of the gradient, which is in agreement with the qualitative description of the effect of the gradient shape on separation as given above in this section.

Chapter 5

CHROMATOGRAPHY WITH COMPOSED GRADIENTS OF THE MOBILE PHASE

5.1. STEPWISE ELUTION CHROMATOGRAPHY

In stepwise elution chromatography, the concentration of the more efficient eluting component in the mobile phase follows a programme consisting of several consecutive isocratic steps. The concentration between the individual steps is changed discontinuously in such a way that the concentration in each subsequent step is increased.

Hence, the retention volume of a sample compound eluted in a given isocratic step consists of contributions of this step and of all preceding steps. In other words, the compound "remembers" the volume and composition of the mobile phase in each step before the elution.

The theory of stepwise elution in thin-layer chromatography was studied by Golkiewicz and Soczewiński[73], who presented a graphical method for the selection of separation conditions. Two calculation procedures for retention volumes in stepwise elution liquid column chromatography were described, the first by Borówko and co-workers[58-60] and the second by Jandera and Churáček[8,49]. The latter approach is simpler than the first, does not require graphical methods for solution and can be easily adapted for application of a computer or a programmable pocket calculator. It is based on the following considerations.

The pump delivers on to the column a volume V_{ei} of the mobile phase of a constant composition in each isocratic step i, where the capacity factor of the sample compound j is k'_i. If the compound j is eluted in one of the following (later) isocratic steps, the step i contributes to the total net retention volume of the compound j, V'_g, by a contribution $V'_{gi} = V_{ei}$.

It is more convenient to use the contributions to the net retention volume, V'_g, for calculation of retention volumes in stepwise elution chromatography than the contributions to V_g, as was used by Borówko and co-workers[58-60]. During the ith step, the centre of the band of the chromatographed compound moves along the column at a distance corresponding to a fraction of the column dead volume, V_{mi}, which represents the contribution of the step i to V_m. The migration of the solute band in stepwise elution chromatography is shown schematically in Fig. 5.1. Consequently, in each isocratic step, the following relationship applies[8]:

References on p. 180.

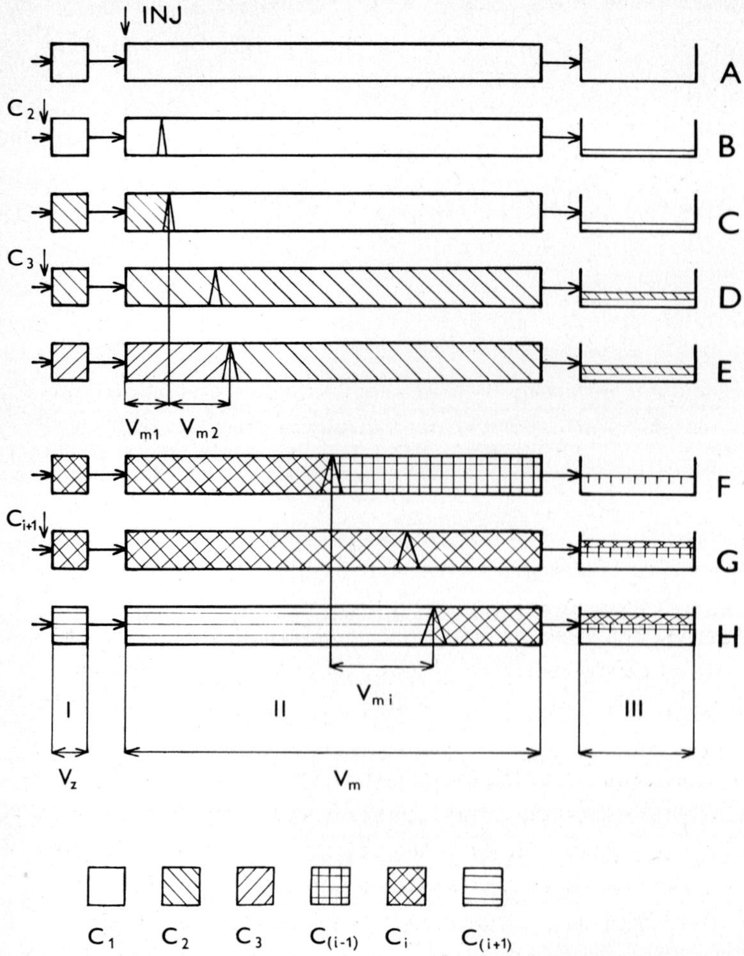

Fig. 5.1. Sample band migration in stepwise elution liquid chromatography at different instances A-H from the sample injection (INJ). I, connection tubing from pump to the column, volume V_z; II, separation column; III, eluate vessel. c_1-c_{i+1} are concentrations of the more efficient eluting component in the mobile phase used in the respective step. V_{m1}-V_{mi} are the contributions of each step to V_m. (From ref. 49.)

$$V'_{gi} = V_{ei} = k'_i V_{mi} \tag{5.1}$$

The net retention volume of the compound j eluted in step n (this is the last elution step with respect to compound j) can be calculated by summation of eqn. 5.1 from the first to the nth step:

$$V'_g = \sum_{i=1}^{n} V'_{gi} = \sum_{i=1}^{n} k'_i V_{mi} = \sum_{i=1}^{n-1} V_{ei} + k'_n V_{mn} \tag{5.2}$$

In all steps from 1 to $n - 1$, the contributions to the net retention volume of compound j are equal to the volumes of the mobile phase in each step. Only the contribution of the nth step to V_g' is not known directly, but it can be calculated from the known value of the capacity factor of compound j in this step, k_n', and from the contribution of this step to V_m, V_{mn}, which can be calculated as[49]

$$V_{mn} = V_m - \sum_{i=1}^{n-1} V_{mi} = V_m - \sum_{i=1}^{n-1} \frac{V_{ei}}{k_i'} \tag{5.3}$$

After introduction of V_{mn} from eqn. 5.3 into eqn. 5.2, the following relationship for the net retention volume in stepwise elution chromatography results[49]

$$V_g' = \sum_{i=1}^{n-1} V_{ei} + k_n' \left(V_m - \sum_{i=1}^{n-1} \frac{V_{ei}}{k_i'} \right) = \sum_{i=1}^{n-1} V_{ei} \cdot \frac{k_i' - k_n'}{k_i'} + k_n' V_m \tag{5.4}$$

From the known relationship between the capacity factor of compound j and the composition of the mobile phase [$k' = f(c)$ function, Table 2.1], we can calculate k_i' in each step from 1 to n and, finally, V_g' can be easily determined from eqn. 5.4[49].

n differs for compounds eluted in different steps in a single stepwise elution experiment, as it represents the number of steps required for elution of a given sample compound under given operating conditions. To determine the step in which a given sample component would be eluted, we can calculate subsequently the contributions V_{mi} for each step with respect to this solute using eqn. 5.1. If we then compare the sum of V_{mi} in all preceding steps and the step under consideration with the column void volume, V_m, we shall find for a certain step $\Sigma V_{mi} > V_m$ and it is in this step that the sample compound would be eluted. This condition can be written as

$$\sum_{i=1}^{n-1} \frac{V_{ei}}{k_i'} < V_m < \sum_{i=1}^{n} \frac{V_{ei}}{k_i'} \tag{5.5}$$

To a first approximation, the bandwidth in stepwise elution chromatography is essentially determined by the instantaneous capacity factor of solute at the moment of the elution of the band maximum, k_f', as in simple gradient elution chromatography, and is given by eqn. 4.35 if the band compression factor, G, is neglected. In stepwise elution chromatography $k_f' = k_n'$ and consequently the following equation can be used to calculate bandwidths, w_g[49]:

References on p. 180.

$$w_g = \frac{4V_m}{\sqrt{n}} \cdot (1 + k_n') \tag{5.6}$$

A more correct, but more complicated way of calculating the widths of peaks in stepwise elution chromatography is to add the contributions to band broadening, w_i, of all parts of the column migrated by the centre of the solute band in all steps i, from 1 to n (ref. 3):

$$w_g = \sqrt{\sum_{i=1}^{n} \frac{V_{mi}}{V_m} \cdot w_i^2} = 4 \cdot \sqrt{\frac{V_m}{n} \cdot \sum_{i=1}^{n} V_{mi}\ (1 + k_i')^2} \tag{5.6a}$$

The last step, n, usually provides the most significant contribution to peak width, as the solute band as a rule migrates the greatest distance through the column in this step. If the contributions of other steps can be neglected, eqn. 5.6a becomes simplified to eqn. 5.6.

Table 5.1 shows the calculated and experimental retention volumes and peak widths in the chromatography of a mixture of barbiturates on an octadecylsilica column using a stepwise concentration gradient of methanol in water, as an example of the verification of the calculation procedure for retention volumes and widths of peaks in stepwise elution chromatography (eqns. 5.4 and 5.6)[49].

TABLE 5.1

CALCULATION OF THE RETENTION VOLUMES AND BANDWIDTHS IN STEPWISE ELUTION CHROMATOGRAPHY (OPTIMIZED TO ACHIEVE R_g = 1.75 BETWEEN THE ADJACENT BANDS)

Column as in Table 4.3. Stepwise gradient: methanol-water, 0.97 ml/min. Compounds: A = barbital, m = 3.203; k_0' = 21.81; B = heptobarbital, m = 3.711, k_0' = 58.44; C = allobarbital, m = 3.547, k_0' = 69.44; D = aprobarbital, m = 3.659, k_0' = 106.96; E = butobarbital, m = 3.776, k_0' = 187.41; F = hexobarbital, m = 3.766, k_0' = 252.29. k_0' and m are the parameters of eqn. 1.28 evaluated by linear regression analysis of the experimental log k_0' = f(c) plots (c = concentration of methanol in the mobile phase, volume ratio x 10^{-2}). V_g' and w_g were evaluated from eqns. 5.4 and 5.6.

Step	c_i	V_{ei} (ml)	Compound eluted	V_g' (ml)		w_g (ml)		R_g	
				Calc.	Exptl.	Calc.	Exptl.	Calc.	Exptl.
1	0.52	2.37	A	1.48	1.54	0.39	0.39		
								1.61	1.29
1	0.52	2.37	B	2.15	2.07	0.44	0.43		
								1.76	1.87
2	0.55	1.78	C	2.95	2.90	0.47	0.46		
								1.77	2.03
3	0.55	1.78	D	3.87	3.84	0.57	0.46		
								1.76	2.65
4	0.64	1.74	E	4.76	5.00	0.44	0.42		
								1.74	1.84
5	0.64	1.74	F	5.62	5.96	0.55	0.63		

5.2. COMPOSED GRADIENTS WITH SIMPLE GRADIENT STEPS

The gradients composed of subsequent steps in which the composition of the mobile phase is changed continuously represent the most complex form of solvent composition programme. A concentration change of the more efficient eluting component in the mobile phase in different steps may follow a common gradient function with different parameters, or different gradient functions. The simplest example of this kind of mobile phase programme is a gradient composed of consecutive linear concentration gradient steps of different steepnesses (multilinear gradients). Such gradients are often used to simulate curved (concave or convex) gradients.

More complex are gradients composed of consecutive steps, where the composition of the mobile phase follows different gradient programmes (linear or curved, convex or concave) controlled by different gradient functions (such as those given in Table 4.1). Modern instrumentation for gradient elution chromatography makes it possible to form gradients of this kind[3,64].

Composed gradients may be useful in the optimization of a given separation problem, where the steepness or shape of the mobile phase composition change should be decreased or increased in certain parts of the gradient in order to improve the resolution or to produce a greater "bunching together" of over-resolved peaks in certain parts of the chromatogram. In practice, this "tailor-made" optimization is most frequently carried out using a trial-and-error method until the required separation is obtained.

However, it is possible in principle to calculate the important parameters of the chromatogram even if such complex gradients are used, even though the calculation procedure is more complicated than in continuous gradient or in stepwise elution chromatography. To calculate retention volumes, the basic equation (eqn. 4.1) should be integrated within each gradient step and consequently the summation of the resulting integrals would yield the equation[8]

$$\sum_{i=1}^{n} \int_0^{V_{mi}} \mathrm{d}V_\mathrm{m} = V_\mathrm{m} = \sum_{i=1}^{n} \int_0^{V'_{gi}} \frac{\mathrm{d}V}{f[\phi_i(V)]} \tag{5.7}$$

The solution of this equation will yield the required relationship for the net retention volume, V'_g, of the compound j eluted in the gradient step n. Here, f describes the dependence of the capacity factor of compound j on the concentration of the more efficient eluting component in the mobile phase, c, and ϕ_i is used for the gradient function $c = \phi(V)$ used in step i.

References on p. 180.

We can apply arguments concerning the summation procedure analogous to that in stepwise elution chromatography. Then,

$$V'_g = \sum_{i=1}^{n} V'_{gi} = \sum_{i=1}^{n} V_{ei} = \sum_{i=1}^{n-1} V_{ei} + V'_{gn} \tag{5.8}$$

As in stepwise elution chromatography, V_{ei} is used here for the volume of the mobile phase that passed through the column in step i and V'_{gi} denotes the contribution of each step i to the net retention volume of compound j, V'_g. The contribution V'_{gi} in each step $i < n$ is directly equal to V_{ei} if compound j is eluted in the nth step. Here again, it remains to calculate the contribution of the step n to V'_g, which can be determined by solving the following equation[4,8]:

$$\int_0^{V'_{gn}} \frac{dV}{f[\phi_n(V)]} = V_{mn} = V_m - \sum_{i=1}^{n-1} V_{mi} = V_m - \sum_{i=1}^{n-1} \int_0^{V_{ei}} \frac{dV}{f[\phi_i(V)]} \tag{5.9}$$

Of course, eqn. 5.9 can be solved only if the corresponding combination of the relationship $k' = f(c)$ (Table 2.1) and of the gradient function (Table 4.1) allows an algebraic solution (cf., Appendix 2). The unknown contribution V_{mn} of the nth step to V_m can be calculated from the contributions V_{mi} of the steps 1 to n-1. These contributions can be determined after the appropriate integration for each step, as indicated in eqn. 5.9. The appropriate equations of V'_g from Appendix 2 can be used for this purpose, but V_{ei} should be used instead of V'_g and V_{mi} instead of V_m.

Obviously, this calculation procedure is complicated and it is possible only for certain combinations of $k' = f(c)$ relationships and of gradient functions that yield algebraic solutions (Appendix 2).

With a known retention volume, V'_g, of compound j eluted in step n (this is the nth step with respect to compound j, but may be step $i < n$ for other compounds), we can calculate easily the composition of the mobile phase corresponding to the elution of the band maximum from the $c = \phi_n(V)$ gradient function applying in step n and the corresponding capacity factor, k'_f, from the relationship $k' = f(c)$ that fits a given chromatographic system. Using this k'_f value, the bandwidth can be calculated from eqn. 4.35 (provided the band compression factor, G, is neglected to a first approximation)[4].

It is open to question whether this complicated calculation procedure is worth carrying out for real complex gradients. However, this is a general method of calculation that may be useful in certain relatively simple kinds of composed gradients, which often occur in practice and are discussed in the following section.

5.3. TWO-STEP GRADIENTS WITH INITIAL GRADIENT DELAY

Two-step elution with the first isocratic step (gradient delay) followed by gradient elution in the second step is often used in practice. It can be very useful in the chromatography of samples that contain weakly retained compounds, the separation of which is possible in a given chromatographic system only if a low concentration of the more efficient eluting component in the mobile phase is used. The use of a two-step gradient with an initial delay allows a good separation of these compounds under near optimum isocratic conditions in the first step, while the retention of more strongly retained compounds is accelerated in the second, gradient, step. Usually, a simple continuous gradient elution can be used in this step with good results[8]. Of course, the concentration of the more efficient eluting component in the first, isocratic, step, c_1, is also the initial concentration in the second, gradient, step, i.e., $A = c_1$.

Most commercial instruments for gradient elution liquid chromatography have an inner volume, V_Z, of a few millilitres between the position where the components of the mobile phase are mixed to form the gradient and the top of the column. If we start the run of the gradient programme at the time of sample injection on the column, the compounds are always chromatographed under the conditions of a two-step gradient, where the first, isocratic, step (the gradient delay) is given by V_Z and the composition of the mobile phase in this step is equal to the initial composition of the mobile phase given by the gradient programme[8].

If sample compounds are relatively strongly retained in the mobile phase of this initial composition, they do not move along the column in the first step and the only influence of V_Z on the chromatographic separation is that the retention volumes of all sample compounds are increased by an equal value, V_Z, but the bandwidths and resolution do not change[43].

However, if sample compounds migrate along a significant part of the column in the first, isocratic, step, the retention volumes of different compounds may be changed to different extents and the resolution and bandwidths may also change. The influence of this first step due to V_Z on separation can be compensated for by delaying sample injection with respect to the start of gradient by a time calculated from V_Z and the volume flow-rate of the mobile phase[50]. Without this correction, it is also possible to calculate retention volumes and other retention characteristics, using the following calculation approach for two-step gradient elution with an initial delay of the gradient[4,8].

The calculation procedure is based on the general calculation of retention volumes in composed gradient elution, as discussed in the preceding section, but it is relatively simple, as only two steps should be considered. The con-

References on p. 180.

tribution of the second, gradient, step to the column dead volume, V_{m2}, which must be known for the calculation of V'_g, can be easily calculated in the same manner as in stepwise elution with isocratic steps (eqn. 5.3):

$$V_{m2} = V_m - V_{m1} = V_m - \frac{V_{e1}}{k'_1} \tag{5.10}$$

where V_{e1} is the volume of the mobile phase that passed through the column in the isocratic step 1 and k'_1 is the capacity factor of compound j (eluted in step 2) in step 1.

Eqn. 5.8 for retention volumes in elution using composed gradients can be adapted in a straightforward manner to two-step gradients with an initial delay:

$$V'_g = V_{e1} + V'_{g2} \tag{5.11}$$

V'_{g2} is the contribution of the second, gradient step to the retention volume, which may be calculated after combination of eqn. 5.9 (adapted to the two-step gradient) with eqn. 5.10 from the following resulting equation:

$$\int_0^{V'_{g2}} \frac{dV}{f[\phi_2(V)]} = V_m - \frac{V_{e1}}{k'_1} \tag{5.12}$$

For this purpose, the appropriate $k' = f(c)$ relationship for a given chromatographic system (Table 2.1) and the gradient function used in step 2, $c = \phi_2(V)$ (Table 4.1) should be introduced into eqn. 5.12. The same possibilities for an algebraic solution and limitations apply for solution of eqn. 5.12 as for the analogous eqn. 4.10 used for calculations of V'_g in simple (continuous) gradient elution (Section 4.4) and the solutions for V'_g from Appendix 2 may be used, but $V_m - (V_{e1}/k'_1)$ should be used instead of V_m and the contribution of the first isocratic step, V_{e1}, should be added to V'_{g2} (ref. 4).

For example, in chromatographic systems where eqn. 1.23 can be used to describe the relationship between k' and c (such as certain systems of adsorption chromatography on polar adsorbents and ion-exchange systems) and the general gradient function given by eqn. 4.15 (Table 4.1) controls the mobile phase gradient in the second step, the following equation can be used for calculation of net retention volumes[4]:

$$V'_g = V_{e1} + \frac{1}{B}\left[(\kappa m + 1)Bk'_0\left(V_m - \frac{V_{e1}c_1^{\,m}}{k'_0}\right) + c_1^{\frac{\kappa m+1}{\kappa}}\right]^{\frac{1}{\kappa m+1}} - \frac{c_1^{\frac{1}{\kappa}}}{B} \tag{5.13}$$

where c_1, the concentration of the more efficient eluting component in the mobile phase in step 1, is also the initial concentration of this component in the second, gradient, step, $c_1 = A$.

Bandwidths can be calculated from the following equation, analogous to eqn. 4.37 (the band compression factor, G, and the contribution of the gradient delay to peak broadening are neglected here):

$$w_g \approx w_{gi} = \frac{4V_m}{\sqrt{n}} \cdot \left\{ 1 + k_0' \left[c_1^{\frac{\kappa m+1}{\kappa}} + (\kappa m + 1)Bk_0' \left(V_m - \frac{V_{e1}c_1^m}{k_0'} \right) \right]^{-\frac{\kappa m}{\kappa m+1}} \right\} \quad (5.14)$$

where w_{gi} is the contribution of the gradient step to the peak width, w_g.

In systems, where the $k' = f(c)$ dependence is described approximately by eqn. 1.28 (such as reversed-phase chromatography on chemically bonded non-polar phases) and where linear concentration gradients are applied in the gradient step 2 (the gradient function is described by eqn. 4.17 in Table 4.1), the following equations for retention volumes and bandwidths can be derived, by analogy with eqns. 4.27 and 4.38 (ref. 4):

$$V_g' = V_{e1} + \frac{1}{mB} \cdot \log \left[2.31mBk_0' \left(V_m - \frac{V_{e1} \cdot 10^{mc_1}}{k_0'} \right) \cdot 10^{-mc_1} + 1 \right] \quad (5.15)$$

and

$$w_g \approx w_{gi} = \frac{4V_m}{\sqrt{n}} \cdot \left\{ 1 + k_0' \left[2.31mBk_0' \left(V_m - \frac{V_{e1} \cdot 10^{mc_1}}{k_0'} \right) + 10^{mc_1} \right]^{-1} \right\} \quad (5.16)$$

The neglect of the contribution of the first, isocratic step to the bandwidths of the early eluted compounds might yield erroneous w_g values calculated from eqns. 5.14 and 5.16 or similar equations and the following more correct calculation is to be preferred:

$$w_g = \sqrt{\frac{V_{m2}}{V_m} \cdot w_{gi}^2 + \frac{V_{m1}}{V_m} \cdot w_{1i}^2} = \sqrt{\left(1 - \frac{V_{e1}}{V_m k_1'}\right) w_{gi}^2 + \frac{16V_{e1}^3}{nV_m} \cdot \frac{(1 + k_1')^2}{(k_1')^3}} \quad (5.16a)$$

where w_{1i} represents the contribution of the isocratic step 1 to w_g and w_{gi} is the contribution of the gradient step, which can be calculated from eqns. 5.14 and 5.16 or similar equations.

Schoenmakers et al.[39] derived an analogous equation to eqn. 5.15 for V'_g in two-step elution, where the three-parameter equation with a quadratic exponent (eqn. 1.26) is used to describe the $k' = f(c)$ dependence in a given system. This should allow more precise calculations in reversed-phase systems; however, the approach using the two-parameter $k' = f(c)$ function, as given above, seems to be satisfactory for most practical reversed-phase gradient elution separations.

5.4. TWO-STEP GRADIENTS WITH THE FIRST GRADIENT STEP FOLLOWED BY A FINAL HOLD-UP AFTER THE END OF THE GRADIENT

In practical gradient elution operations where strongly retained compounds are chromatographed, some of these compounds are not eluted during the continuous gradient step by the time when the maximum possible concentration of the more efficient eluting component in the mobile phase c_{mp}, is reached and the gradient is finished. In such a situation, which often occurs in practice, the elution of these compounds must be accomplished in the following, isocratic step with the concentration c_{mp} of the more efficient eluting component in the mobile phase (this is often 100% of component b).

If a sample compound is to be eluted before the end of the gradient, its retention volume must be lower than the volume of the gradient, V_G. This condition may be formulated, for example for the gradient function given by eqn. 4.15, in the following way:

$$V'_g < \frac{c_{mp}^{1/\kappa} - A^{1/\kappa}}{B} \tag{5.17}$$

We can see that the elution of certain compounds could be achieved in the gradient run if the slope B or the initial concentration of the stronger eluting component A were decreased, but this would lead to longer separation times. A better remedy is to use a stronger solvent b with a higher elution strength, which would allow the elution of strongly retained compounds during the gradient run. This is highly recommended if sample compounds eluted in the "hold-up" step after the end of the gradient are "over-resolved" and the "hold-up" step increases the time of analysis. Otherwise, the elution of strongly retained compounds in the "hold-up" step with adequate resolution is acceptable from the analytical point of view and sometimes is recommended for the separation of mixtures containing compounds with relatively close, strong retentions in a given system.

To calculate the retention volumes of compounds eluted in the isocratic "hold-up" step after the end of the gradient, it is necessary to calculate the contribution of this second, isocratic step to V'_g. The contribution of the gradient step 1 to V'_g is equal to the volume of the mobile phase passed through the column in this step, $V_{e1} = V_G$. Now, it is necessary to calculate the contribution of the gradient step to the column dead volume, V_{m1}. For this purpose, the appropriate equation for retention volume in continuous gradient elution chromatography can be used (Appendix 2), where V_{e1} is introduced instead of V'_g and the unknown value V_{m1} instead of V_m. Eqn. 5.8 in the following form applies for the gradient elution followed by a "hold-up" step:

$$V'_g = V_{e1} + (V_m - V_{m1})k'_m \tag{5.18}$$

where k'_m is the capacity factor of the sample solute eluted in the "hold-up" step, corresponding to $c = c_{mp}$ used in this step.

For systems where eqn. 1.23 describes the dependence between the solute capacity ratio and c (such as certain adsorption and ion-exchange systems) and where the gradient function given by the general eqn. 4.15 is applied, the introduction of the adapted eqn. 4.25 for V_{m1} into eqn. 5.18 yields the following relationship for retention volume[4]:

$$V'_g = V_{e1} + \frac{k'_0 V_m}{c_{mp}^m} - \frac{1}{(\kappa m + 1)Bc_{mp}^m} \cdot \left(c_{mp}^{\frac{\kappa m+1}{\kappa}} - A^{\frac{\kappa m+1}{\kappa}} \right) \tag{5.19}$$

For systems where the $k' = f(c)$ function according to eqn. 1.28 fits the experimental data (reversed-phase chromatography on chemically bonded non-polar phases) and where linear concentration gradients (eqn. 4.17) are applied, a similar approach using a combination of eqns. 5.18 and 4.27 leads to the following equation[4]:

$$V'_g = V_{e1} + \frac{k'_0 V_m}{10^{mc_{mp}}} - \frac{1}{2.31mB} \cdot \left[1 - 10^{m(A-c_{mp})}\right] \tag{5.20}$$

Eqns. 5.19 and 5.20 were derived with the assumption of no or a negligible initial gradient delay. If this delay cannot be neglected, a composed three-step gradient should be considered with a first isocratic ("delay") step followed by a second (main) gradient step and finished by a third ("hold-up") isocratic step. Using the approach described in this and the preceding sections, an equation for retention volumes of compounds eluted in the "hold-up" step may be also derived[4].

The bandwidths of compounds eluted in the "hold-up" step can be calculated very easily using eqn. 5.6 if G is neglected, as we know the composition of the mobile phase in this isocratic step and, consequently, also the capacity factors k'_{m} of these compounds at the moment of their elution. Here, k'_{m} in the isocratic step is equal to k'_{f} at the time when the gradient step is finished. Thus[4],

$$w_{\mathrm{g}}^2 = w_{\mathrm{g}i}^2 \cdot \frac{V_{\mathrm{m1}}}{V_{\mathrm{m}}} + w_{2i}^2 \cdot \frac{V_{\mathrm{m2}}}{V_{\mathrm{m}}} = \frac{16V_{\mathrm{m}}^2}{n}(1 + k'_{\mathrm{m}})^2 \cdot \frac{V_{\mathrm{m1}} + V_{\mathrm{m2}}}{V_{\mathrm{m}}} = \frac{16V_{\mathrm{m}}^2}{n}(1 + k'_{\mathrm{m}})^2 \qquad (5.20a)$$

Chapter 6

TERNARY AND MORE COMPLEX MOBILE PHASE GRADIENTS

6.1. WHEN SHOULD TERNARY GRADIENTS BE USED?

Golkiewicz and Soczewiński[73] suggested the use of ternary mobile phase gradients in chromatography on polar adsorbents as an alternative to the incremental gradient elution introduced by Scott and Kucera[9-13], with the aim of eliminating more easily sudden changes in composition of the adsorbed solvent layer on the surface of adsorbent due to steep changes in polarity of the mobile phase during gradient elution ("solvent demixing effect") and their negative influence on chromatographic separations.

However, the main potential power of ternary and more complex gradients consists in the better possibility of controlling selectivity and retention independently of one another during gradient elution than in binary mobile phase gradients. For example, it should be possible to achieve this in reversed-phase chromatography on non-polar chemically bonded phases by using ternary gradients of two different organic solvents in water.

Bakalyar et al.[6] argued that it could be possible to adjust the selectivity in reversed-phase chromatography by controlling the concentration of one organic solvent in the mobile phase, while the elution strength is changed by the concentration gradient of another solvent during the elution. However, the changes in selectivity and capacity (retention) are always interconnected when the composition of the mobile phase is changed. Therefore, the influence of the ternary gradients on chromatographic behaviour may be much more complicated than with binary mobile phase gradients. For this reason, the content of one component of the mobile phase (one organic solvent) is kept constant during gradient elution, while the concentration of the other solvent in the mobile phase is increased (and that of water is decreased) with time[5,6]. Such gradients can be considered essentially as binary gradients and it is often possible to achieve a similar resolution by using binary gradients only (it is almost always possible when the separation of only one solute pair presents difficulties).

The use of ternary gradients has a good basis for the separation of complex samples where the separation of certain pairs of solutes is difficult in one binary solvent system, while the separation of another solute pair (differing significantly in retention from the first solute pair) cannot be achieved easily

References on p. 180.

in other binary solvent systems. Here, ternary gradients may lead to a considerable simultaneous improvement of selectivity and resolution in different parts of the chromatogram.

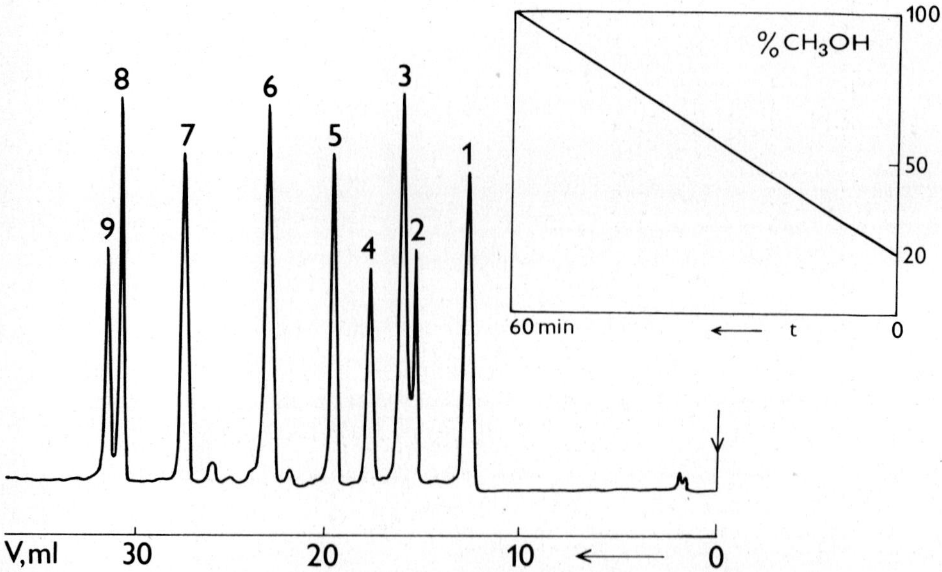

Fig. 6.1. Separation of a mixture of phenols using a linear binary concentration gradient of methanol in water. Column: 300 x 4.2 mm I.D. C18/LiChrosorb Si 100, 5 μm. Flow-rate: 1.0 ml/min. Detection: UV (254 nm). Compounds: 1=4-cyanophenol; 2=2-methoxyphenol; 3=4-fluorophenol; 4=3-fluorophenol; 5=*m*-cresol; 6=4-chlorophenol; 7=4-iodophenol; 8=2-phenylphenol; 9=3-*tert*.-butylphenol. (From ref. 54.)

This can be demonstrated by the example of the separation of a mixture of phenolic compounds, where neither the binary gradient of methanol in water nor the binary gradient of acetonitrile in water could be used to achieve a successful separation of all sample components on an octadecylsilica column (Figs. 6.1 and 6.2). However, the simultaneous use of a concentration change of methanol and acetonitrile in water (a ternary gradient) yielded a good separation of a sample mixture (Fig. 6.3)[54]. More details of this example will be given in Sections 6.3 and 6.4.

6.2. GRADIENT FUNCTIONS IN CONNECTION WITH TERNARY GRADIENTS

Sophisticated instrumentation making possible the mixing of three (or more) components of the mobile phase from the same number of reservoirs containing pure mobile phase components via time-proportioning valves or via auxiliary

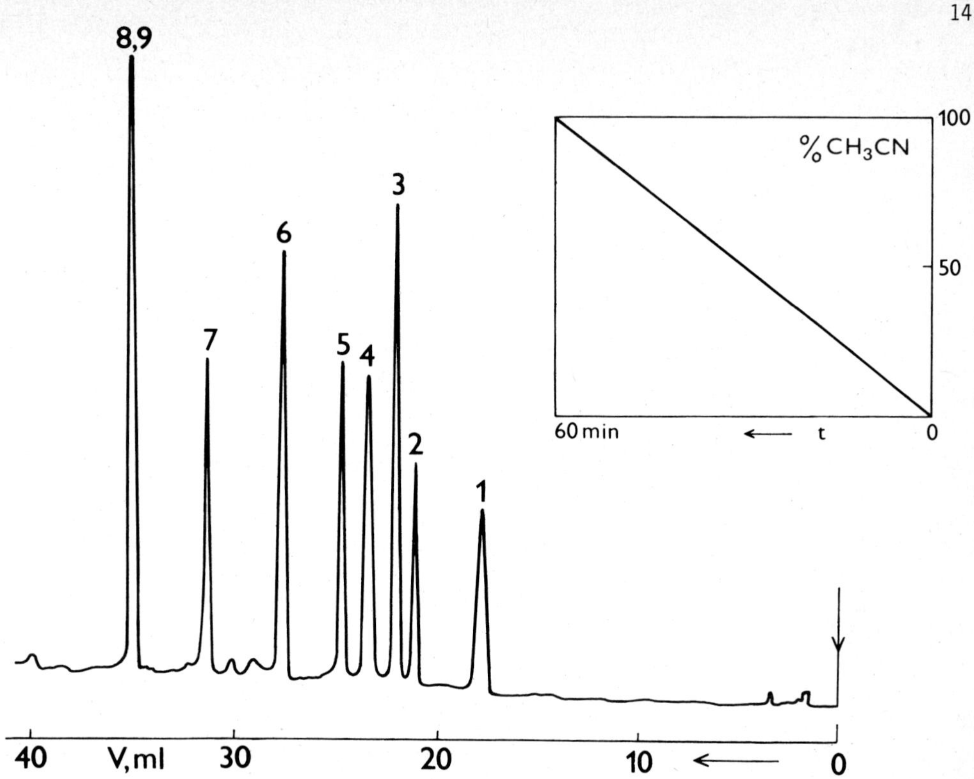

Fig. 6.2. Separation of a mixture of phenols using a linear binary concentration gradient of acetonitrile in water. Conditions and compounds as in Fig. 6.1. (From ref. 54.)

pumping systems are, of course, most convenient and easy to use in performing gradient elution with ternary mobile phase gradients[5-7]. However, a great variety of these gradients are feasible using simple conventional devices with the possibility of mixing two solvents only. This is especially true for linear concentration gradients.

Let each of the two reservoirs contain different concentrations (% v/v) of two organic solvents, x and y, in water, e.g., solution A in reservoir 1 contains the organic solvent x at a concentration c_{Ax} and solvent y at a concentration c_{Ay} and the concentrations of x and y in solution B in reservoir 2 are c_{Bx} and c_{By}, respectively. Let us apply a linear concentration gradient of B in the mixed mobile phase:

$$c_{\mathrm{B}} = A + BV \tag{6.1}$$

(A and B are constants of the gradient function used). c_{B} and c_{A} are the concentrations of B and A in the mixed final ternary mobile phase chosen in such way that

$$c_{\mathrm{A}} + c_{\mathrm{B}} = 1 \tag{6.2}$$

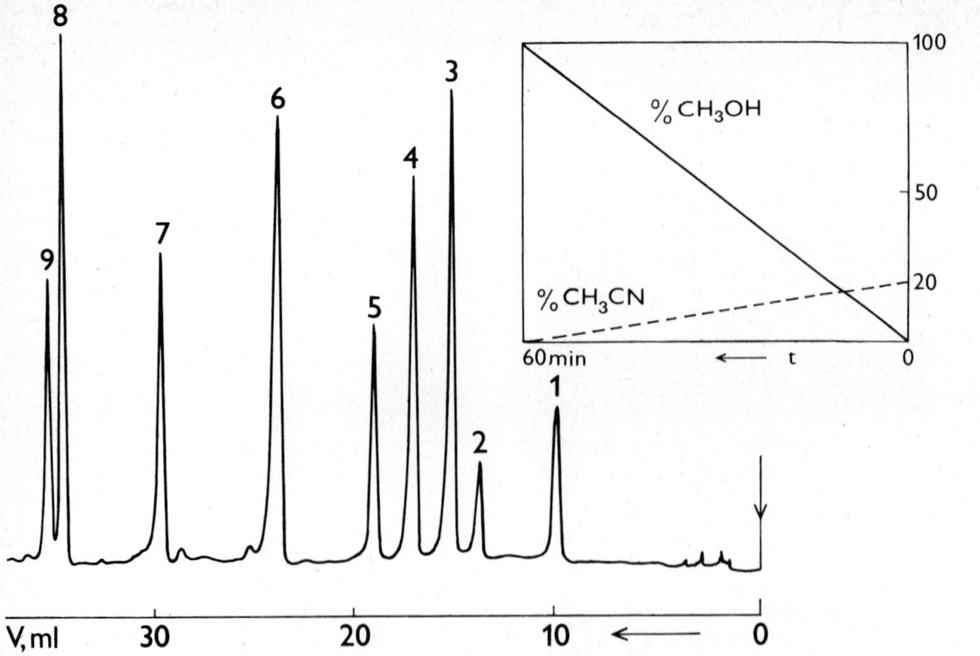

Fig. 6.3. Separation of a mixture of phenols using a linear ternary concentration gradient of methanol and acetonitrile in water. Conditions and compounds as in Fig. 6.1. Optimized ternary gradient. (From ref. 54.)

Then, the concentrations of x and y in the mixed mobile phase at the outlet from the gradient mixer (c_x and c_y, respectively) follow the linear functions

$$c_x = A_x + B_x V \tag{6.3}$$

and

$$c_y = A_y + B_y V \tag{6.4}$$

where A_x, B_x, A_y and B_y are constants of the linear gradient functions for the two solvents x and y, which follow the relationships

$$A_x = c_{Ax} + (c_{Bx} - c_{Ax})A \tag{6.5a}$$

$$B_x = (c_{Bx} - c_{Ax})B \tag{6.5b}$$

and

$$A_y = c_{Ay} + (c_{By} - c_{Ay})A \tag{6.6a}$$

$$B_y = (c_{By} - c_{Ay})B \tag{6.6b}$$

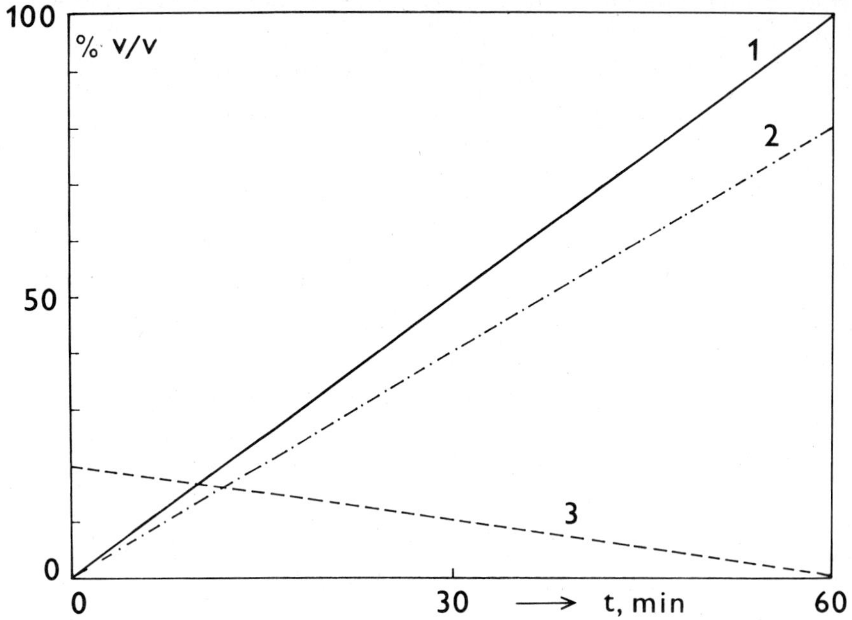

Fig. 6.4. Example of concentration profiles in ternary mobile phase gradients prepared from two binary solvents: A = 20% acetonitrile in water; B = 80% methanol in water. (1) % (v/v) of B in the mobile phase; (2) % (v/v) of methanol in the mobile phase; (3) % (v/v) of acetonitrile in the mobile phase. t = time from the start of the gradient. (From ref. 54.)

This is illustrated by a simple example in Fig. 6.4, where 20% aqueous acetonitrile is used as solvent A in reservoir 1 and 80% aqueous methanol as solvent B and a linear gradient of B (0-100% in 1 h) is run with zero initial concentration (A = 0) (straight line 1). Then, the concentration of methanol in the mixed ternary mobile phase follows the straight line 2 and that of acetonitrile is controlled by the straight line 3[54].

Obviously, in such a way a larger number of various solvents, polar and non-polar, can be mixed to form different ternary or more complex gradients with required profiles. A change in concentration of one or more organic solvents in water can be combined with an adequate concentration gradient of a buffer, salt, acid, base, ion-pair forming compound or complex-forming compound. Not only linear, but also non-linear gradient functions can be utilized to control the concentration of individual mobile phase components during elution. In addition to simple continuous gradients, stepwise or compound ternary gradients may be also used. However, the complexity of the phenomena and difficulties in understanding the influence of the gradient on separation increase with increasing number of mobile phase components and with increasing complexity of the gradient profile.

References on p. 180.

6.3. CALCULATION OF RETENTION VOLUMES FOR ELUTION WITH TERNARY GRADIENTS IN REVERSED-PHASE CHROMATOGRAPHY

Theoretical treatment of gradient elution with ternary gradients can be useful in providing a better understanding of the retention in such systems and can allow the prediction of retention volumes either for better identification of sample solutes or for optimization of the separation of complex mixtures. It would be especially useful for reversed-phase systems, as these systems are the most frequently used in practical liquid chromatography. In addition, theoretical treatment is simplest for reversed-phase chromatography systems and allows the derivation of expressions for retention volumes.

Let us consider chromatographic behaviour in ternary mobile phases used in reversed-phase chromatography on chemically bonded non-polar materials, such as octadecylsilica or octylsilica. Here, the selectivity and solvent strength of the mobile phase can be more or less conveniently adjusted by the simultaneous control of the concentrations of two different organic solvents, x and y, in aqueous solution[6]. If only a binary mobile phase containing one of each of the organic solvents in water is used, well known and widely accepted equations (eqn. 1.28) may be written for the influence of the composition of each of the two solvent binaries on the solute capacity factor, k'[4,8,37-39,50,63]:

$$\log k'_{(x)} = \log k'_{0_x} - m_x c_x \tag{6.7}$$

and

$$\log k'_{(y)} = \log k'_{0_y} - m_y c_y \tag{6.8}$$

where $k'_{(x)}$ and $k'_{(y)}$ are k' in the binary mobile phase containing the organic solvent x or y at concentration c_x or c_y, respectively, and k'_{0x}, k'_{0y}, m_x and m_y are experimental constants that depend on the solute and on the organic solvent but not on the concentration of the organic solvent in the mobile phase. Ideally, k'_{0x} and k'_{0y} should have the meaning of the capacity ratio in pure water ($c_x = 0$ or $c_y = 0$) and, consequently,

$$k'_{0x} = k'_{0y} = k'_0 \tag{6.9}$$

should apply.

If we consider eqns. 6.7-6.9 and assume that the two organic solvents interact with solutes independently of one another, their influence on the retention is additive and we can derive an equation for k' in a ternary mobile phase containing the solvent x at a concentration c_x and the solvent y at a concentration c_y[74]:

$$\log k' = \log k'_0 - m_x c_x - m_y c_y \tag{6.10}$$

where c_x and c_y are the concentrations of x and y, respectively, in the ternary mobile phase. (The exact theoretical derivation yields an additional "cross" term on the right-hand side of eqn. 6.10, which includes both c_x and c_y, but this term can be neglected in many practical reversed-phase systems[74].)

Eqn. 6.10 could theoretically be used to describe the retention in ternary mobile phases. However, in many practical systems significant differences between the values of k'_{0x} and k'_{0y} are often found by linear regressions of the experimental log $k'_{(x)}$ versus c_x and log $k'_{(y)}$ versus c_y plots in x-water and y-water binary solvent systems.

These differences could be probably explained by possible deviations from the linear relationships between log k' and c, as given by eqns. 6.7 and 6.8, respectively, in the regions close to 0% (and to 100%) of the organic solvent in the aqueous binary mobile phase[39,50]. Hence, the linear plots would approximate only a part of these curved plots in the range of practically useful k' values and consequently, even though this approximation may be a very good one in this range (correlation coefficients 0.9999), it would yield false values for k'_{0x} and k'_{0y} as k' in pure water.

To overcome this difficulty, an average value of k'_0, $\bar{k}_0$, in a given ternary mobile phase should be introduced instead of k'_0 into eqn. 6.10, as follows[54]:

$$\log \bar{k}_0 = \frac{c_x \log k'_{0x} + c_y \log k'_{0y}}{c_x + c_y} \tag{6.11}$$

and eqn. 6.10 should be rewritten as

$$\log k' = \log \bar{k}_0 - m_x c_x - m_y c_y \tag{6.12}$$

[It should be noted that a good validity of an empirical relationship

$$\log k'_{(A+B)} = c_A \log k'_A + c_B \log k'_B \tag{6.13}$$

was found for capacity ratios $k'_{(A+B)}$ in ternary mixtures formed by mixing two binary mixtures A and B in the ratio c_A:c_B, while each of the binary mixtures contains a different organic solvent in water and k'_A, k'_B are k' in the original solvent binaries A and B[74].]

Eqn. 6.12 can be used for the calculation of solute retention (k') in various ternary mixtures. Its practical verification for methanol-acetonitrile-water ternary systems is illustrated in Table 6.1[54].

References on p. 180.

TABLE 6.1

CALCULATIONS OF CAPACITY FACTORS IN TERNARY MOBILE PHASES

(I) Methanol-acetonitrile-water (25:20:55); (II) methanol-acetonitrile-water (40:10:50). Column: C18/LiChrosorb Si 100, 10 μm (300 x 4.2 mm I.D.); V_m=3.08 ml. k'_{ox}, m_x, k'_{oy}, m_y = constants in eqns. 6.7 and 6.8 (x=methanol; y=acetonitrile).

No.	Compound	k'_{ox}	m_x	k'_{oy}	m_y	k' in ternary mixture			
						I		II	
						Calc.	Exptl.	Calc.	Exptl.
1	4-Cyanophenol	36.73	3.212	13.94	2.970	0.96	0.91	0.79	0.80
2	Phenol	20.10	2.618	12.90	2.662	1.07	1.16	0.89	0.95
3	3-Methoxyphenol	40.62	3.113	15.87	2.840	1.21	1.22	0.99	0.91
4	2-Methoxyphenol	37.93	3.028	13.87	2.603	1.27	1.25	1.04	1.02
5	4-Fluorophenol	32.25	2.821	20.59	2.881	1.38	1.45	1.13	1.17
6	3-Fluorophenol	46.66	2.919	22.96	2.850	1.71	1.78	1.43	1.43
7	3-Nitrophenol	59.68	3.029	26.17	2.944	1.87	1.91	1.58	1.61
8	*m*-Cresol	81.59	3.215	26.07	2.870	2.06	2.02	1.74	1.77
9	*o*-Cresol	70.17	2.993	27.89	2.807	2.28	2.21	1.94	1.97
10	4-Chlorophenol	142.45	3.268	53.62	3.234	3.17	3.01	2.74	2.52
11	4-Iodophenol	471.81	3.738	136.35	3.666	5.84	5.55	5.06	4.58
12	2-Phenylphenol	1047.60	3.909	487.79	4.307	9.86	10.12	7.97	8.23
13	3-*tert*.-Butylphenol	1775.3	4.021	337.35	4.016	10.94	10.85	9.42	9.03

Eqns. 6.10 and 6.12 can also be used as the basis for the derivation of an equation that would allow calculations of retention volumes in reversed-phase chromatography using ternary gradients, where both the concentration of the solvent x and that of the solvent y are changed simultaneously with time during the elution. Linear concentration gradients have been acknowledged as most useful in reversed-phase chromatography from a number of reasons: (a) they approximate well so-called "linear solvent strength" gradients, which should yield optimum resolution of unknown solute mixtures; (b) they are very simply formed, even with relatively simple instrumentation; and (c) they are easily to be understood, optimized and treated theoretically (see Section 4.2). Therefore, we shall limit our discussion to linear ternary solvent gradients, where the concentrations of the two organic solvents x and y in the ternary mobile phase, c_x and c_y, respectively, are changed linearly with time, or with the volume of the eluate, V, according to the gradient functions described by eqns. 6.3 and 6.4, respectively.

As with binary solvent gradients, the solution for net retention volume of a solute, V'_g, in chromatography with ternary gradients should be based on the fundamental equation of gradient elution chromatography (eqn. 4.2).

If we assume the validity of eqn. 6.10, after introducing eqns. 6.10, 6.3 and 6.4 into eqn. 4.2 we obtain the following relationship for V'_g[54]:

$$V'_g = \frac{1}{m_xB_x + m_yB_y} \cdot \log \left[2.31\ k'_oV_m(m_xB_x + m_yB_y) \cdot 10^{-(m_xA_x + m_yA_y)} + 1\right] \tag{6.14}$$

Eqn. 6.14 is formally almost identical with the analogous equation (eqn. 4.27) for binary linear concentration gradients in reversed-phase chromatography[8,50], but $m_xA_x + m_yA_y$ is used here instead of mA and $m_xB_x + m_yB_y$ instead of mB.

As under isocratic elution conditions, eqn. 6.14 can hardly be used in this form in practice because of differences in the experimental values of k'_{ox} and k'_{oy} for most solutes. If we try to overcome this difficulty by inctroducing $\bar{k}_o$ as defined by eqn. 6.11 into eqn. 4.2 together with eqns. 6.12, 6.3 and 6.4, we obtain an expression that cannot be integrated to yield an algebraic expression and numerical integration should be used for solution. It is much easier to simplify the solution by using eqn. 6.14 for V'_g under gradient conditions and to substitute k'_o by an average value, $\bar{k}_{og}$, which represents an average value of $\bar{k}_o$ during the elution of the solute ($\bar{k}_o$ changes during gradient elution)[54]:

$$\log \bar{k}_{og} = \frac{(A_x + B_x \cdot \frac{V_g}{2}) \log k'_{ox} + (A_y + B_y \cdot \frac{V_g}{2}) \log k'_{oy}}{A_x + A_y + (B_x + B_y) \cdot \frac{V_g}{2}} \tag{6.15}$$

After introducing $\bar{k}_{og}$ from eqn. 6.15 for $\bar{k}_o$ into eqn. 6.14, we obtain an implicit equation for V'_g, which should be solved by an iteration method. The calculation of V'_g, even with the use of a small programmable calculator, such as a TI 58, is very rapid and requires about 1 min for four or five iterations, which are sufficient to obtain a precision of calculation of better than 0.1% relative.

Table 6.2 illustrates the calculation of V'_g using this method for several practical ternary gradients in reversed-phase chromatography. The differences between the calculated and experimental retention volumes are similar to those in chromatography with binary solvent gradients[54].

6.4. SELECTION OF AN APPROPRIATE TERNARY GRADIENT FOR A GIVEN SEPARATION PROBLEM

It is difficult to give generally valid rules for optimization of ternary gradients, as the optimum combination of partial gradient profiles of each mobile phase component depends on the nature of the sample mixture being chromatographed. Rather, a practical example can elucidate a semi-empirical procedure for the selection of gradient elution conditions that would yield an adequate chromatographic separation.

References on p. 180.

TABLE 6.2

VERIFICATION OF CALCULATION OF RETENTION VOLUMES, V_g, IN CHROMATOGRAPHY USING TERNARY GRADIENTS

Column and compound Nos. as in Table 6.1. Gradient from 0 to 100% B in 60 min. (I) A = water, B = methanol-acetonitrile (3:2); (II) A = acetonitrile-water (1:9), B = methanol; (III) A = acetonitrile-water (1:4), B = methanol; (IV) A = acetonitrile-water (3:7), B = methanol. A_x, B_x, A_y, B_y = constants in eqns. 6.3 and 6.4 (x = methanol; y = acetonitrile); concentration of the organic solvent (c_x, c_y) in % (v/v) · 10^{-2}.

Compound No.	Gradient							
	I		II		III		IV	
	(A_x=0, B_x=0.010308, A_y=0, B_y=0.006873)		(A_x=0, B_x=0.017182, A_y=0.1, B_y=-0.0017182)		(A_x=0, B_x=0.017182, A_y=0.2, B_y=-0.0034364)		(A_x=0, B_x=0.017182, A_y=0.3, B_y=-0.0051546)	
	V_g (ml)							
	Calc.	Exptl.	Calc.	Exptl.	Calc.	Exptl.	Calc.	Exptl.
1	22.1	20.8	18.5	16.6	12.6	11.6	8.4	7.7
4	23.2	23.5	19.9	21.1	14.1	13.2	9.7	8.9
5	23.8	24.2	20.2	22.0	14.8	14.7	10.3	9.9
6	25.8	26.1	22.5	24.1	16.6	16.8	11.5	11.2
8	27.8	28.0	25.1	26.3	18.8	18.3	12.9	12.2
10	31.6	31.7	29.5	30.8	23.3	23.2	16.4	16.2
11	36.3	36.1	35.3	36.0	29.7	29.2	22.5	22.2
12	39.4	40.0	38.6	40.0	33.8	34.1	27.3	28.8*
13	40.1	40.1	39.7	40.4	35.1	34.9	28.6	28.8*

*Evaluated from the maximum of a single peak for two compounds; no or poor resolution only.

From eqn. 6.14 it follows that a selectivity change in a ternary gradient with respect to binary gradients can be expected for such pairs of solutes, where the parameters m_x, m_y of eqns. 6.7 and 6.8 in two organic solvents differ at least for one compound of this pair, otherwise the selectivity does not change significantly. Thus, if we inspect the values m_x, m_y for phenols in Table 6.1, we can find that unsubstituted phenol, halogenophenols, 3-nitrophenol and *tert.*-butylphenol have $m_x \simeq m_y$, whereas the largest differences between m_x and m_y occur with 2-methoxyphenol, *m*-cresol and 2-phenylphenol. It follows that we can use

m_x, m_y to compare the selectivities in different solvent systems at least to some extent, and the data from Table 6.1 suggest differences in selective interactions of solvent and the last phenols between methanol and acetonitrile mobile phases (to a lesser extent this applies also to 4-cyanophenol, 3-methoxyphenol and *o*-cresol).

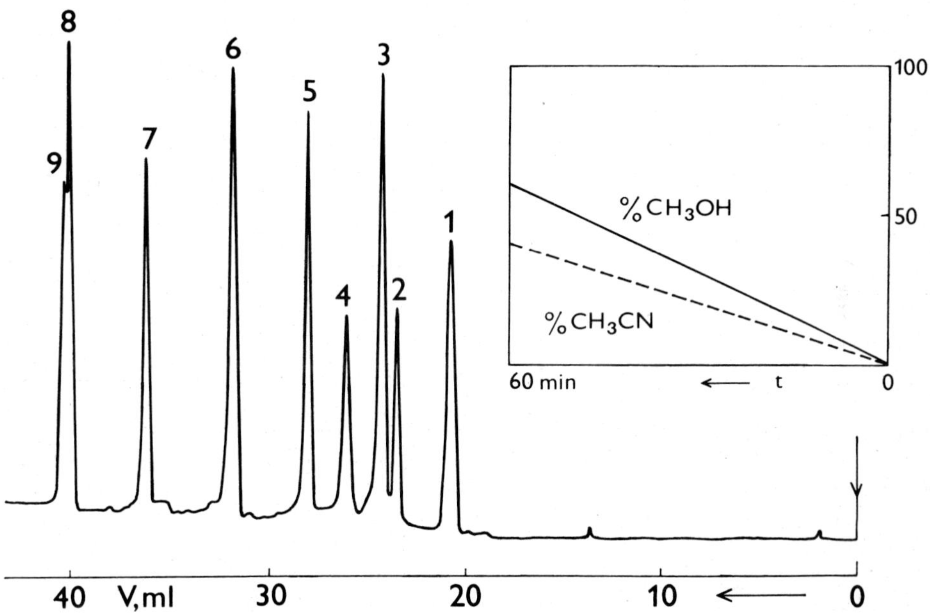

Fig. 6.5. Separation of a mixture of phenols using a linear ternary concentration gradient of methanol and acetonitrile in water. Conditions and compounds as in Fig. 6.1. (From ref. 54.)

As a practical consequence, we find difficulties when attempting the separation of 2-methoxyphenol from 4-fluorophenol in binary mobile phases composed of methanol and water and, on the other hand, the separation of 2-phenylphenol from 3-*tert.*-butylphenol is not possible in water-acetonitrile binary systems. This is illustrated by examples of separations using methanol-water (Fig. 6.1) and acetonitrile-water gradients (Fig. 6.2). The separation of a mixture of nine phenols shown in these examples is suitable for the application of gradient elution, because the retentions of the first and the last compound eluted differ approximately 12-fold[54].

However, the application of a ternary gradient, where the concentrations of methanol and acetonitrile in the mobile phase were simultaneously increased with time in a constant ratio, did not bring about a significant improvement of separation (Fig. 6.5). This might be expected, as we would expect better resolution of 2-methoxyphenol from 4-fluorophenol in a mobile phase relatively rich in acetonitrile and that of 2-phenylphenol from 3-*tert.*-butylphenol in a mobile phase containing methanol in excess over acetonitrile. Consequently, the ternary gradient should be designed in such a way that the concentration of methanol in the mobile phase increases with time, the concentration of acetonitrile decreasing simultaneously.

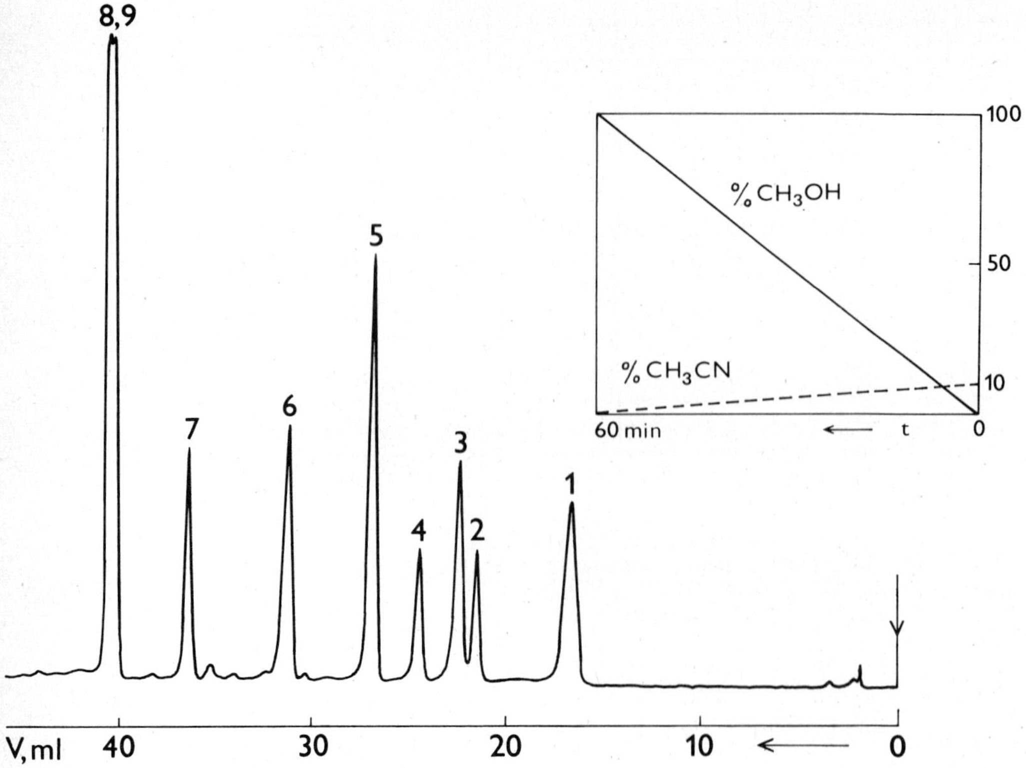

Fig. 6.6. Separation of a mixture of phenols using a linear ternary concentration gradient of methanol and acetonitrile in water. Conditions and compounds as in Fig. 6.1. (From ref. 54.)

It remains for the optimum parameters of the gradient functions (A_x, B_x, A_y, B_y; eqns. 6.3 and 6.4) for each organic solvent to be found. Here, calculations of retention volumes as shown in the preceding paragraph may be useful for saving time and the solvents necessary for trial-and-error experiments, because V'_g for critical pairs of compounds may be calculated for different ternary gradients and from these results the optimum gradient can be chosen.

Three ternary gradients differing in the initial concentration of acetonitrile were tested in this way and, as can be seen from Table 6.2, the gradient starting with 20% acetonitrile (Fig. 6.3) should provide better separations than the gradients starting with 10% or 30% acetonitrile. This was confirmed experimentally, as can be seen from comparison of Fig. 6.3 with Figs. 6.6 and 6.7[54]. The practical utility of ternary gradients designed in such a way that the concentration of one organic solvent increased (here, tetrahydrofuran) and that of the other decreased (methanol) was demonstrated by Roggendorf and Spatz[82] for the separation of hydroxy and methoxy metabolites of cinnamic acid.

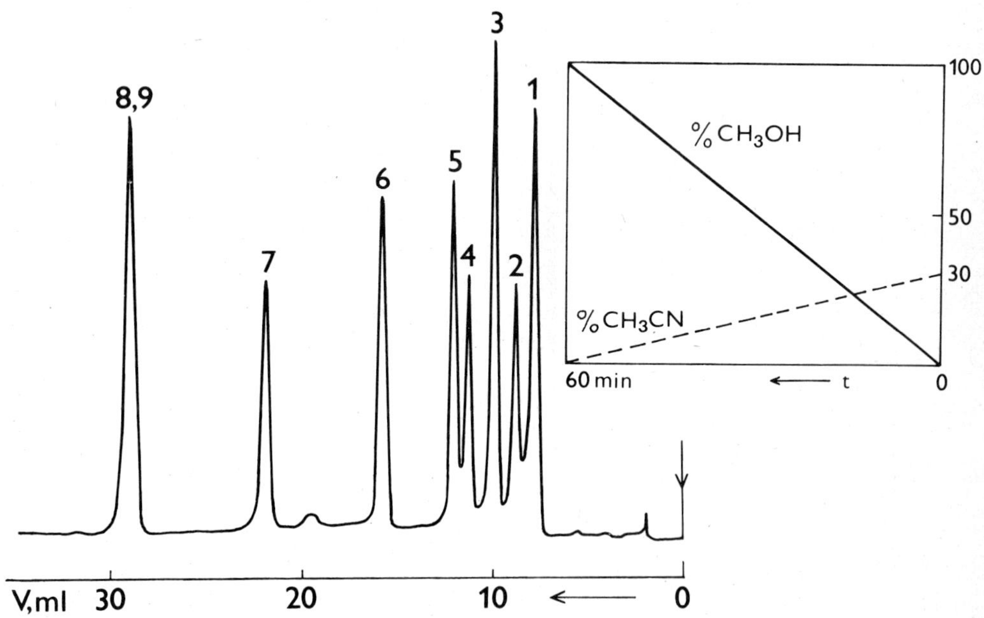

Fig. 6.7. Separation of a mixture of phenols using a linear ternary concentration gradient of methanol and acetonitrile in water. Conditions and compounds as in Fig. 6.1. (From ref. 54.)

Obviously, optimum profiles of ternary gradients exist for various separation problems and the theory can be of a great help in the selection of a "tailor-made" ternary gradient for a given separation problem.

Chapter 7

SELECTION OF THE OPTIMUM GRADIENT

7.1. ASPECTS OF OPTIMIZATION

The "ideal" chromatographic separation can probably be characterized as a separation that provides perfect resolution of all sample compounds of interest in the minimum time, and makes possible a very sensitive detection on the one hand and separations of large amounts of sample for preparative or semi-preparative purposes or for further identification on the other.

It is almost impossible to obtain such an ideal chromatographic separation in practice, because one feature of the separation can be improved only at the cost of other parameters. Thus, the resolution or the amount of sample to be separated can be increased only if the time of analysis is also increased. Similarly, sensitivity can be improved at the cost of resolution; the resolution can be improved, but the time of analysis, sensitivity of detection and the amount of sample that can be loaded on the column are simultaneously adversely affected.

Therefore, an "optimum" separation represents a compromise between these various aspects and such a compromise should be the best fit with respect to the most important requirements, which may differ according to the specific problem to be solved and according to the aim of the separation. We can optimize the separation to obtain one of the following features:

(a) maximum resolution in the minimum time (this is perhaps the most frequent requirement in most routine applications);

(b) maximum resolution, where a short analysis time is of interest but not of primary interest (this applies to the analysis of very complex sample mixtures in natural, biological or environmental samples);

(c) maximum sensitivity of detection (this is required in trace analysis);

(d) maximum amount of sample that can be loaded on the column (separations on a preparative or semi-preparative scale, required for the isolation of sample components for further identification or for other purposes).

It can be seen that the key to optimization in all of the above instances is the resolution, which must be controlled satisfactorily. To this aim, we can optimize the three terms contributing to resolution (I, efficiency; II, selectivity; and III, capacity). With given instrumentation, the optimization consists in a suitable choice of the column and of the mobile phase. With some experience,

References on p. 180.

TABLE 7.1

USEFUL COMBINATIONS OF MOBILE PHASE COMPONENTS (WEAKER ELUTION STRENGTH COMPONENT, *a*; STRONGER ELUTION STRENGTH COMPONENT, *b*), THE SHAPE AND THE STEEPNESS LIKELY TO YIELD GOOD SEPARATIONS OF UNKNOWN MIXTURES TO A FIRST APPROXIMATION (CONDITION FOR THE FIRST EXPERIMENT IN EMPIRICAL OPTIMIZATION OF GRADIENT ELUTION CONDITIONS)

The initial composition of the mobile phase at the start of the gradient should be selected according to the absolute retention of chromatographed compounds in sample mixtures.

Each gradient may be started from the mobile phase containing a certain amount of component *b* and use component *b* diluted by component *a* or to end the gradient before 100% *b* is achieved. Gradients from 5 to 95% of *b* are recommended by Snyder.

Column packing	Mobile phase components		Gradient shape	Gradient steepness (V_G/V_m)
	a	*b*		
(I) Normal-phase chromatography on polar adsorbents:				
Silica, alumina, porous, 5-10 μm	*n*-Hexane or *n*-heptane	*n*-Propanol, isopropanol, ethanol	Concave (or linear - less adequate)	15-25
		Chloroform, diisopropyl ether, diethyl ether*, methylene chloride*		10-20
	Diisopropyl ether	Methanol, acetonitrile	Concave	15-25
	1-Chlorobutane	Methanol, acetonitrile		15-25
(II) Reversed-phase chromatography, aqueous:				
C18/silica (C8/silica less frequently), chemically bonded; carbon adsorbents, 5-10 μm	Water**	Methanol, acetonitrile	Linear, slightly convex	10-20
		Tetrahydrofuran (dioxan)		15-25
(III) Reversed-phase chromatography, non-aqueous:				
As in (II)	Acetonitrile, methanol	Tetrahydrofuran	Linear	5-10
		Diethyl ether*, diisopropyl ether		10-15

TABLE 7.1 (continued)

Column packing	Mobile phase components		Gradient shape	Gradient steepness (V_G/V_m)
	a	*b*		
(IV) Normal-phase chromatography on chemically bonded phases of medium polarity:				
NH_2/silica, CN/silica, Diol/silica, NO_2/silica, etc., 5-10 μm	*n*-Hexane, *n*-heptane	*n*-Propanol, isopropanol, ethanol, chloroform, diisopropyl ether, diethyl ether*, methylene chloride*	Linear or concave	5-10
(V) Ion-exchange chromatography:				
Cation, anion exchanger***	Water	0.1-0.5 *M* salt or buffer§	Concave or linear	10-15
Cation, anion exchanger	pH 2 pH 6-8	pH 6-8 pH 2	Linear pH gradient	15-25
(VI) Ion-pair chromatography in reversed-phase systems:				
As in (II)	Ion-pairing ion§§, 0.0005-0.05 *M*		Linear	10-20
	In water	In CH_3OH or CH_3CN		
	0.1 *M* ion-pairing ion in CH_3OH-H_2O§§§ or in CH_3CN-H_2O§§§	CH_3OH-H_2O§§§ CH_3CN-H_2O§§§	Concave	5-10

*Volatile solvent, difficulties may occur when reciprocating pumps are used.

**A salt or a buffer at a concentration of 0.01-0.4 *M* may be added to improve the separation of strongly polar and ionized compounds (attention must be paid to non-miscibility of solvents a and b!).

***Resinous exchangers, exchangers chemically bonded on porous material (silica).

§E.g., $NaNO_3$, Na_2SO_4, KH_2PO_4, K_2HPO_4, CH_3COONa, sodium citrate, borate, lactate, formate, oxalate, NH_4Cl, KCl (halide ions are usually not recommended owing to corrosion of the equipment), acetate, phosphate or citrate buffers.

§§Tetrabutylammonium phosphate, sulphate or perchlorate for chromatography of acidic compounds; pentane- or heptanesulphonic acid for chromatography of basic compounds, pH buffered to between 3-7 and/or ionic strength adjusted.

§§§This concentration is kept constant during the gradient.

a suitable column packing material can be found relatively easily for a given separation problem, as there are far fewer commercially available useful column packing materials for liquid chromatography than stationary phases for gas chromatography. On the other hand, with each column packing material, we can combine binary or more complex mobile phases. However, a few combinations of mobile phases for each packing material have proved useful for a number of practical separation problems and are listed in Table 7.1. Of course, numerous other combinations may be more suitable for specific separation problems. A more detailed discussion of this aspect of separation would be beyond the scope of this book, and can be found in other literature on liquid chromatography[75,76].

Once the column packing material has been selected, the efficiency of separation can be controlled by using appropriate particle size of the material, column dimensions, mobile phase flow-rate and, to a lesser extent, temperature.

Columns with an inner diameter between 3 and 6 mm and a length between 100 and 300 mm packed with 5-10 μm particle diameter porous material have become standard for most routine analytical separations. Recently, more efficient columns packed with 3 μm particles have become available commercially and may be used; there are also less efficient columns packed with 15-25 μm material.

In ion-exchange chromatography and in semi-preparative chromatography, columns of length 500-1000 mm and I.D. 8-10 mm are often used in connection with materials of particle diameter 20-50 μm. Here, a large column loading and a lower cost of the packing material are of primary interest.

Pellicular and controlled surface porosity packing materials containing a thin layer (1-5 μm) of the stationary phase coated, grafted or sintered on an impervious core, 30-80 μm in diameter, were very popular in the early 1970s, but they are used only occasionally now, as they combine low sample loadability with a relatively lower efficiency of separation in comparison with standard analytical columns. Their only advantage, the possibility of high-speed analysis at a relatively low pressure, does not represent a significant benefit with modern liquid chromatographic instrumentation.

In recent years, microbore columns of I.D. 0.5-2.0 mm, 1 to several metres long, packed with 5-20 μm particle diameter material, have been introduced[77] and have recently become commercially available. These columns offer a very high efficiency (up to several hundred thousands of theoretical plates), but the separation times are longer than with commercial analytical columns. Microbore columns are useful in the chromatography of very complex samples and seem promising for direct coupling with mass spectrometers. Similar advantages would be offered by open-tubular columns, but their capacity is low and extreme care must be devoted to the miniaturization of other parts of the instrumentation (sample injectors, detector cells).

When the efficiency of separation has been established by the appropriate choice of column geometry, particle size of the packing material and flow-rate of the mobile phase (which is usually between 0.5 and 4.0 ml/min in work with standard analytical columns), the selectivity and capacity contributions to resolution can be optimized (the efficiency can be assumed not to change significantly with changing composition of the eluent). This can be done much more easily in isocratic elution chromatography (as is discussed in Sections 2.1 and 2.2) than in gradient elution chromatography. There, the optimum gradient profile has to be selected, which means that the shape and steepness of the gradient and the initial composition of the mobile phase should be optimized (this also holds true for each step in composed gradients). The optimization is complex and has usually been done using a trial-and-error method. However, the theory of gradient elution can be used to increase the efficiency of and to speed-up the optimization procedure.

The rational optimization of conditions for gradient elution chromatography is based on the optimization of resolution. Two different situations should be distinguished, each of which requires different methods for solution[4,36]:

(1) specific case, where the sample compounds and the relationship between their retention and the composition of the mobile phase are known;

(2) general case, where the parameters of the $k' = f(c)$ relationship for the chromatographed compounds and often their exact number are not known.

7.2. SPECIFIC CASE

The optimized profile of the gradient ideally should yield a separation in which all sample compounds of interest are sufficiently well resolved and their elution is achieved in the minimum time. This means that the peaks of compounds should be spaced as regularly as possible with the required resolution (e.g., $R_g = 1.5$) on the chromatogram from its beginning to its end. A higher resolution than that just required would only increase the time of analysis and therefore is undesirable.

It is naturally only rarely possible to achieve such an ideal, exactly regular spacing of the peaks of all sample compounds in practical systems, where the capacity factors usually increase irregularly between the individual sample compounds. This leads to "bunching together" of peaks at certain places on the chromatogram and to "over-resolution" at other places (homologous, benzologous and oligomeric series are exceptions). However, the gradient profile can be optimized in order to improve the spacing of peaks in the chromatogram while maintaining the required resolution and to approximate the "ideal" separation. Gradient elution offers more efficient means for this purpose than elution under isocratic conditions.

References on p. 180.

To optimize the profile of the gradient, the steepness (*B*), shape (κ) and the initial concentration of the more efficient eluting component in the mobile phase (*A*) should be chosen so as to yield optimum contributions of the selectivity and capacity terms to resolution. [The efficiency term (number of theoretical plates) is assumed not to depend significantly on the composition of the mobile phase and, consequently, on the profile of the gradient.]

Let us recall briefly the influence of the three parameters of gradient profile on separation, as it appears on the chromatogram. If the initial concentration of the more efficient component in the mobile phase, *A*, is not too high and the steepness of the gradient, *B*, not too low (otherwise the separation conditions are close to isocratic elution), an increase in *A* would have a much greater influence on retention volumes (decrease) than on the widths of the peaks, which usually decrease to some extent also.

Resolution and bunching together of peaks are usually not influenced much by minor changes in *A*, with the exception of the early eluted compounds. An increase in *B* leads to a significant decrease in the retention volumes of sample compounds. The differences in the retention volumes of individual compounds and (to a lesser extent) the peak widths also decrease with increasing *B*. So does the resolution, if the selectivity of separation (α) does not depend significantly on the composition of the mobile phase. In the opposite case, the plots of R_g versus *B* may be complex (see Fig. 4.14 and the discussion in Section 4.9).

If the shape of gradient is changed from convex to linear and further to concave (while increasing the parameter κ of the gradient function, eqns. 4.15-4.18), the retention volumes and compression of the chromatogram increase, and α_g decreases. At a certain curvature of the gradient, maximum resolution can be observed. Generally, an increase in κ would lead to an increase in the resolution of earlier eluted compounds and, at the same time, the resolution of later eluted compounds may be impaired (the tendency for bunching together of peaks is moved from the earlier to the later parts of the gradient).

The above general rules apply if only one parameter of the gradient function is changed while the remainder are kept constant and, further, if the compounds are eluted under true gradient conditions, i.e., if the compound "senses" sufficiently the change of the mobile phase composition during the migration along the column before the elution. Otherwise, with early eluted compounds and with "flat" gradients, the solute band is eluted under quasi-isocratic conditions and the differences in the gradient profile can hardly have a significant influence on separation.

The best way of optimizing a given gradient elution separation seems to begin with the selection of the gradient shape[3,4]. In most instances, we select either linear concentration gradients (in reversed-phase systems) or a certain curvature of the gradient by analogy with other experiments in similar systems (concave gradients are to be preferred in chromatography on polar adsorbents and in ion-exchange chromatography). The curvature may be calculated with respect to the approximate compression of the gradient (E) required (see Section 4.10).

The second step is to choose a convenient steepness of the gradient in order to achieve the required resolution of the pair of sample compounds that are most difficult to separate. Of course, the separation of such compounds is feasible in gradient elution chromatography only if it can be accomplished in a given system under isocratic conditions with the same components of the mobile phase.

Finally, the initial concentration of the more efficient eluting component in the mobile phase can be conveniently adjusted with the aim of keeping the time of separation as short as possible.

The optimization approach described may be performed using a trial-and-error method, by calculation or by combination of calculations and experiment, and the method selected will depend on the nature of the separation problem and on the experience of the chromatographer. In the following, a method will be described for optimization of the steepness of the gradient (B) and of the initial concentration of the more efficient eluting component in the mobile phase (A) by calculation[52]. As both of these parameters influence to a certain extent both the resolution and the retention volumes of sample solutes, which control the time of separation, simultaneous optimization of A and B is the most correct theoretically.

First, it is necessary to choose appropriately two compounds j and k, with adjacent peaks, the resolution of which should be kept at a certain required level, $R_{\mathrm{g,o}}$. These are usually the two components of a sample mixture that are most difficult to separate, which should guarantee adequate resolution over the whole chromatogram. Further, another sample compound, i, is chosen, the retention volume of which, $V'_{\mathrm{g}i}$, should be minimal, in order to keep the separation time as short as possible. This is usually the most strongly retained sample component. An appropriate shape of the gradient (parameter κ) is pre-selected.

From the mathematical point of view, the optimization of A and B consists in the calculation of the minimum of the function $V'_{\mathrm{g}i} = \mathrm{f}(A,B)$ at a given value of κ, for which A and B are further interrelated by the condition that the required value $R_{\mathrm{g,o}}$ should be obtained. As the usual Lagrange method for the solution of the problem would suffer from severe difficulties, a modified method of solution was suggested, the details of which, together with the calculation program, are given elsewhere[52]. The calculation requires a knowledge of the parameters of the $k' = \mathrm{f}(c)$ dependences for the sample compounds i, j and k and is feasible for combinations of these functions and gradient functions that yield an explicit solution for retention volume (Appendix 2). A schematic diagram of the optimization program is given in Appendix 3.

References on p. 180.

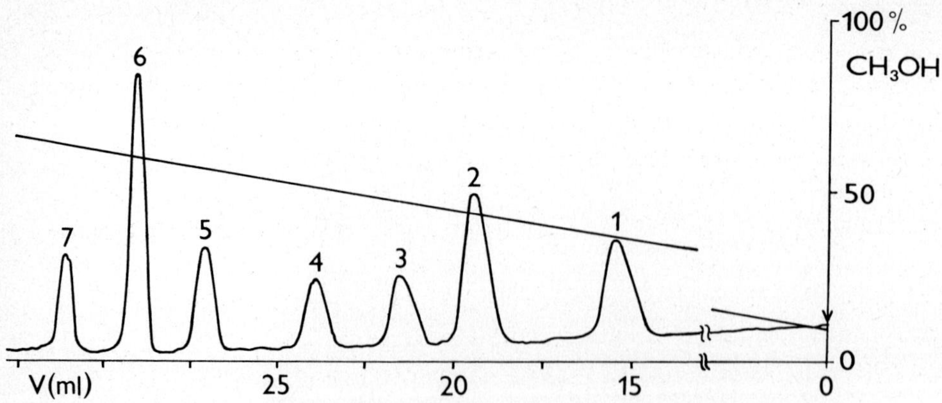

Fig. 7.1. Optimized reversed-phase gradient elution separation of a mixture of seven barbiturates using a linear concentration gradient of methanol in water. Conditions and Nos. of compounds as in Fig. 4.7; compound 7 = amobarbital. Resolution optimized with respect to the compounds 6 and 7 and minimum retention volume of compound 1 required. Gradient function: c = 0.523 + 0.00846 V [c = %(v/v) methanol x 10^{-2}; V in ml]. (From ref. 52.)

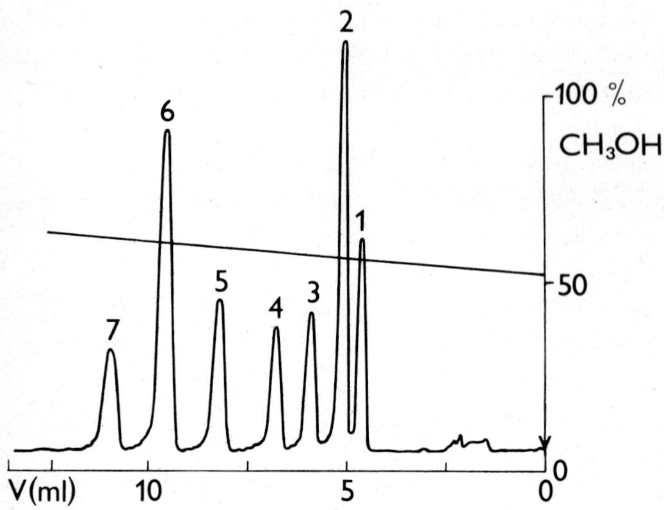

Fig. 7.2. Reversed-phase separation of a mixture of seven barbiturates using an empirically selected gradient (linear gradient function: c = 0.1 + 0.0172 V). Conditions and Nos. of compounds as in Figs. 4.7 and 7.1.

A practical example of the separation of a mixture of seven barbiturates on an octadecylsilica column can illustrate this optimization procedure[52]. A linear concentration gradient of methanol in water was optimized in order to achieve a resolution of 1.5 for the separation of the two most strongly retained compounds, 6 and 7, while the retention volume of the earliest eluted compound, 1 should be kept to a minimum. The optimization calculation yielded a linear gradient function of the form $c = 0.523 + 0.00846\ V$. The chromatogram obtained using the optimized gradient function is shown in Fig. 7.1. For comparison, the chromatogram obtained originally using the gradient function selected by a trial-and-error procedure is shown in Fig. 7.2. A significant reduction in the time of separation is apparent for the gradient optimized by calculation.

Obviously, such a relatively simple separation could be achieved under isocratic conditions in an equal or even shorter analysis time if we take into account the time necessary to re-equilibrate the column after the end of the gradient. However, optimized gradient elution separations may result in significant shortening of the time of analysis of more complex mixtures.

To elucidate the influence of the selected shape (curvature) of the gradient on the optimization procedure, the calculation approach was repeated for the same column and components of the mobile phase and the same optimization criteria, but a logarithmic (convex) gradient function given by eqn. 4.20 (Table 4.1) was used. The $c = f(V)$ plot according to the calculated optimized logarithmic gradient function was almost identical with that in the previous instance of the calculated optimized linear gradient function. This suggests that the pre-selected form of the gradient function is not very critical for the optimization procedure[52].

To verify each calculated optimized gradient function, it is advisable to calculate the retention volumes and resolution for all chromatographed compounds. Owing to minor changes in selectivity with changing composition of the mobile phase, which are difficult to forecast without calculation, sometimes the pair of compounds expected to be the most difficult to separate is well resolved under calculated optimized gradient conditions, but the resolution of another pair of sample compounds is impaired. This is apparent from the calculated V'_g and R_g values. In this instance, the selection of the compounds i, j and k should be changed and the optimization calculation repeated.

This was the case with the above example of the separation of barbiturates, where originally the gradient function was optimized for the resolution $R_{g,o}=1.5$ between compounds 1 and 2 and the minimum retention volume of the last eluted compound, 7; under these conditions, however, the resolution of compounds 6 and 7 would be insufficient. Therefore, the criteria of optimization were changed and the calculation was repeated as indicated above[52].

Further, we shall bear in mind that in each stationary-mobile phase system the resolution of each pair of sample compounds is limited by certain minimum and maximum values, in both isocratic and gradient elution, which cannot be surpassed with a given column (the number of plates is assumed to be constant and optimized) and a combination of the mobile phase components (see Section 4.9).

Therefore, the required value of the resolution in the optimization calculations should be kept within these practical limits (the efficiency of the column could be changed to yield the resolution required), otherwise the calculation would fail.

The optimized gradient conditions can be simply transformed to other columns of different geometry and efficiency and to other flow-rates of the mobile phase following the rules outlined in Section 4.5.

Regular spacing of the peaks of sample compounds on the chromatogram can usually be better approximated in stepwise elution chromatography and in chromatography using composed gradients than in simple continuous gradient elution chromatography. Here, the conditions can be optimized step by step in order to achieve a "tailor-made" separation of two (or more) neighbouring peaks in each step. Three approaches have been suggested for the optimization of stepwise elution chromatography.

The relatively simple graphical method introduced by Golkiewicz and Soczewiński[73] is used for the construction of a stepwise gradient in such a way that each compound eluted in a given isocratic step as the last peak should have a certain, pre-determined capacity factor in this step (e.g., $k' = 1$). The composition of the mobile phase in this step necessary for this aim is found from plots of $k' = f(c)$ for the individual compounds. On the other hand, the $k' = f(c)$ plots are used to determine the capacity factors for compounds that are still retained on the column after the end of this step and the corresponding R_F values that give the migration of these compounds along the column. However, this approach does not take into account the efficiency of the column necessary to achieve a certain resolution.

The optimization approach suggested by Borówko and co-workers[58,60] consists in the calculation of the volume and composition of the mobile phase in each isocratic step, which is necessary to keep the resolution of the adjoining peaks of compounds within certain limits.

The approach of Jandera and Churáček[49] is similar, but much simpler owing to another definition of the column sequences corresponding to the individual isocratic elution steps (see Section 5.1). It works as follows. The composition and the volume of the mobile phase in each isocratic elution step are calculated provided a single sample compound is eluted in each step and the resolution between the neighbouring peaks is kept at a required value $R_{g,o}$. The width of the peak of the compound eluted in step n is controlled by the capacity factor of the compound in this step (at the time of elution) (eqn. 5.6) (this is somewhat simplified). Then, the retention volumes of the pair of compounds with adjacent peaks that are eluted in two neighbouring steps, n-1 and n, are interrelated as follows:

$$V'_{gn} = V'_{g(n-1)} + R_{g,o} \cdot \frac{w_{gn} + w_{g(n-1)}}{2} = V'_{g(n-1)} + \frac{2R_{g,o}V_m}{\sqrt{n}} \cdot [k'_{(n-1)} + k'_n + 2] \quad (7.1)$$

where the subscripts n-1 and n relate to V'_g, k' and w_g of compounds eluted in steps n-1 and n, respectively.

The capacity factors, k'_i, of the solute eluted in step n, and the volumes of the mobile phase passed through the column, V_{ei}, in each step i from 1 to n-1 are known. Then, it is possible to calculate the capacity factor, k'_n, necessary to elute the solute in step n so that the required resolution $R_{g,o}$ between this compound and the solute eluted in step n-1 is achieved. For this purpose, eqn. 7.1 is combined with eqn. 5.4 as follows:

$$k'_n = \frac{V'_{g(n-1)} - \sum_{i=1}^{n-1} V_{ei} + \frac{2V_m R_{g,o}}{\sqrt{n}} \cdot [k'_{(n-1)} + 2]}{V_m - \sum_{i=1}^{n-1} \frac{V_{ei}}{k'_i} - \frac{2\,V_m R_{g,o}}{\sqrt{n}}} \quad (7.2)$$

After introducing k'_n calculated from eqn. 7.2 into the appropriate $k' = f(c)$ function that fits a given chromatographic system (Table 2.1), the necessary concentration of the more efficient eluting component in the mobile phase, c_n, is calculated. Finally, the volume of the mobile phase in step n, V_{en}, is determined, e.g., if step n is finished just at the time when the elution of the sample compound is accomplished:

$$V_{en} = V'_{gn} + \frac{w_{gn}R_{g,o}}{2} - \sum_{i=1}^{n-1} V_{ei} = V'_{gn} + \frac{2V_m R_{g,o}}{\sqrt{n}} \cdot (1 + k'_n) - \sum_{i=1}^{n-1} V_{ei} \quad (7.3)$$

This sequence of calculations of k'_n, c_n and V_{en} is consequently repeated for each step and compound eluted, beginning with step 1 until the step in which the last sample compound is eluted. It is convenient to consider the elution of two compounds in step 1 while the composition of the mobile phase in this step is calculated as in chromatography under isocratic conditions after introducing the

References on p. 180.

corresponding $k' = f(c)$ functions for the two compounds into the definition equation for resolution (see Appendix 1).

This calculation approach may be easily adapted to the construction of stepwise gradients where two compounds instead of a single one are eluted in each isocratic step. Fig. 7.3 shows an example of the separation of a mixture of barbiturates on an octadecylsilica column using a stepwise concentration gradient of methanol in water optimized for the resolution $R_{g,o} = 1.75$ between the adjacent bands[49].

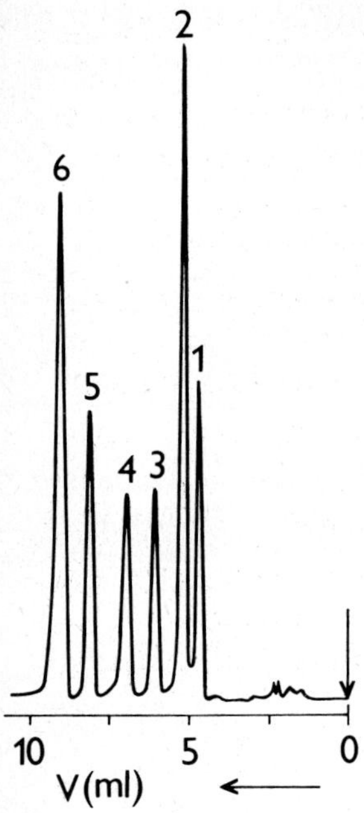

Fig. 7.3. Reversed-phase separation of a mixture of barbiturates using a stepwise gradient optimized by calculation to achieve a resolution $R_g = 1.75$ between neighbouring peaks. Step 1: 2.37 ml, 52% (v/v) methanol, compounds 1 and 2 eluted. Step 2: 1.78 ml, 55% (v/v) methanol, compounds 3 and 4 eluted. Step 3: 64% (v/v) methanol, compounds 5 and 6 eluted. Conditions and Nos. of compounds as in Fig. 4.7. (From ref. 52.)

Of course, the optimization calculation approach described is subject to certain limitations. It is possible to obtain the resolution within certain practical limits, as under isocratic conditions and as in chromatography using simple continuous gradients (see above).

Sometimes the concentration of the more efficient eluting component in the mobile phase for step n, c_n, calculated using the above optimization approach may be lower than the corresponding concentration in step n-1. If this calculated concentration were realy used in step n, difficulties might occur in the form of broad and sometimes even unsymmetrical peaks. In other instances, the calculation may yield a retention volume of the compound eluted in step n lower than the sum of V_{ei} for steps 1 to n-1, which means that the concentration of the more efficient eluting component in the mobile phase in step n-1 is too high to allow the required resolution $R_{g,o}$ between the adjacent peaks eluted in steps n-1 and n. If one or the other situation arises, it is necessary to re-calculate the conditions for step n-1 so that the elution of the two compounds is accomplished in step n-1 with the required resolution $R_{g,o}$. This condition is easy to establish in a program of optimization calculation[49].

For more precise calculations, eqn. 5.6a for peak widths w_g should be used instead of eqn. 5.6 in combination with eqns. 7.1 - 7.3. In principle it would be possible to optimize gradient elution using composed gradients consisting of several subsequent continuous steps in an analogous way to stepwise elution. Of course, the calculation procedure would be much more complex. In such calculations, linear concentration changes in the individual steps are sufficient, because the curvature of the gradient in a relatively short part of the gradient has an almost negligible influence on separation. It is possible to calculate the steepness of the gradient necessary for the elution of one or two sample compounds in each step in order to achieve the resolution required. The contribution of each step to V'_g and to V_m should be calculated using the appropriate equations from Appendix 2 (see Section 5.2). The solution would be possible only with the aid of a computer or at least a programmable calculator.

A similar optimization approach was applied by Svoboda[69] to calculate optimum composed gradients for anion-exchange separations of mixtures of nucleotides.

7.3. GENERAL CASE

For most practising chromatographers, it is of great interest to have the possibility of optimizing gradinet elution without the isocratic experiments necessary to determine the constants of the $k' = f(c)$ functions for the individual sample compounds in a given system. Such a generally applicable gradient optimization is usually performed using a trial-and-error method, until the chromatographer is satisfied with the result of the separation. Some commercial gradient elution liquid chromatographs make it possible to pre-programme a sequence of trial-and-error experiments to be performed subsequently by the instrument[5,6]. After the end of this unattended series of experiments, the chromatographer selects the "best looking" separation and the corresponding profile of the gradient for routine repeated analyses.

References on p. 180.

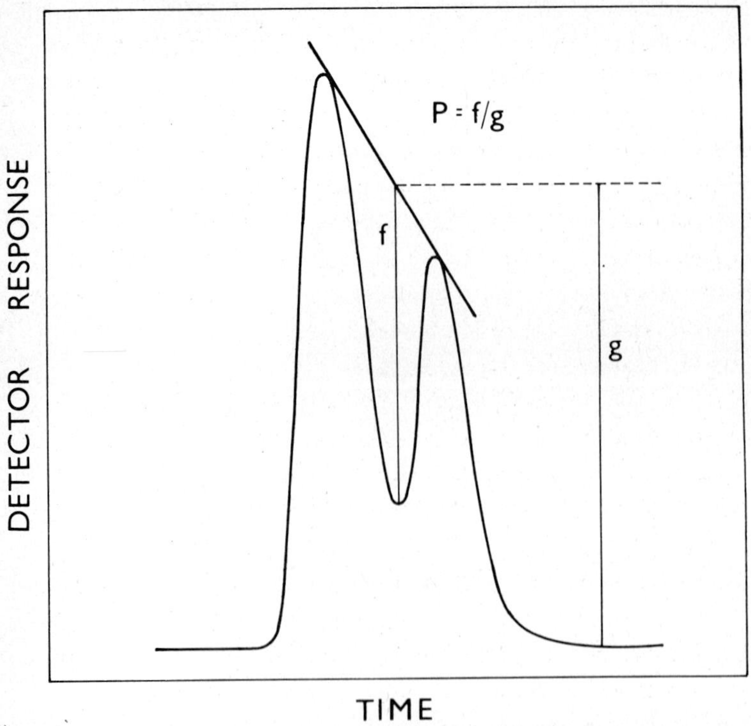

Fig. 7.4. Method for calculation of the peak separation function P. (Adapted from ref. 79.)

This empirical optimization method may be aided and controlled using a computerized simplex algorithm method suggested by Watson and Carr[78]. In this method, the peak separation function, P, as originally defined by Kaiser[79] (see Fig. 7.4), is used as a measure of the quality of separation and is integrated into the chromatographic response function (*CRF*). This function involves the experimental peak separation for each pair of adjacent peaks (P_i), the desired peak separation (P_0), the actual analysis time (t_L) and the acceptable analysis time (t_M):

$$CRF = \Sigma \ln \frac{P_i}{P_0} + \bar{\alpha}(t_M - t_L) \tag{7.4}$$

where $\bar{\alpha}$ is an arbitrary weighting factor. P_0 and t_M should be consistent with the achievable signal-to-noise ratio or with the desired resolution. Then, the optimization consists in the search of a separation that yields the *CRF* optimization function closest to zero. In subsequent optimization experiments, the peak separation functions, P_i, for each pair of peaks must be evaluated as indicated in Fig. 7.4. This optimization procedure was verified in the practical separation of a five-component phenylthiohydantoin-amino acid mixture[78] and, in a further refined form, on the separation of antioxidants[85] by reversed-phase chromatography. A similar simple statistical approach was suggested for optimization of gradient elution conditions both in reversed-phase and in normal-phase chromatography, but this approach took into consideration a rational choice of solvents from the point of view of possible effects on selectivity[84].

It is possible to give rules for "optimum" gradients with general applicability with considerably simplified assumptions only as "hints" that cannot be expected to provide the best separation of a given sample mixture, but rather to offer the general probability of yielding acceptable separations in a wide variety of different separation problems.

According to Snyder[1-3], the "linear solvent strength" (LSS) gradients are especially suitable for providing the optimum separation conditons in the above sense. Snyder and co-workers[37,38] suggested an optimization approach for determining the optimum steepness of LSS gradients that would yield maximum resolution per unit time over the whole chromatogram. Let us recall that the peak widths and resolution are approximately equal in different parts of the chromatogram for solutes with equal relative retentions (isocratic), provided that the separation factors, α, of compounds in the sample mixture do not change significantly with the composition of the mobile phase. In this instance, an optimum steepness, β, of an LSS gradient can be derived by analogy with the optimum capacity factor under isocratic conditions (eqn. 4.9): β = 0.1-0.3 for different column packing materials, provided there is a constant column length and a variable separation time. For example, this corresponds in practice to an increase of 1-3% in the amount of organic solvent in the mobile phase per millilitre of the eluate from the column in reversed-phase chromatography on an octadecylsilica column with $V_m \simeq 3$ ml. This is in good agreement with the empirically suggested optimum gradient steepness in reversed-phase chromatography[61].

In reversed-phase chromatography, it is possible to predict the steepness of a linear concentration gradient of the organic solvent necessary to achieve approximately the resolution required for such mixtures, where the selectivity of separation between the individual compounds (under isocratic conditions) is approximately constant and either is known or can be estimated. For example, the selectivity is approximately constant between the neighbouring members of a homologous or of a benzologous series in reversed-phase chromatography ($\alpha \simeq$ constant).

Further, the widths of all peaks are approximately constant ($w_g \simeq$ constant) under the gradient conditions, so that the differences of the retention volumes, $\Delta V'_g$, and the resolution between the adjacent peaks are approximately constant also[52]:

References on p. 180.

$$\Delta V'_g = V'_{g2} - V'_{g1} = R_g w_g = \frac{1}{mB} \cdot \log\left(\frac{2.31mBk'_{o2}V_m + 10^{mA}}{2.31mBk'_{o1}V_m + 10^{mA}}\right) \simeq \frac{\log \alpha}{mB} \tag{7.5}$$

Provided that the elution is started at a low concentration of the organic solvent in the mobile phase, the term 10^{mA} in eqn. 7.5 can be neglected to a first approximation. If $\alpha \simeq$ constant, the constants m in the $k' = f(c)$ functions (eqn. 1.28) are approximately equal for different sample compounds.

From eqn. 7.5, the steepness of the linear concentration gradient, B, necessary to yield the required resolution, R_g, is given as follows:

$$B = \frac{\log \alpha}{m w_g R_g} = \frac{\sqrt{n} \log \alpha}{4mV_m R_g (1 + k'_f)} \tag{7.6}$$

If we use $k'_f \simeq 1$, we obtain

$$B = \frac{\sqrt{n} \log \alpha}{8mV_m R_g} \tag{7.7}$$

or, introducing the more correct eqn. 4.30c for k'_f in gradient elution using LSS gradients:

$$B = \frac{\sqrt{n} \log \alpha - 1.73\, R_g}{4mV_m R_g} \tag{7.8}$$

The calculation of optimum gradient steepness for separation in a homologous series is illustrated by Table 7.2. Fig. 4.5(B) shows the separation of a homologous series of fluorescent derivatives of aliphatic amines using a linear gradient, with near to optimum conditions (B = 0.015)[70].

In practice, the parameters m increase with increasing retention and the separation factors, α (isocratic), of the neighbouring members in a homologous series decrease to a certain extent with increasing content of the organic solvent in the mobile phase. To compensate for this moderate influence of concentration on α, the mean values of m and of the separation factors (corresponding to the composition of mobile phase in the middle part of the gradient) should be considered in calculations according to eqns. 7.6-7.8.

A practical consequence of the dependence of α on mobile phase composition is that slightly convex gradients sometimes approximate better a linear solvent strength change than linear gradients of concentration of the organic solvent in the mobile phase[3,80].

TABLE 7.2

OPTIMUM GRADIENT STEEPNESS, B_{op}, OF A LINEAR CONCENTRATION GRADIENT OF AN ORGANIC SOLVENT AS A FUNCTION OF RESOLUTION, R_g, IN REVERSED-PHASE CHROMATOGRAPHY OF HOMOLOGOUS SERIES

Column, C18/LiChrosorb Si 100 (10 μm), 300 x 4.2 mm I.D.; V_m = 3.1 ml; n = 2000. Calculation based on separation of a homologous series of fluorescent derivatives of C_1 - C_{10} n-alkylamines using (a) eqn. 7.7 and (b) eqn. 7.8 (B in gradient function given by eqn. 4.17; V_G = volume of the gradient (ml) if the elution is started with zero concentration of the organic solvent; A = 0).

Organic solvent	m*	log α*	Parameter	R_g					
				1.0		1.5		2.0	
				a	b	a	b	a	b
Methanol	5.25	0.12	B_{op}	0.041	0.056	0.027	0.028	0.021	0.015
			V_G	24	18	37	36	48	67
Acetonitrile	2.95	0.13	B_{op}	0.079	0.111	0.053	0.059	0.040	0.032
			V_G	13	9	19	17	25	31
Tetrahydrofuran	5.44	0.13	B_{op}	0.043	0.061	0.029	0.032	0.022	0.017
			V_G	23	16	34	32	45	59

*Mean values.

7.4. GRADIENT ELUTION AS A "SCOUTING" TECHNIQUE FOR ISOCRATIC ELUTION LIQUID CHROMATOGRAPHY

In certain chromatographic systems, the gradient elution technique can be used for the rapid selection of a suitable composition of the mobile phase in liquid column chromatography under isocratic conditions.

According to Snyder and co-workers[37,38], optimized elution using "linear solvent strength" (LSS) gradients can be used with advantage for this purpose. If optimized LSS gradients (where β = 0.2 in eqn. 4.4) are used, all sample compounds should elute with approximately equal instantaneous capacity factors, $k'_f \approx 2.2$ (cf., eqn. 4.30c).

For example, to achieve $k' \approx 4$ under isocratic conditions, it would be necessary to select a composition of the mobile phase corresponding to $k'_c \approx 4$ in eqn. 4.4. Eqn. 4.4 can then be used to calculate the time, t_c, from the start of gradient elution, when the compound eluted in time t_g had the capacity factor $k' = k'_c = 4$ at the inlet of the column, i.e.,

$$t_c = t_g - \Delta t_{c/f} - t_m - t_z = t_g - t_m \cdot \frac{\log k'_c - \log k'_f}{\beta} - t_m - t_z \simeq$$

$$\simeq t_g - 2.5\ t_m - t_z \tag{7.9}$$

where $\Delta t_{c/f}$ is the difference between the times corresponding to the instantaneous capacity factors $k' = k'_f$ and $k' = k'_c$ at the outlet of the column and t_z is the delay time corresponding to the void (dead) volumes between the outlet from the gradient-forming device and the top of the column. The composition of the mobile phase corresponding to the time t_c can be easily found from the gradient function used to control the mobile phase composition programme.

Snyder and co-workers[37,38] found good agreement between the experimental and expected chromatographic data in chromatography on chemically bonded non-polar stationary phases using linear concentration gradients of acetonitrile in water. A more extensive study by Elgass[61] showed that nine compounds on an octadecylsilica column had instantaneous capacity factors at the time of elution of $k'_f \simeq 2.4$-3.2 for linear concentration gradients of acetonitrile in water and $k'_f \simeq 0.9$-2.5 for linear concentration gradients of methanol in water ($\beta \simeq 0.2$ in these examples), which is in satisfactory agreement with Snyder's assumptions ($k'_f \simeq 2.2$ for $\beta = 0.2$). Based on these experiments, the following empirical rules were derived[61].

Capacity factors $k' = 1$ can be expected under isocratic conditions when working with a mobile phase with a composition corresponding to the time of elution of peak maxima for compounds that yield the following retention volumes in gradient elution when eluted with linear concentration gradients beginning at zero concentration of the more efficient eluting component in the mobile phase:

(a) $V'_g \simeq 12\ V_m$ for linear gradients of methanol concentration in water, starting with a zero concentration of methanol, on a C18 column;

(b) $V'_g \simeq 9\ V_m$ for linear gradients of acetonitrile concentration in water, starting with a zero concentration of acetonitrile, on a C18 column;

(c) $V'_g \simeq 11$ to $12\ V_m$ for linear gradients of methylene chloride concentration in *n*-heptane, starting with a zero concentration of methylene chloride, on a silica column.

The precision of these estimates should be within ± 20-40% relative.

Another possibility for optimization of isocratic conditions is to calculate the constants of the $k' = f(c)$ functions in a given system from the data obtained in gradient elution chromatography. Under gradient conditions, we can determine these constants for a larger number of compounds with significant differences in retention than is possible under isocratic conditions. These constants can be used to calculate the composition of the mobile phase necessary to achieve the required resolution for given pairs of compounds, as indicated in Part I.

Jandera and Churáček[8,43] showed this possibility for systems where eqn. 1.23 describes the $k' = f(c)$ relationships (such as chromatography on polar adsorbents or on ion exchangers) and where the general gradient function given by eqn. 4.18 controls the concentration programme of the more efficient eluting component in the mobile phase. Here, the constants of the two-parameter eqn. 1.23, k'_0 and m, may be calculated from the regression lines of log V'_g versus log B plots obtained from the data under gradient conditions (eqn. 4.34). The following relationships between the constants of the regression lines, $\bar{a}$ and $\bar{b}$, and the parameters k'_0 and m apply:

$$m = \frac{\bar{b}}{\kappa(1 - \bar{b})} \tag{7.10}$$

and

$$k'_0 = \frac{1}{(\kappa m + 1)V_m} \cdot 10^{\bar{a}(\kappa m + 1)} \tag{7.11}$$

where κ characterizes the shape (curvature) of the gradient (eqn. 4.18). The accuracies of the constants k'_0 and m calculated in this way are comparable to those determined under isocratic conditions.

A similar method was suggested by Schoenmakers et al.[41] for optimization of isocratic conditions in reversed-phase liquid chromatography. They observed an empirical correlation between the parameters m and k'_0 of the $k' = f(c)$ relationship applying in reversed-phase systems (eqn. 1.28) if aqueous methanolic mobile phases are used:

$$m = p' + q' \log k'_0 \tag{7.12}$$

This correlation may be introduced into eqn. 4.27 for retention volumes in reversed-phase gradient elution liquid chromatography to yield

$$V'_g = \frac{1}{mB} \cdot \log \left[2.31 mBV_m \cdot 10^{\frac{m(1-q'A)-p'}{q'}} + 1 \right] \tag{7.13}$$

With the known column dead volume, V_m, and the parameters of the linear concentration gradient function, A and B, the only unknown quantity on the right-hand side of eqn. 7.13 is the constant m of eqn. 1.28, which can be determined from the experimental retention volume using eqn. 7.13 and iteration calculations, provided the parameters p' and q' of eqn. 7.12 are known (from the data of Schoenmakers et al.[41], $p' \simeq 2.86$ and $q' \simeq 1.77$ for an octadecylsilica column and

References on p. 180.

aqueous methanolic mobile phases). The correlation is poorer with mobile phases that contain other organic solvents. To calculate the constant k'_0, the value of m determined from eqn. 7.13 is introduced into eqn. 7.12 and, once the constants of the $k' = f(c)$ function according to eqn. 1.28 are known, the composition of the mobile phase necessary to achieve the required resolution in a given mixture may be calculated as in the previous optimization method for normal-phase chromatography. The method has been used to find optimum isocratic conditions for the separation of mixtures of phenolic compounds and of phenylthiohydantoin amino acids[41]. However, one must be careful when using the correlation according to eqn. 7.12, as the constants p' and q' were evaluated using only a limited number of compounds and exceptions are likely to occur (it is interesting that the correlation given by eqn. 7.12 can be predicted theoretically using the interaction indices model for reversed-phase separation mechanism, as mentioned in Section 1.1[83]).

7.5. EMPIRICAL APPROACH FOR SELECTION OF OPTIMUM GRADIENT CONDITIONS

With a basic knowledge of the theory of gradient elution, it is possible to place on a rational basis an empirical design and optimization of gradient elution conditions. This empirical optimization approach may be understood in terms of a subsequent approach to the conditions that would yield a resolution in the chromatogram as near to the required "best" values as possible in a reasonably short separation time. For this purpose, the three contributions to resolution (efficiency, selectivity and capacity) should be suitably adjusted.

The efficiency, i.e., the number of plates in the column used, can be increased and the resolution improved in much the same way as under isocratic conditions, i.e., by (a) increasing the column length, (b) decreasing the particle diameter of column packing material, (c) decreasing the flow-rate of the mobile phase and (d) increasing the temperature or choosing components of the mobile phase of lower viscosity. However, changes (a) and (c) or their simultaneous combination are connected with an increase in the time of analysis; (a) and (b) lead to an increase in the operating pressure, which is increased most if the column length and the flow-rate of the mobile phase are increased simultaneously to hold the analysis time constant. The change (d) usually produces minor improvements.

As both the time of analysis and the operating pressure have practical upper limits, the optimization of efficiency is limited in practice. Currently used standard analytical columns packed with materials with particle diameters of 5-10 μm are usually sufficiently efficient for most practical separation problems.

Selection of an appropriate column packing material and combination of solvents for a given separation is also controlled by the same rules as in isocratic elution chromatography[1-6]. Some useful combinations of column packing materials and components of the mobile phase are given in Table 7.1[2-4,61].

Once having selected the appropriate column and the components of the mobile phase, the optimization of gradient elution separation consists in a search for the best gradient profile.

For this purpose, we run a gradient corresponding to a change from 0 to 100% of the more efficient eluting component in the mobile phase in a volume of gradient $V_G \simeq$ 10-15 V_m (generally, a 15-45 min gradient at a flow-rate of 1 ml/min, when using common analytical columns). A linear concentration gradient is selected for work in reversed-phase systems, whereas a concave gradient would be generally preferred in ion-exchange chromatography and in chromatography on polar adsorbents[3,4]. The hints concerning the shape and steepness of the initial gradient profile are also given in Table 7.1[2-4,61].

Of course, for certain separations, the selectivity may be modified by addition of a ligand-forming, ion-pairing compound or another suitable compound into the mobile phase which offers specific interactions with the sample compounds.

If the peaks of all sample compounds are eluted in the part of the gradient volume corresponding to a ca. 20-25% change in the concentration of the more efficient eluting component b in the mobile phase, it is likely that isocratic conditions could be found for a successful separation of the sample mixture. If the compounds are eluted over a larger part of the gradient, gradient elution usually should be used to speed up the elution. Of course, this rule is only approximate and the above indicated range of composition is influenced by the system used and by the differences in the elution strengths of the solvents a and b. It should approximately fit reversed-phase systems with methanol or acetonitrile as the solvent b. In normal-phase chromatography on polar adsorbents, this rule can be used if the compounds are not eluted in the early part of the gradient, where the changes in concentration of the more polar solvent have a much more dramatic influence on retention.

(a) The peaks are regularly spaced on the chromatogram, but they are not sufficiently resolved and overlap:

The steepness of the gradient, B, should be decreased or the number of plates increased to improve resolution. The time of analysis and the peak widths are thus increased and the peak heights lowered. A decrease in the initial concentration of the more efficient eluting agent in the mobile phase, A, also leads to a slight increase in resolution, but also to a large increase in the analysis time, and it should be tried only if the first peaks eluted are close to V_m and poorly resolved, otherwise a decrease in B is usually a more efficient remedy.

References on p. 180.

(b) The peaks are regularly spaced at large distances on the chromatogram - they are "over-resolved", which leads to unnecessarily long analysis times:

The steepness of the gradient should be increased to yield the resolution required; the time of analysis is thus shortened and the peaks become higher and narrower. A shorter column or a higher flow-rate of the mobile phase may be used.

(c) The peaks are resolved satisfactorily but the first compound has a large elution volume:

The concentration of the more efficient eluting component b in the mobile phase at the beginning of the gradient, A, should be increased. As an empirical rule, a starting concentration A selected as ca. 50% of the concentration of the component b at the instant of elution of the first sample band in the initial gradient run is likely to provide a satisfactory improvement. This usually slightly impairs the resolution (to compensate for this effect, B can be lowered), but decreases significantly the time of analysis.

(d) Sample compounds are eluted too early and are badly resolved:

Both A and B should be decreased; in certain instances the solvent b is too strong and should be changed for a weaker one. If the slope of the gradient is low and the resolution is still poor and the peaks are relatively early eluted at $A = 0$, the less efficient eluting component in the mobile phase (solvent a) has too high an elution strength and should be changed for a weaker one. If this is not possible (e.g., if a is water in reversed-phase chromatography or an n-alkane in normal-phase chromatography), a change of the whole chromatographic system would be necessary. The retention may be increased using a certain form of additional interactions in the mobile phase, such as a change in the ionic strength of pH or the addition of ion-pair-forming or complex-forming agents in reversed-phase chromatography of strongly polar or ionized solutes.

(e) A number of compounds can be eluted only after the end of the gradient, in pure component b of the mobile phase:

If the resolution is adequate and all sample solutes are eluted in a reasonable time, it is not necessary to change the profile of the gradient, which is a two-step gradient with a final hold-up (Section 5.4). However, if sample peaks are "over-resolved" and the elution volumes of the most strongly retained compounds are excessive, the initially chosen solvent b has too low an elution strength and should be changed for a stronger one.

(f) The later eluted peaks are sufficiently resolved or "over-resolved", but the early eluted compounds are poorly resolved; the peaks are "bunched together" towards the beginning of the chromatogram; the peak widths increase towards the end of the chromatogram:

The shape (curvature) of the gradient should be changed in such a way that the gradient is less steep at the beginning and steeper towards the end, i.e., use a more concave gradient. If the peaks of the early eluted compounds still remain badly resolved and are eluted close to V_m, try the use of a two-step elution with an initial gradient delay using pure component *a* (with a low content of the component *b* in the first, isocratic step).

(g) The earlier eluted peaks are sufficiently resolved or "over-resolved", but the later eluted compounds are poorly resolved; the peaks are "bunched together" towards the end of the chromatogram; the peak widths decrease towards the end of the chromatogram:

The shape (curvature) of the gradient should be changed to become steeper at the beginning and less steep towards the end, i.e., use a more convex gradient. A two-step gradient with a final hold-up, either at 100% or at a lower content of the solvent *b* in the mobile phase, may improve the separation.

(h) The resolution and spacing of most peaks on the chromatogram are satisfactory, but there are one or more pairs of compounds insufficiently resolved at different parts of the chromatogram:

If an increase in column efficiency does not help, it is necessary to adjust the selectivity of separation for these "critical" pairs of compounds. This may be attempted by changing the steepness of the gradient (changes in the shape of the gradient or in the initial concentration of the more efficient eluting component in the mobile phase may sometimes also improve the selectivity, but the influence of *B* on selectivity is more significant), which can be successful for pairs of compounds with different coefficients *m* of the two-parameter $k' = f(c)$ functions. A change in *B* may either improve or impair the selectivity (for more details, see Section 4.8). If a change in *B* is not successful, we can try a composed gradient with gradient delays at positions of elution of the unresolved pairs of peaks. If the separation is still not satisfactory, it is necessary to change the more efficient eluting component in the mobile phase (such as methanol for acetonitrile or tetrahydrofuran in reversed-phase chromatography); a change of the less efficient eluting component is usually not of much help. This means that the solvent *b* from another selectivity group (as defined by Snyder[75]) is to be preferred. The use of a ternary gradient may often solve the problem efficiently (see Chapter 6). Another possibility for improving the selectivity involves the addition of a further component to the mobile phase that can undergo specific interactions with some of the badly resolved compounds (a change in pH, addition of a complex-forming compound, etc.).

If the separation problem is not solved by any of these means or by increasing the efficiency, the chromatographic system (column packing material) should be changed.

Some further details of points (a)-(h) should be considered. For certain pairs of compounds, where the selectivity depends significantly on the composition of the mobile phase, a change in *B* or *A* may lead to opposite results to those pre-

References on p. 180.

dicted above owing to changes in selectivity, but this does not ocurr very frequently. Further, we should recall that the steepness of the gradient, *B*, is defined as the concentration change of component *b* in the mobile phase per unit *volume* of the eluate from the column. Consequently, if we are changing the flow-rate of the mobile phase, we should not forget to change correspondingly the time of the gradient in order to keep *B* constant. If we decrease the flow-rate of the mobile phase to improve the efficiency of separation (and resolution), but we do not change the time of the gradient, an increase in the gradient steepness, *B*, would result, which usually would greatly predominate over the influence of the mobile phase flow-rate and the resolution would not be increased, but lowered!

Further details concerning the possibilities of using the optimum gradient profile, once determined, with columns of other geometries or with other mobile phase flow-rates are given in Section 4.5.

Not only the time of the gradient, but also the time necessary for re-equilibration of the column back to the initial gradient conditions should be included in the analysis time.

The specific requirements for separation, such as the detection sensitivity or semi-preparative or preparative separation purposes, should be considered when developing the optimum separation (they would largely determine the resolution necessary).

With certain gradient profiles, additional difficulties may occasionally occur, such as band tailing and solvent demixing effects.

With certain gradient elution instruments, the gradients are poorly reproducible in the initial and final parts. Here, it is recommended to start the gradient at ca. 5% of component *b* and to end the gradient at a maximum of 95% of component *b*[38]. The compositions of the two liquids mixed during the gradient elution should be adjusted correspondingly.

REFERENCES TO PART II

1 L.R. Snyder, *Chromatogr. Rev.*, 7 (1965) 1.
2 L.R. Snyder and J.J. Kirkland, *Introduction to Modern Liquid Chromatography*, 2nd ed., Wiley-Interscience, New York, 1979, Ch. 16.
3 L.R. Snyder, in Cs. Horváth (Editor), *High-Performance Liquid Chromatography, Advances and Perspectives*, Vol. 1, Academic Press, New York, 1980, p. 208.
4 P. Jandera and J. Churáček, *Advan. Chromatogr.*, 19 (1981) 125.
5 L.B. Sybrandt and E.F. Montoya, *Int. Lab.*, July/August (1977) 51.
6 S.R. Bakalyar, R. McIlwrick and E. Roggendorf, *J. Chromatogr.*, 142 (1977) 353.
7 D.J. Popovich, J.B. Dixon and W.A. McKinley, *Amer. Lab.*, May (1981).
8 P. Jandera and J. Churáček, *J. Chromatogr.*, 91 (1974) 223.
9 R.P.W. Scott, *J. Chromatogr. Sci.*, 9 (1971) 385.
10 R.P.W. Scott and P. Kucera, *J. Chromatogr. Sci.*, 11 (1973) 83.
11 R.P.W. Scott and P. Kucera, *J. Chromatogr.*, 83 (1973) 257.
12 R.P.W. Scott and P. Kucera, *Anal. Chem.*, 45 (1973) 749.

13 R.P.W. Scott, *Analabs Res. Notes*, (1975) 15.
14 L.R. Snyder, *Anal. Chem.*, 46 (1974) 1384.
15 W.W. Pitt, Jr., *J. Chromatogr. Sci.*, 14 (1976) 396.
16 C. Liteanu and S. Gocan, *Gradient Liquid Chromatography*, Wiley, Chichester, New York, 1974.
17 B. Drake, *Ark. Kemi*, 8 (1955) 1.
18 C. Mader, *Anal. Chem.*, 26 (1954) 566.
19 L.M. Marshall, K.O. Donaldson and F. Friedberg, *Anal. Chem.*, 24 (1952) 773.
20 E.C. Freiling, *J. Amer. Chem. Soc.*, 77 (1955) 2067.
21 E.C. Freiling, *J. Phys. Chem.*, 61 (1957) 543.
22 A.S. Said, *AIChE J.*, 2 (1956) 477.
23 H. Schwab, W. Rieman and P.A. Vaughan, *Anal. Chem.*, 29 (1957) 1357.
24 M.A. Jermyn, *Aust. J. Chem.*, 10 (1957) 55.
25 G.B. Maslova, P.P. Nazarov and K.V. Khmutov, *Ionoobm. Sorbenty v Prom., Akad. Nauk SSSR, Inst. Fiz. Khim.*, (1963) 103.
26 F. Molnár, A. Horváth and V.A. Khalkin, *J. Chromatogr.*, 26 (1967) 215.
27 D.L. Massart and W. Bossaert, *J. Chromatogr.*, 32 (1968) 195.
28 K. Koguchi, H. Waki and S. Ohashi, *J. Chromatogr.*, 25 (1966) 398.
29 S. Ohashi and K. Koguchi, *J. Chromatogr.*, 27 (1967) 214.
30 S. Ohashi, N. Tsuji, Y. Veno, M. Takeshita and M. Muto, *J. Chromatogr.*, 50 (1970) 349.
31 P.R. Rony, *Separ. Sci.*, 3 (1968) 425.
32 J. Carles and G. Abravanel, *J. Chromatogr.*, 56 (1971) 231.
33 J. Inczédy, *Magy Kém. Lapja*, 24 (1969) 232.
34 L.R. Snyder, *J. Chromatogr.*, 13 (1964) 415.
35 L.R. Snyder and H.D. Warren, *J. Chromatogr.*, 15 (1964) 344.
36 L.R. Snyder and D.L. Saunders, *J. Chromatogr. Sci.*, 7 (1969) 195.
37 L.R. Snyder, J.W. Dolan and J.R. Gant, *J. Chromatogr.*, 165 (1979) 3.
38 J.W. Dolan, J.R. Gant and L.R. Snyder, *J. Chromatogr.*, 165 (1979) 31.
39 P.J. Schoenmakers, H.A.H. Billiet, R. Tijssen and L. de Galan, *J. Chromatogr.*, 149 (1978) 519.
40 P.J. Schoenmakers, H.A.H. Billiet and L. de Galan, *J. Chromatogr.*, 185 (1979) 179.
41 P.J. Schoenmakers, H.A.H. Billiet and L. de Galan, *J. Chromatogr.*, 205 (1981) 13.
42 P. Jandera and J. Churáček, *J. Chromatogr.*, 91 (1974) 207.
43 P. Jandera and J. Churáček, *J. Chromatogr.*, 93 (1974) 17.
44 P. Jandera and J. Churáček, *J. Chromatogr.*, 104 (1975) 9.
45 P. Jandera and J. Churáček, *J. Chromatogr.*, 104 (1975) 23.
46 P. Jandera and J. Churáček, *J. Chromatogr.*, 104 (1975) 257.
47 P. Jandera, M. Janderová and J. Churáček, *J. Chromatogr.*, 115 (1975) 9.
48 P. Jandera, M. Janderová and J. Churáček, *J. Chromatogr.*, 148 (1978) 79.
49 P. Jandera and J. Churáček, *J. Chromatogr.*, 170 (1979) 1.
50 P. Jandera, J. Churáček and L. Svoboda, *J. Chromatogr.*, 174 (1979) 35.
51 P. Jandera and J. Churáček, *J. Chromatogr.*, 192 (1980) 1.
52 P. Jandera and J. Churáček, *J. Chromatogr.*, 192 (1980) 19.
53 P. Jandera, J. Churáček and L. Svoboda, *J. Chromatogr.*, 192 (1980) 37.
54 P. Jandera, J. Churáček and H. Colin, *J. Chromatogr.*, 214 (1981) 35.
55 P. Jandera and J. Churáček, *Sb. Ved. Pr. Vys. Sk. Chemickotechnol., Pardubice*, 35 (1976) 107.
56 P. Jandera, *Sb. Ved. Pr. Vys. Sk. Chemickotechnol., Pardubice*, 44 (1981) 87.
57 P. Jandera and F. Král, *Sb. Ved. Pr. Vys. Sk. Chemickotechnol., Pardubice*, 44 (1981) 103.
58 M. Borówko, M. Jaroniec, J. Narkiewicz, A. Patrykiejew and W. Rudziński, *J. Chromatogr.*, 153 (1978) 309.
59 M. Borówko, M. Jaroniec, J. Narkiewicz and A. Patrykiejew, *J. Chromatogr.*, 153 (1978) 321.
60 M. Jaroniec, M. Borówko, J. Narkiewicz, A. Patrykiejew and W. Golkiewicz, *Chromatographia*, 12 (1979) 29.

61 H. Elgass, *Ph.D. Thesis*, Saarbrücken, 1978.
62 H. Engelhardt and H. Elgass, *J. Chromatogr.*, 158 (1978) 249.
63 R.A. Hartwick, C.M. Grill and P.R. Brown, *Anal. Chem.*, 51 (1979) 34.
64 M. Martin and G. Guiochon, in J.F.K. Huber (Editor), *Instrumentation for High-Performance Liquid Chromatography*, Elsevier, Amsterdam, 1978, Ch. 3.
65 E.A. Peterson and H.A. Sober, *Anal. Chem.*, 31 (1959) 857.
66 J. Inczédy, *Analytical Applications of Complex Equilibria*, Akadémiai Kiadó, Budapest, 1976.
67 S.R. Abbott, J.R. Berg, P. Achener and R.L. Stevenson, *J. Chromatogr.*, 126 (1976) 421.
68 J. Inczédy, *J. Chromatogr.*, 154 (1978) 175.
69 V. Svoboda, *Radioisotopy*, 18 (1977) 775.
70 P. Jandera, H. Pechová, D. Tocksteinová, J. Churáček and J. Královský, *Chromatographia*, 16 (1982) 275.
71 J.C. Giddings, *Anal. Chem.*, 39 (1967) 1027.
72 Cs. G. Horváth and S.R. Lipsky, *Anal. Chem.*, 39 (1967) 1993.
73 W. Golkiewicz and E. Soczewiński, *Chromatographia*, 11 (1978) 454.
74 H. Colin, G. Guiochon and P. Jandera, *Anal. Chem.*, 55 (1983) 442.
75 L.R. Snyder, *J. Chromatogr. Sci.*, 16 (1978) 223.
76 L.R. Snyder and J.J. Kirkland, *Introduction to Modern Liquid Chromatography*, Wiley-Interscience, New York, 2nd ed., 1979, Ch. 6-11.
77 R.P.W. Scott and P. Kucera, *J. Chromatogr.*, 169 (1979) 51.
78 M.W. Watson and P.W. Carr, *Anal. Chem.*, 51 (1979) 1835.
79 R. Kaiser, *Gas Chromatographie*, Geest and Portig, Leipzig, 1960, p. 33.
80 F.P.B. van der Maeden, M.E.F. Biemond and P.C.G.M. Janssen, *J. Chromatogr.*, 149 (1978) 539.
81 H. Poppe, J. Paanakker and M. Bronckhorst, *J. Chromatogr.*, 204 (1981) 77.
82 E. Roggendorf and R. Spatz, *J. Chromatogr.*, 204 (1981) 263.
83 P. Jandera, H. Colin and G. Guiochon, *Anal. Chem.*, 54 (1982) 435.
84 J.J. Kirkland and J.L. Glajch, *J. Chromatogr.*, 255 (1983) 27.
85 J.C. Berridge, *J. Chromatogr.*, 244 (1982) 1.
86 I. Molnár, Cs. Horváth and P. Jatlow, *Chromatographia*, 11 (1978) 260.

PART III

INSTRUMENTAL AND OTHER PRACTICAL ASPECTS OF CHROMATOGRAPHY WITH PROGRAMMED COMPOSITION OF THE MOBILE PHASE

Chapter 8

INSTRUMENTATION FOR PRODUCTION OF MOBILE PHASE GRADIENTS

As with other analytical techniques, the success of gradient elution liquid chromatography is determined above all by the instrumentation used. The reproducibility and precision of the retention characteristics and of the quantitative results depend on the performance of the equipment generating gradients of the mobile phase. For accurate results, it is necessary to control very precisely both the composition and the flow-rate of the mobile phase so that the pre-set values of these variables can be accurately reproduced in repeated experiments, and their random fluctuations should be minimized. For a number of reasons, which will be discussed later, the reproducibility and accuracy of the experimental conditions can be far more easily controlled in isocratic elution than in gradient elution chromatography, but the latter technique can yield results as reproducible and accurate as those obtained under isocratic conditions if appropriate attention is paid to the instrumental design and to the operation of the chromatographic system. Therefore, it is very useful for practising chromatographers to become acquainted with the basic principles of the construction of various instruments for gradient elution liquid chromatography and with possible instrumental errors connected with various types of construction of gradient elution systems.

The earlier devices for gradient elution chromatography were constructed from common laboratory glassware, slightly adapted, and used gravity flow to drive the mobile phase through the column. These devices, used in the early stage of "classical" liquid column chromatography with inefficient columns, were usually constructed in each users laboratory and numerous less or more ingenious variants and modifications of several basic types of construction were described. For each device, it was necessary to derive the appropriate relationship between the time and the composition of the mixed mobile phase at the outlet of the device.

Comprehensive descriptions of these devices and of the corresponding time-composition equations were given in a book by Liteanu and Gocan[1] and in reviews by Snyder[2] and others[3,4]. Devices of these types are now used for certain biochemical applications in ion-exchange or affinity chromatography on larger diameter columns packed with ion exchangers and gels based on relatively "soft" (non-rigid) matrices. Of course, it is possible to use them as a low-pressure gradient mixer connected to the inlet of a high-pressure pump in modern high-efficiency

References on p. 282

high-performance liquid chromatography; however, the precision of the concentration gradients generated in these types of devices is usually too poor for the demanding requirements of accuracy of quantitative results common in modern chromatographic analysis.

To meet these requirements, commercial gradient elution liquid chromatographs are used almost exclusively in modern analytical applications of liquid column chromatography. These instruments are usually able to generate a great variety of linear and curved (both concave and convex) gradients, either of several profiles pre-programmed in the electronic system, or drawn on a sheet of paper or on special cards, or entered as a sequence of linear or non-linear gradient steps into the memory of a microprocessor-controlled instrument. Commercial gradient elution instruments have been reviewed by Veening[5], Martin and Guiochon[6], Snyder[7], Snyder and Kirkland[8] and, concisely, by Jandera and Churáček[9], but new or improved versions of these instruments are continuously being introduced by the numerous manufacturers.

In modern gradient elution liquid chromatographs the concentration gradient is controlled by the electronic part of the system, which is usually located in a separate module. These instruments can be classified according to the hydraulic system into two groups:

(a) devices in which the solvents forming gradient are mixed in a low-pressure part at the inlet of a single high-pressure pump; and

(b) apparatus in which the components of the mobile phase are mixed in the high-pressure part by means of two high-pressure pumps.

In the following sections, the operating principles of various devices for gradient elution chromatography are outlined prior to a discussion of possible instrumental sources of errors in the gradient profile and flow-rate of the mobile phase.

8.1. SIMPLE CLASSICAL GRADIENT DEVICES

In the early period after the initial introduction of the gradient elution technique, a large number of devices capable of forming various gradient profiles were designed. The apparatus was utilized in classical column liquid chromatography, where the mobile phase usually flowed through the column by virtue of gravity. These simple instruments can be subdivided into two basic groups, "exponential" and "proportional" gradient devices, which were used in numerous variations and modifications[2].

8.1.1. "Exponential" gradient devices

These were the first devices used for gradient elution liquid chromatography. The basic principle is shown in Fig. 8.1a. The reservoir R containing the more efficient eluting solvent b at a concentration c_R (%, v/v) is connected to a closed mixing chamber M of a given volume V_1 placed below the reservoir. The mixing chamber contains initially the more efficient eluting component of the mobile phase at a concentration $c_0 < c_R$. Its bottom is connected via a stop-cock to the top of the column. When the stop-cock is opened, the solvent b flows into the mixing chamber, the contents of which should be stirred vigorously to ensure good mixing. The mixed mobile phase flows from the mixing chamber through the column by virtue of gravity. In this apparatus, the flow-rate of the solvent b between the reservoir and the mixing chamber is the same as the flow-rate of the mixed mobile phase between the mixing chamber and the column, $Q_1 = Q_2 = Q$. Then, the concentration c of the more efficient eluting solvent b at the outlet of the gradient device in time t from the start of the gradient elution is given by the equation

$$c = c_R - (c_R - c_0)e^{-tQ/V_1} = A - Be^{-V\kappa} \tag{8.1}$$

where $V = tQ$ is the volume of the mobile phase delivered from the start of the gradient run (cf., Table 4.1).

A modification of this type of gradient devices is shown in Fig. 8.1b. It consists of the connection of an additional mixing chamber M_2 of a constant volume $V_2 = V_1$ between the reservoir and the original mixing chamber M_1. This mixing chamber contains originally the more efficient eluting component of the mobile phase b at a concentration c_{RA} and its contents are also stirred. Thus, the liquid flows from the reservoir R first into the additional mixing chamber M_2, then into the mixing chamber M_1, and through the column, at a constant volume flow-rate Q, due to gravity. The concentration of the more efficient eluting component b in the mobile phase at the outlet of the gradient device in time t is given by

$$c = c_{RA} + \left(\frac{c_{RA} - c_R}{V_1} \cdot tQ - c_{RA} + c_0 \right) e^{-tQ/V_1} = A + (DV - B)e^{-V\kappa} \tag{8.2}$$

The exponential devices produce exponential, i.e., convex, concentration gradients, which are relatively less useful in practice[2]. The profile of the exponential gradient may be controlled to a certain extent by varying the volume of the mixing chamber(s) and the composition of solution in the reservoir and

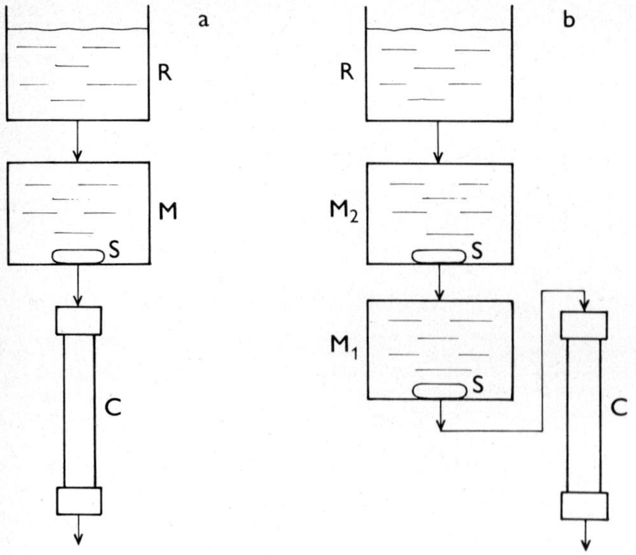

Fig. 8.1. Exponential gradient elution device with (a) one or (b) two mixing chambers. R = reservoir; M, M_1, M_2 = mixing chambers; S = stirrer; C = chromatographic column.

in the chamber(s)[1,2]. If only the initial part of the exponential gradient is used, the gradient shape is close to the linear profile. The apparatus using two mixing chambers is able to produce gradients that approximate better the linear profile than the devices with one mixing chamber. However, the popularity and use of these earliest devices decreased considerably with the introduction of proportional gradient devices and modern gradient liquid chromatographs, so that they are hardly used any longer.

8.1.2. "Proportional" gradient devices

These devices utilize two or more open mixing chambers interconnected in series at the bottom of each vessel by means of a tube with a stop-cock or by a siphon connection.

The simplest apparatus of this type (Fig. 8.2) employs a reservoir R containing the more efficient eluting component of the mobile phase at a concentration c_R (%, v/v) connected to one mixing chamber M containing this component at a concentration c_0. The bottom of the mixing chamber is connected to the top of the column and the contents of the chamber are stirred to ensure good mixing. The profile of the gradient formed depends on the concentrations c_0 and c_R and on the volumes of the reservoir and the mixing chamber. In the simplest form, these are two cylindrical vessels, which may have different diameters.

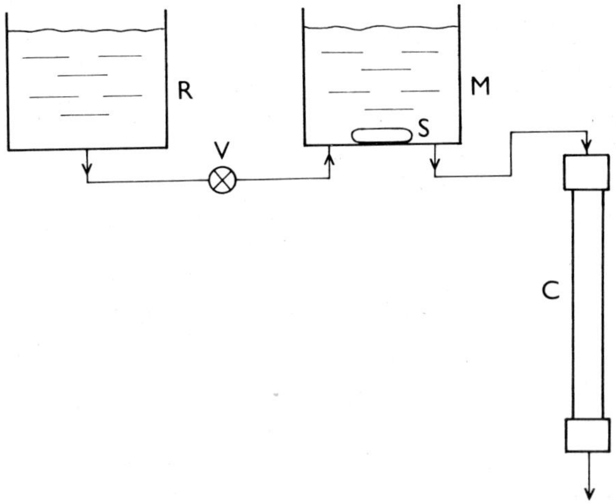

Fig. 8.2. Proportional volume gradient elution device. R = reservoir; M = mixing chamber; S = stirrer; V = valve; C = chromatographic column.

Before the start of the gradient, the reservoir and the mixing chamber are filled with the appropriate solvents so that the levels of the liquids are in hydrostatic equilibrium. The initial volumes of the liquids in the reservoir and in the mixing chamber are V_{o2} and V_{o1}, respectively, and the instantaneous volumes after the start of the gradient (by opening the stop-cocks) are V_2 and V_1 respectively. If the two liquids in the reservoir and in the mixing chamber have equal densities, the volume flow-rate between the mixing chamber and the column, Q_2, is twice the flow-rate between the reservoir and the mixing chamber, Q_1 (gravity is the driving force), i.e., $Q_2 = Q_1$ $(V_2/V_1 = V_{o2}/V_{o1} = \kappa)$. Gradient devices of this type may be termed "proportional volume devices".

The concentration of the more efficient eluting component b in the mobile phase at the outlet of the gradient device, c, in time t is controlled by the following equation:

$$c = c_R - (c_R - c_0)\left[1 - \frac{Q_2 t}{V_{o2} + V_{o1}}\right]^{\frac{V_{o2}}{V_{o1}}} = D - \left(A^{\frac{1}{\kappa}} - BV\right)^{\kappa} \qquad (8.3)$$

(cf., Table 4.1), where

$$D = c_R \tag{8.3a}$$

$$A = c_R - c_0 \tag{8.3b}$$

$$B = \frac{(c_R - c_0)^{\frac{1}{\kappa}}}{V_{o2} + V_{o1}} \tag{8.3c}$$

If the two cylindrical vessels used as the reservoir and the mixing chamber have identical diameters, then $V_{o1} = V_{o2} = V_1 = V_2$ and $\kappa = 1$. In this instance, the device produces linear concentration gradients. Such an apparatus is commercially available from Pharmacia (Uppsala, Sweden) (GM-1 gradient mixer). If $V_{o2} > V_{o1}$ concave gradients are produced, and if $V_{o2} < V_{o1}$ convex gradient profiles result. Hence, for a given pair of vessels, it is not possible to vary the gradient profile in such a simple device.

Eqn. 8.3 fails to describe the gradient profile if the solutions in the mixing chamber and in the reservoir have different densities, ρ_1 and ρ_2. Here, the hydrostatic pressure forces the denser liquid to flow at a higher volume flow-rate. The effect of densities can be partially accounted for in eqn. 8.3 by using

$$\kappa = \frac{V_{o2}\,\rho_1}{V_{o1}\,\rho_2} \tag{8.4}$$

instead of $\kappa = V_{o2}/V_{o1}$. However, this correction does not account for the continuous change in the density in the mixing chamber, ρ_1, with time from the start of the gradient run. Only if the gradient volume, V_G, is sufficiently low with respect to V_{o1}, does the density ρ_1 not change significantly during the gradient run and eqn. 8.3 may be used satisfactorily to calculate the gradient profiles. Otherwise, the experimental gradient profiles would deviate from the calculated values. In practice, however, these deviations are relatively unimportant, with the exception of the later part of the gradient.

The reservoir and mixing chamber may have conical or other irregular shapes to produce gradients of the required curvature, which is fixed by the vessels used. If siphon connectors are used between the reservoir and the mixing chamber, various vessels can be easily interchanged, but if the two liquids differ in density, the performance of the device may be inadequate[2].

It is possible to adapt the devices using two fixed vessels in such a way that they are able to produce gradients of different shapes. The simplest way is to immerse closed glass or plastic cylindrical or conical vessels of various dimensions in the reservoir to vary the volume V_{o2} at constant V_{o1} and thus to control the exponent κ in eqn. 8.3.

More sophisticated proportional gradient devices employ auxiliary pumps, P_1 and P_2, between the reservoir and the mixing chamber and between the mixing chamber and the column. A schematic diagram of such a "proportional flow" gradient device is shown in Fig. 8.3a. Inexpensive low-pressure pumps are suitable for this purpose, such as, e.g., a P-3 peristaltic pump (Pharmacia) with three individual channels, each of which may be operated at a different flow-rate at the same time. A single multi-channel peristaltic pump does the work of the two auxiliary pumps P_1 and P_2 and this pump, with only two beakers and a magnetic stirrer, form the gradient device (Fig. 8.3b).

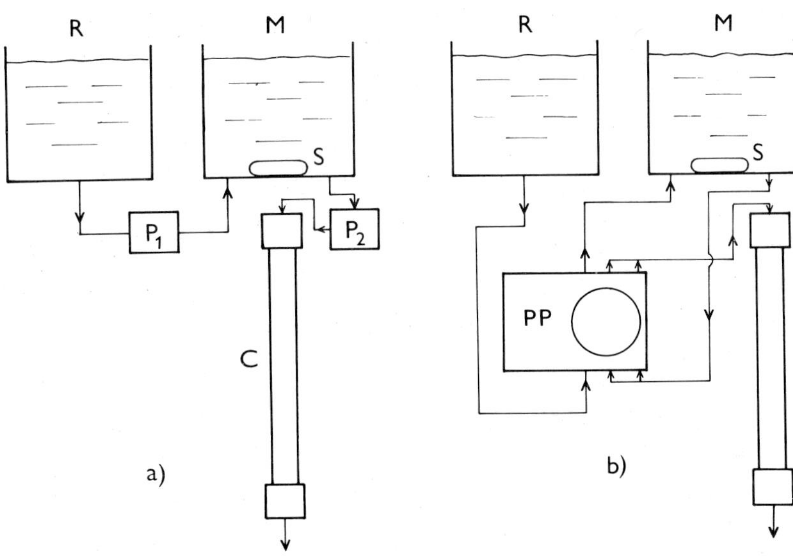

Fig. 8.3. Proportional flow gradient elution device using (a) two auxiliary pumps or (b) a single multi-channel peristaltic pump. R = reservoir; M = mixing chamber; S = stirrer; P_1, P_2 = auxiliary pumps; PP = multi-channel (three-channel here) peristaltic pump; C = chromatographic column.

In this type of equipment, the flow-rates of the two pumps, Q_1 and Q_2, may be selected independently and thus the shape of the gradient is controlled more efficiently[10]. The concentration of the more efficient eluting component at the outlet of the gradient device at time t is given by

$$c = c_R - (c_R - c_0)\left[1 - \frac{Q_2 t}{V_{02} + V_{01}}\right]^{\frac{Q_1}{Q_2 - Q_1}} = D - \left(A^{\frac{1}{K}} - BV\right)^K \tag{8.5}$$

References on p. 282.

(cf., Table 4.1). This equation applies for $Q_1 \neq Q_2$. If $Q_1 = Q_2$, eqn. 8.1 for exponential gradients can be used to describe the gradient profile.

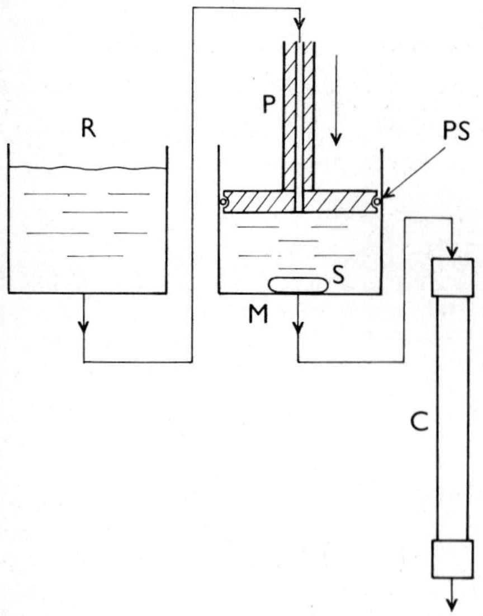

Fig. 8.4. Modified syringe-type proportional gradient elution device. R = reservoir; M = mixing chamber of variable volume (syringe); P = motor-driven flow-through plunger; PS = piston seal; S = stirrer; C = chromatographic column.

A modification of this type of gradient device consists in the use of only one pump delivering the mixed mobile phase from the mixing chamber on the column. The mixing chamber is closed and combines its role with the operation of the second pump. This is achieved using a motor-driven plunger, which can move up and down in the cylindrical mixing chamber. The plunger provides a leak-proof moveable limit of the mixing chamber volume, which is changed continuously during the elution, as the plunger moves down or up[11]. Eqn. 8.5 or 8.1 can be used to describe the gradient profile formed using this device, provided that the decrease in mixing-chamber volume per unit time is used for $Q_2 - Q_1$ and the flow-rate between the mixing chamber and the column for Q_2. A device of this type is commercially available from Analabs (North Haven, CT, U.S.A.) (Analabs concentration profile modulator), where the stronger eluting solution flows into the mixing chamber via a flow-through plunger (Fig. 8.4).

8.1.3. Other gradient devices

These are the gradient devices that make it possible to achieve different gradient shapes when using one device without auxiliary pumps. Certain devices use two vessels with two solutions from which the mobile phase gradients are formed. These are improved versions of the two-vessel proportional volume gradient devices, where various closed vessels are inserted to vary the shape of the gradient. Such is the device introduced by Brown[12] and Watt[13] (Fig. 8.5). Here, two open reservoirs, R_1, R_2, are used instead of the one reservoir and one

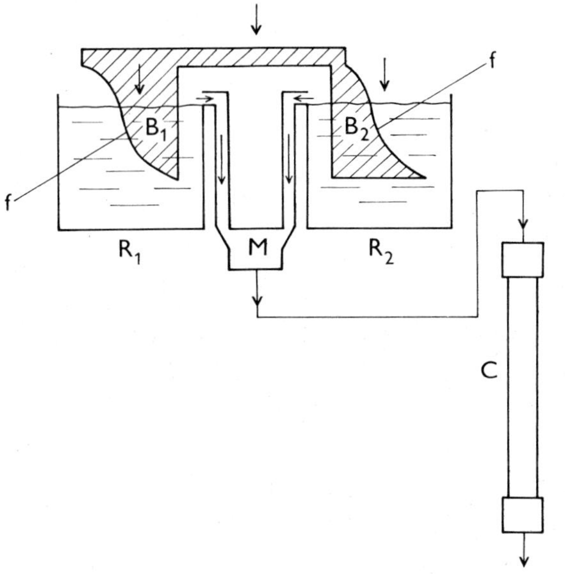

Fig. 8.5. Gradient elution device of the type suggested by Brown[12] and Watt[13]. R_1, R_2 = reservoirs with overflow tubings; B_1, B_2 = two parts of the plastic block cut according to the gradient function f, moving slowly down; M = mixing chamber; C = chromatographic column.

mixing chamber usual with proportional gradient devices. The gradient function giving the volume ratio of the solvents in the two reservoirs as a function of time is drawn on a plastic block, the block is cut in two along the curve, the two parts are fastened on two motor-driven screws and are continuously moved and immersed each in one reservoir. The liquids displaced from the reservoirs in this way overflow and are mixed together in a small-volume chamber in front of the top of the column. Another possibility is to use an orthogonal glass or plastic chamber divided vertically into two parts by means of a plastic spacer foil.

References on p. 282.

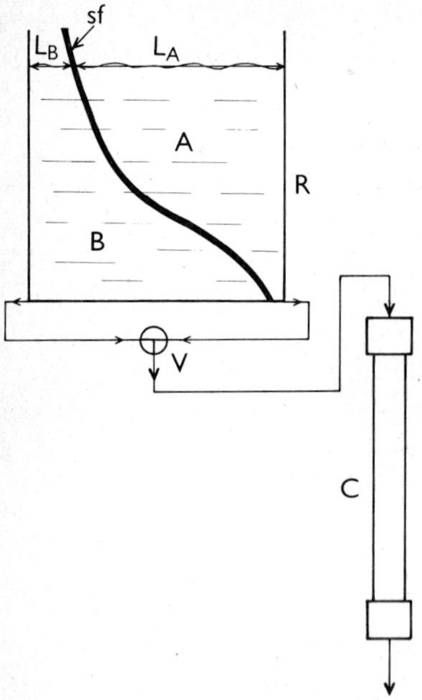

Fig. 8.6. Gradient device with rectangular chamber divided into two compartments by a flexible sealing foil. R = rectangular chamber divided by a sealing foil(sf) into two compartments (A, B) containing the components of the mobile phase; L_A, L_B = width of the two compartments at the air-liquid interface; V = valve; C = chromatographic column.

This foil provides a leak-proof seal between the two compartments which are interconnected at the bottom, but it is flexible enough to be formed and bent to provide various gradient shapes (Fig. 8.6). For two solvents of equal densities in the two compartments A and B, the actual concentration of the more efficient eluting agent b at the outlet of the gradient device is given by the ratio of the widths L_A and L_B of the two compartments at the air-liquid interface level at time t:

$$c = c_B + \frac{L_A}{L_A + L_B} \cdot (c_A - c_B) \tag{8.6}$$

where c_A and c_B are the concentrations of b in compartments A and B, respectively. This equation fails if the two solutions mixed differ in density. Commercial devices of this type are the Universal Scientific gradient former and Reeve Angel gradient former.

Of other gradient devices using more than two vessels to prepare mobile phase gradients of various shapes, the Varigrad system according to Peterson and Sober[14] is the most popular. It is composed of several cylindrical or rectangular open vessels (chambers) of equal dimensions, interconnected at the bottoms, and their contents are stirred, as with proportional volume gradient devices. The outlet from the first chamber is connected to the column (Fig. 8.7). Usually nine cham-

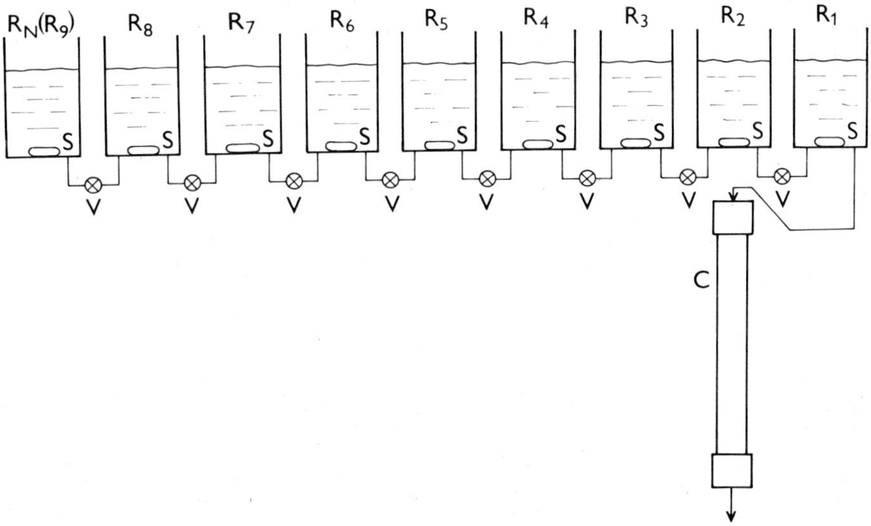

Fig. 8.7. Varigrad gradient device. R_1 - R_N (R_9) = equal-volume reservoirs (nine in original Varigrad system). S = stirrers; V = valves; C = chromatographic column.

bers are used, but it is possible to use as few as two vessels. The liquids in all of the chambers are in hydrostatic equilibrium during the gradient run and gravity drives the liquid between the chambers and through the column. In this device, a wide variety of gradient shapes, even those with concentration maxima, may be generated depending on the initial volume and composition of the solutions in the individual chambers. In practical work with the Varigrad system, often only the last chamber (the most distant from the column) is filled with the solution of the more efficient eluting component b of concentration c_R, while the remaining chambers are filled with pure less efficient eluting component of the mobile phase. In such an instance, the concentration of b at the outlet from the Varigrad system varies with time approximately according to the following equation:

References on p. 282.

$$c = c_{\mathrm{R}}\left(\frac{F_{\mathrm{m}}t}{V_{\mathrm{t}}}\right)^{(N-1)} = (BV)^{\kappa} \tag{8.7}$$

(cf., Table 4.1), where F_{m} is the flow-rate of the mobile phase at the outlet from the Varigrad system and V_{t} is the total volume of the liquid contained in N chambers at the beginning of the gradient elution (equal volumes in each chamber).

As with other gradient devices, accurate functioning of the Varigrad system is possible only with the solutions of approximately equal densities, otherwise it fails. Chase[15] reported a modification of this system using tightly closed reservoirs, which should allow work not only with solutions of different density, but also with volatile solvents, but his system has not been widely applied in practice.

8.1.4. Performance of classical gradient devices

The shape of gradients generated by the exponential devices is limited to the convex form; with proportional devices and the Varigrad system it is possible in principle to produce a wide variety of gradient shapes. The profile of the gradient is controlled by the geometry of the vessels used as the reservoirs and mixing chambers and by the composition of solutions used in these vessels, and also by the flow-rates of the liquids between the reservoir and the mixing chamber, delivered by the auxiliary pumps. These quantities have to be calculated from the equation describing the gradient profile required (using eqns. 8.1-8.7). To change the gradient profile, it is necessary to exchange the solutions in the reservoirs and mixing chambers, or to use other vessels, which is not very convenient.

The accuracy and reproducibility of the gradient profiles formed with these gradient devices are usually poor, because the flow-rate through the column and between the reservoir and mixing chambers depends on the column back-pressure, which may change with gradually changing density and viscosity of the mobile phase flowing through the column and with temperature fluctuations. Further, the performance of these devices may be disturbed by the accidental formation of gas bubbles in the connecting tubing, by a decrease in the column permeability due to the precipitation of certain sample components or impurities or due to the deposition of microparticulate material formed by erosion of the column packing, etc.

These difficulties are partially eliminated with proportional flow gradient devices equipped with auxiliary pumps, but the flow-rate of simple pumps, such as peristaltic pumps, is also influenced by variations in the density of liquids and the permeability of the column.

The main advantage of these devices is their low cost in comparison with other, more sophisticated instruments. They may be conveniently used in applications where the accuracies of the gradient profile and of the flow-rate are not of primary interest, such as in ion-exchange or affinity chromatography using large-diameter columns packed with large-particle materials.

High flow-rates can be easily achieved by virtue of gravity in such systems, which makes possible preparative or semi-preparative applications.

In principle, it would be possible to connect simple classical gradient devices at the inlet of a high-pressure reciprocating pump to give the possibility of gradient elution chromatography on modern analytical liquid chromatographic columns, requiring high operating pressures. For example the Varigrad system would offer a wide variety of gradient profiles. However, the accuracy of gradient profiles generated in such systems would generally be insufficient for the demanding requirements of precision and reproducibility of retention data and quantitative results, unless precise electronically controlled metering pumps or valve systems are used to mix the components of the mobile phase in the low-pressure part at the inlet of the high-pressure pump (see Section 8.3).

8.2. DEVICES GENERATING GRADIENTS AT A HIGH PRESSURE

8.2.1. High-pressure exponential and proportional devices

The simplest type of this device utilizes a high-pressure mixing chamber, the contents of which are stirred, between the high-pressure pump and the column[16]. In principle, this is a high-pressure exponential device and only convex exponential gradients may be formed, which are described by eqn. 8.1.

The Tracor Chromatec Model 6000 gradient elution accessory is based on this principle, but it is modified by using a high-pressure blend valve, which divides the flow of the more efficient eluting solution into two parts. One part enters the stirred mixing chamber and the other passes directly into the mixing port before the top of the column. By varying the proportions of the two flows by means of the solvent blend valve, it is possible to modify the gradient profile[19].

The principle of a high-pressure proportional device is basically the same as that of low-pressure proportional devices (Section 8.1) and was used in the Nester/Faust Model 1200 early liquid chromatograph, in which two positive-displacement (large-volume syringe) pumps are used (Fig. 8.8). Pump A delivers the more efficient eluent into the other pump, B, originally containing the weaker eluent. The contents of pump B are stirred to ensure perfect homogenization and the flow-rates of the two pumps can be controlled independently of one another by changing the speed of the motors driving the plungers in each syringe pump.

Thus different linear and curved gradients may be formed, which are controlled by eqn. 8.5 if $Q_1 \neq Q_2$ and by eqn. 8.1 (exponential gradients) if $Q_1 = Q_2$ (Q_1 is the flow-rate of pump A and $Q_2 - Q_1$ is the flow-rate of pump B).

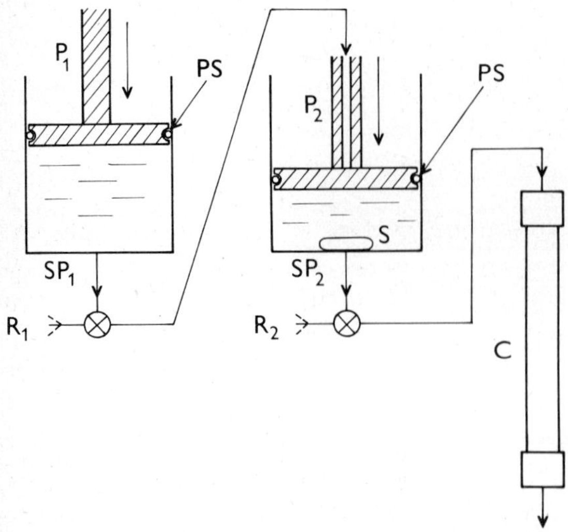

Fig. 8.8. High-pressure proportional flow device using two large-volume syringe pumps. SP_1, SP_2 = high-pressure syringe pumps; P_1 = motor-driven plunger; P_2 = motor-driven flow-through plunger; PS = piston seals; S = stirrer; C = chromatographic column; R_1, R_2 = inlets from solvent reservoirs.

In these types of pumps, both exponential and proportional, both the gradient profile and the flow-rate deviate considerably from the expected and calculated values at high pressures, owing to the compressibility of the liquid inside the mixing chamber. As the composition of the mobile phase changes, the column pressure drop and the compressibility of the liquid in the mixing chamber (or in pump B) also change, which causes the above deviations. These deviations depend on the character of the components of the mobile phase used and increase with increasing operating pressure and with increasing inner volume of the mixing chamber (or with increasing actual inner volume of syringe pump B). The influence of the compressibility of liquids on the actual gradient profile is discussed in more detail later in this section in the part dealing with the devices for preparing mobile phase gradients from the independent flow streams using two syringe pumps.

8.2.2. Devices using two (or more) independent flow streams

Far more sophisticated and more frequently used are high-pressure gradient devices, in which two pumps are used to deliver independently the solutions to be mixed in a homogenization chamber of small volume or directly in the connecting tubing before the column. Each pump operates at an independently selected flow-rate, but the sum of the flow-rates of the two pumps is constant. The ratio of the flow-rates is changed with time by means of an electronic control unit to provide the gradient profile required. The two liquids are either continuously mixed "on-stream" at the same time, or alternately allowed to flow into a mixing chamber of small volume ahead of the column.

In these systems, the electronic control unit is usually located in a module independent of the hydraulic unit where the two solutions composing the mobile phase are mixed and it is convenient to treat the hydraulic part of the system separately from the electronic part.

The performance of the hydraulic part of this type of gradient device depends on the type of pumps used. These can be classified according to their operating principles in the following order of increasing utility for the gradient elution technique:

(a) pneumatic amplifier pumps;

(b) hydraulic amplifier pumps;

(c) positive displacement (motor-driven syringe) pumps;

(d) single-head reciprocating pumps;

(e) dual-head sinusoidal reciprocating pumps;

(f) dual-head special-drive reciprocating pumps; and

(g) dual-head reciprocating pumps with flow-rate control by means of a feed back system or microprocessor-controlled reciprocating pumps.

The reciprocating pumps are of either a plunger or a diaphragm type.

8.2.2.1. Devices using pneumatic and hydraulic amplifier pumps

These pumps use a pressurized gas or liquid as the driving force to provide the flow of mobile phase through the column. The force of the pressurized medium is applied on the piston with a larger surface area, connected rigidly with the small-surface-area piston in contact with the mobile phase, which is pumped on the column at a constant pressure. This pressure is equal to the pressure of the pressurized medium amplified by a factor given by the reciprocal of the surface-area ratio (Pascal law). As the pumps operate under constant pressure, the flow-rate of the mobile phase through the column depends on the permeability of the column and on the density and viscosity of the mobile phase and changes with temperature fluctuations and with changing composition of the mobile phase. The consequence is that these pumps cannot be used to deliver simultaneously the two components of the mobile phase. An alternate flow of the two liquids must be used, controlled by two high-pressure solenoid valves, which are alternately switched on and off in short cycles. Further, a mixing chamber must be used between the solenoid valves and the

References on p. 282.

column to ensure perfect homogenization of alternately delivered liquids[17]. The mixing chamber should have an inner volume small enough not to distort the gradient profile required. The Applied Chromatography Systems Model 750 gradient elution digital programmer was designed to control two pneumatic amplifier pumps in gradient elution operation.

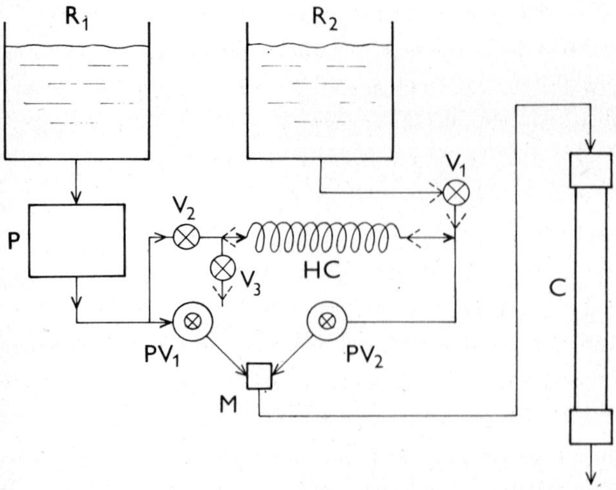

Fig. 8.9. High-pressure gradient device using a single high-pressure pneumatic pump and a holding coil. R_1, R_2 = solvent reservoirs; P = high-pressure pneumatic pump; HC = holding coil for solvent B; PV_1, PV_2 = high-pressure proportioning valves; M = mixing chamber; V_1-V_3 = valves; C = chromatographic column. Full arrows indicate the flow of the solvents during the chromatographic operation; broken arrows show the flow of solvent B in the filling cycle of the holding coil.

An ingenious modification of the gradient elution device was used in one of the first commercial gradient elution liquid chromatographs, the DuPont Model 830 gradient elution accessory[18]. A simplified schematic diagram of this device is shown in Fig. 8.9. This instrument makes use of a single pneumatic pump instead of two pumps. A holding coil of volume 140 ml is used instead of the second pump, which is filled with liquid b before the start of the experiment. Then, the high-pressure valve between the pneumatic pump and the holding coil is opened and that between the holding coil and the reservoir of liquid b is closed. The pump starts to deliver liquid a into two streams, one to the tubing with proportioning valve PV_1 and the other to the holding coil, from which the liquid b is pushed to the tubing with proportioning valve PV_2. These two solenoid valves open and close alternately in a timing cycle of 7.5 sec and let the two liquids a and b flow into the 2-ml mixing chamber prior to entering the chromatographic column. The volume proportions of the two liquids mixed are given by the ratio of the opening times of the respective proportioning valves during one time cycle. This ratio is controlled electronically to produce various linear and curved gradients.

Gradient elution devices using pneumatic pumps operating at constant pressure cannot reproduce exactly the gradient profile required, as the flow-rate changes continuously and the composition of the mixed mobile phase deviates from the preset gradient profile as the viscosity of the mobile phase changes with time. Therefore, in order to obtain meaningful quantitative results, it is necessary to use a flow feedback control, such as the DuPont 833 precision flow controller, which adjusts the pressure of the air actuating the pump so as to keep the flow-rate constant. However, even with such precautions, the actual gradient profiles can deviate from the expected curves owing to differences in the compressibilities of liquids a and b and to imperfect mixing of liquids with large differences in densities and viscosities[6].

The earlier types of Micromeritics pumps, Model 7000 and 7000B, used the principle of two hydraulic amplifier pumps creating the gradients on the high-pressure side. The pressurized liquid medium (oil) is less compressible than the air used in pneumatic amplifier pumps, but the compressibility could change the flow-rate significantly at high pressures if the viscosity of the mobile phase and the operating pressure change significantly during the gradient run. Therefore, a flow-feedback system is necessary to achieve a constant flow-rate. These systems are subject to similar errors to the pneumatic pumps and are expensive.

8.2.2.2. Devices using two positive-displacement (motor-driven syringe) pumps

The operating principle of these devices is shown schematically in Fig. 8.10. The two syringes, of large volume (200-500 ml), contain different liquids, a and b. The pistons of the pumps are mechanically driven by electric motors with variable speeds. The speed of each of the two motors is programmed with time in such a way that the sum of the velocities of the two pistons remains constant during the gradient run. So does the total flow-rate delivered by the two syringes, while the ratio of the flow-rates from the syringes is equal to the composition ratio of the two solvents in the mixed mobile phase. It is programmed with time by an electronic device to form the gradients required. Because the two liquids are mixed "on-stream", i.e., at the same time, the mixing chamber to which they are led from the pumps may have a very low volume or a capillary tee-union may even be used instead of the mixing chamber. The syringe pumps are expensive, but they deliver a very smooth, virtually pulseless flow-rate. As they have a limited capacity, it is necessary to refill them from time to time. Commercial gradient elution liquid chromatographs based on two syringe-type pumps are available from Applied Chromatography Systems (Model 750/30, using Model 750/02 pumps), ISCO

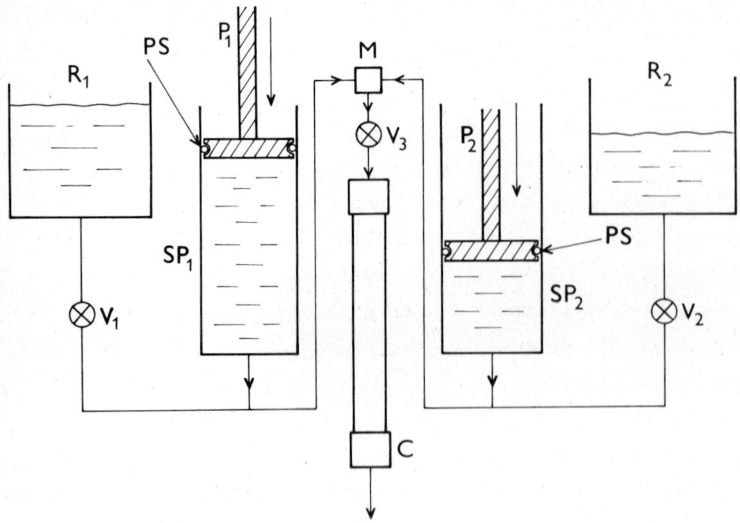

Fig. 8.10. High-pressure gradient device using two large-volume syringe pumps. R_1, R_2 = solvent reservoirs; SP_1, SP_2 = high-pressure syringe pumps; P_1, P_2 = motor-driven plungers; PS = piston seals; M = mixing chamber; V_1 - V_3 = valves; C = chromatographic column.

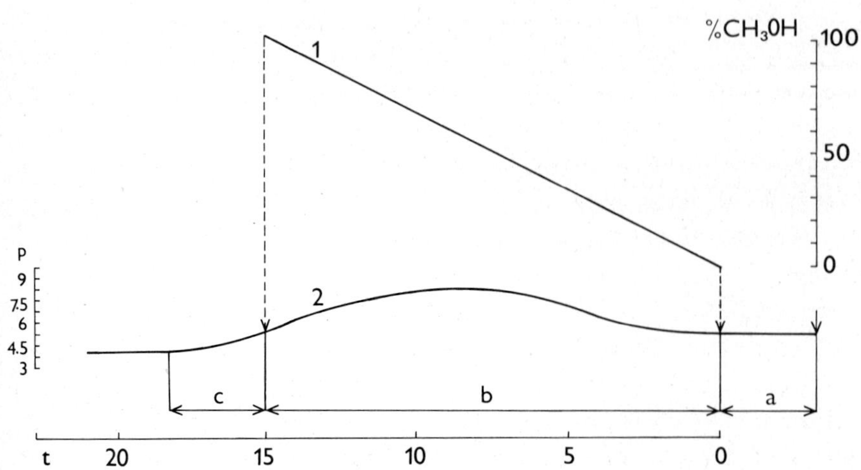

Fig. 8.11. Operating pressure profile in reversed-phase gradient elution chromatography using a linear gradient of methanol concentration in water. Column: C 18 on LiChrosorb Si 100 (10 μm), 300x4.2 mm I.D. Flow-rate of the mobile phase: 1 ml/min. 1 = Gradient profile (a 15-min linear gradient from 0 to 100% methanol in water); 2 = pressure profile. p = Pressure in MPa; t = time in min. a = Delay corresponding to the volume between the mixing chamber and the column inlet; b = gradient line; c = time corresponding to the migration of the gradient end along the column.

(Dialgrad Model 384), Perkin-Elmer (Series 1210 and 1220 Modular Analytical Liquid Chromatograph; Model 601) and Varian (Model 8500 - perhaps the best known instrument of this class).

The syringe-type instruments are generally expensive, but they are not very good gradient devices. Their poor performance is due to compressibility effects, as with high-pressure exponential and proportional devices[20-23]. Because the electronic control unit controls only the speed of the piston and not the flow of the solvent from each pump, the device can reproduce the exact profile of the gradient only at zero operating pressure.

Most of the solvents used as components of the mobile phase in liquid chromatography will be compressed by 0.5-2% of their volume on increasing the pressure by 10 MPa. The exact compressibility varies from one solvent to another and it changes to a certain extent with the operating pressure for a given solvent.

Suppose we are mixing two liquids with different viscosities and approximately equal compressibilities, such as *n*-hexane and isopropanol (these two solvents are useful in normal-phase liquid chromatography on polar adsorbents). *n*-Hexane, used as the weaker solvent a, has a viscosity of ca. 0.32 cP and isopropanol ca. 2.3 cP. During the gradient from pure *n*-hexane to pure isopropanol, the viscosity of the mobile phase increases approximately linearly with the content of isopropanol in the mobile phase and at the end of the gradient it is ca. seven times as high as at the start of the gradient run. The pressure drop across the column increases in the same ratio and the difference in operating pressure between the end and the start of the gradient may be as high as 10-15 MPa under normal working conditions with standard analytical columns. The compressibilities of the two solvents are approximately 1.5% by volume per 10 MPa. Thus, starting with 200 ml of each solvent in the corresponding pump and assuming a 10 MPa pressure difference between the end and the starting conditions, the volume of the solvent in each pump is reduced by about 3 ml. This means that the corresponding part of the forward movement of the piston was lost by this compression and the actual volume of the gradient delivered by this gradient device is lower by ca. 6 ml than the expected value. With a gradient volume of 30 ml (a 30-min gradient at a flow-rate of 1 ml/min), this corresponds to as much as 20% of the total volume of the mobile phase. Necessarily, both the concentration profile and the flow-rate would deviate considerably from the pre-set values, i.e., the steepness of the gradient would be ca. 20% higher and a negative flow-rate gradient will be produced simultaneously.

The situation is even worse if solvents with different viscosities and compressibilities are used. This is the situation in reversed-phase chromatography using water-acetonitrile and water-methanol gradients. Here, the viscosity decreases from ca. 1 cP for water to ca. 0.37 cP for acetonitrile, and if methanol (0.6 cP) is mixed with water a viscosity maximum (1.4 cP) is observed at ca. 40-50% of

References on p. 282.

methanol. Generally, organic solvents are more compressible than water. This means that with changing operating pressure (decreasing or following a curve with a maximum, Fig. 8.11), the organic solvent is more decompressed than water if the pressure drops during the gradient (and vice versa). Consequently, complex flow-rate gradients combined with significant deviations from the expected concentration profile result and, under certain circumstances, one solvent may even flow into the syringe pump containing the other solvent.

As can be understood from the above example, the magnitudes of the errors in flow-rate and in the composition of the mobile phase depend not only on the nature of the solvents used and on the operating pressure, but also on the volume of the liquids in the syringe pumps. This means that these deviations are irreproducible from run to run, unless the initial volumes in the pumps are always set at the same values. Consequently, irreproducible retention times (volumes) and erroneous quantitative data result. The errors in the flow-rate and the concentration profile can be approximately estimated from the initial volumes of the liquids in the two pumps, with known compressibilities of the liquids and the dependence of viscosity on the composition of the mobile phase[22].

Reversal of the liquid flow from one syringe pump into the other pump may be avoided using outlet check-valves for each pump[22]. The influence of the compressibility effects on the profile of gradient and on the flow-rate can be suppressed to a considerable extent by using a pressure back-valve between the mixing chamber and the column, which allows the two pumps to be operated at constant pressure, independent of the column operating pressure[21]. Unfortunately, this valve is not used in commercial gradient devices based on two syringe pumps, as far as we know.

An interesting possibility of gradient formation for gradient elution on microbore columns (i.e., columns of I.D. 0.5-1.0 mm) was suggested which makes use of two positive-displacement (syringe) micropumps (250 μl capacity each) in connection with the principle of the proportional gradient device[24]. The pistons of the two equal dimension syringe pumps are driven in such a way that, in syringe 1, the plunger moves down and pushes the liquid into the stirred open vessel (beaker) with the second component of the mobile phase, b. At the same time, the plunger in syringe 2 moves up and sucks the liquid from the beaker into this syringe via capillary tubing with a total inner volume of ca. 200-250 μl (0.5 mm I.D., 1 m long). After the end of this step, this capillary contains the total volume of mobile phase necessary to achieve the separation on a microbore column 5-30 cm long. This means that the whole "stock" mobile phase gradient is contained in this capillary, which is then connected to the microbore column and, after application of the sample, the syringe micropump 2 is used to push the prepared mobile phase gradient through the column. Of course, the accuracy and reproducibility of the gradient profile formed in such a way are not very high, but this simple method makes it possible to apply gradient elution at a low flow-rate (ca. 10 μl/min with 0.5 mm diameter columns) necessary in work with microbore columns[24].

8.2.2.3. Devices using two reciprocating pumps

In these devices, two (or more) reciprocating pumps are used to mix the solvents in the high-pressure part in a small-volume mixing chamber or in a capillary tee-union connected to the top of the chromatographic column (Fig. 8.12). These pumps deliver the liquids from reservoirs of unlimited volume and thus need not be refilled periodically. To form gradients of various shapes, the ratio of the flows from the two pumps is changed with time, while the sum of the flow-rates from the individual pumps remains constant during the gradient run.

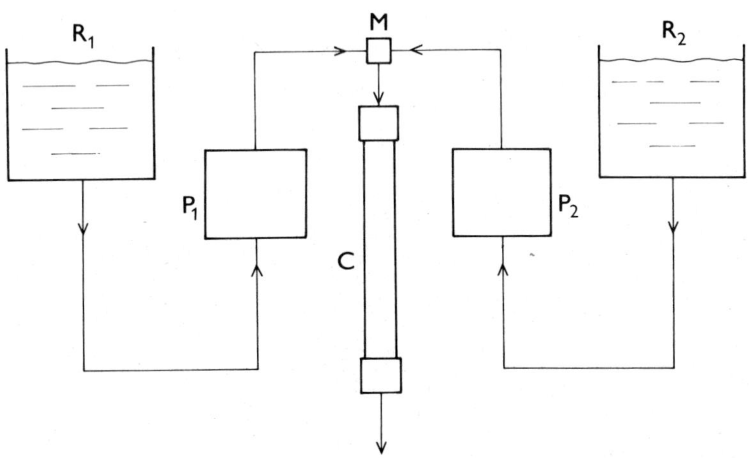

Fig. 8.12. High-pressure gradient device using two reciprocating pumps. R_1, R_2 = solvent reservoirs; P_1, P_2 = high-pressure reciprocating pumps; M = mixing chamber; C = chromatographic column.

Two methods may be used to control the flow ratio from the two pumps. The first method works with a constant driving motor (and consequently plunger) velocity and the volume of the liquid delivered by each pump in a single stroke is changed according to a pre-set programme. This is achieved by using an additional motor to adjust the stems limiting the path through which each piston moves back and forth during a single stroke in each pump. The motor is controlled using a programmable electronic device and a special gear system ensures that the sum of the piston paths is constant and only the ratio of the paths in each pump changes with time according to the selected gradient function[25,26].

The other way, which is more frequently used in commercial instruments, is to operate with constant volumes delivered in each stroke, but to change the velocities of the plungers by means of the electronic (frequency) control of the rotation of driving stepping motors used in each pump.

Gradient devices of this type are less subject to the errors originating from compressibility effects than the syringe-type devices, but the influence of compressibility of liquids at high pressures is not fully eliminated and these instruments suffer from other problems, originating from the operating principle of reciprocating pumps[20].

In gradient devices using two (or more) pumps to mix the components at high pressure, it is necessary for the two pumps to be very accurately matched together. Each pump should deliver exactly the same flow of liquid as the other, if operated independently at the same flow-rate. If not, the flow-rate decreases or increases continuously during the gradient run, as indicated later in Fig. 8.14, and the actual concentration gradient deviates from the pre-set profile[27].

The core of a reciprocating pump (the pump head) consists of a small-volume chamber (usually 0.05-0.1 ml), connected to inlet and outlet high-pressure check valves. A motor-driven piston moves back and forth in the chamber and alternately fills the chamber with a liquid during the retract period and delivers the liquid from the pump during the forward period. Thus only half of each piston cycle is actually used to deliver the liquid. The motor drives the piston by means of a circular excentric cam, so that the piston velocity is zero at the extreme positions in the chamber and is maximum in the middle position. Consequently, a sinusoidal piston position time profile results (Fig. 8.13.A, a) and the resulting flow-rate, which is given by the slope of the piston position time profile, is half-sinusoidal (Fig. 8.13.A, b, c). Thus, single-head reciprocating pumps operate with large periodic flow (and pressure) pulsations, which should be damped. An interesting possibility of damping the flow pulsations is used in the ACS 300 pump, where a reciprocating piston operates in a single pump head at a rapid and fixed speed of 23 cycles per second, which is sufficient to damp the flow pulses without additional flow dampers in the system. However, cavitation may give rise to flow problems with certain solvents if the piston operates at a high speed.

In dual-head reciprocating pumps, the same motor drives two pistons in two pump heads of an identical design. The movement of the pistons is 180° out of phase, i.e., during the retract motion of one piston the other moves forward and delivers the liquid. Thus, the pump delivers the liquid in the two half-cycles. However, as is illustrated in Fig. 8.13B, a, b, c, the flow-rate still oscillates between the maximum value and zero (at the moment of the piston changeover). This pulse flow profile can be smoothed by using a third pump head in a triple-head pump, where the heads operate 120° out of phase with each other.

However, it is possible to smooth the flow profile of a dual-head pump in a more economical and efficient way by using a pair of specially shaped cams to transduce the motor rotation to the movement of the pistons in such a way that the flow profiles from the two pumps are superimposed during a part of the cycle, which results in an essentially constant flow from the pump, as shown in Fig. 8.13C, a, b, c (in reality, a short-term flow decrease is observed at the moment of the piston changeover owing to the inertia of the check valves, as shown in Fig. 8.13C, c). When working at a high pressure, this flow ripple becomes more significant owing to the compressibility of the solvent in the piston chamber and the short-term flow decrease may amount to about 20% of the maximum flow[28]. Most commercial reciprocating pumps are based on this principle.

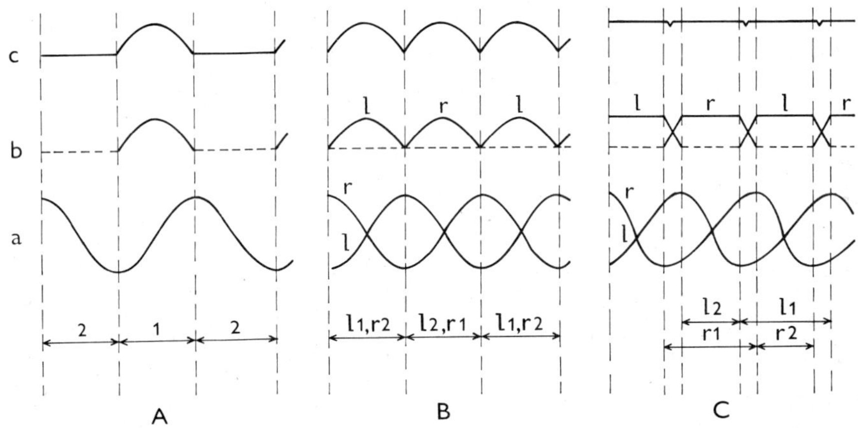

Fig. 8.13. Piston position and flow profiles of a reciprocating pump. A = single-head pump; B = dual-head, sinusoidal movement pump; C = dual-head pump using specially shaped cams. a = Piston position profiles; b = flow-rate profiles of the individual piston heads (l = left; r = right); c = composite flow-rate profile delivered by the pump. Piston movement periods: 1, piston advances; 2, piston retracts (l = left piston; r = right piston).

The dual-pump gradient devices can usually be easily adapted for work with microbore columns, with which much lower flow-rates are used than with standard analytical columns. For this purpose, it is necessary to decrease electronically the speed range of the motors driving each of the reciprocating pumps; For example, a ten-fold decrease would yield a total flow-range of 0.01-0.99 ml and would make possible gradient elution with a flow-rate of 0.04-0.1 ml/min, which is adequate with microbore columns. However, special attention must be paid to the mixing of the two liquid streams in order to minimize the mixing volume. A mixing T-union was described for 0.01 mm I.D. tubing, which has a mixing volume of less than 2 μl[29]. Microbore-column gradient elution chromatography is possible, for example, with commercially available Waters equipment using two M 6000 A reciprocating pumps [or one M 6000 A and one M 45 (G)] controlled by an M 680 or by a 720 microprocessor.

References on p. 282.

The compressibility effect affects the performance of the reciprocating pumps to a lesser extent than with syringe pumps but, as has just been mentioned, it cannot be neglected. The flow delivered by the reciprocating pump decreases with increasing operating pressure.

The reason is that a certain fraction of the stroke volume is required to compress the liquid in the pump chamber until it reaches the operating pressure and the outlet check valve of the pump head opens. Consequently, the liquid volume delivered by the pump head is equal only to the remaining fraction of the stroke volume.

The magnitude of the flow decrease is determined by the solvent compressibility and by the ratio of the stroke volume to the total volume of the piston chamber of the pump (the maximum liquid volume in the pump chamber), and it should be reproducible from run to run as long as the column permeability does not change.

Some reciprocating pumps are equipped with a potentiometer which makes it possible to eliminate the compressibility effects by speeding up the stepping motor by an increment proportional to the operating pressure, so that the correct volume is delivered over the entire range of the operating pressure[30]. The amount of this correction should be found by a trial-and-error method for each solvent. This method, however, is not exact because it cannot account for the dependence of compressibility on pressure, but this has a relatively minor effect. This compressibility control is used, e.g., with the Waters M 6000 and M 6000 A pumps and the Altex 110 A pump and makes it possible to adjust the correct average solvent flow. However, this compressibility control does not eliminate the short-term ripple due to the compressibility effects, which may cause errors in the quantitation of the peak areas and, to a lesser extent, of the retention times.

A more efficient way of controlling the delivery of a reciprocating pump is to use a flow feedback system[31,32]. This is most often achieved by measuring the pressure drop across a restrictor, such as a piece of capillary tubing or a small column packed with glass beads. This pressure drop, which is proportional to the average flow-rate through the restrictor, is compared with the reference calibration value and an error signal is created and used to control the speed of the motor driving the pump. The flow feedback control is a very efficient means of damping the flow and pressure oscillations produced by the reciprocating pumps. It also makes it possible to eliminate the deviations of the flow-rate from the pre-set value caused by the solvent compressibility effects and to maintain not only a good average flow-rate, but also to correct the regular short-term flow decrease (flow ripple). Therefore, it is indispensable for modern instruments using single-head reciprocating pumps and very useful for those using double-head pumps.

The Beckman (Altex) Model 332 microprocessor-controlled gradient system makes use of two single-head piston reciprocating pumps (Model 110 A). Here, the electronic pulse compensation circuitry reduces pressure pulsations at high pressure by speeding up the pump motor during the refill and part of the compression periods[30].

Several commercial systems, such as the Siemens S 112, Orlita WKE 500-3, Touzart et Matignon Chromatem 38 and Hewlett-Packard 1084, use a dual-head reciprocating diaphragm pump in which each pump head delivers a different solvent to form concentration gradients at high pressure. These instruments are essentially based on a two single-head reciprocating pump principle. Without a flow feedback control, these pumps would deliver a sinusoidal flow as shown in Fig. 8.13A. The compressibility effects are far more significant with diaphragm reciprocating pumps than with piston reciprocating pumps. In a diaphragm pump, the piston moves in an oil chamber, which is separated from the solvent chamber by a thin diaphragm, usually made of stainless steel. As the piston moves forward, the oil in the oil chamber is pressurized and displaces the diaphragm, which pushes the liquid from the solvent chamber equipped with an inlet and outlet check-valve system. During the backward motion of the piston, the diaphragm returns to its original position and the solvent chamber is refilled. With a diaphragm pump, part of the piston forward stroke is used to compress not only the liquid in the solvent chamber but also the oil in the oil chamber to a volume corresponding to the operating pressure. As the sum of the volumes of the oil and solvent chambers is greater than the volume of the piston chamber in piston reciprocating pumps, the flow delivered by the diaphragm pumps at high pressures is decreased much more by the compressibility effects than the flow from the piston pumps. Therefore, all gradient devices using diaphragm pumps utilize a flow feedback control, which is usually based on the measurement of the pressure difference across a flow restrictor[6].

The Hewlett-Packard Model 1084 liquid chromatograph employs a different method of pump flow control. This is based on a hydraulic capacitor, which is an elastic vessel that changes its volume, V_c, if the operating pressure, p_s, changes. The flow-through capacitor is designed in such a way that there is a linear relationship between V_c and p_s. By measuring Δp_s during a time period Δt, the liquid volume ΔV_c delivered on to the column during this time period can be measured as long as the capacitance $C = \Delta V_c/\Delta p_s$ is independent of the operating pressure, p_s. Δp_s is measured by integrating the pressure over the refill part of the pump cycle and the flow-rate F_m is calculated in the electronic part of the instrument as

$$F_m = \frac{2C}{\Delta t^2} \cdot \int_0^{\Delta t} p_s \mathrm{d}t \qquad (8.8)$$

References on p. 282.

Owing to the design of the capacitor, C is almost independent of the nature of the liquid and temperature, so that it is not necessary to re-calibrate the measurement on changing the conditions.

The measured value is compared with the pre-set flow-rate and any difference is compensated for by adjusting the volume of the liquid delivered during a single stroke. This is achieved using a servo-motor, which adjusts the position of the pump head (piston chamber with check-valve systems) with respect to the piston, while the velocity of the piston is constant. In the Model 1084 liquid chromatograph two such pumps are used, each controlled independently by a microprocessor, which changes the pre-set flow-rate of each pump according to the pre-selected multilinear gradient programme, consisting of a set of linear sequences[32].

Gradient devices using two single-head pumps deliver each component of the mobile phase at different times, i.e., each pump head delivers the solvent during the refill period of the other head. Therefore, a high-pressure mixing chamber should be used to ensure good mixing of the mobile phase components. The content of the mixing chamber is usually stirred electromagnetically (dynamic mixing chamber), which enables a mixing chamber with an inner volume, V_c, of 1-2 ml to be used. Alternatively, a static chamber (an empty tube with a ca. 2:1 ratio of length to diameter) or capillary laminar-flow blending narrow-bore tubing of volume 1-2 ml is used[106].

It can take a considerable amount of time to sweep the mixing volume to achieve the desired mobile phase composition at the outlet from the mixing chamber after each change in the inlet composition. A dynamically stirred mixing chamber behaves in the same way as the mixing chamber in exponential gradient devices (with a much lower inner volume, of course) and the time dependence of the mobile phase composition at the outlet of the mixing chamber can be accordingly described by eqn. 8.1, where c_0 is the original concentration of solvent b in the mixing chamber at the moment when it is changed to c_R according to the gradient programme. A low volume of the mixing chamber is essential in order to achieve a rapid response to programmed compositional changes and not to distort the actual gradient profile. The time t necessary to change the original concentration c_0 in the dynamic mixing chamber to the actual concentration c_t may be calculated as[30]

$$t = \frac{V_c}{F_m} \cdot \ln \frac{c_0}{c_t} \tag{8.9}$$

where F_m is the flow-rate of the liquid flowing through the mixing chamber. However, the volume of the mixing chamber should be larger than the sum of the volumes of the piston chambers of the two pumps used. A good practical compromise used in most instruments is to have mixing chambers of inner volume 1-2 ml, where the original content is exchanged from 95% within 1-3 min at a flow-rate of 1-2 ml/min. However, if steep gradients are used, the profiles are flattened at the beginning and at the end. Of course, the mixing chambers would also introduce the corresponding gradient delay.

Gradient devices using two reciprocating pumps driven by stepping motors are subject to compositional errors at those parts of the gradient where the composition is close to either 100% of solvent a or 100% of solvent b. This is caused by the fact that the reciprocating stepping motor driven pumps are flow-rate limited because the stepping motors have an accurate speed range of only about two orders of magnitude. Then, if the maximum flow-rate is 10 ml/min, as with most commercial reciprocating pumps, the minimum accurate flow-rate is ca. 0.1 ml/min and, moreover, the pumps do not deliver the liquid at all at a flow-rate lower than 0.03-0.04 ml/min[6,20,23].

On going from 100% solvent a to 100% solvent b in gradient elution chromatography at a given flow-rate, the pump delivering solvent b speeds up slowly to provide an increased contribution to the total constant flow-rate, while at the same time the pump delivering solvent a slows down until it stops completely at 100% b. However, the pump delivering solvent b starts to deliver the liquid first when the gradient profile corresponds to ca. 3-4% of solvent b (at a total flow-rate of 1 ml/min). Then, a sudden compositional jump follows and the pump still does not work correctly until ca. 10% of solvent b has been delivered. The same is true at the other end of the gradient. Therefore, it is recommended that these devices be used for gradient elution within the range 10-90% of solvent b rather than over the full composition range.

It is possible to overcome this problem by using pulse-width-modulated signals in the gradient system controller. Here, the pump delivering a low flow-rate is controlled by the microprocessor in such a way that it works only for part of the pulse cycle at a piston speed that ensures mechanically correct operation and it is stopped for the remainder of the cycle[30]. This makes it possible to operate the two reciprocating pumps even at the parts of the gradients where less than 0.5% of solvent a or b should be delivered. Such a control system is used in the Altex (Beckman) 332 and 322 gradient elution instruments, where the microprocessor automatically begins to generate pulse-width-modulated signals to control either of the two pumps whenever its flow-rate drops below 0.1 ml/min.

Other flow and compositional errors common to all gradient devices generating gradients at a high pressure originate in the thermodynamic volume changes (volume contractions) connected with the mixing of the components of the mobile phase[27]. The volume contractions can be significant, especially if an organic solvent is mixed with water. For example, the volume contraction in 50% aqueous methanol represents ca. 4% relative, which means that the volume of the mixed mobile phase

References on p. 282.

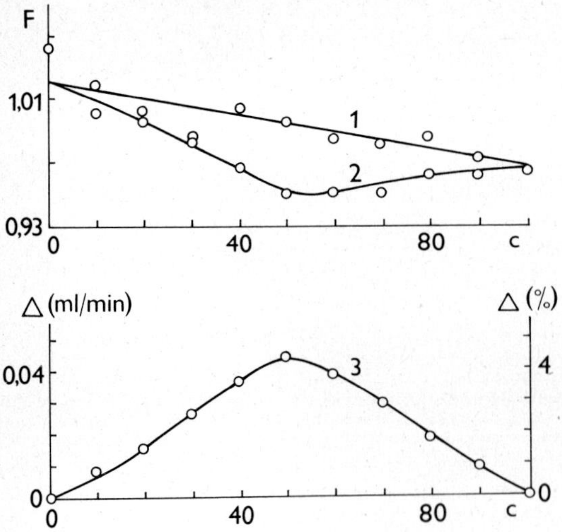

Fig. 8.14. Dependence of the flow-rate (F) on the pre-set composition of the mobile phase [c, in % (v/v) of component B] in systems with mobile phase composed of water (component A) and methanol (component B) using a high-pressure gradient device with two reciprocating pumps. Gradient device: Waters M 6000 + M 6000 A high-pressure reciprocating pumps controlled by Waters M 600 solvent programmer. Curves: 1 = $F_A + F_B$ versus c; 2 = F_C versus c; 3 = Δ versus c. F_A and F_B = flow-rates at the outlet of the individual pumps delivering components A and B, respectively; F_C = actual total flow-rate at the outlet from the gradient device; $\Delta = F_A + F_B - F_C$. The systematic shift of $F_A + F_B$ with c is due to inaccurate fit of the two pumps in the device used. (Reproduced from ref. 27 with permission.)

is significantly lower than the sum of the volumes of the individual mobile phase components (Fig. 8.14). Consequently. lower flow-rates result at intermediate compositions of the mobile phase, i.e., in the middle part of the gradient, than at the beginning and at the end of the gradient and the actual composition of the mobile phase is consequently also changed with respect to the pre-set gradient profile. These deviations are reproducible from one run to another. Gradient devices of this type, where only the flows delivered by the individual pumps are controlled, do not provide the possibility of eliminating the errors due to volume contractions.

8.2.3. Performance of high-pressure gradient devices

The use of an individual pump for each mobile phase component in the more sophisticated types of these devices is the reason for the relatively high price of the instrumentation. Usually, only two-component gradients are possible with this class of instruments, but certain electronic gradient controllers or micro-processors have a provision that enables a third high-pressure pump to be connected to form gradients from three liquid components. In principle, it is possible to form any required gradient profile using an adequate control unit for the two pumps (see Section 8.4). The devices using two reciprocating pumps are superior to those using two large-volume syringe pumps, where large deviations of the actual flow-rate and gradient profile from the pre-set values may result owing to the liquid compressibility effects.

The devices using two reciprocating pumps offer an unlimited reservoir capacity and the compressibility effects are less important. These effects, however, may still be significant and may lead to erroneous gradient profiles (and flow-rates). Therefore, it is advisable to use devices equipped with compressibility correction possibilities or, better, with flow feed-back control. However, the feed-back control of the flow-rates of the individual pumps cannot fully eliminate the deviations of the actual gradient profile (and of the total flow-rate) from the expected values owing to the thermodynamic contractions connected with the mixing of the liquids.

The devices using two reciprocating pumps driven by two specially formed cams (Fig. 8.13C) are superior to those using sinusoidal double- or single-head pumps (Fig. 8.13A, B), because they yield much more steady flows and usually do not require further hydraulic flow and pressure dampers. These dampers have relatively large internal volumes (and often off-line "dead" spaces), which would contribute to the gradient delay and may distort the shape of the gradient formed. The damping problems do not exist with the devices using two large-volume syringe pumps.

The problems connected with the poor performance of reciprocating pumps at very low flow-rates lead to systematic errors in gradient profiles at the beginning and end of the gradient. The most sophisticated instruments overcome this problem by using pulse-width-modulated signals to control the reciprocating pumps in the low flow-rate region.

The reciprocating pumps usually require good degassing of the mobile phase components and may fail to deliver certain highly volatile organic solvents (cavitation problems).

It is essential that the two high-pressure pumps used in the gradient device be mechanically fitted to one another to deliver exactly the same flow-rates of the liquid when used separately under identical conditions. Sometimes (rarely), wear particles of seal material can be flushed into the outlet check valves, where they can cause malfunctions. More frequent are inlet check valve malfunctions due to the microparticulate material in the solvents, which should therefore be carefully filtered (e.g., with 0.5-μm filters). The use of filtered solvents is very helpful in improving the performance of reciprocating pumps.

With appropriate precautions, the functioning of the mechanical parts of the reciprocating pumps is usually sufficiently reliable and does not lead to significant errors in flow-rates and gradient profiles.

8.3. DEVICES GENERATING GRADIENTS AT A LOW PRESSURE AT THE INLET OF THE HIGH-PRESSURE PUMP

Devices of this type make use of a single high-pressure pump to deliver the mixed mobile phase on to the column. Only high-pressure pumps of the reciprocating type can be used. The components of the mobile phase are mixed at a low pressure in a mixing chamber connected to the inlet of the high-pressure pump. Obviously, the number of components of the mobile phase can be increased easily.

In the simplest type of these devices, classical gradient elution devices, such as exponential, proportional or Varigrad gradient formers, are connected to the inlet of the high-pressure reciprocating pump. The principle of operation and performance of classical gradient formers is described in Section 8.1. The errors of these classical gradient formers originating from random or systematic changes in the column operating pressure are essentially eliminated by use of a high-pressure reciprocating constant-flow pump between the gradient former and the column. However, this does not reduce the errors in the mixing of the mobile phase components caused by different solvent densities, the relatively large-volume connecting tubing and valves and irregularities in the vessels used (usually common or slightly modified laboratory glassware).

The main advantage of these gradient elution devices consists in their low price, they may be obtained (or assembled) at a cost only slightly higher than that of a single reciprocating pump. The precision and reproducibility of the gradient profile formed in this type of devices are usually too low for analytical high-pressure applications using columns of standard dimensions[33], but their use may be adequate in applications using ion-exchange and other columns of larger diameters and lengths, mainly if accuracy of the retention data is not of primary interest.

For most analytical gradient elution applications, precision metering instrumentation is required to mix the components of the mobile phase in the low-pressure part of the gradient elution device. Precision metering of the solvents in the low-pressure parts may be achieved either by using alternate flows of the individual mobile phase components or by mixing these components at the same time, "on-stream", in analogy with high-pressure gradient elution devices using two (or more) independent flow streams.

8.3.1. Devices using alternate flows of the individual mobile phase components

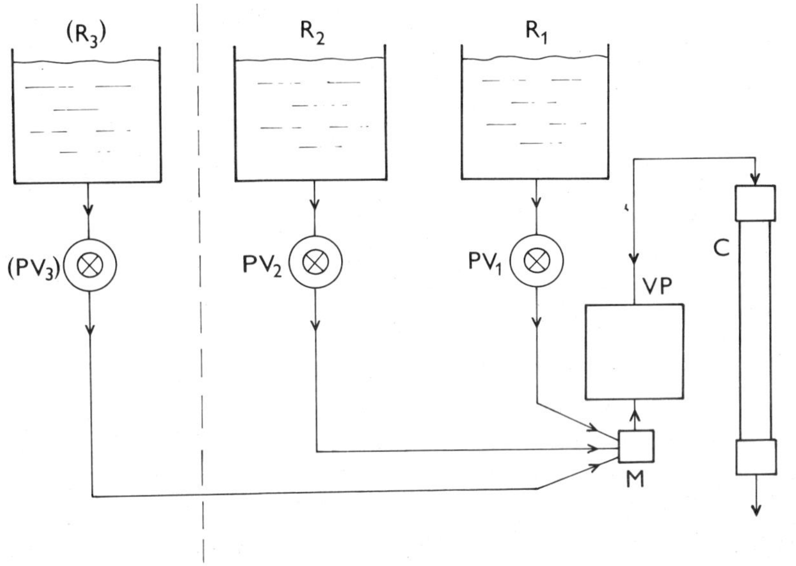

Fig. 8.15. Gradient device using alternate-flow mixing of the mobile phase components in the low-pressure part at the inlet of the high-pressure pump. R_1, R_2, R_3 = solvent reservoirs; PV_1, PV_2, PV_3 = time-proportioning electrovalves; M = mixing chamber; VP = high-pressure reciprocating pump; C = chromatographic column. If the parts denoted by letters in parentheses are used, three solvents can be mixed; otherwise the mobile phase is prepared from two components.

The operating principle of this type of gradient elution device is shown schematically in Fig. 8.15. The solvents fed to the high-pressure pump from the individual reservoirs are passed through time-proportioned electrovalves before they reach the mixing chamber at the inlet port of a high-pressure reciprocating pump. The composition of the mixed mobile phase is determined by the ratio of the switching period of the individual proportioning valves. This ratio is programmable with time and is controlled by an electronic programmable unit or by the central microprocessor of the liquid chromatograph[6,27,34-39]. Commercial instruments of this type prepare the mobile phase gradients from two or three (usually an optional possibility) components and make use of either a single-head reciprocating pump (Varian LC-5000 Series, Models 5060 and 5061), or more often of a dual-head reciprocating pump with pistons driven by specially shaped cams (Fig. 8.13C) such as the Micromeritics 7500 liquid chromatograph, the Spectra Physics SP-8000 or the Pye Unicam LC-XP liquid chromatograph using an LC-XPD pump in connection with an LC-XP gradient programmer.

References on p. 282.

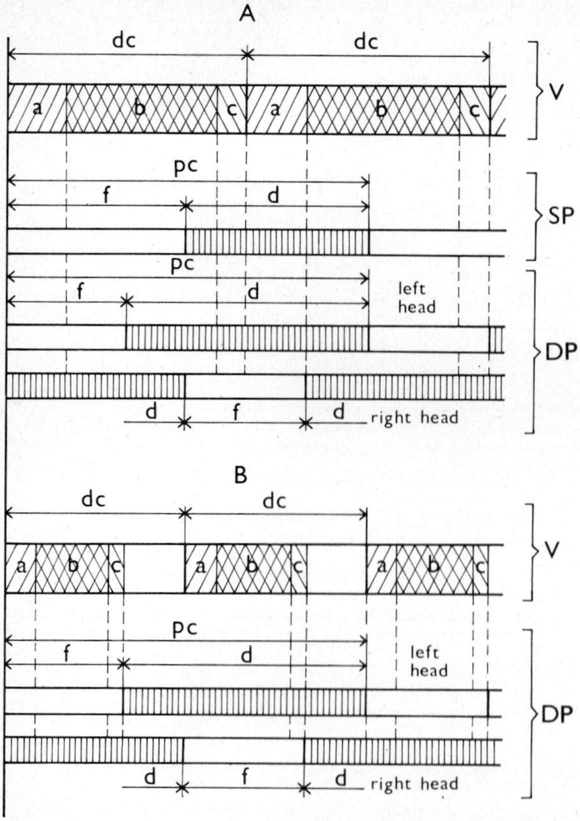

Fig. 8.16. Schematic diagram of duty cycle (dc) and pump cycle (pc) in gradient devices using alternate-flow mixing of the mobile phase components at the inlet of the reciprocating pump. Isocratic performance, ternary mixture a-b-c (2:5:1). A = non-synchronized duty and pump cycles; B = synchronized duty and pump cycles; V = proportioning valve working cycle-delivery periods of solvents a, b, c (volume ratio 2:5:1). SP = reciprocating single-head pump working cycle; DP = reciprocating dual-head pump working cycle. d = Delivery period (pump head inlet valve closed); f = filling period (pump head inlet valve open).

The total time of the solenoid valve on-off cycle is called the duty cycle (Fig. 8.16). On the other hand, each piston head of a reciprocating pump works with its own periodic solvent delivery-filling cycle (pump cycle). If the duty cycle of the valves is not synchronized with the pump cycle, serious oscillations of the mobile phase composition may occur because the components of the mobile phase enter the pump head only during the filling period and the opening of the proportioning valve has no effect during the solvent delivery period, when the pump head inlet check valve is closed[37,39]. The result is the oscillation frequency, f_0, which is equal to the difference in the pump frequency and the proportioning valve frequency:

$$f_o = 1/T_p - 1/T_c \tag{8.10}$$

where T_p is the time of the pump cycle and T_c is the time of the duty cycle[37].

It would seem that the composition of the mobile phase delivered by the pump would be correct if a dual-head pump is used or if the time of the pump cycle is twice that of the valve duty cycle, as the pump inlet check valves would be opened to each solvent during the whole duty cycle. In reality, this is not the case because most reciprocating pumps make use of specially shaped driving cams and consequently the filling period of each pump head is considerably shorter than its solvent delivery period (Fig. 8.16A). An example of flow oscillations that may occur with a dual-head reciprocating pump is demonstrated in Fig. 8.17.

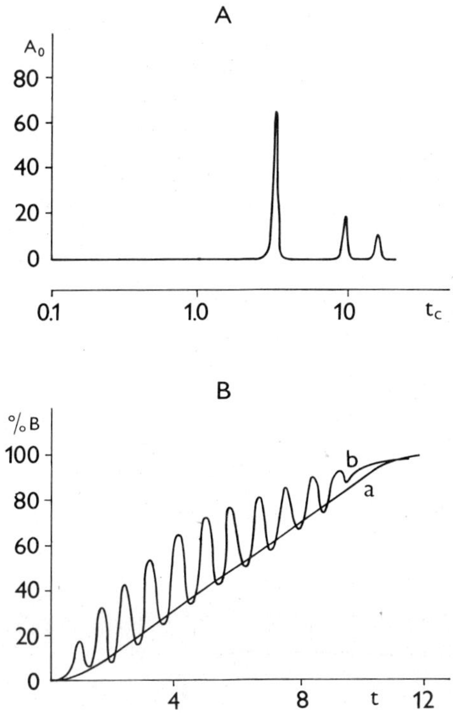

Fig. 8.17. Examples of concentration oscillations in non-synchronized devices using proportioning valves to mix solvents at the inlet of the high-pressure reciprocating pump. (A) Isocratic operation: t_c = Time of duty cycle, sec; A_o = amplitude of concentration oscillations in % of solvent B. (B) Gradient elution operation: 10-min linear concentration gradient of solvent B; (a) without concentration oscillations; (b) with concentration oscillations in a non-synchronized system with small mixing chamber; t = time (min). (Adapted from refs. 37 and 39.)

When the pump cycle is much longer than the duty cycle, the concentration oscillations should disappear because all solenoid valves open and close many times during the filling period of the pump. However, it is not possible to decrease the duty cycle much below 1 sec, because errors in solvent composition can occur at short duty cycles owing to the finite time it takes to open and close the proportioning valve (e.g., the delivery of solvent b in mobile phase containing 1% of b requires 10 msec at a duty cycle of 1 sec).

The concentration oscillations may be considerably smoothed if a mixing chamber is inserted between the proportioning valves and the reciprocating pump[37]. The volume of the mixing chamber should be large enough to allow for good mixing of the solvents delivered in a number of duty cycles (e.g., 2 ml). A number of commercial gradient liquid chromatographs of this type make use of such a mixing chamber, usually stirred. On the other hand, the mixing chamber would lead to a decrease in the capability of the instrument to reproduce exactly rapid minor concentration changes, if these are required. The gradient profiles would be distorted and flattened to a certain extent and the gradient delay of the instrument would be increased.

The column itself contributes to damping of concentration oscillations to a certain extent.

Obviously, the most efficient means of eliminating the concentration oscillations is to synchronize the proportioning valves and the pump in such a way that the duty cycle coincides exactly with the filling period of each individual pump piston head[39] (Fig. 8.16B). The movement of the piston itself can be used to trigger the valve switching period, so that the device would be automatically adapted to all flow-rates, or the job can be done by the microprocessor. Such a synchronization is used in Micromeritics 7500 liquid chromatograph or the Pye Unicam PU 4001 liquid chromatograph. Here, the volume of the mixing chamber at the pump inlet can be decreased, but still a considerable volume is necessary to ensure good mixing of "slugs" of the individual solvents before they reach the column.

However, the simple synchronization of the valve duty cycle with the pump cycle on a time basis can give rise to serious inaccuracies in the composition of the solvent at the outlet from the high-pressure pump, in spite of suppression of concentration oscillations. These inaccuracies originate from the changing velocity of the piston movement with time during the refill period. Therefore, if the duty cycle corresponds to the refill period of one piston chamber, the resulting composition of the mobile phase in this chamber does not correspond to the ratio of the valve switching times, because of the different average piston velocities during the opening time of each valve. As a remedy, the synchronization of the duty cycle with the pump cycle should not use a time basis, but the position of the piston in the chamber. In this way, accurate solvent ratios in both isocratic and gradient elution can be achieved[40].

The reciprocating pump should be designed in such a way that additional pulse dampers could be avoided or small-volume flow-through dampers are used. This is achieved using specially driven dual-head or triple-head pumps. An interesting possibility is represented by the Varian LC 5000 liquid chromatograph with a single-head reciprocating pump[34]. Here, the stepping motor is accelerated to its maximum speed during the filling period of the pump, so that the 90-μl pump chamber is filled within 0.2 sec. This is a very short time relative to the delivery period, where the motor speed is constant and depends on the flow-rate delivered. Thus, at a flow-rate of 1 ml/min the delivery period is 5.4 sec, i.e., 96% of the pump cycle. Therefore, the flow-rate is nearly constant and is interrupted only with short fill spikes, which are damped in a special flow-through damper.

The pump inlet check-valve is replaced by a mechanically controlled inlet valve that is synchronized to the fill cycle of the pump. This eliminates cavitation problems with volatile organic solvents. Two or three microprocessor-controlled proportioning valves are connected between the reservoirs and the pump inlet to allow the formation of gradients from two or three individual solvents.

Because the pump filling period is very short in comparison with the duty cycle, the system must involve a high-pressure static mixer between the pump and the pulse damper.

To avoid the effects of compressibility on the reciprocating pump performance, three methods are used: (a) the pump is operated at a constant pressure, higher than the column back-pressure, such as in the Varian Model 5000, or (b) electronic compressibility compensation is used (see Section 8.2) or, best, flow feed-back control is employed. The flow feed-back control with this type of gradient devices not only makes it possible to eliminate the effects of solvent compressibility, but also to eliminate the effect of the volume contractions connected with the mixing of different solvents. It is possible to use this type of instrumentation also for work with microbore or capillary columns, but the hydraulic parts of the pump should be carefully minimized. A single 2150 pump controlled with a 2152 microprocessor, which can be used for this purpose, is commercially available from LKB.

In certain instruments, different solvent densities and viscosities can affect the performance of the proportioning solenoid valves and cause errors in the flows through the valves and in the composition of the mixed mobile phase, especially at a low percentage of one of the solvents. Further, the solenoids generate heat that may, under certain circumstances, either partially evaporate highly volatile liquids or outgas poorly degassed solvents. The gas bubbles formed in this way may cause problems in the functioning of electrovalves and of the high-pressure check-valves. Consequently, errors in solvent composition and flow-rate of the mobile phase may occur[41].

References on p. 282.

8.3.2. Devices using "on-stream" mixing of the individual mobile phase components

Gradient elution devices of this type use auxiliary precision metering pumps instead of proportioning electrovalves (Fig. 8.18). This makes it possible to mix the individual solvents "on-stream", at the same time, so that mixing chambers are unnecessary and zero dead volume tee-unions can be used to collect the individual solvent streams from the outlets of the auxiliary pumps before they reach the high-pressure reciprocating pump. This provision ensures perfect mixing of the individual solvents and at the same time avoids the possibility of gradient distortion (flattening of the gradient) and significantly decreases the gradient delay. Further, the possibilities of concentration oscillations are completely eliminated.

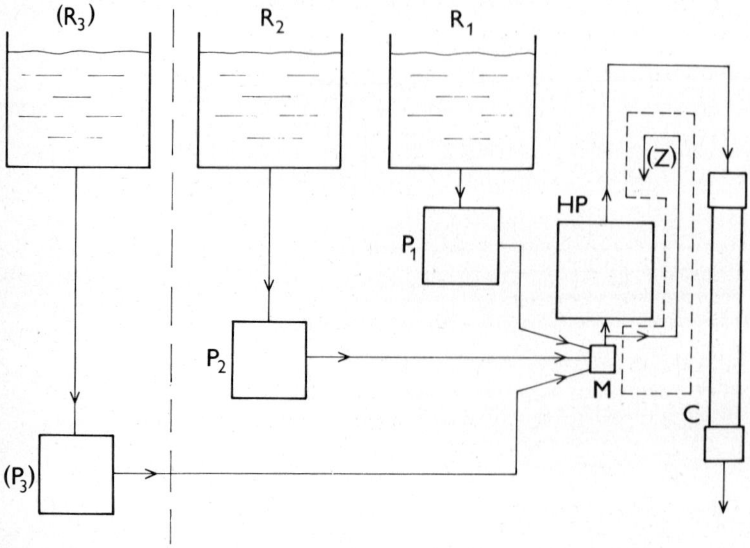

Fig. 8.18. Gradient device using "on-stream" mixing of the mobile phase components in the low-pressure part at the inlet of the high-pressure pump. R_1, R_2, R_3 = solvent reservoirs; P_1, P_2, P_3 = auxiliary precision metering pumps; M = small-volume mixing chamber or tee-union; HP = high-pressure reciprocating pump; C = chromatographic column; Z = overflow to waste (used in overflow type instruments); R_3, P_3 are used only in ternary solvent mixing devices.

The auxiliary pumps delivering the individual solvents are of the reciprocating type, like the high-pressure pump, which offers an unlimited solvent reservoir capacity. They need not withstand high pressures, as the solvents are mixed at atmospheric pressure. Therefore, the auxiliary pumps can be efficiently miniaturized.

In gradient devices of this type, it is necessary to match the total flow from the auxiliary pumps to the demand of the high-pressure pump. This can be simply done by adjusting this flow to be 10-20% higher than that delivered by the high-pressure pump. This solvent overflow is sufficient to compensate for eventual volume contractions when mixing the individual solvents. The remaining overflow is led to waste.

A gradient elution device of this type constructed from modified commercial instruments has been used in the authors laboratory for several years with good accuracy of gradient profiles and of mixed mobile phase flow-rates[27]. The auxiliary pump system is represented by a dual-head reciprocating pump where each pump head delivers one solvent to a mixing chamber of very small volume. The total flow-rate is fixed by adjusting the plunger stroke frequency by means of a gear system. The ratio of the volume in each piston chamber is continuously adjustable by a special motor driving the two stems limiting the distance travelled by each piston during one stroke cycle in such a way that the sum of these dis-

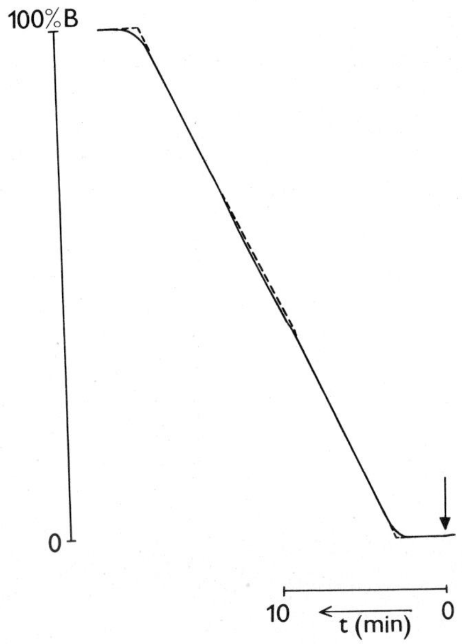

Fig. 8.19. Comparison of the experimental (full line) and expected (broken line) gradient profiles at the outlet of the M 6000 pump connected in series to a PPM 68005 gradient-generating device. Component A = methanol; component B = 0.3% (v/v) solution of nitromethane in methanol. Gradient: $c = 0.06872\ V$, where c represents the volume ratio of B to A+B; V (ml) = volume of the mobile phase delivered from the mixing chamber. The experiemntal profile is measured as the signal of the R-401 RI detector (Waters Assoc., sensitivity 32). (Reproduced from ref. 27 with permission.)

References on p. 282.

tances remains constant (and so does the total flow-rate), while their ratio follows a pre-set programme[25]. Strictly, this represents an alternate flow system, but the total volume delivered by the two pump heads during one stroke cycle was adapted to as a low value as 19 μl (the piston displacement volume in each pump head of the dual-head reciprocating high-pressure pump used in this device is 100 μl), which means that the components of the mobile phase are well mixed together and the device actually behaves as an "on-stream" mixing type. The inlet port of the high-pressure pump (Waters M 6000) is adapted to minimize its volume and is equipped with a side overflow tube. Fig. 8.19 shows the required gradient profile at the outlet of the reciprocating high-pressure pump using this device to produce a linear gradient from pure methanol to methanol with 0.3% nitromethane as a tracer. The actual gradient profile (full line) is corrected for the gradient lag in the high-pressure pump (3.08 ml).

The other possibility of matching the flows of the high-pressure pump and auxiliary pump system makes use of electronic microprocessor control[42]. This system is used in the Tracor 985 liquid chromatograph (Fig. 8.20A). Here, a Tracor 951 dual-head reciprocating pump is used to deliver the mixed mobile phase on to the column. The auxiliary pumping system employs three specially designed dual-head reciprocating precision metering pumps, each driven by its own stepping motor (ternary solvent mixing possibility). The stepping motors allow the control of each metering pump in 0.064-μl increments. The three pumps run simultaneously, with the ratio of their speeds equalling the ratio of solvents delivered. The mixing of the three solvents occurs in a cross-union with an internal volume of only 10 μl. Fig. 8.20B shows the exposed auxiliary pumps and mixing cross-union (on the extreme right). The motors are placed below the pump heads (a 1 : 1 gear system is used to drive the pistons) and the check-valve systems are placed in special stainless-steel blocks (foreground of the photograph).

The matching of the output of the metering pumps to the demand of the high-pressure pump is accomplished through the use of a specially designed flow sensor, which monitors the solvent demand of the high-pressure pump and feeds the information back to the microprocessor to adjust the speed of the stepping motors of the metering pumps and to regulate in this way the delivery from the auxiliary pumps. Therefore, no overflow system and no electronic interface between the gradient former and the high-pressure pump are necessary. The gradient delay in this system is about 2 ml, which is comparable to the previously described authors' gradient device.

Devices with "on-stream" solvent mixing in the low-pressure part virtually eliminate the composition and flow errors of the gradinet formed, if the high-pressure pump is provided with a flow feed-back control or, at least, with the possibility of compressibility compensation or is equipped with an on-line constant overpressure back-valve.

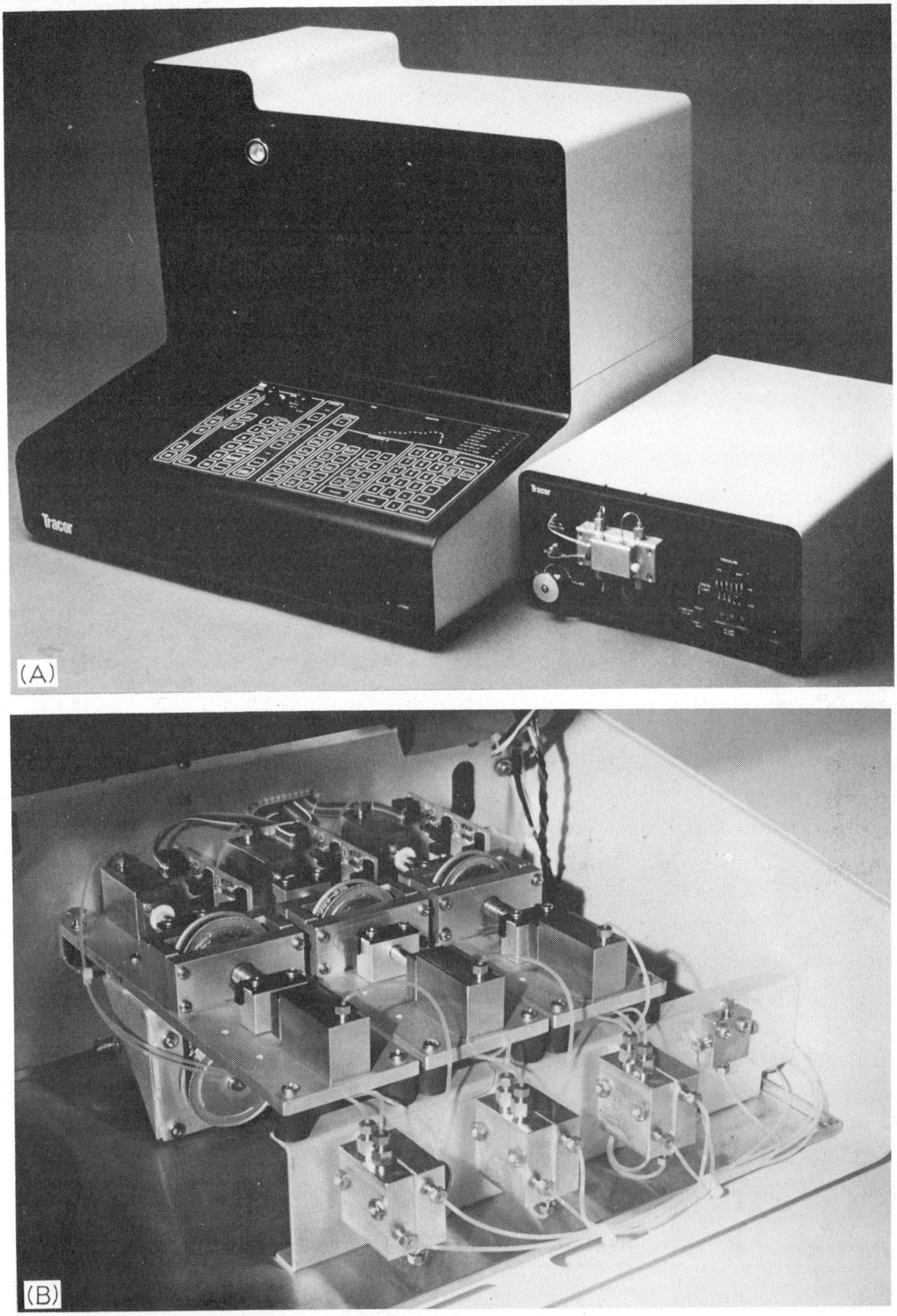

Fig. 8.20. Tracor 985 liquid chromatograph. (A) Overall view; (B) the heart of the gradient former - the auxiliary pumping system using three low-pressure specially designed precision metering dual-head reciprocating pumps. (Reproduced by Courtesy of Tracor, Inc., Austin, TX, U.S.A.)

References on p. 282.

8.3.3. Performance of gradient devices mixing the solvents before reaching the inlet of the high-pressure pump

With this type of gradient elution equipment, it is not necessary to use two costly high-pressure reciprocating pumps. As with high-pressure gradient devices, it is possible to form almost any required gradient profile, using the appropriate electronic control of the low-pressure gradient mixer at the inlet of the high-pressure reciprocating pump. If the high-pressure pump is equipped with compressibility correction or with a flow feed-back control, the gradient profile formed in this type of device is essentially free from the errors caused by the liquid compressibility effects and by thermodynamic volume changes on mixing of the liquid components of the mobile phase. However, more perfect degassing of the mobile phase components is required than with high-pressure gradient devices.

In devices using alternate flows of the individual mobile phase components to mix the gradient at the inlet of the high-pressure pump, synchronization of the duty cycle of the proportioning electrovalves with the filling-delivery period of the high-pressure reciprocating pump is required to avoid possible periodic oscillations in the mobile phase composition[39]. Further, additional mixing chambers (usually stirred dynamic mixing chambers) are necessary to ensure perfect mixing of the "slugs" of pure liquids delivered from the proportioning electrovalves. The internal volume of this mixing chamber (a few millilitres), together with the inner volume of the high-pressure reciprocating pump, is usually considerably higher than the volume of the mixing chamber or of the tee-union used in high-pressure gradient devices. The increased mixing volume in this type of gradient device leads to an increased gradient delay compared with the high-pressure gradient devices and can cause destortion (flattening) of the actual gradient profile at the points of sudden concentration changes[43]. This flattening of the gradient profile in its initial and terminal parts may ressemble the analogous flattening with high-pressure gradient devices, where it originates mainly in the inaccurate functioning of the individual reciprocating pumps at very low flow-rates.

The problems with large mixing volumes and corresponding gradient distortion, with malfunctions of proportioning electrovalves and with possible concentration oscillations, are eliminated in gradient devices using "on-stream" mixing of the individual mobile phase components at the inlet of the high-pressure reciprocating pump. Gradient devices of this type do not require mixing chambers and essentially are not influenced by liquid compressibility effects or by thermodynamic volume changes connected with the mixing of different liquid components of the mobile phase[27,42]. Consequently, the principle of their design seems to be the most sophisticated and to offer the most accurate formation of actual gradient profiles in comparison with all other currently available types of instruments for analytical gradient elution liquid chromatography. However, the necessity to use precision metering low-pressure auxiliary reciprocating pumps instead of the proportioning electrovalves in this type of gradient device increases their price in comparison with that of low-pressure gradient devices using alternate flows of the individual mobile phase components. The gradients formed by this type of device are necessarily delayed by a few millilitres with respect to the gradient programme and the flattening of the gradient profile at its extreme parts is given only by the inner volume of the reciprocating high-pressure pump (usually ca. 2 ml), which is comparable to or even better than the delay and flattening with high-pressure gradient devices using a dynamically stirred mixing chamber.

8.4. ELECTRONIC PART OF THE GRADIENT DEVICE USED TO CONTROL THE FORMATION OF THE GRADIENT PROFILE

The electronic control of gradient formation is used to give the appropriate information to the electrovalves for selecting the opening sequence of the two solvents during the duty cycle or to the motors that control the speeds of the two high-pressure pumps or to the motor that controls the distance travelled by the pistons in constant-motor-speed driven reciprocating pumps. This control unit is pre-programmed in order to provide the gradient profile required. This does not mean that the gradient really obtained corresponds exactly to the desired gradient profile, because it can be influenced by the performance of the mechanical and hydraulic parts of the gradient device, as discussed in preceding sections. The aim of the electronic part of the gradient device discussed here is only to "tell" the hydraulic part the required composition of the mobile phase as a function of time.

The electronic part of the gradient elution instrument is either located in a separate module or is integrated in a central microprocessor that controls the complete gradient elution liquid chromatograph. We can classify the electronic control units essentially into three groups: (a) modules that make it possible to reproduce gradient profiles of any shapes drawn on sheets of paper or on special cards; (b) electronic modules that make it possible either to form gradients according to several continuous curves of a given shape or to form so-called multi-linear gradients composed of a sequence of several linear segments whose slope and time can be programmed in such a manner that the sum of these segments would approximate a curved gradient; and (c) a central microprocessor controlling the whole liquid chromatograph including the control of the gradient profile in a similar manner to the special-purpose electronic modules (b).

Some electronic devices make it possible to reproduce gradient profiles of any shape according to hand-drawn curves. A photoelectric curve follower can be used, which makes use of a photoelectric head following exactly the black trace of the gradient curve drawn on a white chart sheet fastened on a slowly rotating drum[25]. The photoelectric head consists of a light source and a photocell, to which the light from the source is reflected from the surface of the paper. The head is sensitive to deviations of its position from the boundary between the white paper surface and the black trace of the gradient curve. This boundary moves slowly as the drum rotates and the photoelectric head generates an electric signal that actuates a servo-motor, which always corrects the position of the photoelectric head exactly on the new boundary. The electric signal from the photoelectric head is simultaneously used to adjust the flow outputs of the two pumps delivering the individual mobile phase components. Usually, a dual-head reciprocating pump is used, driven by a constant-speed motor, where the pumping ratio of the two heads is adjusted by two rods limiting the distance travelled by each piston during one stroke. The positions of these two rods are controlled by the servo-motor actuated by the photoelectric head in such a way that the total flow-rate delivered by the two pump heads is constant and does not depend on the position of the distance-limiting rods and the ratio of the flow-rates from the two heads corresponds to the instantaneous position of the photoelectric head over the gradient curve. For example, the Touzart et Matignon Chromatem 38 gradient system and the PPM-68005 gradient pump from the Workshops of the Czechoslovak Academy of Sciences operate on this principle.

This principle of gradient control was modified for use in combination with devices forming the gradient in a low-pressure part from alternate liquid flows at the inlet of the high-pressure reciprocating pump. These gradient mixers can also be used directly in low-pressure liquid chromatography applications. Here, the gradient curve is hand-drawn on a sheet of ordinary millimetre or special-purpose chart paper, and the profile is cut out with a pair of scissors and fastened on a special rectangular frame of the electronic device. During the gradient run the photoelectric sensing head moves vertically in this frame and generates an electric signal which controls the open-shutdown cycle of electrovalves on the inlets of the individual liquids composing the gradient. The ratio of the distance covered by the cut sheet of paper to the blank distance gives the ratio of the opening times of the valves. One photoelectric head sensing period agrees with the duty cycle of the electrovalves and is usually between 6 and 12 sec. After each sensing cycle the photoelectric head moves an adjustable distance from left to right and then repeats the sensing of the gradient profile. Thus, the gradient profile is scanned in the form of a number of discrete points corresponding to each duty cycle of the electrovalves. In this way, either two liquids may be mixed to form binary gradients, such as in the Gilson Mixograd instrument, or three solvent flows may be combined, such as in the LKB Ultrograd gradient mixer. These two instruments are equipped with the possibility of automatic insertion of a gradient hold-up period when the feedback information from the detector indicates that a certain level of a signal has been reached and the gradient run is continued from the time when the detector signal again drops below this pre-set value. The automatic insertion of such a detector-signal-controlled hold-up period is essentially possible also with microprocessor-controlled instruments[84]. These modified gradients may help to improve the resolution of certain poorly separated peaks (see Fig. 8.21).

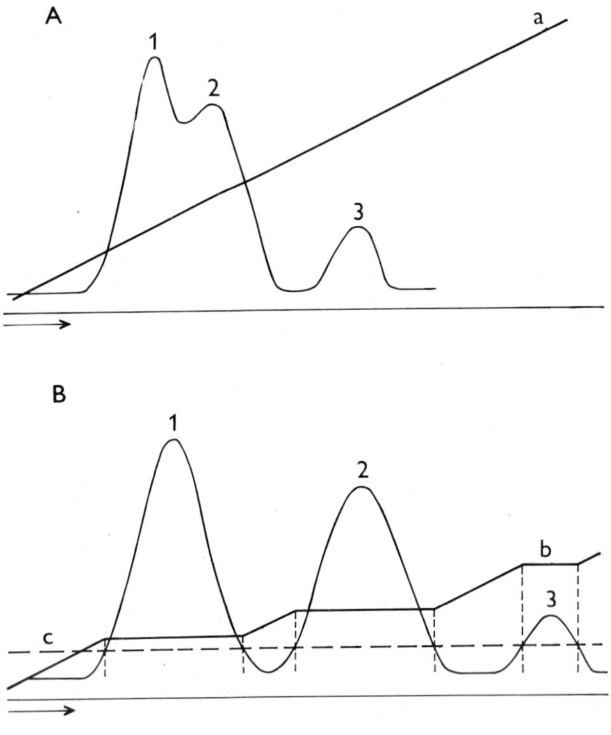

Fig. 8.21. Use of automatically adapted gradients to improve the resolution. (A) Non-adapted gradient run; a = gradient profile. (B) Automatically adapted gradient run; b = gradient profile with inserted hold-up periods; c = adjustable level control of detector signal used to introduce the hold-up periods; 1,2,3 = numbers of peaks. (Adapted from ref. 84)

The DuPont 838 programmable gradient elution accessory employs essentially the above-mentioned system of electronic gradient control, but the electrovalves close and open the streams of liquids under high pressure (see Section 8.2)[18].

References on p. 282.

The chart sheets with drawn or cut gradient profiles can be filed in order to ensure exact reproducibility of the gradient profile in future experiments. Electronic gradient programmers of this type offer unlimited possibilities of any gradient profile as complex as the user wishes; however, there is some inconvenience in the necessity for hand-drawing and cutting out the gradient profile on the chart sheets.

Another group of modular gradient elution liquid chromatographs employ electronic gradient controllers generating the electric control signal according to a few pre-programmed fixed gradient shapes. The desired gradient curve is selected using a curve selector switch. For each continuous gradient programme chosen, the initial and final compositions of the mobile phase, the total flow-rate of the mobile phase and the desired time of the gradient run are set up either using digital dials or by means of turnable switches that make possible the continuous selection of these parameters.

The shape of the gradient curves is usually described by the following equation:

$$c = B't^{\kappa} = (BV)^{\kappa} \tag{8.11}$$

where c is the concentration (%, v/v) of the stronger eluting component in the mobile phase, t is the time and V is the volume of the mobile phase from the start of the gradient run; B, B' and κ are experimental constants. The constant B' (slope of the gradient) depends on the time of the gradient run and on the initial and final conditions selected on the programmer. The exponent $\kappa = 1$ for linear gradients, $\kappa > 1$ for concave gradient curves and $\kappa < 1$ for convex gradient curves. Eqn. 8.11 is identical with eqn. 4.18 and the retention characteristics in experiments using these gradient curves can be calculated theoretically (see Part II).

It should be noted that sometimes the equation

$$c = B' \left[1 - (1 - t)^{\kappa}\right] \tag{8.12}$$

is used to describe the convex gradient curves, in order to keep the exponent $\kappa > 1$, but eqn. 8.11 can also be used satisfactorily for the correct description of such curves[6,7].

Most commercial gradient programmers of this type employ a certain number of selectable gradient functions with fixed values of κ, for example κ = 1/5, 1/4, 1/3, 1/2, 1, 2, 3, 4, 5 (Fig. 8.22), with a Waters M 660 solvent programmer, LDC Model 1601 gradient master, DuPont 820 and 830 gradient elution accessory and Micromeritics Model 752 gradient programmer. In the Perkin-Elmer Series 3B liquid

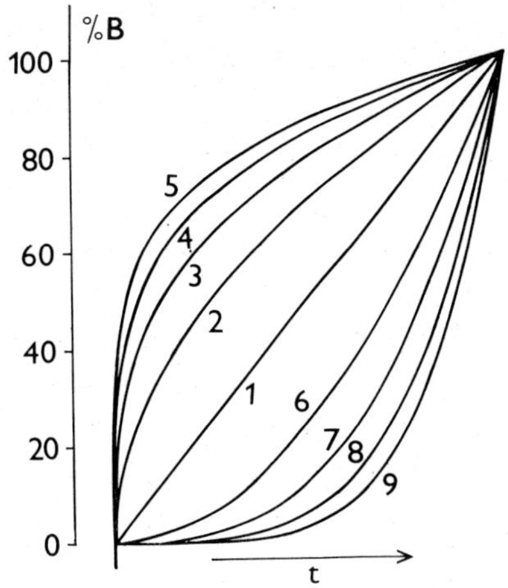

Fig. 8.22. Gradient shapes provided by Waters M 660 gradient programmer. Gradient function according to the equation $y = A + BV^{\kappa}$ (eqn. 4.16). (1) κ = 1; (2) κ = 1/2; (3) κ = 1/3; (4) κ = 1/4; (5) κ = 1/5; (6) κ = 2; (7) κ = 3; (8) κ = 4; (9) κ = 5. t = Time elapsed; B = more efficient eluting component in the mobile phase.

chromatograph and the Waters Model 720 system controller, the liquid chromatograph is controlled by a central microprocessor. Here, the solvent gradients according to the above nine different gradient functions may be formed and, moreover, the gradient programme may group continuous gradients with any of these gradient curves into a sequential composed gradient. Similar possibilities are also offered by other microprocessor-controlled liquid chromatographs, such as the Spectra-Physics SP 8000, DuPont Model 850 and Tracor 985. The Pye Unicam LC-XP gradient programmer enables the exponent κ to be chosen in steps of 0.1 from 0.1 to 9.9, whereas with the Perkin-Elmer Series 2 liquid chromatograph only linear gradients (κ = 1) are possible. All of these gradient devices have the possibility of incorporating an initial gradient delay and hold-up periods and of runing reversed gradients, which are useful in re-equilibration of the column to the initial conditions. Usually, step changes from one mobile phase composition to another are also possible.

Another group of gradient programmers make use of so-called multi-linear gradients, i.e., a succession of linear segments whose slope and time can be programmed[26]. These programmers are available commercially as modular units, such as the Applied Chromatography Systems Model 750/36 decilinear gradient elution programmer (ten segments), the Applied Research Laboratories 750/30, 31 and 32 gradient elution digital programmer (4, 7 and 10 segments) and the Knauer 9100 gradient former (8 segments).

References on p. 282.

In certain instruments, such as the Hewlett-Packard 1084, Varian LC 5000 and Beckman 421 ORT system controller, the multi-linear gradient programming is done by a central microprocessor, which controls all functions of the liquid chromatograph. The gradient profile is programmed from the microprocessor keyboard as a sequence of data, giving the composition of the mobile phase at different times from the start of the gradient. The time intervals between neighbouring pairs of data (break points) give the lengths of the individual gradient steps, in which the composition of the mobile phase changes linearly.

Using a suitable succession of linear gradient segments, it is possible to approximate more or less well continuous curved gradient profiles, depending on the number and length of the individual segments. The mobile phase gradients may be combined with flow-rate programming. Several microcomputer-controlled instruments offer the possibility of linear and curved (concave or convex) composition changes in each segment (see above).

The gradient programming can be carried out easily and conveniently with a few keyboard touches and can be stored either in the memory of the microprocessor or, with some instruments, on magnetic tape. Certain instruments (e.g., the ACS 750/36 and ARL 750/30-32) make use of thumb-wheel switches and others (e.g., the Varian 8500, 16 segments maximum) a pin-board to enter the programme, which cannot be stored.

Of course, gradient delays, hold-up periods and reversed gradients can be programmed with this type of instrument, using the necessary number of gradient segments for this purpose. A very useful feature of microprocessor-controlled gradient elution instruments is the ready possibility of forming and controlling ternary and even more complex gradients. Microprocessor-controlled instruments working with ternary gradients (such as the Beckman 345, Spectra Physics SP 8000, 8100, 8700, Tracor Model 985, DuPont 850, Micromeritics 7500, Varian 5000) or with quaternary gradients (such as the DuPont 8820 or Perkin-Elmer Series 4) are now commercially available.

8.5. DEVICES FOR STEPWISE AND MULTI-SOLVENT GRADIENT ELUTION CHROMATOGRAPHY

Any gradient elution device can be used to form stepwise gradients of a mobile phase consisting of two components, where the individual isocratic steps differ from one another only in the content of these components in the mobile phase. This is not the case if the mobile phase programme used during the separation should consist of segments with several different pure liquids. With instruments that have a ternary gradient capability, it is of course possible to use three different solvents in one elution run. This may be useful for enlarging the range of the solvent polarity change in a single gradient run, e.g., to combine aqueous and non-aqueous reversed-phase chromatography on a chemically bonded alkylsilica packing when starting with water as the initial solvent, going to methanol or acetonitrile and then to methylene chloride or di-*n*-propyl ether. In such a way, it is possible to use even starting and final solvents which are insoluble or poorly soluble in one other. It is also possible to use the gradient from solvent a to solvent b to separate compounds of one class and then proceed with the gradient from solvent b to solvent c to achieve the separation of compounds of another class, for example, to separate nucleosides and bases with a gradient formed from one pair of mobile phase components on an ion-exchange column and then to continue with the gradient formed from another pair to separate the nucleotides.

A Varigrad system (Section 8.1) can be used to form multi-component gradients and can be connected to the inlet of a high-pressure reciprocating pump. Of course, using this system it is not possible to achieve gradients from pure liquid a to pure solvent b, then to pure solvent c and so on, but all of the solvents would be mixed simultaneously in the mobile phase after a certain initial period and the concentration of all of the solvents (with exception of the initial one, a) first increases, until it reaches a maximum, and then decreases again. The form of the gradient would depend on the flow-rate of the mobile phase, the initial volume of the individual solvents in each mixing chamber and the size and form of the chambers, if they are not equidimensional.

In normal-phase adsorption chromatography on silica or alumina, more than three solvents may be necessary to cover the polarity range from an *n*-alkane to water, because there is a risk of displacement or solvent-demixing effects. It has been claimed that as many as twelve solvents are necessary for this purpose and elution employing such a variety of solvents was called "incremental gradient elution". The solvents should be used in a well defined order to make sure that the stepwise change from one solvent to the next in this series would result in a constant incremental change in the logarithm of the capacity factor of a solute[44,46]. To use such a large number of solvents would be impractical from the points of view of both detection and manipulation the solvents. It seems that a smaller number of solvents, four to six (or possibly even three), would be sufficient (see discussion in Chapter 10).

Originally, a rather complicated device was used for incremental gradient elution, where the most polar solvent was pumped by the pump on to the column via a series of auxiliary columns packed with glass beads, each column containing another solvent in order of decreasing polarity from the pump to the column. The gradient formed with such a device should ideally consist of subsequent steps of pure solvents, but, by virtue of the mixing of solvents in the auxiliary columns, the actual gradient involved continuous concentration changes in subsequent steps between neighbouring solvents rather than a sudden discontinuous changeover. It is clear that it is very inconvenient to work with such a device, the precision of the gradient formed is poor and it is hardly possible to influence the length and steepness of the individual gradient steps.

Much more practical is the use of a multi-port valve at the inlet of the high-pressure reciprocating pump, connected to the desired number of solvent reservoirs[45,46]. Here the time of opening of the individual solvent ports can be programmed and controlled by a relatively simple electronic device. The actual gradient would depend on the volume of the mixing chamber, usually inserted between the multi-port valve and the pump. The use of such a mixing chamber would allow polarity changes to be "smoothed" during the individual gradient steps, which would be of a more or less sharp exponential continuous profile.

The Analabs Anasol-20 solvent programmer is based on this principle and can be used for incremental gradient elution in high-performance liquid chromatography. It employs 20 solvent reservoirs, a 20-port rotary valve controlled by a digital timing programmer control unit and a dynamically stirred mixing chamber with an adjustable volume. Sixteen solvents may be used for the incremental gradient elution and the other four for the regeneration of the column. The time of opening of each solvent port in the rotary multi-port valve may be programmed from 0 to 99 min.

With the Siemens S 100 liquid chromatograph, it is possible to form only stepwise gradients composed of three steps. The instrument uses two holding coils filled with different solvents, while the third solvent is pumped by a membrane pump. A high-pressure switching valve[47] allows each of the three solvents to be selected and to be directed on to the column by pressing one of three push-buttons. Because the coils holding the liquids have large inner volumes, there is a strong influence of solvent compressibility at high pressure on the performance of the instrument and it takes a long time to achieve steady flow and composition conditions.

8.6. TESTS OF THE PERFORMANCE OF GRADIENT ELUTION INSTRUMENTS

A wide variety of instruments for gradient elution liquid chromatography are commercially available. These instruments differ from one another in the design of the hydraulic and electronic parts, which utilize different operating principles described in previous sections. The properties of some frequently used commercial instruments are summarized in Table 8.1. This table cannot give complete up-to-date information on all commercially available equipment, because it is based on the information that was obtained from the various manufacturers at the time of completion of the manuscript. Omission of certain instruments does not indicate an adverse appraisal by the authors.

The quality and performance of the instruments for gradient elution can be evaluated according to the following criteria: (a) possibility of formation a variety of linear, curved and composed gradient profiles; (b) reproducibility of the gradient profile in repeated runs; (c) accuracy of gradient formation, i.e., agreement between the actual and the required gradient profiles; (d) speed of response in the composition of the mobile phase to a change in the composition programme; (e) stability and constancy of flow-rate at different compositions of the mobile phase during the gradient run.

It is difficult to compare the quality and performance of different instruments from the manufacturers' literature because often the operating principles are not clearly explained. Therefore, it was felt that a set of control experiments should be performed to show clearly the performance of instruments for gradient elution under operating conditions and the possible influence of various possible sources of errors in flow-rates and gradient profiles, which are decisive for the reproducibility and accuracy of retention data and quantitative results[27,34,39,48].

The manufacturers usually demonstrate the quality of their instrumentation for gradient elution chromatography by a record from the UV detector during a blank gradient run with two components, A and B (usually a linear gradient from 0 to 100% of B in 10-40 min), where component A is a solvent that does not absorb at the wavelength used and B is the same solvent with the addition of a small amount of a UV-absorbing substance. However, this test cannot show what occurs when two different solvents are used to form the gradient, as is usual in practical chromatographic work, because these solvents usually differ more or less in density and viscosity and their mixing may result in volume changes. All of these effects can have an adverse influence on the precision of the flow-rate and the gradient profile.

The effect of these possible sources of errors originating in solvent "non-idealities" on the gradient formation would appear most clearly if water and methanol or acetonitrile are used as the components of the mobile phase and the mixed mobile phase is pumped at a high pressure, e.g., 10-30 MPa. This would simulate the usual conditions in reversed-phase gradient elution liquid chromatography[27,34,43,48]. To perform this test, a "blank" linear gradient should be run from pure water to 100% methanol containing 0.1-0.3% of a UV-absorbing substance, such as nitromethane, acetone or benzene, in an instrument with a chromatographic column and a UV detector. The signal of the detector is recorded and the flow-rate is measured in different parts of the gradient. However, precautions should be taken in order to avoid preferrential sorption of a mobile phase component or of the UV tracer on the column. An analytical column packed with 5-10 μm glass beads would be ideally suited for this purpose, because it would only produce a

TABLE 8.1

SURVEY OF CERTAIN COMMERCIAL GRADIENT ELUTION INSTRUMENTS

Manufacturer and instrument model	Design type*	Gradient possibilities		Mixing chamber**	Flow control***	Max. pressure (psi)	Note§
		Number of components	Profiles				
Analabs Anasol 20	B	Up to 16	Convex steps	Dy (50)	-	-	1
ACS (ARL) 750/30,31 or 32	G,F	2	4,7 or 10 linear steps	St	-		1a
750/36	H (r)	2	10 linear steps	St	c	6000	1a
750/37	H (r)	2	13 linear steps	St	c	6000	1a
APPLE II	H (r)	2	25 linear steps	St	c	6000	1b
Beckman (Altex) 322	H (sd)	2	Multilinear	Dy (2)	b	10,000	2
324	H (sd)	2	Multilinear	Dy (2)	b	10,000	3
332	H (s)	2	Multilinear	Dy (2)	b	6000	2
334	H (s)	2	Multilinear	Dy (2)	b	6000	3
342	H (s)	2	Multilinear	Dy (2)	c	6000	2a
344	H (s)	2	Multilinear	Dy (2)	c	6000	3a
345	H (s)	3	Multilinear	Dy (2)	c	6000	3b
Bruker LC 21B	B	3	Multilinear	St (1)	c	7000	28
DuPont 830	F	2	Linear, convex, concave	St (0.4)	Optional	3000	4
850	B	2-3	Linear, convex or concave steps	Dy (1)	b	6000	5
8820	B	2-4	Linear, convex or concave steps	Dy	b	7000	5a
Hewlett-Packard 1084	H (d)	2	Multilinear	Dy (0.4)	c	6000	6
1090	C	2-3	Multilinear		c	6000	6a
Gilson Mixograd		2	Any	-	-	-	7

References on p. 282.

Knauer LC 250/2	H (sd)	2	10 linear steps	St	-	6000	
50/B	H (s)	2-3	Multilinear	St	-	6000	29
Kontron Model 200	H (s)	2	Multilinear	Not specified	b	5000	7a
Kratos Series 250		2	Multilinear				25
Laboratory Data Control			Linear, convex				
GLC 7800,401,403	H (sd)	2	concave	Dy	-	5000	8
CCM		2	Multilinear	-	-	-	8a
LKB Ultrograd		2-3	Any	-	-	-	9
2152 Controller	B	2	Linear,	-	-	5000	26
	H (sd)	2	multilinear	St	-	5000	27
Micromeritics 7500	B	2-3	Linear, convex concave	Dy (1.5)	c	6000	10
Perkin-Elmer Series 2/2	H (sd)	2	Linear	Dy (2.5)	-	6000	
Series 3B	H (sd)	2	Linear, convex, concave steps	Dy (2.5)	-	6000	11
Series 4	B	2-4	Linear convex, concave steps	Dy	c	6200	11a
Pharmacia							
Gradient Mixer	A	2	Linear	Large	-	-	12
Peristaltic Pump P-3	A	2	Linear (convex, concave)	Large	-	-	13
FPLC System	H	2	Multilinear	Dy	-	600	13a
Pye Unicam LC-XP	B	2	Linear, convex, concave steps	Dy (0.8)	c	10,000	14
PU 4001	B	2	Linear, convex, concave steps		c	6000	14a
PU 4002	H (sd)	2	Linear convex, concave steps		b	6000	14b
Reeve Angel Gradient Former	A	2	Any	St (tee tap)	-	-	15
Siemens S 100	Special	2-3	Step (max. 3)	3-port valve	-	5000	16
S 111	H (d)	2	Linear, convex, concave	St (0.5)	-	5000	17

(Continued on p. 236)

Table 8.1 (continued)

Manufacturer and instrument model	Design type*	Gradient possibilities: Number of components	Gradient possibilities: Profiles	Mixing chamber**	Flow control***	Max. pressure (psi)	Note§
Spectra Physics SP 8000	B	2-3	Linear, multilinear, curved	Dy (1.6)	c	6000	18
SP 8100	B	3	Linear, multilinear, curved	St + Dy (0-10 ml)	c	6000	18a
SP 8700	B	3	Multilinear		c	6000	18b
Tracor Model 8000	D	2	Linear, concave, convex	Large	-	2000	19
Model 980A, 981	B	2	Linear, concave, convex		-		19a
Model 985	C	3	Linear, multilinear, curved	St (0.01)	c	10,000	20
Universal Scientific Gradient Former	A	2	Any	St (tee tap)	-	-	15
Varian 8500	G	2	Multilinear	St	-	8500	21
5020, 5060	B	2-3	Multilinear	St (0.7)	a	5000	22
5500	B	3	Multilinear	St	a	6000	22a
Waters M 660	H (sd)	2	Linear, convex concave	St (tee)	b	6000	23
M 680	H (sd)	2	Linear, convex concave steps	St (tee)	b	6000	24
M 720, M 721	H (sd)	2 (3)	Linear, convex, concave steps	St (tee)	b	6000	24a

*Design type: A, classical low-pressure proportional device; B, low-pressure gradient former using alternate flows of liquids at the inlet of the high-pressure pump; C, low-pressure gradient former using simultaneous flows of liquids at the inlet of the high-pressure pump; D, high-pressure gradient former of the exponential type; E, high-pressure gradient former of the proportional type; F, high-pressure gradient former using two pneumatic or hydraulic amplifier pumps; G, high-pressure gradient former using two syringe pumps; H, high-pressure gradient former using two-reciprocating pumps of a diaphragm (d) or special driven dual-head (sd) or single-head (s) type, or using a single dual-head rapid-stroke pump (r).

**Mixing-chamber: St, static; Dy, dynamic (stirred); volume of the chamber in millilitres in parantheses.

***Flow control: a, solvent compressibility compensation using a constant pump overpressure; b, electronic compensation of solvent compressibility by adjustment of the speed of pump motor; c, flow feed-back control.

§Notes:

1. Instrument for incremental gradient elution.
1a. Gradient controller; should be used in connection with two higher-pressure pumps or with a single a dual-head ACS 400/02 pump.
1b. Gradient system as in 1a, but microprocessor controlled.
2. Instrument controlled by a Model 420 microprocessor; special pump design assuring accurate function at low flow-rates of the individual pumps (Model 110A in 332 and Model 100A in 322).
2a. As in 2, but two improved Model 112 single-head pumps used.
3. As in Note 2, but Model 421 microprocessor used to control the instruments; ternary gradient possibility using three reciprocating pumps. (Model 110A in 334 and Model 100A in 324).
3a. As in 3, but two improved Model 112 single-head pumps used.
3b. As in 3, but three improved Model 112 single-head pumps used for ternary gradients.
4. Special design; one pneumatic pump and a solvent holding coil; flow feed-back control unit 833 (optional) necessary for precise flow and gradient profile; nine gradient shapes pre-programmed.
5. Triple-head high-pressure reciprocating pump, microprocessor controlled; five gradient segments, each with seventeen pre-programmed gradient shapes.
5a. As in 5, but improved triple-head pump design.
6. Microprocessor controlled two single-head diaphragm pumps with a common drive.
6a. Gradient capability optional; microprocessor controlled; simultaneous low-pressure gradient mixing using 2-3 special auxiliary reciprocating dual-head precision metering pumps that feed a diaphragm high-pressure reciprocating pump; both conventional and microbore column operation possible.
7. Gradient controller in connection with a low-pressure peristaltic pump.
7a. Microprocessor control - in linear steps, up to 150.
8. One Consta Metric I + one Consta Metric II G (or Consta Metric III) reciprocating pump controlled by a Model 1601 Gradient Master.
8a. Microprocessor control unit only; controls two high-pressure reciprocating pumps as in 8.
9. Gradient controller only.
10. A 752 Gradient Programmer (nine gradient shapes pre-programmed) or a 740 Control Module in connection with a 753 Ternary Solvent Mixer and a 750 reciprocating pump; solenoid valves synchronized with the pump.
11. Microprocessor controlled; five segments maximum; each can use any of nine pre-programmed gradient shapes.
11a. Microprocessor controlled as in 11; quaternary gradients standard; improved low-pressure solvent mixing and flow-rate control.

(Continued on p. 238)

Table 8.1 (continued)

12. Low-pressure gradient mixer, two cylindrical vessels, one stirred; gravity flow.
13. A large-volume stirred mixing chamber is necessary for a complete low-pressure gradient device.
13a. Two P-500 pumps controlled by a microprocessor control unit (Liquid Chromatography Controller LLC 500 or Gradient Programmer GP-250), specially designed for analytical and preparative chromatography of proteins, high chemical resistance of the pumps against salt solutions (even those containing halides) used in ion-exchange chromatography; chromatofocusing is possible with this system.
14. Microprocessor controlled; program in nine segments; 99 various pre-programmed shapes can be used in each segment.
14a. Microprocessor (PU 4030 controller) controls a single dual-head reciprocating pump (PU 4011); proportioning valves synchronized with pump; gradient programs as in 14.
14b. Microprocessor (PU 4800 video chromatography control centre) controls two dual-head reciprocating pumps (PU 4010); gradient programs as in 14.
15. Low-pressure gradient former; rectangular chamber divided into two compartments by a Neoprene gasket according to the gradient profile required; gravity flow; mixing in tee-tap.
16. Diaphragm pump + two holding coils + three-way high-pressure switching valve; step gradients only (three steps maximum); no longer produced.
17. Two diaphragm pumps + gradient controller; eight pre-programmed gradient shapes; no longer produced.
18. Instrument controlled by a central microprocessor; ternary gradients optional.
18a. Central microprocessor controls the instrument, ternary gradients standard, seven gradient segments maximum.
18b. Ternary gradients standard, up to 52 gradient steps; microprocessor controlled; reciprocating pump with two check valves and two pistons working in series.
19. Large-volume stirred mixing chamber with solvent a between an auxiliary and a high-pressure pump; low gradient shape precision.
19a. Solvent programmer, controls one dual-head reciprocating pump such as Tracor Model 950, 951 (10,000 psi) or 955 (6000 psi); sequential gradients from solvents A to B and then from B to C are possible.
20. Microprocessor controlled; simultaneous low-pressure gradient mixing using three special auxiliary reciprocating precision metering pumps; their feed synchronized with the high-pressure reciprocating pump.
21. Ten linear gradient segments maximum; programmed via a pin-board.
22. Microprocessor controlled; ternary capability optional; proportioning electro-valves connected directly to pump chamber.
22a. As in 22, ternary gradients standard, microstepped control of flow-rate; microprocessor controlled; microbore column chromatography possible.
23. M 660 Solvent Programmer controls two M 6000 A pumps, or one M 6000 A and one M 45(G) pump; or two M 510 pumps; nine pre-programmed gradient curvatures.
24. Microprocessor M 680 controls the same pump combinations as in 23; multi-step gradients - segments with nine pre-programmed gradient shapes; flow-rates down to 10 μl/min are possible for microbore column operation.

References on p. 282.

24a. Central microprocessor - 720 or 721 System Controller - controls two or three M 6000A or an M 6000A and an M 45(G) pump, or two or three M 510 pumps; multi-step gradients - segments with nine pre-programmed gradient shapes; ternary gradients are possible with three high-pressure pumps; flow-rates down to 10 µl/min for microbore column operation.

25. Microprocessor control unit only. Kratos KLIC 1 microprocessor or LC 250/2 gradient programmer controls two high-pressure reciprocating pumps. In connection with MM 700 memory module, automatic detector baseline drift correction is possible.

26. A single 2150 pump is controlled (2150 pump: special piston seals, ceramic liquid ends, miniaturized check valves and liquid paths - compatible with salt solutions and low flow required in chromatography on microbore and capillary columns, 10 µl/min flow possible).

27. Two 2150 pumps are controlled.

28. A single triple-head pump (double-feedback control), microprocessor controlled.

29. Two single-head pumps (Model 64), microprocessor controlled.

pressure drop in the system, but the column packing material would be inert[31,43]. Because such columns are not available commercially, silica-packed columns could be also used. If we wish to use a reversed-phase column, nitromethane is a better choice than benzene or acetone as a UV tracer.

To perform a more complete test of the gradient elution equipment, which can detect and differentiate various potential sources of error, we can use the following approach[20,27,34,39]:

(1) The column is disconnected from the instrument and the UV detector is connected directly to the outlet of the gradient former. A pure solvent is used as component A of the mobile phase and the same solvent with the addition of a UV-absorbing substance as component B (for example, the combinations methanol / methanol + 0.1-0.3% of nitromethane, acetone or benzene; *n*-heptane + 0.1-0.3% of tetrachloroethylene; water / water + 0.01 *M* potassium bromide). The record from the UV detector and the flow-rate of the mobile phase under isocratic conditions are measured at several different compositions of the mobile phase covering the range from 100% A to 100% B. A stepwise gradient can be used for this purpose. Errors indicate poor matching of the pumps or imprecise functioning of the pumps (check valves) or of the proportioning electrovalves (less often of the electronic gradient programmer). This test can be particularly recommended with instruments without feed-back flow control.

(2) Using the same connection as in test (1), a "blank" linear gradient is run from 0 to 100% of B in 10-15 min at 1-2 ml/min and the signal from the UV detector is recorded. This test will show the ability of the instrument to reproduce the programmed gradient profile at zero working pressure and the speed of the response in the mobile phase composition to the imposed value. Large deviations (flattening) of the gradient profile in its initial and final parts and a slow response in the composition of the mobile phase can indicate either poor functioning of the pumps at low flow-rates or mixing chambers with excessive internal volumes (these can also be recognized from a large gradient delay).

(3) The column is again connected between the gradient former and the detector. Then, the same gradient is run as in test (2) and the results of the two tests are compared. The two records of "blank" gradients should not differ significantly; only the gradient delay would increase in test (3) of the column void volume. (Of course, the UV tracer should be chosen such that it is not retained on the column under these conditions.) The flow-rate of the mobile phase is measured at different parts of the gradient. Deviations of the actual gradient profile and flow-rate from the required values indicate the influence of non-compensated (or poorly compensated) compressibility effects, or possible malfunctions of the pump seals, check valves or electrovalves at elevated pressures.

(4) Water is used as component A and methanol containing 0.1-0.3% of UV tracer as component B of the mobile phase with the column connected in the instrument.

The series of measurements or a stepwise gradient from test (1) are repeated and the results of the two tests compared. The conditions should be chosen in such a way that the UV tracer is not retained on the column (see above). Deviations of the actual flow-rates and compositions in this test of the mobile phase from the expected values show (a) the influence of the volume changes connected with the mixing of the two solvents or (b) the influence of varying compressibility, viscosity and density during the gradient run. To differentiate between (a) and (b), it is possible to repeat test (4) with the column disconnected; the volume changes on mixing the solvents are far less influenced by the operating pressure than the effects in (b). To discriminate the effects of density from those of viscosity, it is possible to run a gradient of *n*-heptane-methylene chloride + 0.1-0.3% of tetrachloroethylene using a silica column. Here, the density changes far more significantly than the viscosity. The changes in density may significantly influence the performance of the column check valves, electrovalves and mixing chamber. Then, the gradient is run from *n*-heptane to isopropanol + 0.1-0.3% of tetrachloroethylene using a silica column, as with *n*-heptane-methylene chloride. Here, the viscosity changes far more significantly than the density during the gradient run. As the operating pressure increases with increasing viscosity, the flow-rate may decrease if the liquid compressibility effects are not compensated correctly.

(5) The "blank" gradient as in test (2) is run with the column and mobile phase components as in test (4) to demonstrate the effects of the volume changes due to solvent mixing and changing compressibility, density and viscosity on the gradient profile and flow-rate.

(6) A test mixture of sample compounds is injected on to a reversed-phase column using a water-methanol gradient and the retention times (volumes) and peak areas and heights are measured in several repeated runs. Here the influence of various potential error sources on the reproducibility of these data can be determined. It should be noted that this test cannot verify the accuracy of retention times and peak areas, which can be subject to systematic errors due to the design and properties of the given instrument for gradient elution and these data may acquire different values with another instrument.

A practical note to the test experiments can be given. High-purity solvents should be used in experiments in which chromatographic columns are connected in the instruments to avoid the concentration of the UV absorbing solvent impurities on the column in the earlier part of the gradient and their later elution, which would interfere with the UV signal of increasing concentrations of the UV tracer in component B of the mobile phase. It is preferable to use as low a sensitivity of the detector as possible, but a linear response over the range of concentration change of the UV tracer should be guaranteed. The flow-rate of the

References on p. 282.

mobile phase can be measured from the time necessary to collect a certain volume of the eluate in a calibrated cylinder or flask. It is also possible to connect the outlet from the detector to graduated pipettes in a vertical position either by means of a small flexible tube or by using a three-way valve to allow easy emptying and re-use of the pipette for further measurements[34].

Chapter 9

DETECTION AND QUANTITATION IN GRADIENT ELUTION LIQUID CHROMATOGRAPHY

9.1. DETECTOR COMPATIBILITY WITH GRADIENT ELUTION

Various principles of operation have been proposed for the continuous monitoring of the column effluents in liquid chromatography, but only a few have so far been utilized in widely used commercial detectors.

Universal detectors produce a signal proportional to a certain bulk property of the effluent, such as refractive index or dielectric permitivity, which means that such detectors respond to all substances in a more or less equivalent manner. These detectors are usually very sensitive to temperature changes and, of course, to each change in composition of the mobile phase. For this reason, it is not possible to use universal detectors in gradient elution liquid chromatography. In the only published example[49] where a differential refractometer was used for detection in gradient elution chromatography, two columns were used. The analytical column was connected to the measuring cell of the refractometer and another column with the same packing and dimensions was connected to the reference cell. Then, when the mobile phase gradient passed through the analytical column and was used to separate sample compounds, the same "blank" gradient passed through the other column, where the sample was not injected. In this way, the compositions of the mobile phase in the measuring and reference cells of the refractometer should be equal at any time. However, it is very difficult to achieve good compensation of mobile phase composition to allow work at a reasonable detection sensitivity, even when the mobile phase components are carefully selected in such a way that they have almost equal refractive indices as in the separation example reported. For this reason, it is almost impossible to use this compensation technique in practice.

Another obvious way to use universal detection in gradient elution liquid chromatography is to remove all of the mobile phase from the column effluent prior to detection. This can be done in the simplest way if the components of the mobile phase are sufficiently volatile, in so-called mass transport detectors[50-58]. Here the column effluent is fed on to a moving wire[52,53], chain[54], belt or disc[56], where the relatively volatile mobile phase evaporates and the less volatile sample components are left on the moving transporting system, with which they are introduced into a small pyrolytic oven. The products of the pyrolysis of the sample compounds are flushed by a stream of an inert gas or hydrogen into a flame-ionization or other gas chromatographic detector.

Scott and Lawrence[59] modified the detector in attempting to increase the detection sensitivity for a number of oxygenated and other compounds by using an oxidation oven instead of pyrolysing the solutes. Here, the organic solutes are completely combusted in an oxygen or air stream so that all of the carbon in the solute is converted into carbon dioxide, which is converted into methane on a nickel catalyst in a stream of excess hydrogen and then led to the flame-ionization detector.

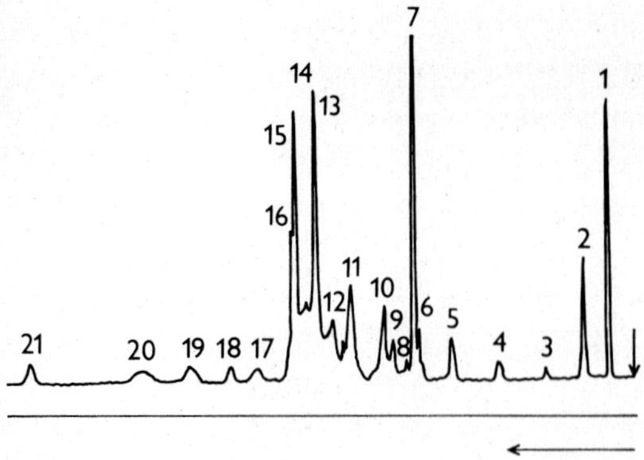

9.1. Chromatogram of an artificial mixture of solutes differing widely in polarities using incremental gradient elution suggested by Scott and Kucera[44] and a transport-wire flame-ionization detector. Column: Bio-Sil A, 500x5 mm I.D., Solvent sequence: *n*-heptane, carbon tetrachloride, chloroform, ethylene dichloride, 2-nitropropane, nitromethane, propyl acetate, methyl acetate, acetone, ethanol, methanol, water (20 ml each); 0.5 ml/min, mixing chamber volume = 6 ml. Sample solutes: 1 = squalane; 2 = anthracene; 3 = methyl stearate; 4 = benzophenone; 5 = chloroaniline; 6 = nitroaniline; 7 = *p*-dinitrobenzene; 8 = *o*-nitrophenol; 9 = dihydrocholesterol; 10 = catechol; 11 = phenacetin; 12 = adenine; 13 = phenolphthalein; 14 = EEDQ; 15 = quinine; 16 = acetylsalicylic acid; 17 = benzoic acid; 18 = *tert*.-BOC-leucine; 19 = *tert*.-BOC-glycine; 20 = alanine; 21 = glucose. (Reprinted with permission from ref. 44. Copyright 1973 American Chemical Society.)

Transport detectors are the only detection systems that make possible the detection in incremental gradient elution using a series of twelve solvents, as suggested by Scott and Kucera[44,46] (see Section 8.5), because certain solvents from this series absorb light in the UV region, which precludes the use of UV detectors. Use of the incremental gradient elution technique with transport detectors allows the separation of sample compounds with very large differences in polarities and the detection of almost any sufficiently non-volatile organic compound, as is illustrated by the example in Fig. 9.1. However, the practical utility of these detectors is rather limited and their popularity and use have decreased in recent years, owing to their expense, complicated construction, manipulation and maintenance, usually a significantly high noise level and, above all, a low sensitivity in comparison with other detectors. The low sensitivity is due to a low yield of volatiles on pyrolysis and to a low percentage of the column effluent carried into the pyrolyser or into the combustion oven by the transport system, in spite of a number of attempts to modify the moving wire, chain or belt in such a way that they could hold a larger volume of the column effluent on its surface (e.g., ref. 53). Further, the moving transport system together with the gas stream between the pyrolyser (or combustion oven) and the flame-ionization detector contribute significantly to broadening of the detected solute peaks.

Another detector, in which the mobile phase is separated from the solutes and no moving transport system is used, has been described recently and seems to offer some possibility of universal detection in gradient elution chromatography[60]. The detector can be used for the detection of non-volatile solutes if the mobile phase is composed only of volatile components. The eluate from the chromatographic column passes into a special heated column, where it is nebulized and the solvent is evaporated. Finely divided solute particles are produced which pass through a light beam. A photomultiplier placed at an angle of 120^0 to the light beam detects the light scattered from the solute particles and yields a signal that is amplified and recorded. The noise level of the detector depends on the perfect evaporation of the mobile phase and therefore the temperature of the heated column should be carefully adjusted. Under properly selected conditions, the sensitivity and detection limit should be better than with refractive index detectors.

The principle of the wire transport detector was used for coupling of a liquid chromatograph with a mass spectrometer. The interface between the two instruments described by McFadden et al.[61] and manufactured by Finnigan makes use of a moving continuous stainless-steel ribbon transporting the column effluent through the solvent evaporator using infrared radiation and through vacuum locks into the ion source of a quadrupole mass spectrometer, equipped with a flash solute vaporizer. With this system, 25-40% of the amount of solute in the column effluent would be utilized for the formation of spectra in the mass spectrometer.

Such a detection system can be used in both isocratic and gradient elution liquid chromatography. The on-line connection of mass spectrometry with gradient elution chromatography can be extremely useful in the analysis of complex natural samples, where gradient elution offers the possibility of separation of a large number of solutes in a single run, and mass spectra can be used for the identification or structure elucidation of the individual unknown compounds.

Quadrupole mass spectrometers also make possible direct coupling of liquid chromatographic columns to the ion source using either electron impact or chemical

References on p. 282.

ionization. However, this connection is feasible in practice when using low flow-rates of the mobile phase. For this reason and because of the relatively very high efficiency of microbore columns, these seem to be particularly suitable for direct coupling with mass spectrometers. If gradient elution were applied in such a system, it should be possible to separate and to obtain mass spectra of several hundred solutes contained in very complex samples during a single chromatographic run, of course at the cost of an increased analysis time (possibly several hours for a single analysis). Thus, the combination of microbore columns with gradient elution and on-line mass spectrometry of column effluents seems to offer tremendous prospects for the analysis of complex naturally occurring samples, biochemical materials, samples containing environmental pollutants and their metabolites, etc.

Detailed information about the on-line combination of liquid chromatography and mass spectrometry can be found elsewhere[62,63].

Specific detectors monitor the total column effluent, but they measure only some specific property of a solute or of a class of solutes. These detectors can detect only solutes that have this specific property and they restrict the choice of the mobile phase to solvents or components that do not possess this specific property. The most sensitive and versatile detectors with a wide linear dynamic range used in liquid chromatography today belong to this class. Therefore, specific detectors have been most widely used not only for isocratic but also for gradient elution liquid column chromatography. It follows from the nature of these detectors that they cannot be used for all samples, and several detectors of this class are necessary in order to meet the necessity of sensitive detection for various compounds and there are a number of compounds for which sufficiently sensitive specific detectors have not yet been developed.

For the application of specific detectors in gradient elution liquid chromatography it follows that the components of the mobile phase used for the formation of mobile phase gradients should not have the specific property detected. This requirement is often difficult to meet in practice, where, for example, practically all solvents absorb certain, sometimes very small, amounts of light in the UV region depending of the wavelength used (this may also be caused by trace amounts of UV-absorbing impurities present in commercial solvents). In such an instance, the solvent used as a component of the mobile phase should not differ significantly from the other components in the property detected, otherwise a serious baseline drift occurs during gradient elution, which would often make work at higher detector sensitivity ranges impossible.

The most frequently used specific detectors in modern liquid chromatography are the UV detector and, to a far lesser extent, the fluorescence detector.

Photometric detectors measure the absorption of radiation (light) by the column effluent in the measuring flow-through cell of small volume, usually 8-20 μl. The detector cell usually consists of a small block of stainless steel with a bore closed at the ends by windows transparent to the radiation used. The column effluent is introduced into and led out of the cell via capillary tubing connected (usually soldered) to the ends of the bore forming the inner volume of the cell. A beam of radiation passes through the cell and is focused on a photoelectric cell or on a photomultiplier. Single-beam detectors, using only one cell, are not very frequent in practice. Most detectors make use of two flow-through cells, one the measuring cell and the other the reference cell. A dual photometric sensing element is used to form an output signal, which is amplified and usually fed to a logarithmic converter to ensure that the signal is directly related to the concentration of solute in the measuring cell.

According to the wavelength region of the radiation used for measurement, UV, visible and infrared detectors are available. The commonest and most useful are the UV detectors. Simple UV detectors may operate at a single wavelength (usually 254 nm) or at two or more wavelengths (254, 280 nm, etc.) selected by an appropriate filter and use a low-pressure mercury bulb as the light source. More expensive are detectors using a deuterium and a tungsten lamp and a monochromator (usually a grating), which allows the required wavelength to be selected from a continuous UV - visible spectrum. This makes it possible to increase the sensitivity of detection for any particular solute by selecting the wavelength at which the compound shows maximum absorption. Some of these detectors offer the possibility of scanning the UV-visible spectrum of a solute in the measuring cell (e.g., Varian Variscan LC/UV spectrometer). For this purpose, the flow-rate is interrupted and the elution is started again on completion of the spectrum scanning.

In the so-called diode array detectors, the light from the measuring cell passes to the grating monochromator and is then reflected on the diode array, which gives electrical output signals corresponding to light absorption at a number of different fixed wavelengths (one diode for one wavelength). The data from the diode array are stored and processed by a computer, which provides direct access to the spectra at any time during the elution without interrupting the flow of the mobile phase and offers smoothing procedures to improve the signal-to-noise ratio and thus to increase the sensitivity. Rapid scanning spectrometers can be also used for this purpose.

The main problem in gradient elution liquid chromatography with UV detection is baseline drift, which depends on the wavelength used for detection and may originate from three different sources:

(a) Changes in the refractive index during gradient elution. As the composition of the mobile phase is changed during gradient elution, the refractive index in the detector cell also changes. A certain amount of the light passing through the narrow-bore cell of the detector is deflected to the walls of the cell and cannot reach the photocell or other sensing element. The amount of the deflected light depends on refractive index and changes during gradient elution, which may cause baseline drift. This problem can be remedied, at least partly, by using a special inner geometry of the detector cell (taper cell), e.g., as in the Waters M440 UV detector[72] (see Fig. 9.2).

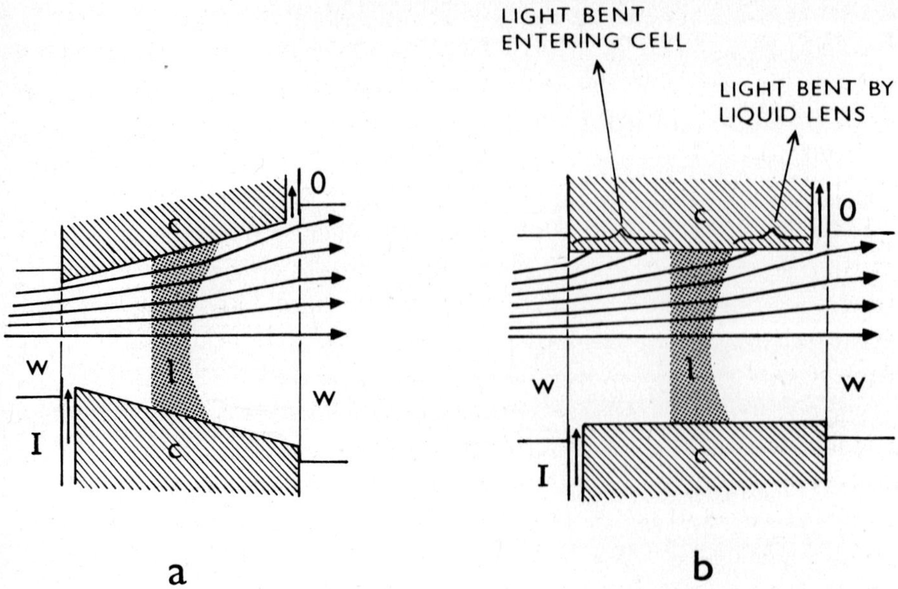

Fig. 9.2. (a) Taper cell and (b) conventional cell of a UV detector. Effect of the bending of a light beam on a liquid lens (l) formed by the eluted sample compound in the detector cell (dotted area) is indicated. I = Sample inlet; O = sample outlet; w = quartz window; c = cell wall. (Reproduced from ref. 72 with permission.)

(b) UV-absorbing impurities. These are present at different levels in commercial solvents and in other chemicals used as components of the mobile phase and changes in the ratio of the mobile phase components during gradient elution cause corresponding changes in the content of these impurities in the mobile phase and thus baseline drift. Therefore, careful purification of the solvents and the use of spectroscopic or LC-grade solvents are recommended (see Chapter 10).

(c) Differences in the absorption of light between the liquids used as components of the mobile phase. All pure solvents or salt solutions used to prepare mobile phase gradients differ more or less each in the amount of the light absorbed and transmitted in a given detector cell. Consequently, baseline drift of a UV detector is observed during gradient elution, which is proportional to the difference in absorption between the mobile phase components at the wavelength used. This difference (and the drift) is wavelength-dependent and usually increases with decreasing wavelength used for detection. In the visible region of light, the baseline drift is usually negligible, but at 254 nm, the wavelength most often used for detection, it may often make it difficult to use higher sensitivity ranges of the detector. The use of a short UV wavelength between 200 and 210 nm would be extremely useful for the detection of a number of compounds, including those of biochemical interest, which do not exhibit significant absorption at 254 or 280 nm, but which absorb at short UV wavelengths. It is necessary to take great care with the purity of solvents used, even for isocratic elution, when working at wavelengths lower than 254 nm and special precautions should be taken to eliminate or reduce baseline drift when attempting to use these wavelengths for detection in gradient elution chromatography. This can be achieved either by running simultaneously a "blank" gradient on the column connected to the reference cell of the UV detector or by blending the solvents or buffer solutions to achieve approximately equal absorbances of the liquid components of the gradient.

The latter approach is more practical and has had some success[64,65]. Here, a UV absorbing compound is added to the component of the gradient with a lower absorbance value (usually the component with a lower eluting strength). For example, 5-6% (v/v) of acetone or butanone-2 was added to water or to a lower concentration ammoniacal buffer as component A, with a phosphate buffer or a higher concentration ammoniacal buffer as component B of the mobile phase in the gradient elution anion-exchange chromatography of peptides or enzymes monitored at 206 nm[65]. However, such a blending compound should not be retained on the column or interfere with the chromatographic system used. When blending the solvents for gradient elution, it is important that the absolute light absorption in the detector cell does not exceed a certain value (such as 40-50%), which depends on the dimensions of the cell and the type of detector, otherwise the detector would not function correctly. Therefore, the solvent blending approach to low-wavelength UV detection in gradient elution chromatography has only limited applicability in practice and its use should be considered with great care.

Baseline drift in gradient elution chromatography may be corrected electronically in certain microprocessor-controlled instruments, such as these using the

References on p. 282.

Kratos KLIC 1 microprocessor and MM 700 memory module. Here, the "blank" gradient is run first and the background signal as a function of time from the start of the gradient is stored in the microprocessor memory. In the repeated, analytical, gradient run, the background signal is electronically subtracted from the signal of the detector. Exact reproducibility of the gradient profile, of the flow-rate and of the delay time is essential for the successful application of this convenient automatic drift supression method[66].

Infrared detectors are commercially available from several manufacturers (e.g., DuPont). They have mainly been used in liquid column chromatography for monitoring separations of polymers performed by gel (size-exclusion) chromatography where the composition of the mobile phase is not changed during the separation. However, they may be used in other chromatographic systems, if a lower detection sensitivity is acceptable. These detectors are specific like UV detectors, but may be used to monitor a wide variety of organic compounds. By selecting the wavelength of detection, it is possible to detect selectively compounds with certain structural (functional) groups in their molecules, such as CH_2, C=O and OH. This makes it possible to monitor certain compounds that cannot be detected by a UV detector at 254 or 280 nm or other wavelengths. Of course, it is necessary to find suitable solvents providing spectral "windows", i.e., a sufficiently high optical transmission at the IR wavelength used for detection.

With certain column packing materials, it is difficult to find suitable solvents for IR detection, but a number of lipophilic substances with CH_2, C=O and other groups may be satisfactorily detected in reversed-phase non-aqueous chromatography using mobile phases formed from methanol, acetonitrile, tetrahydrofuran or methylene chloride (carbon tetrachloride or carbon disulphide could be used, but they are toxic). However, these solvents show significant differences in transmittance at the IR wavelengths suitable for detection and consequently, blending of the mobile phase is necessary in order to avoid a large baseline drift, as with UV detection at a low wavelength (see above). In contrast to UV detection, either component A or component B of the gradient may be blended. However, it is important to select very precisely the composition of the two components of the mobile phase gradient to minimize the baseline drift. For this purpose, the following operating conditions have been recommended[67].

For CH_2 monitoring (almost universal, as most organic compounds possess methylene groups): IR wavelength, 3.4 μm; solvent A, 40% methylene chloride-60% acetonitrile; solvent B, 100% dichloroethane. For C=O monitoring (e.g., esters, acids, aldehydes and ketones): IR wavelength, 5.75 μm (this does not yield maximum sensitivity for all compounds, as the exact wavelength of absorption maximum of the carbonyl group is variable); solvent A, 100% acetonitrile; solvent B, 52.25% tetrahydrofuran-47.75% methylene chloride.

The range of elution strengths in the above suggested gradients is limited, but it makes possible separations of a number of compounds, such as saturated triglycerides or higher alkanes (C_{10}-C_{20}) on an octadecylsilica column[67].

As lower alcohols and other strongly polar solvents and the small amounts of water contained in these solvents may partially dissolve alkali metal halides, the cell windows cannot be made from sodium chloride or potassium bromide (the most frequently used window materials in infrared spectroscopy) and calcium fluoride is used as the window material, but this precludes the use of higher IR wavelengths.

Fluorescence and radioactivity detectors belong among the most sensitive LC detectors. They are very specific, so that they make possible the monitoring of only a limited number of compounds (fluorescent or labelled). This means, on the other hand, that these detectors can be used very well in gradient elution chromatography and usually have far less difficulty with baseline drift than UV or IR detectors. However, some solvents may yield more scattered light than others at certain emission wavelengths, which could cause baseline drift. Therefore, care should be taken with the selection of the emission wavelength (longer wavelengths are preferable, if they can excite the fluorescence with sufficient sensitivity). Certain limitations on the choice of the components of mobile phase gradients with fluorimetric detection are imposed by the fact that some solvents, such as halogenated hydrocarbons, quench the fluorescence, which may lead to a decreased sensitivity of detection. The quantum yield of fluorescence may depend strongly on the polarity of the solvent used and it usually increases with decreasing polarity of the mobile phase. Therefore, it is essential for quantitative gradient elution applications that the mobile phase programme is reproducible from one run to another and calibration must be carried out under the same operating conditions (with identical gradient and flow-rate) as those used for the determination of the content of sample compounds.

Conductivity detectors monitor the electrical conductivity in the column effluents; the mobile phase used should not conduct, but should dissolve fully or partially ionized compounds, the only ones that can be detected. Consequently, the use of conductivity detectors is limited to water or aqueous solutions of strongly polar organic solvents as the mobile phase. This means that conductivity detection can be used in the reversed-phase gradient elution liquid chromatography of acids, bases or salts using gradients composed of water and methanol or acetonitrile. However, this approach seems to be limited as conducting compounds usually have to be added to the mobile phase to obtain the full selectivity advantages of reversed-phase systems for the chromatography of ionized compounds.

Electrochemical detectors use a system of two or three electrodes at a given applied potential to monitor the current produced if oxidizable or reducible compounds appear in the column effluent. More details about the principle, design and function of these detectors may be found elsewhere[63,68]. Detection is possible only if the eluate conducts an electric current, i.e., in ion-exchange or reversed-phase chromatography, where it can be used with gradient elution. The electrolyte, if it is not contained in the mobile phase, may be added to the column effluent before it enters the detector cell. The sensitivity of detection is very high, but the detection is limited to certain classes of compounds. The selectivity of detection is increased by the possibility of adjusting the working potential necessary for the detection of a given solute or of a group of solutes.

For a number of organic compounds, such as amino acids, amines, lipids and other biochemically important compounds, there are no adequate specific detectors at present. Transport detectors can be used, but their low sensitivity and other disadvantages make them rather impractical. For these compounds, a derivatization step is usually necessary to allow sufficiently sensitive detection. In this step, derivatives are formed that either strongly absorb light in the UV or visible region (such as phenacyl esters of fatty acids) or possess strong fluorescence (such as dansyl derivatives of amines and amino acids) and can be monitored in column effluents using common UV or fluorescence detectors. Of course, only reactions that yield 100% of derivatives from the sample compounds and are sufficiently fast are adequate.

The derivatives may be formed directly in the sample before injection on to the chromatographic column (pre-column derivatization), which often allows the use of a more efficient chromatographic system for separation, such as reversed-phase chromatography for dansyl amino acids, while less efficient ion-exchange chromatography should usually be used for non-derivatized amino acids.

The so-called post-column derivatization is the other approach to the chromatography of derivatives. This is a procedure where suitable derivatives are formed "on-line", between the chromatographic column and the detector. One or more reagents are continuously added to the column effluent by means of an auxiliary pumping system (usually peristaltic pumps) and the combined streams of the effluent and reagent(s) from the mixing tee union(s) are led into the flow-through microreactors, which have the form of small inner diameter glass or plastic coils, microcolumns packed with glass beads, etc., and are usually placed in a heating bath, which allows the necessary reaction temperature to be achieved. The dimensions (length and diameter) of the reactors should be carefully designed to allow for a sufficiently long dwelling time of the reaction mixture in the reactor to achieve completion of the reaction, but at the same time not to cause significant broadening of the eluted peaks and a loss of

TABLE 9.1

SOME USEFUL REAGENTS FOR DERIVATIZATION AND POST-COLUMN REACTION DETECTION APPLICABLE IN GRADIENT ELUTION LIQUID CHROMATOGRAPHY

Reagent	Functional group	Product	Type of detection*
Ninhydrin	$-NH_2$ + -COOH (amino acid)		Vis
Phenyl isothiocyanate	$-NH_2$ + -COOH (amino acid)	Phenylthiohydantoin-amino acids	UV
3,5-Dinitrobenzoyl chloride	$-NH_2$, -OH	3,5-Dinitrobenzoyl amides and esters	UV
5-Dimethylaminonaphthalene-1-sulphonyl chloride (dansyl chloride)	$-NH_2$ -OH (phenolic) Amino acids	Dansyl amides and esters	Fl
Fluorescamine	$-NH_2$ Amino acids	Pyrrolinones	Fl
o-Phthalaldehyde	$-NH_2$ Amino acids		Fl
1,2-Naphthoylenebenzimidazole-6-sulphonyl chloride (NBISyl chloride)	$-NH_2$	NBISyl amides	Fl
N-Succinimidyl-*p*-nitrophenyl acetate	$-NH_2$	*p*-Nitrophenylacetamide	UV
p-N,N-Dimethylamino-*p*'-benzeneazobenzoyl chloride	$-NH_2$, -OH	*p*-N,N-Dimethylamino-*p*'-benzeneazobenzoyl-amides and esters	Vis
Phenyl isocyanate	-OH	Alkyl phenyl urethanes	UV
Benzoic anhydride	-OH (sugars)	Sugar perbenzoates	UV
1-Chloromethylisatin	-COOH	Isatinylmethyl esters	UV
o-Phenylenediamine	-COCOOH	2-Quinoxalinol derivatives	UV
9,10-Diaminophenanthrene	-COOH	2-Phenanthrimidazoles	Fl
p-Methoxyaniline	-COCl	*p*-Methoxyanilides	UV
p-Bromophenacyl bromide	-COOH	*p*-Bromophenacyl esters	UV
2-Naphthacyl bromide	-COOH	2-Naphthacyl esters	UV
1-*p*-Nitrobenzyl-3-*p*-tolyltriazene	-COOH	*p*-Nitrobenzyl esters	UV

(Continued on p. 254)

Table 9.1 (continued)

Reagent	Functional group	Product	Type of detection*
4-Bromomethyl-7-methoxy-coumarin	-COOH		Fl
2,4-Dinitrophenylhydrazine	-C=O	2,4-Dinitrophenyl-hydrazones	UV, Vis
7-Chloro-4-nitrobenzo-2-oxa-1,3-diazole	-SH		Fl
p-Nitrobenzyl-*n*-propylamine hydrochloride	-NCO	Ureas	UV

*Vis = visible-range absorption; UV = ultraviolet absorption; Fl = fluorescence.

separation efficiency. Each reactor should be matched to the flow-rate of the mobile phase. More details about derivatization in liquid chromatography can be found in a book by Lawrence and Frei[69].

Some useful reagents for derivatization that can be utilized in gradient elution liquid column chromatography are listed in Table 9.1.

A useful possibility of post-column reactors is to utilize the selective capabilities of certain compounds to react with the reagents added to the eluate to form a certain absorbing (coloured) or fluorescent product.

The amino acid analysers were the first commercial instruments where the principle of post-column reaction was used (reaction with ninhydrin followed by monitoring of the absorption in the visible region). These and analogous instruments (such as carbohydrate analysers) make use of relatively low-efficiency large-diameter ion-exchange columns, where post-column peak broadening is usually not critical.

Another example of reaction detectors of this type is the use of enzymatic catalysed reactions to detect selectively certain types of enzymes that catalyse the same reaction (isoenzymes). For example, if a mixture of lactate and nicotinamide adenine dinucleotide (NAD) buffered to pH 9.0 is added continuously to the eluate from the chromatographic column, certain enzymes catalyse the reaction yielding pyruvate and the reduced form of NAD (NADH). NADH can be detected either by a UV detector, using its strong absorption at 340 nm or by an electrochemical detector, using the possibility of its electrochemical oxidation[70].

The derivatization approach is especially useful in connection with the gradient elution technique, where the choice of detectors is much more restricted than under isocratic elution conditions. The peculiarities of the use of UV and fluorescence detectors to monitor the derivatives are discussed above.

9.2. SENSITIVITY OF DETECTION; DETECTION LIMIT

The sensitivity of a detector is the change in its signal per unit concentration change of a solute in the measuring cell, and the minimum detectable concentration of a solute in the detector cell is usually defined as twice the detector noise (i.e., of average random fluctuations of the detector baseline). Analysts are usually most concerned with (a) the least *amount* (mass) of sample solute and (b) the least *concentration* of solute in the sample than can be detected. These quantities are commonly, but rather incorrectly, termed the "detection sensitivity"; however, in reality they refer to the detection limit for a given solute, which is the property of the whole chromatographic system including the detector and not of the detector alone.

Three factors control the detection limits in liquid column chromatography and there are significant differences between their role under isocratic conditions and in the gradient elution technique:

(a) peak broadening as the solute migrates in the column;

(b) noise level and baseline drift, which determine the sensitivity range of the detector and the minimum detectable concentration in the detector cell;

(c) amount of sample that can be injected on to the column.

Migration of the solute in the column leads to dilution of the solute in the eluate with respect to its concentration in the sample injected on to the column. The concentration of the solute in the eluate at the time of elution of the peak maximum is approximately twice as high as the mean concentration in the peak volume (peak width in volume units, i.e., the volume of the eluate, containing the eluted solute). Hence both the mean and the maximum peak concentrations are inversely proportional to the peak volume, i.e., to the bandwidth. The height of the peak recorded as the detector signal is proportional to the concentration of the solute in the peak maximum and inversely proportional to the bandwidth.

Consequently, the ratio of the peak height in gradient elution chromatography to the peak height under isocratic conditions, $S_{g/i}$, may be used as a measure of the gain in sensitivity (or, better, in detection limit) in gradient elution chromatography in comparison with isocratic conditions. Then, using eqns. 1.5 and 4.35, we obtain

$$S_{g/i} = \frac{w_{is}}{w_g} = \frac{1 + k'_{is}}{1 + k'_f} \qquad (9.1)$$

where w_{is} and w_g are the peak widths under isocratic and gradient, respectively, conditions, k'_{is} is the capacity factor under isocratic conditions and k'_f is the instantaneous capacity factor of the solute at the time of elution of the peak maximum in gradient elution chromatography (see Part II). Because k'_f is lower than k'_{is} under the usual operating conditions, the peak widths are diminished and peak heights increased when working with mobile phase gradients in comparison with the isocratic elution.

It is possible to calculate $S_{g/i}$ after introducing the necessary equations for k'_{is} and k'_f that apply in a given chromatographic system with the gradient profile used (Part II). To obtain a better understanding of the influence of the gradient parameters on peak height, we can use an example of a linear gradient of the concentration of an organic solvent in the mobile phase, c:

$$c = A + BV \tag{4.17}$$

applied to a reversed-phase system, where the dependence of the capacity factor, k', on c can usually be described as

$$k' = k'_0 10^{-mc} \tag{1.28}$$

where V = volume of the eluate, A and B = parameters of the gradient, and k'_0 and m = characteristic constants of a solute in a given reversed-phase system (see Part II).

In this example, eqn. 9.1 can be rewritten in the following form, using the equations for w_i and w_g from Parts I and II:

$$S_{g/i} = \frac{1 + \dfrac{k'_0}{10^{cm}}}{\left(1 + \dfrac{k'_0}{2.31\, mBk'_0V_m + 10^{mA}}\right) G} \tag{9.2}$$

where G is the so-called band-compression factor.

From eqn. 9.2 we can see that the peak heights under gradient conditions relative to isocratic elution increase with increasing slope of the gradient, B, with increasing concentration of the organic solvent, A, in the mobile phase at the start of the gradient, with increasing retention of the solute, k'_0, and with decreasing concentration of the more efficient eluting component in the mobile phase used under isocratic conditions. Thus, for example, a solute with k'_0 = 1000 and m = 3 would be eluted in 66% methanol in ten column void volumes and would be eluted in an approximately equal retention time using a 30 min gradient from 0 to 100% methanol (at 1 ml/min; B = 1/30) on a column with V_m = 3 ml; however, the height of peak would be increased almost 5-fold in the gradient experiment.

Owing to the additional band compression (factor G), even narrower and higher peaks could theoretically be achieved in gradient elution than the peak of an unretained compound, but at a cost of impractically steep gradients.

Unfortunately, the gain in the detection limit due to the band sharpening in gradient elution chromatography can be used fully only with fluorescence and UV detectors at higher wavelengths, but it is at least partly lost by the lower sensitivity range of UV detectors that must be used because of the baseline drift at 254 nm and lower wavelengths, which would allow more general detection.

The gradient elution technique offers another possibility of improving the minimum detectable concentration in samples, namely by increasing the volume of sample injected on to the column. In isocratic elution chromatography, it is important to keep the volume of the injected sample smaller than one quarter to one third of the peak volume of the compound that would be eluted with the column void volume, otherwise larger sample volumes would contribute significantly to additional peak broadening. Usually, 25-50 μl is a tolerable maximum sample volume with modern analytical columns.

This is not the case in gradient elution chromatography, where usually much larger sample volumes may be tolerated. The elution strength of the mobile phase at the time of injection is usually relatively very low in gradient elution chromatography and sample compounds become retained in a narrow layer of the packing material on the top of the column. Because they begin to migrate along the column only after a high enough elution strength has been achieved in the mobile phase, the solute bands are not broadened more than if a few microlitres of sample were injected.

Surprisingly high sample volumes can sometimes be injected, so that gradient elution allows sample pre-concentration to be connected with chromatographic separation on a single column. This technique was successfully applied, for example, to the analysis of UV-absorbing trace impurities in samples of water[71,72]. However, two conditions must be fulfilled to allow this approach: (a) a sufficient amount of sample must be available and (b) the sample must not contain large amounts of other compounds that would overload the column.

A pre-column packed with the same material as the analytical column may be used for the enrichment step to increase the lifetime of the analytical column[73].

9.3. QUANTITATIVE ANALYSIS BY THE GRADIENT ELUTION TECHNIQUE

When considering quantitative analysis in connection with the gradient elution technique, we should first of all be aware of the limitations imposed on the use of detectors and their sensitivity. Usually, far more significant baseline drift is observed in gradient elution experiments than under isocratic conditions,

References on p. 282.

which must be taken into account in evaluation of the chromatograms. Therefore, it is advisable to use integrators with the capability of correction for baseline drift.

High-precision quantitative analysis by gradient elution chromatography requires precise control of the flow-rate and of the composition of the mobile phase, as under isocratic conditions[21,74].

The precision in the flow-rate of modern liquid chromatographic instrumentation is usually 0.5-1.0%, whereas the ultimate precision reported is 0.2-0.3%[74,75]. With earlier pumping systems, the flow-rate precision may be worse than 1%.

Under isocratic conditions, the composition of the mobile phase can usually be controlled precisely. The precision of the mobile phase composition programme that can be obtained in the gradient elution technique is usually determined by the performance of the hydraulic part of the gradient device. It depends on the precision of the flow-rates of the pumps used, is influenced by the pump ripple[28], by compressibility effects[20-23], imperfect mixing of the mobile phase components[30] or compositional oscillations[37,39] and other effects discussed in Chapter 8. The precision may be worsened by virtue of the solvent demixing effect in adsorption chromatography (see Chapter 10). For purposes of quantification, good reproducibility of flow-rates and of the composition (or programme) of the mobile phase is more essential than the absolute accuracy of pre-set operating conditions. Short-term noise is more detrimental for accurate quantitative data than long-term baseline drift.

Quantitative analysis may be performed by either peak-height or peak-area measurement. Both of these quantities are influenced by the precision of the flow-rate and of the gradient profile. The precision of quantitative results is more influenced by run-to-run fluctuations in the slope of the gradient than by fluctuations of the initial mobile phase composition and/or of the shape (curvature) of the gradient.

When selecting the appropriate method for the quantitative evaluation of results in gradient elution chromatography, it should be noted that a change in the flow-rate of the mobile phase usually leads to a greater change in the peak areas than in the peak heights. Consequently, peak-height measurement yields better quantitative results if the flow control is poor but the composition of the mobile phase or the gradient programme can be controlled and reproduced precisely[74]. This was the case with earlier liquid chromatographic systems operated under isocratic conditions.

The peak area is relatively less dependent on the composition of the mobile phase and is less influenced than the peak height by fluctuations in the gradient programme at a constant flow-rate. A change in the composition of the mobile phase leads to a change in the capacity factor of a sample compound and conse-

TABLE 9.2

PRECISION OF PEAK AREAS (NORMALIZED) IN GRADIENT ELUTION EXPERIMENTS

Compounds (peaks): 1 = fluorene; 2 = phenanthrene; 3 = fluoranthene. Instrument, Varian LC 5000; column, 30 cm x 4 mm I.D., Micro-Pak MCH-5 (reversed phase); gradient from 65 to 85% (v/v) acetonitrile in 12 min.

Run No.	Peak No. 1 (%)	Peak No. 2 (%)	Peak No. 3 (%)
1	18.74	55.73	25.53
2	18.86	55.76	25.37
3	18.74	55.84	25.43
4	18.61	55.82	25.57
5	18.84	55.66	25.51
6	18.91	55.18	25.91
7	19.02	55.48	25.50
8	18.77	55.32	25.91
9	18.68	55.44	25.88
10	18.93	54.90	26.18
11	18.64	55.11	26.25
12	18.75	55.61	25.64
13	18.71	55.73	25.56
14	19.11	55.39	25.50
15	18.58	55.76	25.66
$\bar{x}$	18.79	55.52	25.69
σ	0.151	0.287	0.270
$\sigma_{rel.}$	0.81%	0.52%	1.05%

quently to a change in peak broadening. An altered bandwidth thus corresponds to a changed concentration of sample compound at the peak maximum and thus to a changed peak height. Therefore, it is preferable to use peak-area measurements if the flow control is good, but the gradient profile (or composition of the mobile phase under isocratic conditions) cannot be maintained precisely. Thus, whether peak-height or peak-area measurement yields more precise results in gradient elution chromatography depends on the instrumentation and system. In modern gradient elution instruments with flow feedback control, peak-area measurement seems to offer distinct advantages over peak-height determination[74,76].

The reproducibility of peak-area measurements in repeated runs in gradient elution chromatography using modern commercial gradient elution chromatographs in well defined chromatographic systems is given by manufacturers as relative standard deviations between 0.5 and 1.0% while the reproducibility of retention times is usually characterized by relative standard deviations better than 0.5%[30,42,119]. There are not very significant differences between the data from different manufacturers either producing equipment with gradient formation in the high-pressure part using two reciprocating pumps, or instruments with gradient formation using alternate flows of mobile phase components at the low pressure at the inlet of a high-pressure reciprocating pump. Table 9.2 shows an example of the reproducibility of the peak areas of three polyaromatic hydrocarbons when chromatographed on a Micro-Pak MCH-5 column (300 x 4 mm I.D.) with a linear gradient from 65 to 85% (v/v) of acetonitrile in water in 12 min using a Varian Model LC 5000 single reciprocating pump liquid chromatograph, where the gradient is formed at the inlet of the high-pressure pump.

Chapter 10

COLUMN PACKING MATERIAL AND MOBILE PHASE IN GRADIENT ELUTION LIQUID CHROMATOGRAPHY: PRACTICAL ASPECTS

10.1. PREFERENTIAL SORPTION OF A CERTAIN COMPONENT OF THE MOBILE PHASE: SOLVENT DEMIXING EFFECT

The columns packed with fine particle material used in modern liquid column chromatography do not cause significant deformation of the concentration gradient profile applied on the top of the column by virtue of the column void (dead) volume[27]. However, a special type of discrepancy between the experimental and the pre-set gradient profile may be caused by the column packing material in certain systems. Under certain conditions, the stronger eluting agent from the binary mobile phase may be preferentially retained by the column packing material that has not been equilibrated previously with the mobile phase used. This phenomenon is well known as the "solvent demixing effect" from thin-layer and dry-column chromatography on polar adsorbents[77,78]. In modern liquid column chromatography, solvent demixing does not occur under isocratic conditions if the column has been allowed to come to equilibrium with the mobile phase for a sufficiently long time. However, it may be occasionally observed if a column not sufficiently equilibrated with the mobile phase is used.

In gradient elution chromatography, the column is initially in equilibrium with the mobile phase containing no or a low concentration only of the stronger eluting agent. This concentration is increased continuously with time as the gradient elution progresses and preferential uptake of the stronger eluting agent may occur from the mobile phase by the column packing material. Thus the stronger eluting agent is retained almost completely from the mobile phase during the initial period of the elution, where the mobile phase has only a very low eluting strength. After the saturation of the column packing material by the stronger eluting agent, the concentration of this component in the mobile phase increases rapidly. Consequently, a front of the mobile phase rich in the strong eluting agent is created, which moves along the column with a relatively high elution strength. The rapid increase in the elution power at this front may displace weakly retained sample compounds, which migrate together in the column and are eluted as a sharp unresolved band or as a few poorly resolved bands. The retention volumes of other relatively poorly retained compounds may be strongly influenced, even if they are eluted as resolved peaks. The retention volumes of more strongly retained solutes are relatively little influenced by the solvent demixing effect[7,78,79]. This phenomenon is illustrated in Fig. 10.1, where the deviations from the linear gradient profile and the peak of weakly retained compounds (impurities in the weak solvent) eluted by the front of the strong solvent are apparent from the trace of the UV detector and from the results of the GC analysis of the solvent components in narrow fractions of the eluate (system: silica, gradient 0-10% of n-propanol in n-heptane in 30 min)[79].

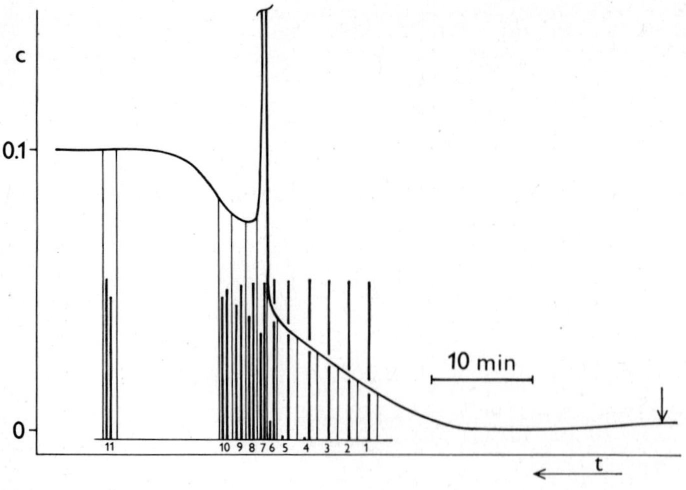

Fig. 10.1. An example of the solvent demixing effect in gradient elution chromatography on polar adsorbents. Comparison of the concentration profile in gradient elution chromatography [linear blank gradient from 0 to 10% (v/v) of n-propanol in n-heptane accomplished in 30 min on a column packed with Porasil A, 37-75 μm, 900 x 2 mm I.D.] with the results of gas chromatographic analysis of the individual fractions. Full curve, trace of the Waters (LDC) UV 254 nm detector, 0.02 a.u.f.s.; below the curve, 11 fractions (0.4-0.8 ml each) taken for gas chromatographic analysis are indicated; for each fraction, the concentration of n-heptane (first vertical line) and that of n-propanol (second vertical line) are traced inside the corresponding interval (the sensitivity of the gas chromatograph for n-propanol was 64 times greater than that for n-heptane). c = % (v/v)·(10^{-2}) of n-propanol in n-heptane; t = time elapsed (min). (Reproduced from ref. 79 with permission.)

Of course, the solvent demixing effect is highly undesirable and may cause difficulties in the prediction of elution behaviour and in the interpretation of chromatograms obtained in gradient elution chromatography. As has been already mentioned, the solvent demixing effect occurs mainly in chromatography on polar adsorbents. Its relative importance increases with increasing differences in the polarities of the solvents used as the components of the mobile phase.

To eliminate the undesirable influence of this effect on separation, some workers have suggested the use of a set of several individual solvents with gradually increasing polarities instead of two solvents only with a large difference in polarities to form concentration gradients. To cover a wide polarity range of the mobile phase, Scott and Kucera[44] introduced so-called "incremental gradient elution" using as many as twelve different solvents, which would allow the elution of compounds with such large polarity differences as squalane and glucose to be accomplished in a single chromatographic run. The practical use of incremental gradient elution is connected with great experimental difficulties, the main one being the use of UV-absorbing solvents, so that special instrumentation with a moving-wire flame-ionization detector must be used (for further details concerning instrumentation see Sections 8.5 and 9.1).

Snyder and Saunders[80,81] reduced the set of solvents to six that do not absorb in the UV region and make possible the use of common UV detectors.

Gradients composed of more than two or three different mobile phase components are difficult to generate in practice with most commercial equipment (Section 8.5) and are extremely complex to treat theoretically. Therefore, gradients formed from only two or three solvents are preferable and have been used almost entirely in practical applications. However, the solvent strength range is limited and one should be aware of the possible effect of solvent demixing on separation, as the necessity for a large polarity change during gradient elution contrasts with the requirement of a small difference in polarities between the mobile phase components to decrease the solvent demixing effect.

Another complicating aspect in practice is the presence of water in all solvents used in adsorption chromatography. Even the least polar solvent (such as a saturated hydrocarbon) contains a trace amount of water, which is very difficult to control and varies with respect to the purity, previous treatment and time of storage of the solvent and fluctuations of the ambient temperature and humidity of the air. Even a short period of contact of the solvent with the ambient air is sufficient for absorption of trace amounts of water by the solvent. The capacity factors of sample compounds in non-polar solvents may change dramatically by an order of magnitude in a non-polar hydrocarbon if the content of water is changed by only a few parts per million[33,48,82,83] and the time of equilibration of the water content between the non-polar solvent and the adsorbent may take several hours. However, the relative effect of the water content on the capacity factors of sample compounds and the time necessary for equilibration decrease with increasing polarity of the organic solvent.

In chromatography on polar adsorbents, gradient elution is usually started from a non-polar solvent, which contains a different amount of water (usually lower) than the more polar solvent used as the second component of the mobile

References on p. 282.

phase gradient. Thus, in practice, we have a ternary mobile phase gradient with changing concentration of water as the third component instead of the binary solvent system. This additional unintended gradient of water concentration may give rise to problems analogous to those with the use of the gradient formed from two solvents with considerably different polarities, especially significant irreproducibility in the retention behaviour in the initial part of the gradient elution.

This additional undesirable water concentration gradient can be eliminated, the reproducibility of the elution improved and the equilibration time shortened to a few minutes if so-called "isohydric" solvents are used, i.e., those solvents which have the water content adjusted to equilibrium with the water content in the adsorbent[83]. The adjustment of the water content in a series of solvents is tedious and it is virtually impossible to keep the equilibrium water content in the solvents over a long time period.

If we consider the effects of solvent demixing and of the additional water gradient in gradient elution chromatography on polar adsorbents (the first effect would be suppressed by using solvents with smaller differences in polarities while the opposite holds true for the second effect) together with the requirements on the range of polarity change during the gradient elution and limitations imposed by the instrumentation, binary or ternary gradients seem preferable to incremental gradient elution. If we are careful, even gradients formed from solvents with relatively largely differing polarities, such as *n*-heptane or *n*-hexane as component a and *n*-propanol as component b, may be used successfully in chromatography on polar adsorbents[9].

To suppress the influence of solvent demixing effect in such systems, it is preferable to start the elution with the mobile phase containing the stronger eluting component at a low concentration (e.g., 0.5-5%, depending on the polarity of solvent b) rather than with the pure non-polar solvent. This would allow the adsorbent to be saturated at least partially with the more polar component of the mobile phase and the preferential adsorption of this component would be greatly diminished in the gradient run. Further, it is advisable to avoid too steep an increase in concentration of the polar component in the mobile phase, especially in the initial part of the gradient. Therefore, convex gradients are unsuitable for chromatography on polar adsorbents, because these gradients are steepest just in their initial parts. Linear, or better, concave gradients are to be preferred (moreover, concave gradients approximate the best linear solvent strength gradients in chromatography on polar adsorbents)[7].

If the sample contains some very weakly retained compounds, the solvent demixing effect can be suppressed and the separation improved by using two-step elution with an isocratic step first, with the mobile phase containing the polar component at the maximum concentration that would just allow the separation of weakly retained sample solutes. Then the gradient step follows, to speed up the separation of more strongly retained compounds, the retention behaviour of which is not influenced significantly by the solvent demixing effect[9].

10.2. REGENERATION OF CHROMATOGRAPHIC COLUMNS AFTER THE END OF THE GRADIENT

Restoring the initial equilibrium on the column after the end of the gradient elution may be subject to certain problems, especially in chromatography on polar adsorbents. It is very difficult to achieve complete re-equilibration of the column after the end of the gradient if the gradient elution is started in a pure non-polar hydrocarbon with a low water content. Therefore, it is important to standardize the conditions of the column re-equilibration and to use the same regeneration procedure after the end of each gradient elution run. This is the best way to achieve good reproducibility of the results.

Sometimes it is useful to allow the pure stronger eluting component of the mobile phase to pass through the column after the end of the gradient elution for some time to accomplish the elution of strongly retained impurities present originally either in the sample or in the weaker eluting component of the mobile phase. Otherwise, these impurities may be eluted during the later gradient runs and may interfere with the interpretation of the results.

It has been acknowledged that the initial equilibrium in gradient elution chromatography can be restored much faster by running a "reverse" gradient after the completion of the gradient elution and after the elution of strongly retained impurities rather than by switching directly to the initial composition of the mobile phase. After the end of the "reverse" gradient, which is usually much steeper than the "working" gradient, it is necessary to let a sufficient volume of the starting mobile phase (at least ca. 10 column dead volumes) pass through the column before the start of the next experiment[48]. If a sample loop injection valve is used in the instrument, we must not forget also to flush the injection loop with the starting mobile phase. The regeneration of the column can be accelerated by using an increased flow-rate in comparison with the gradient run.

It is important to standardize the conditions for column re-equilibration not only in work with polar adsorbents, but also with non-polar chemically bonded phases. Here, too, the complete elution of the organic solvent from the column containing a non-polar bonded phase may take a long time, which makes impractical the achievement of perfect equilibrium between the column and pure water, often used as the starting eluent in reversed-phase gradient elution chromatography.

Another problem consists in the formation of void volumes in columns packed with chemically bonded phases, which is increased by changes in mobile phase

References on p. 282.

composition during gradient elution and column regeneration procedures and is further aided by high pressures built up if regeneration is performed at an elevated flow-rate. Therefore, it is necessary to renew the packing layer at the top of the column from time to time. These problems seem to be improved with so-called radially compressed columns, consisting of a flexible-walled cartridge packed with a microparticulate column packing material and placed in a special compression chamber, where a hydraulic liquid pressure is applied on the outer walls of the flexible column tube. Thus, the packing material is compressed towards the flexible column walls and creation of void spaces is largely avoided, which improves the long-term stability of the column even at the elevated flow-rates and pressures used during column re-equilibration procedure[120].

10.3. PURITY OF THE MOBILE PHASE COMPONENTS

The purity of the mobile phase components is far more critical in gradient elution chromatography than in isocratic operation. Under isocratic conditions, after a certain time equilibrium is established between the content of the impurities in the mobile phase and on the column. Consequently, the presence of trace impurities in isocratic elution chromatography leads only to an increase in the baseline response of the detector, but does not cause significant baseline drifts or other difficulties. Of course, this holds true only if the impurities in the mobile phase cannot influence significantly the retention of sample compounds.

In gradient elution chromatography, trace impurities in the mobile phase components usually increase the baseline signals to a different extent for each component. Impurities in the stronger eluting component of the mobile phase usually lead to baseline drift that may make it difficult or impossible to work at higher sensitivity ranges of the detector (for further details, see Chapter 9). The situation is even worse with impurities in the less efficient eluting component of the mobile phase. Such impurities are often retained on the column. Because relatively large amounts of the mobile phase component with a lower elution strength flow through the column in the equilibration step and at the beginning of the gradient run, relatively large amounts of the impurities accumulate on the column during these periods and are eluted later during gradient elution as a group of badly resolved bands or, at least, give rise to baseline fluctuations. This effect may interfere severly with the interpretation of the chromatogram and its magnitude depends on the volume of the less efficient component of the mobile phase that passed through the column prior to the start of the gradient. This occurs when polar impurities are present in saturated hydrocarbons used as the starting component of the mobile phase in gradient elution chromatography on polar adsorbents or if organic impurities are present in water, usually used as the starting mobile phase component in reversed-phase chromatography.

Consequently, special care should be devoted to the purification of the solvents used in gradient elution chromatography. Very often, the purity of spectroscopic-grade solvents is insufficient for this purpose. Ultrapure solvents, such as HPLC grade, are commercially available. If these costly solvents are not available, it is necessary to purify lower purity solvents. However, even repeated distillation in glass is often not sufficient. To examine whether the purity of solvents is sufficient for use in gradient elution chromatography, it is useful to run a blank gradient experiment without injecting a sample. This approach is also used by producers of HPLC solvents as the strictest test of quality[85,86]. In spite of the commercial availability of a number of solvents with purity generally adequate for gradient elution applications, there is no a priori guarantee that such solvents will be pure enough for the specific application and method of detection used. In such an instance, it would be necessary to purify the solvents before use. Another reason is the relatively high price of HPLC- and spectroscopic-grade solvents, which may make it necessary to use and purify common reagent-grade solvents[48].

Non-polar and low-polarity solvents such as hydrocarbons, dichloromethane, chloroform, diethyl ether and tetrahydrofuran may be purified in the laboratory by extracting them with concentrated sulphuric acid to remove basic impurities. Then, distillation on a rectification column is carried out and the purfication may be finished by passing the solvent through a large-diameter column packed with a highly active adsorbent (silica and alumina). Water may be doubly distilled in glass with addition of potassium permanganate and potassium hydrogen carbonate to oxidize organic impurities and the remaining traces of organics may be removed by adsorption on a preparative chemically bonded reversed-phase column. The purified water should be stored in glass bottles for no longer than 2 days, because it absorbs organics from the air. Contact with plastic bottles and, if possible, with plastic tubing should be avoided if a high-sensitivity detector setting is intended. In Fig. 10.2, chromatograms are compared from experiments using de-ionized water stored for 1 day in a polyethylene bottle (A) and water doubly distilled with addition of potassium permanganate-potassium hydrogen carbonate and passed through an octadecylsilica purification column (B).

It is more difficult to purify medium-polarity solvents usually used as component b of the mobile phase gradient, such as methanol or acetonitrile in reversed-phase chromatography and propanol in chromatography on polar adsorbents. Here, suitable adsorbents for purification by adsorption are not available and thus usually only repeated distillation on a rectification column can be used. If salts are used in the mobile phase in ion-exchange chromatography and some-

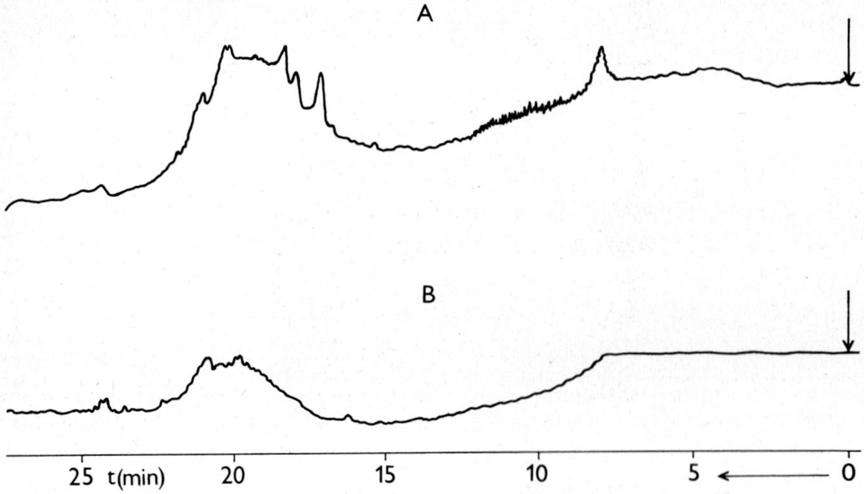

Fig. 10.2. UV record of a blank gradient using deionized water stored for one day in a polyethylene flask (A) versus the record of a blank gradient using deionized water doubly distilled with the addition of potassium permanganate-potassium hydrogen carbonate and then passed through an octadecylsilica purification column (B). Column, 300 x 4.2 mm I.D., C 18/silica (10 μm); gradient, 0-100% methanol in water in 15 min; 1 ml/min. Detection: UV at 254 nm, 0.08 a.u.f.s.

times in reversed-phase chromatography, they can be recrystallized if they are not available commercially in a sufficiently pure grade. Recrystallization may be combined with purification on ion-exchange columns[87]. Thus, it is recommended to purify potassium dihydrogen phosphate (which is very suitable for the formation of gradients in ion-exchange and in salting-out reversed-phase chromatography) from UV-absorbing impurities by passing it through a column packed with Dowex 1-X8 anion-exchange resin in the phosphate form. Other salts that can be used are sodium, potassium and ammonium citrates, acetates, borates and sulphates. Formate buffers can also be employed, but they may corrode stainless-steel parts of the instrument, which may give rise to leaks.

The difficulties with baseline drift and with impurities in the mobile phase components increase with decreasing wavelength of detection if UV detectors are used. At low wavelengths, the presence of oxygen dissolved in different concentrations in the mobile phase components may interfere. In this instance, it is necessary to degas the components of the mobile phase, preferably by using an inert gas, such as helium.

Degassing of the mobile phase components is highly advisable for another reason - the air dissolved at different concentrations in the mobile phase components may be set free when they are mixed, especially when mixing water with organic solvents in reversed-phase gradient elution chromatography. Air bubbles formed at the inlet of the high-pressure reciprocating pump would interfere with the functioning of the pump inlet check valve, which may lead to failure of the functioning of the devices forming gradient at the low-pressure side. Furthermore, air bubbles that appear in the detector cell may cause malfunctioning of the detector.

Of course, it is necessary to filter the mobile phase components used in gradient elution to remove fine solid particles, which could plug the column inlet frits or cause malfunctioning of the column check valves, especially if reciprocating pumps are used, as in isocratic operation; 0.5-2 μm filters, such as those from Millipore, are suitable for this purpose.

10.4. PRACTICAL ASPECTS OF THE SELECTION OF MOBILE PHASE COMPONENTS FOR GRADIENT ELUTION CHROMATOGRAPHY

Theoretical aspects of an appropriate choice of the components of the mobile phase gradient are discussed in Part II. However, we should bear in mind several practical points important for this selection. First, if binary gradients are used, it is necessary that the mobile phase components are completely miscible over the whole range of mobile phase composition used in the gradient run. Further, the stronger eluting component of the mobile phase should dissolve the sample well. The sample can be dissolved either in this component of the gradient or in some mixture of the two components used to prepare the gradient. The components of the mobile phase should be compatible with the detector and sufficiently pure not to cause problems with baseline fluctuations or drift and with impurity peaks at the sensitivity range of the detector used for analysis.

The components used for the formation of the gradient should allow a sufficiently large change of the elution strength during the gradient.

If reciprocating pumps are used, the components of the mobile phase should not be very volatile, otherwise they are evaporated partially by virtue of the heat evolved in the pump or in the solenoid valves at the inlet of the pump and bubbles are formed during the suction period of the pump which could cause failure of the pump inlet check valves.

It is further important to degas completely the two components of the mobile phase and to filter solid particles (see Section 10.3).

Finally, the price of the solvents, their toxicity and the possibility of potential explosions should be considered when selecting appropriate solvents for gradient elution chromatography.

A survey of certain mobile phase components generally useful in gradient elution chromatography is given in Table 7.1, but of course other solvents may be also used, such as those in various application examples given in Part IV.

References on p. 282.

Chapter 11

SPECIAL GRADIENT ELUTION TECHNIQUES AND OTHER TECHNIQUES OF PROGRAMMING OPERATING CONDITIONS IN LIQUID CHROMATOGRAPHY

In liquid chromatography, techniques for programming the operating conditions other than gradient elution may be also used. Gradient elution will be briefly compared with these techniques here. It can be also combined with some of them. In certain special instances, mobile phase gradients may be formed directly in the column itself without the need for gradient elution instrumentation.

11.1. "SEQUENTIAL ISOCRATIC STEP" LIQUID CHROMATOGRAPHY

This is not a technique for programming operating conditions, but it can be used as an alternative to gradient elution for the separation of complex mixtures that differ widely in retention. It consists in a series of successive isocratic experiments with a different composition of the mobile phase in each run[88]. Sample material is injected and chromatographed in each step. If the composition of the mobile phase is changed from one run to another in a systematic manner, a set of chromatograms results in which a few sample compounds appear as resolved bands while the remainder are either not resolved or are not eluted from the column. However, the position of the peaks changes from one run to another so that theoretically all sample solutes appear in the part of the chromatogram with good resolution at least in one of the runs in the set.

The advantage of this technique is better detection possibilities than with gradient elution. Virtually all LC detectors that can be used under isocratic conditions may be employed here. Further, they may be operated at a high sensitivity range, as there are no problems with baseline drift or solvent impurity accumulations. However, the method also has serious drawbacks in comparison with gradient elution chromatography: (1) it is lengthy as the set of "sequential isocratic steps" takes more time than a gradient run; this disadvantage may be partially overcome with microprocessor-controlled liquid chromatographs that can be pre-programmed for unattended operation; (2) the identification of peaks is more difficult than in gradient elution chromatography; (3) there is a danger of the elution of certain more retained sample compounds in one of the following sequential isocratic steps, which would interfere with the interpretation of the chromatogram. In our opinion, the application possibilities of this technique are more limited than those of the gradient elution technique.

11.2. FLOW PROGRAMMING IN LIQUID CHROMATOGRAPHY

In the flow-programmed elution technique, the composition of the mobile phase does not change during the elution, i.e., we use isocratic elution operation, but the flow-rate of the mobile phase is changed (usually increased) either continuously or stepwise with time. For this purpose, a single-pump instrument employed under isocratic conditions can be used and it is possible to change the flow-rate either manually (in the stepwise technique) or automatically with the pump controlled by a central microprocessor or an electronic programmer. Thus, the instrumentation necessary for flow programming is essentially cheaper than the gradient elution liquid chromatograph. However, this is the only advantage of flow programming in comparison with the gradient elution technique.

Flow programming is of much less value than gradient elution as a solution to the "general elution problem". Under flow-programmed conditions, the capacity factors of sample compounds do not change and nor do the retention volumes; only the retention times of later eluted peaks are shortened. As an increased flow-rate of the mobile phase causes a larger broadening of the eluted peaks in volume units, the resolution and the peak heights decrease from the start to the end of the flow-programmed elution, while the operating pressure increases in the same direction.

It is simple to calculate retention times under flow-programmed conditions. As the retention volumes, V_R, and capacity ratios, k', are constant, the retention times, t_R, may be predicted by solving the equation

$$V_R = \int_0^{t_R} F_m dt \tag{11.1}$$

after introducing the respective equation for the time change of the volume flow-rate, F_m. For example, if $F_m = B't$, then

$$t_R = \sqrt{\frac{2\,V_R}{B'}} \tag{11.2}$$

The utility of flow programming is rather limited and this technique is relatively rarely used. It is not suitable for very complex mixtures, but it may be useful in instances where all sample compounds are separated with satisfactory resolution in the early part of the chromatogram and only one or two compounds are "over-resolved" and eluted with higher retention volumes. The retention times of these compounds may be shortened to 25-50% of the original values using a flow programme, at the cost of increased column pressure. Consequently, flow programming is potentially more useful with less efficient columns than with high-efficiency columns packed with small-diameter particles, where the pressure drop is higher and usually cannot be increased very much owing to the instrumental limitation.

Sometimes, flow programming can be combined with gradient elution to decrease the retention times of "over-resolved" last eluted compounds.

More details about this programming technique may be found elsewhere[89,90].

11.3. TEMPERATURE PROGRAMMING

Working at an elevated but constant temperature has proved useful in certain liquid chromatographic separations, mainly those employing ion-exchange columns. Here, the separations at higher temperatures usually yield better resolution and sharper bands of eluted solutes by virtue of decreased viscosity. However, temperature programming is relatively rarely used in liquid chromatography, in contrast to gas chromatography.

As in gas chromatography, the capacity factors and retention volumes in liquid chromatography usually decrease with increasing temperature. However, a change in temperature in liquid chromatography has a far smaller effect on retention than a change in the composition of mobile phase. Under certain circumstances, nearly constant capacity factors are observed at different temperatures; sometimes the capacity factors even increase with increased temperature.

From the technical point of view, temperature programmes are easy to form, if the column is placed in a heated oven where the temperature may be controlled and programmed in much the same way as in gas chromatography. It is also possible to use a temperature control system based on a circulating water-bath. The temperature in the bath can be programmed by programming the voltage of the immersion heater. The insulated bath is connected to the column jacket, through which the water circulates as the column heating medium[105].

A significant shortening of the analysis time (but inferior to that which could be achieved in gradient elution chromatography) can be obtained in reversed-phase chromatography on non-polar chemically bonded phases, where reproducibility of retention data comparable to that in gradient elution chromatography was reported[105].

In chromatography on polar adsorbents using mixed mobile phases composed of a non-polar and a polar or moderately polar solvent, an increase in temperature leads to an activation of the adsorbent surface, i.e., some of the adsorbent sites set free molecules of the preferrentially adsorbed more polar solvent and thus become available for the adsorption of sample compounds. Consequently, the capacity factors of sample compounds should increase with increasing temperature and an inverse temperature programme, i.e., a continuous decrease in temperature

References on p. 282.

with time should be used to achieve acceleration of the elution of later eluting peaks.

However, if a pre-column pre-saturated with the more polar solvent in the mobile phase is used, the content of this solvent in the mobile phase can reach equilibrium with that in the adsorbent in the pre-column while the temperature is increased and the capacity factors of solute compounds decrease with increasing temperature, as in other chromatographic systems.

If the temperature gradient is applied to systems employing columns packed with polar adsorbents, it is never a "pure" temperature gradient, but is always combined with a mobile phase composition gradient, because the concentration of the stronger eluent in the mobile phase changes as the equilibrium distribution of this more polar eluent between the mobile phase and the adsorbent is changing with temperature and consequently with time[92].

If the column with a pre-column pre-saturated with a binary solvent is placed in a heated oven or in a water-jacket, the temperature of which is increased with time, the concentration of the more polar solvent in the mobile phase will increase uniformly throughout the length of the pre-column and column. Thus a gradient of the concentration of the more polar solvent in the mobile phase with "axial equilibrium" can be produced[93].

Another possibility of programming column temperature is offered by "dynamic" temperature gradients, where the temperature gradient is created by a moving oven over the column or by a heating spiral fixed to the column. Here, only part of the column is heated and as the position of heating is moving along the column towards its end, a pulse of a stronger concentration of the more polar solvent in the mobile phase is created, which moves along the column in the same direction and thus, again, a combination of temperature and mobile phase composition gradient is accomplished[94,95].

Of course, it is necessary to re-equilibrate the column (and the pre-column) with the mobile phase at the initial temperature after the end of each experiment. If the temperature gradient is applied with an instrumental device using a constant-pressure pump, a flow-rate gradient is generated at the same time, as the viscosity decreases and the column permeability increases with increasing temperature.

In practice, temperature gradients are very rarely used in liquid column chromatography, even less frequently than flow programming. More details on this topic may be found elsewhere[89,90].

11.4. PROGRAMMING OF THE STATIONARY PHASE; COLUMN COUPLING AND COLUMN SWITCHING TECHNIQUES

It is not easily possible to achieve a continuous change in the properties of the column packing material (stationary phase) in a given column without a simultaneous change in the properties of the mobile phase. In normal-phase chromatography on polar adsorbents, it is possible to achieve a continuous change of the activity of the adsorbent in the column by adding continuously increasing amounts of water to a non-polar solvent, such as a saturated hydrocarbon, used as the mobile phase[96]. However, as the polarity of the mobile phase changes significantly during this procedure, this method of programming the conditions of elution (not frequently used) may be understood as a special example of gradient elution with a very strong solvent demixing effect (due to very large polarity differences between hydrocarbons and water).

More freqently used and far more promising seems to be the use of several columns connected in series or in parallel[89-91,96-102,107-118]. These may be either packed with materials that have different chemical functionalities on the surface or with those with the same functionalities, but differing in capacity or specific surface area (e.g., porous layer versus microparticulate totally porous material).

The mobile phase used in separations on these "coupled columns" usually has a constant composition (isocratic elution). It is possible to separate sample compounds by chromatographing all of them on two or more columns connected in series. More frequently, each column is used to separate only some of sample components, while the remaining compounds are chromatographed on other columns. For this purpose, it is necessary to connect the columns by means of switching valves that make possible the selection of the appropriate column for the separation of a certain number of the sample components. This procedure, so-called "column switching", may speed up the separation of complex sample mixtures and thus provide an alternative to gradient elution for the solution of the "general elution problem"[90].

Several possibilities of experimental arrangements for column switching technique are shown in Fig. 11.1. In arrangement A, three columns are connected in series and the mobile phase may flow either through all three columns, or through columns 1 + 2 or only through column 1 to the detector. During the chromatographic run, all three columns are operated in series first until the sample compounds are distributed in all three columns, but are not yet eluted. Then, column 1 contains the most strongly retained solutes, column 2 moderately retained compounds and column 3 the least retained sample components. In the following step, the outlet from column 1 is switched directly to the detector and the most strongly retained

References on p. 282.

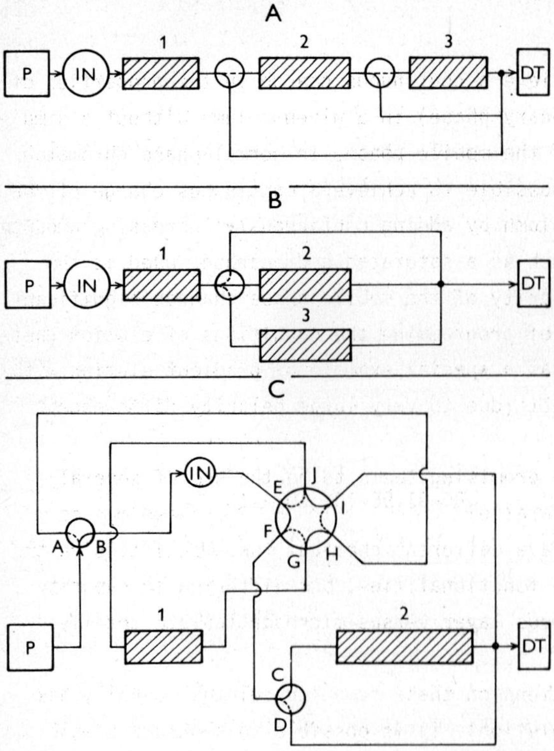

Fig. 11.1. Examples of column coupling - column switching arrangement. 1, 2, 3 = Chromatographic columns; P = pump; IN = injector; DT = detector. (A) Coupling in series; (B) parallel coupling; (C) coupling of two columns making possible operation in series or single operation of each column and back-flushing of column 1. A-J = Inlet and outlet ports of two three-way and one six-way switching valves.

compounds are separated and eluted. Next, the series of columns 1 + 2 is switched to the detector and the less strongly retained solutes are separated and eluted. Finally, the series of three columns is connected to the detector to achieve the separation and elution of the least retained compounds. Thus, the least retained compounds have migrated along the longest column distance, while the most retained solutes have migrated along the shortest part of the column system, which makes it possible to speed up the elution of the strongly retained compounds and to achieve a good resolution of the weakly retained ones. Usually, columns packed with the same material are used, but different packing materials may also be used.

Example B shows another possibility of a three-column arrangement, where the mobile phase may flow through column 1, or through columns 1 + 2 or 1 + 3. Here again, after appropriate switching, column 1 may be used to separate more strongly retained compounds, while some of less retained solutes are allowed to pass into column 2 and the other weakly retained compounds into column 3, where their separation and elution may be accomplished[107].

In example C, the use of a two-column arrangement is demonstrated in connection with one six-port and two three-way valves. This arrangement allows the following four operation techniques with different flow connections of the valve ports: (a) connection B-E-J-G-F-C allows the chromatography of sample compounds on columns 1 (pre-column) and 2 in series; (b) connection B-E-J-G-F-D is used when column 1 (pre-column) should be connected directly to the detector; (c) connection B-E-F-C joins the pump directly to column 2 and then to the detector; (d) connection A-H-G-J-I allows the direction of the flow in column 1 to be reversed and the solute to be back-flushed from this column directly to the detector. By appropriate column switching, it is possible to divide sample compounds into two groups. Each group may be chromatographed on a separate column or one group on column 1 and the other group on columns 1 + 2. The column back-flushing technique may be used for chromatographic separations of samples where the analysis of only a few moderately retained compounds of interest is important and the more retained compounds may be eluted in a much shorter time by flow reversal from column 1 (pre-column). Certain sample mixtures where moderately retained solutes only are of analytical interest contain large amounts of non-retained compounds, which could interfere with the quantification of the compounds of interest and which are eluted near the column dead volume. Here, so-called "pre-column venting" may be used to elute the non-retained compounds directly from the pre-column to the detector and then the compounds of interest are chromatographed on columns 1 + 2 connected in series.

More generally, column switching as in example C in Fig. 11.1 would allow one to "extract" any group of sample compounds of interest from column 1 to column 2, where they will be subjected to further separation, while other compounds would be eluted from column 1 directly to waste. In this way, only the interesting compounds are "cut out" from the mixture for further processing, and the technique is called "front cut" if weakly retained compounds at the front of the chromatogram are of interest, "heart cut" if moderately retained compounds in the middle part of the chromatogram are further chromatographed on column 2 and "end cut" if only the separation of strongly retained compounds at the end of the chromatogram is to be performed.

In these arrangements, columns of different lengths but packed with the same packing material may be used[97]. Here, the programming of the column length for different sample compounds is equivalent to the programming of separation efficiency, i.e. of number of plates, which should be lower for more retained compounds than for those less retained, in order to shorten the time of separation.

References on p. 282.

It is possible to use columns packed with different materials with the same functional groups on the surface, but with different capacities. Thus, polar adsorbents with different specific surface areas may be used for this purpose[98]. More frequently and more generally, column 1 (pre-column) can be packed with porous surface layer material, which has a low capacity due to a low phase ratio (volume of the stationary phase to the volume of the mobile phase in the column), while the other column(s) would contain totally porous microparticulate material with a considerably higher phase ratio and capacity[90,99]. This technique allows greater shortening of the retention times of strongly retained compounds than with column length programming, because these compounds would be allowed to enter only the column containing porous surface layer packing material, where their elution is much more rapid than it could be with totally porous packing materials. Obviously, the column capacity (together with column efficiency) is programmed here.

Selectivity programming is achieved when the columns used are packed with materials containing different functionalities, which makes it possible either to utilize selective or specific interactions between packing materials and solutes or to take advantage of different chromatographic mechanisms in each column. For example, coupled anion- and cation-exchange columns may be used for the chromatographic analysis of complex biochemical mixtures[100]. Another application example is the use of the mechanism of steric exclusion in column 1 packed with a hydrophilic gel, while in the second column, packed with a chemically bonded non-polar material, a reversed-phase mechanism is operative[101]. It is also possible to couple two columns with alkyl bonded phases of different alkyl-group lengths[108].

In the selectivity programming technique, often very complex mixtures may be separated, even those which are difficult or impossible to separate using gradient elution. However, it is essential that all of the sample compounds of interest should have capacity factors in the range that makes possible their elution from either of the columns used in a reasonable time with a mobile phase of constant composition used for the elution in all of the columns of the arrangement. This is a strict requirement, which significantly limits the application possibilities of the very attractive looking selectivity programming technique.

From the experimental point of view, it is necessary to take certain precautions when attempting to use column coupling and column switching techniques. First, the compatibility of the mobile phase with a reasonably rapid elution of sample components from all of the columns used must be guaranteed, as mentioned above. Further, the connection tubing and switching valves used should have as low an internal volume as possible in order to avoid extra-column contributions to band broadening. With certain columns, the reversal of the flow direction in the column back-flushing technique could result in a serious decrease in the efficiency of the column. Finally, columns of approximately equal diameters and HETP should be used for column coupling, otherwise the overall separation efficiency would be seriously deteriorated[102]. Of course, it is necessary to expect a certain decrease in efficiency even for coupled columns with approximately equal plate heights in comparison with the hypothetical simple addition of the plate numbers for each of the columns used in series.

In comparison with gradient elution, column coupling and column switching techniques do not require a costly second pump or programmable valve system and gradient programmer, but only a less expensive system of high-pressure switching valves of low internal volume. On the other hand, this technique is less generally applicable than gradient elution and more demanding on the method development step, because usually the column lengths and the packing materials used have to be "tailor made" for each individual separation problem. However, in certain specific instances, column coupling and switching can yield extremely good results and it seems that much can be expected from this technique in the future.

In particular, column coupling and switching techniques could be useful in trace analysis and in the analysis of complex samples, such as biological materials. Here, the compounds of interest should be concentrated and interfering compounds should be removed. Some common sample clean-up or enrichment techniques may be expensive and time consuming, may yield incomplete recovery of compounds of interest or certain unstable compounds may decompose during the sample pre-treatment steps. Using column coupling and switching techniques, it is often possible to shorten the total analysis time (sample pre-treatment and separation) as the concentration step or the removal of interferents can be achieved on-line to the liquid chromatographic separation system and the recovery of the compounds of interest may be improved, the risk of their decomposition lowered and the total analysis better suited to automation.

To further increase the power of the column switching technique, it is possible to couple it with gradient elution and to speed up the separation[108].

11.5. COLUMN SELF-GENERATED GRADIENTS

Under certain circumstances, gradients of the composition of the mobile phase may be generated in the column itself, without the need for gradient elution instrumentation. For example, a gradient of the polarity of the mobile phase is formed in a column packed with a polar adsorbent if the column temperature is changed, as is discussed in Section 11.3.

Another example is the so-called "chromatofocusing", which was first described by Sluyterman and co-workers[103,104]. This technique has been used with success

References on p. 282.

in the ion-exchange chromatography of proteins, where the separation is based on the elution of proteins from ion-exchange columns in the order of their isoelectric points when the pH of the eluting buffer is continuously changed. For this purpose, a conventional gradient elution device may be used. However, in the chromatofocusing technique, the pH gradient is generated in the column itself as follows. The anion-exchange column used is first equilibrated with a "start" buffer with a pH in the alkaline region, which gives the upper limit of the pH gradient. However, the pH of the buffer used for the elution is selected so as to give the lower limit of the pH gradient. The sample is equilibrated with eluent, applied to the column and the column is eluted with the eluting buffer. As the eluting buffer, containing a large number of differently charged species, moves down the column, the most acidic components of the buffer bind to the basic anion exchanger and the pH of the buffer is thus increased, but the pH at each point in the column is continuously lowered as the elution progresses. Consequently, the pH gradient is formed in the solution leaving the column as if the gradient were prepared in a mixing device connected to the top of the column.

The sample protein applied in the eluting electrolyte originally migrates in a medium where the pH is lower than its isoelectric point and thus it carries a positive charge. However, after migrating a certain distance down the column, the protein reaches the point where the pH is greater than its isoelectric point and here the protein reverses its charge and is retained by the anion exchanger until the pH at this point decreases during the elution below its isoelectric point, where the charge of protein again reverses and the protein is released from the exchanger. This procedure is repeated until the protein is eluted. Thus the sample proteins migrate down the column at certain distances given by the differences in their isoelectric points and will be eluted as separated bands if the differences are sufficiently great. Therefore, the pH interval limited by the starting and the eluting buffers should be chosen in such a way that the isoelectric points of the proteins fall approximately in the middle of the pH gradient. A multi-component buffer should be used that would give an even buffering capacity across the pH range used during the elution. Such buffers and suitable ion exchangers are commercially available from Pharmacia (Uppsala, Sweden). It is recommended that the gradient volume is about 10-15 bed volumes of the exchanger in the column.

As each protein migrates exactly with this point of the gradient, where the pH is equal to its isoelectric point, the band sharpening (focusing) effect is observed, which allows relatively large sample volumes to be applied. The focusing effect can be explained by the fact that the proteins migrate in the lower pH region at the same velocity as the mobile phase, but at an essentially higher velocity than the speed of migration of the pH gradient and thus the protein molecules will reach later or earlier the point where the pH equals the isoelectric point, depending on the volume of sample applied. The proteins originally contained in a large volume would become concentrated at the pH boundaries given by their isoelectric points and eluted as bands whose widths would be given only by the usual band-broadening effects in the column used.

Some examples of the application of this technique are mentioned in Chapter 21 dealing with proteins.

References on p. 282.

REFERENCES TO PART III

1 C. Liteanu and S. Gocan, *Gradient Liquid Chromatography*, Wiley, London, New York, 1974.
2 L.R. Snyder, *Chromatogr. Rev.*, 7 (1965) 1.
3 C. Liteanu and S. Gocan, *Pure Appl. Chem.*, 31 (1972) 455.
4 O. Mikeš, *Chem. Listy*, 54 (1960) 576.
5 H. Veening, *J. Chem. Educ.*, 50 (1973) 429.
6 M. Martin and G. Guiochon, in J.F.K. Huber (Editor), *Instrumentation for High-Performance Liquid Chromatography*, Elsevier, Amsterdam, 1978, Ch. 2 and 3.
7 L.R. Snyder, in Cs. Horváth (Editor), *High Performance Liquid Chromatography. Advances and Perspectives*, Vol. 1, Academic Press, New York, 1980, p. 207.
8 L.R. Snyder and J.J. Kirkland, *An Introduction to Modern Liquid Chromatography*, 2nd ed., Wiley-Interscience, New York, 1979, Ch. 3 and 4.
9 P. Jandera and J. Churáček, *Advan. Chromatogr.*, 19 (1981) Ch. 5, p. 125.
10 M. Del R. Huang and I.S. Fagerson, *J. Chromatogr. Sci.*, 13 (1975) 347.
11 M. Kamiński, J. Klawiter and J.S. Kowalczyk, *J. Chromatogr.*, 176 (1979) 171.
12 J.R.C. Brown, cited in ref. 2.
13 P.R. Watt, *Chem. Ind. (London)*, (1963) 76.
14 E.A. Peterson and H.A. Sober, *Anal. Chem.*, 31 (1959) 857.
15 M.A. Chase, *Anal. Chem.*, 35 (1963) 1457.
16 E.H. Pfadenhauer, T.E. Lynes and T.V. Updyke, *Anal. Chem.*, 46 (1974) 623.
17 P.A. Bristow, *Anal. Chem.*, 48 (1976) 237.
18 S.H. Byrne, J.A. Schmit and P.E. Johnson, *J. Chromatogr. Sci.*, 9 (1971) 592.
19 K.J. Bombaugh, *J. Chromatogr.*, 107 (1975) 201.
20 *Characteristics of Gradient Elution Liquid Chromatography*, DuPont, Wilmington, DE, 1976.
21 S.R. Abbott, J.R. Berg, P. Achener and R.L. Stevenson, *J. Chromatogr.*, 126 (1976) 421.
22 M. Martin and G. Guiochon, *J. Chromatogr.*, 151 (1978) 267.
23 J.C. Helmer, *Anal. Chem.*, 48 (1976) 1741.
24 D. Ishii, K. Asai, K. Hibi, T. Jonokuchi and M. Nagaya, *J. Chromatogr.*, 144 (1977) 157.
25 P. Jandera and J. Churáček, *J. Chromatogr.*, 93 (1974) 17.
26 B.D. Warner, G. Boehme and J.I. Legg, *J. Chromatogr.*, 210 (1981) 419.
27 P. Jandera, J. Churáček and L. Svoboda, *J. Chromatogr.*, 192 (1980) 37.
28 J. Rock, *Chromatography Review (Spectra Physics)*, 5, No. 1 (1979) 8.
29 R.P.W. Scott and P. Kucera, *J. Chromatogr.*, 188 (1979) 27.
30 M. Savage, *Int. Lab.*, May/June (1979) 191.
31 M.T. Jackson and R.A. Henry, *Int. Lab.*, November/December (1974) 57.
32 H. Schrenker, *Int. Lab.*, July/August (1978) 67.
33 H. Engelhardt and H. Elgass, *J. Chromatogr.*, 112 (1975) 415.
34 P. Bünter, M. Schönmann and H. Kern, *Application Notes*, No. 13, Varian, Zug, 1980.
35 H. Kern and K. Imhof, *Int. Lab.*, January/February (1978) 65.
36 L.B. Sybrandt and E.F. Montoya, *Int. Lab.*, July/August (1977) 51.
37 D.L. Saunders, *J. Chromatogr. Sci.*, 15 (1977) 129.
38 A.M.C. Davies, E.H.A. Prescott and R. Stansfield, *J. Chromatogr.*, 171 (1979) 117.
39 H.A.H. Billiet, P.D.M. Keehnen and L. de Galan, *J. Chromatogr.*, 185 (1979) 515.
40 C. Laurent, H.A.H. Billiet, H.C. van Dam and L. de Galan, *J. Chromatogr.*, 218 (1981) 83.
41 R.D. Conlon, *A Discussion of Gradient Systems Pre-pump Versus Dual Pump*, TN-79, Perkin-Elmer, Norwalk, CT, 1979.

42 D.J. Popovich, J.B. Dixon and W.A. McKinley, *A Unique Solution to the Problems of Gradient Formation in a New Microprocessor Based HPLC System*, Tracor, Austin, TX, 1980.
43 S.T. Balke and R.D. Patel, *J. Liq. Chromatogr.*, 3 (1980) 741.
44 R.P.W. Scott and P. Kucera, *Anal. Chem.*, 45 (1973) 749.
45 R.P.W. Scott, *J. Chromatogr. Sci.*, 9 (1971) 385.
46 R.P.W. Scott and P. Kucera, *J. Chromatogr. Sci.*, 11 (1973) 83.
47 J.F.K. Huber, R. van der Linden, E. Ecker and M. Oreans, *J. Chromatogr.*, 83 (1973) 267.
48 H. Elgass, *PhD Thesis*, Saarbrücken, 1978.
49 K.J. Bombaugh, R.N. King and A.J. Cohen, *J. Chromatogr.*, 43 (1969) 332.
50 K. Aitzetmüller, *J. Chromatogr.*, 79 (1973) 329; 83 (1973) 461.
51 K. Aitzetmüller, *J. Chromatogr. Sci.*, 13 (1975) 454.
52 A.T. James, J.R. Ravenhill and R.P.W. Scott, *Chem. Ind. (London)*, (1964) 746.
53 A. Stolyhwo and O.S. Privett, *J. Chromatogr. Sci.*, 11 (1973) 20.
54 E.O.A. Haahti and T. Nikkari, *Acta Chem. Scand.*, 17 (1963) 2565.
55 M. Popl and Z. Havel, *Fette-Seifen-Anstrichm.*, 77 (1975) 51.
56 H. Dubský, *J. Chromatogr.*, 71 (1972) 395.
57 W. Scholtan and F.K. Woll, *J. Macromol. Chem.*, 151 (1972) 33.
58 P. Vandamme, B. Declercq, A. Denco, V. Blaton and H. Peters, *Anal. Biochem.*, 69 (1975) 29.
59 R.P.W. Scott and J.F. Lawrence, *J. Chromatogr. Sci.*, 11 (1973) 263.
60 R. Macrae and J. Dick, *J. Chromatogr.*, 210 (1981) 138.
61 W.M. McFadden, H.L. Schwartz and S. Evans, *J. Chromatogr.*, 122 (1976) 389.
62 E. Kenndler and E.R. Schmid, in J.F.K. Huber (Editor), *Instrumentation for High-Performance Liquid Chromatography*, Elsevier, Amsterdam, 1978, Chap. 10.
63 R.P.W. Scott, *Liquid Chromatography Detectors*, Elsevier, Amsterdam, 1977.
64 A.S. Brown, J.E. Mole, A. Weissinger and J.C. Bennett, *J. Chromatogr.*, 148 (1978) 532.
65 J. Sjödahl, *Sci. Tools*, 26, No. 1 (1979) 9.
66 *Correction of a Rising Baseline of a Chromatogram Generated by a Gradient Elution HPLC System*, Kratos, 1981.
67 N.A. Parris, *J. Chromatogr. Sci.*, 17 (1979) 541.
68 H. Poppe, in J.F.K. Huber (Editor), *Instrumentation for High-Performance Liquid Chromatography*, Elsevier, Amsterdam, 1978, Ch. 8.
69 J.F. Lawrence and R.W. Frei, *Chemical Derivatization in Liquid Chromatography*, Elsevier, Amsterdam, 1976.
70 R.R. Schroeder, P.J. Kudirka and C. Torren, Jr., *J. Chromatogr.*, 134 (1977) 83.
71 C.G. Creed, *Res./Dev.*, 27 (1976) 40.
72 J.N. Little and G.J. Fallick, *J. Chromatogr.*, 112 (1975) 389.
73 F. Erni, R.W. Frei and W. Lindner, *J. Chromatogr.*, 125 (1976) 265.
74 S.R. Bakalyar and R.A. Henry, *J. Chromatogr.*, 126 (1976) 327.
75 K. Asai, Y.I. Kanno, A. Nakamoto and T. Hara, *J. Chromatogr.*, 126 (1976) 369.
76 P.R. Brown, *J. Chromatogr.*, 57 (1971) 383.
77 L.R. Snyder, *J. Chromatogr.*, 13 (1964) 415.
78 L.R. Snyder, *Principles of Adsorption Chromatography*, Marcel Dekker, New York, 1968, Ch. 8.
79 P. Jandera, M. Janderová and J. Churáček, *J. Chromatogr.*, 115 (1975) 9.
80 L.R. Snyder and D.L. Saunders, *J. Chromatogr. Sci.*, 7 (1969) 195.
81 L.R. Snyder, *Anal. Chem.*, 46 (1974) 1384.
82 W. Boehme and H. Engelhardt, *J. Chromatogr.*, 133 (1977) 67.
83 J.P. Thomas, A. Brun and J.P. Bounine, *J. Chromatogr.*, 129 (1977) 21.
84 D. Berger and B. Gilliard, *J. Chromatogr.*, 210 (1981) 33.
85 D.W. Bristol, *J. Chromatogr.*, 188 (1980) 193.
86 N.E. Spingran, G.T. Garvie-Gould and L.L. Vuolo, *Anal. Chem.*, 53 (1981) 565.
87 H.W. Shmukler, *J. Chromatogr. Sci.*, 8 (1970) 581.
88 V.V. Berry, *J. Chromatogr.*, 199 (1980) 219.
89 L.R. Snyder, *J. Chromatogr. Sci.*, 8 (1970) 693.

90 L.R. Snyder and J.J. Kirkland, *An Introduction to Modern Liquid Chromatography*, 2nd ed., Wiley-Interscience, New York, 1979, Ch. 16.
91 H. Engelhardt, *Z. Anal. Chem.*, 277 (1975) 267.
92 R.P.W. Scott and J.G. Lawrence, *J. Chromatogr. Sci.*, 7 (1969) 65.
93 R.P.W. Scott and J.G. Lawrence, *J. Chromatogr. Sci.*, 8 (1970) 619.
94 M. Krejčí and D. Kouřilová, *J. Chromatogr.*, 91 (1974) 151.
95 D. Kouřilová and M. Krejčí, *J. Chromatogr.*, 138 (1977) 329.
96 G. Hesse and G. Roscher, *Chromatographia*, 2 (1969) 512.
97 R.J. Dolphin, L.P. Hoogeveen and F.W. Willmott, *J. Chromatogr.*, 122 (1976) 259.
98 J.F.K. Huber and R. v. d. Linden, *J. Chromatogr.*, 143 (1977) 125.
99 C.D. Scott and N.E. Lee, *J. Chromatogr.*, 83 (1973) 383.
100 C.D. Scott, D.D. Chilcote and N.E. Lee, *Anal. Chem.*, 45 (1972) 85.
101 F. Erni and R.W. Frei, *J. Chromatogr.*, 149 (1978) 561.
102 J. Kwok, L.R. Snyder and J.C. Sternberg, *Anal. Chem.*, 40 (1968) 118.
103 L.A.E. Sluyterman and O. Elgersma, *J. Chromatogr.*, 150 (1978) 17.
104 L.A.E. Sluyterman and J. Wijdenes, *J. Chromatogr.*, 150 (1978) 31.
105 E.J. Kikta, Jr., A.E. Stange and S. Lam, *J. Chromatogr.*, 138 (1977) 321.
106 J. Korpi and B.A. Bidlingmeyer, *Amer. Lab.*, June (1981) 110.
107 H. Hulpke and V. Werthmann, *Chromatographia*, 12 (1979) 390.
108 F. Erni, M.P. Keller, C. Morin and M. Schmitt, *J. Chromatogr.*, 204 (1981) 65.
109 A. Wehrli, *Chimia*, 28 (1974) 690.
110 J.F.K. Huber and F. Eisenbeiss, *J. Chromatogr.*, 149 (1978) 127.
111 J.F.K. Huber and R. Vodenik, *J. Chromatogr.*, 122 (1976) 331.
112 F.W. Willmott, I. Mackenzie and R.J. Dolphin, *J. Chromatogr.*, 167 (1978) 31.
113 R.J. Dolphin and F.W. Willmott, *J. Chromatogr. Sci.*, 14 (1976) 12.
114 J.J. Kirkland, *Analyst (London)*, 99 (1974) 859.
115 H. Straub and E. Ecker, *GIT Fachz. Lab.*, 19 (1975) 1.
116 G.C. Davis and P.T. Kissinger, *Anal. Chem.*, 51 (1979) 1960.
117 D.D. Koch and P.T. Kissinger, *Anal. Chem.*, 52 (1980) 27.
118 J.C. Gfeller and M. Stockmeyer, *J. Chromatogr.*, 198 (1980) 162.
119 F. Klink, *Analusis*, 9 (1981) 291.
120 S.P. Assenza and P.R. Brown, *J. Liq. Chromatogr.*, 3 (1980) 41.

PART IV

APPLICATIONS OF GRADIENT ELUTION IN LIQUID CHROMATOGRAPHY

In this part, examples are presented of applications of the gradient elution technique in liquid column chromatography. The examples discussed cannot exhaust all application possibilities, nor it is possible to discuss or even refer to all published gradient elution separations. The aim of this part is rather to demonstrate the gradient elution technique as a practical analytical tool and to illustrate the wide possibilities for its application by means of practical examples of the separation of various classes of compounds, and therefore most attention is paid to the choice of mobile phase components.

Each of the chapters treating the individual classes of organic compounds is sub-divided according to the chromatographic systems used (chromatography on polar adsorbents, reversed-phase chromatography, chromatography on polar bonded phases, ion-exchange chromatography, etc.) that generally determine the range of useful mobile phase components. The detection possibilities are more limited in gradient elution applications than in separations under isocratic conditions and UV detection at 254 or 280 nm is most frequently used. Other detection methods applied in gradient elution chromatography are mentioned in connection with the particular separation examples.

Pre-column and post-column derivatization are frequently used in the gradient elution liquid chromatography of non-absorbing compounds. If pre-column derivatization is employed, it usually not only improves the possibilities of detection, but also determines the choice of the stationary and mobile phase, owing to the changed chemical nature of the compounds to be separated. Post-column derivatization requires the design of a specific reaction detector fitted to the reaction and mobile phase used. These techniques are mentioned in the chapters on individual compounds and again the main concern is in the choice of the components for the formation of mobile phase gradients.

Chapter 12

HYDROCARBONS

12.1. ALIPHATIC HYDROCARBONS

Liquid chromatography is not a very suitable method for the separation of aliphatic hydrocarbons. Small and medium-sized aliphatic hydrocarbons are either gaseous or volatile compounds and therefore gas chromatography is the method of choice for their separation. Higher alkanes can be separated using liquid chromatography and the application of the gradient elution technique would be very efficient. However, this technique has not been frequently used for the separation of these compounds because of difficulties connected with detection. UV and fluorescence detection, commonly applied in the gradient elution technique, cannot be used owing to the absence of a π-electron system in saturated hydrocarbons. There are also no suitable derivatives that could be prepared to enhance the detection possibilities. Mass-transport detectors or an on-line liquid chromatograph-mass spectrometer combination are detection systems that could be used in the gradient elution chromatography of saturated hydrocarbons.

Another possibility mentioned in the literature is the application of an infrared detector at 2950 cm^{-1}. Microgram amounts of 2,2,4-trimethylpentane, decane, dodecane, hexadecane, octadecane and eicosane were separated on an octadecylsilica column (Zorbax ODS, 4.6 x 250 mm I.D.) using a linear gradient from methylene chloride-acetonitrile (40 : 60) to 100% dichloroethane in 30 min using this method of detection[1]. However, in spite of careful matching of the infrared absorbance of the initial and final mobile phases, the sensitivity of detection is low in comparison with UV detection for absorbing compounds.

12.2. AROMATIC HYDROCARBONS

Gradient elution liquid chromatography has become very popular for the analysis of aromatic hydrocarbons, mainly polycyclic compounds. It has been applied in the separation and determination of aromatic hydrocarbons in oil fractions and in the analysis of polycyclic hydrocarbons as environmental pollutants with possible carcinogenic properties in air, drinking water and waste waters.

References on p. 293.

12.2.1. Chromatography on polar adsorbents

Early work on the analysis of aromatic hydrocarbons in oil fractions, heavy distillates and residues, shale oil and other fuels used columns packed with polar adsorbents. It is possible to use gradient elution chromatography to separate oil samples into fractions for further characterization using capillary gas chromatography or other methods. A column, 250 x 10 mm I.D., packed with Chromosorb LC-8 with an exponential gradient from 0 to 100% of *n*-butanol in *n*-hexane in 15 ml was used for this purpose[2]. Another possibility is the application of coupled columns packed with alumina and silica or a dual column half packed with alumina and half with silica. A stepwise gradient of benzene, diethyl ether and methanol in *n*-pentane was applied on such a column for the fractionation of high-boiling petroleum distillates[3].

Aromatic hydrocarbons have been separated on columns packed with alumina using concave, linear and multilinear gradients of diethyl ether in isooctane[4] or in *n*-pentane[5-7] or a gradient of cyclohexane in *n*-hexane followed by a gradient of methanol in chloroform[8].

In more recent applications, columns packed with microparticulate silica have been far more often used than columns packed with alumina. Linear gradients of chloroform or dichloromethane in *n*-hexane are frequently used here[9-11]. The selectivity for biphenyl and substituted biphenyls can be increased by using a column packed with silica coated with thallium nitrate and elution with a linear gradient of ethanol in *n*-heptane (0-3% in 9 min)[12].

12.2.2. Chromatography on polar and non-polar bonded phases and on other sorbents

Aromatic hydrocarbons have been separated on polar chemically bonded phases, such as on the ethereal phase Permaphase ETH using a linear gradient of isopropanol in water[13], on the nitrile phase Cyano-Sil-X-1 or on the amino phase Amino-Sil-X-1 using a concave gradient of methylene chloride in *n*-hexane[14] and on the nitro phase Nucleosil 5 NO_2 using a linear gradient of chloroform in *n*-hexane[15].

Selective interactions between the π-electron systems and metals were utilized in chromatography on a chemically bonded organomercuric phase prepared on LiChrosorb Si 60 by bonding phenyl rings first to the silica surface and introducing a mercuriacetyl group on to the phenyl ring. Polyaromatic hydrocarbons were separated using a linear gradient from 1 to 30% of dioxane in *n*-hexane in 40 min[50].

Chromatography of polyaromatic hydrocarbons on polar bonded phases offers a valuable complement to chromatography on alkyl bonded phases. Because the retention of polyaromatic hydrocarbons on an alkyl bonded phase increases both with increasing number of aromatic rings and with increasing number and length of alkyl side chains, it is virtually impossible to achieve the fractionation of polyaromatic hydrocarbons into groups containing hydrocarbons with a given number of aromatic rings. This is in principle possible, however, by chromatography on a bonded amino, nitrile or nitro phase. Gradient elution separation of aromatic hydrocarbons on cation-exchange columns (Aminex 50W-X4-Fe) using a linear gradient of acetonitrile in water has also been described[16].

Chemically bonded non-polar phases have been most frequently used for separations of aromatic hydrocarbons with the application of reversed-phase gradient elution. Aromatic hydrocarbons in petroleum fractions, shale oils, oil distillates and polystyrene oligomers are usually separated on columns 200-300 mm long of 2.5-6.5 mm I.D., packed with 5-10 μm microparticulate octadecylsilica[17-22]. Columns 500-2000 mm long of 2-2.5 mm I.D., packed with larger diameter material have been occasionally used, mainly in earlier applications[23,24]. Linear or convex gradients of methanol or acetonitrile in water can be applied, usually started at 30-65% of the organic solvent in the mobile phase[17-32,51]. A microbore column, 150 mm long of 0.5 mm I.D., packed with Permaphase ODS (30 μm) was applied for the separation of polycyclic aromatic hydrocarbons using the above-described gradients[33]. Parris[34] applied non-aqueous reversed-phase liquid chromatography on an octadecylsilica column (Zorbax ODS, 6 μm, 250 x 4.6 mm I.D.) with linear gradients of methylene chloride in acetonitrile and of tetrahydrofuran in methanol for the separation of polynuclear aromatic hydrocarbons and of polystyrene oligomers (Fig. 12.1).

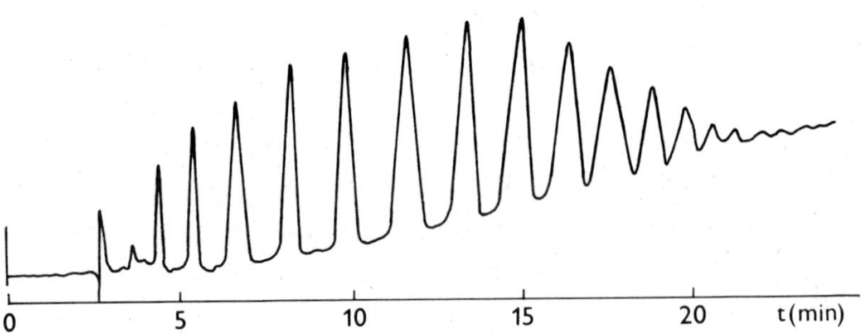

Fig. 12.1. Separation of polystyrene 800 oligomers by non-aqueous reversed-phase gradient elution liquid chromatography. Column: Zorbax ODS, 150 x 4.6 mm I.D. Gradient: linear, 1.4% B/min. Solvents: A, methanol; B, tetrahydrofuran. Flow-rate: 1.5 ml/min. Detection: UV, 254 nm. (Reproduced from ref. 34 with permission.)

References on p. 293.

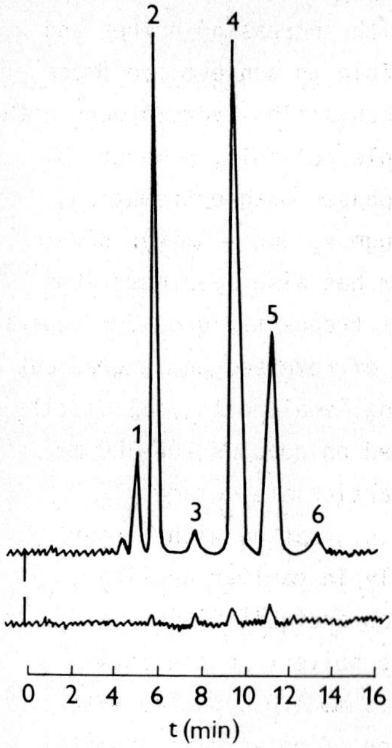

Fig. 12.2. Separation of a mixture of several environmentally important polyaromatic hydrocarbons by reversed-phase gradient elution liquid chromatography. Column: ODS SIL-X-I, 250 x 2.6 mm I.D. Gradient: linear, 1.0% B/min. Solvents: A, 45% (v/v) acetonitrile in water; B, acetonitrile. Flow-rate: 1 ml/min. Detection: fluorimetric, λ_{ex} = 338 nm, λ_{em} = 385 nm (upper trace), and UV at 280 nm, 0.02 a.u.f.s. (lower trace). Peaks: 1 = anthracene; 2 = pyrene; 3 = chrysene; 4 = benzo[a]pyrene; 5 = benzoperylene; 6 = coronene. (Reproduced from ref. 29 with permission.)

Great attention has been paid to analysis of polycyclic aromatic hydrocarbons as carcinogenic pollutants in drinking, sea and waste waters, air particulate material, cigarette smoke and other samples. With a few exceptions[10,11], reversed-phase chromatography on octadecylsilica[53] or octylsilica[52] columns has been used almost exclusively for this purpose. Here again, gradients of acetonitrile or of methanol in water as described above have been applied[35-46], but a gradient from 25 to 40% of isopropanol in water can also be used[13]. Reversed-phase chromatography is especially suitable for the analysis of pollutants in water, because it simplifies considerably the treatment of samples. This makes it possible to use direct on-line enrichment of aqueous samples by passing large amounts (even several litres) of water through a pre-column packed with octadecylsilica or directly through the analytical column before the start of the gradient elution[47,53].

UV detection is convenient for aromatic hydrocarbons, but fluorimetric detection is more sensitive and more specific and therefore it is usually the method of choice in the trace analysis of polycyclic aromatic hydrocarbons. In such analyses, reversed-phase gradient elution liquid chromatography with fluorimetric detection is far superior to gas chromatography, with which it would be very difficult to distinguish polycyclic aromatic hydrocarbons from biogenic olefins and small amounts of aromatic hydrocarbons in environmental samples could be overlooked in gas chromatograms. The strong fluorescence of polycyclic aromatic hydrocarbons makes possible their selective detection in liquid chromatography, which can provide much more valuable information concerning possible petroleum pollution than gas chromatographic analysis, which would also detect a number of non-aromatic and non-petroleum hydrocarbons[48]. An example of the reversed-phase gradient elution separation of polycyclic aromatic hydrocarbons with fluorimetric detection is shown in Fig. 12.2.

Fixed excitation (360 or 280 nm, for example) and emission wavelengths may be used in fluorimetric detection[43,49], but the use of various combinations of excitation and emission wavelengths during a single gradient run may yield improved sensitivity of determination[46]. Colmsjö and co-workers[31,41] applied the Shpolskii effect to improve the possibilities of identification of polycyclic aromatic hydrocarbons in fractions of the eluate. This effect makes use of fluorescence measurements in rigid media at a low temperature (63-77° K), where a considerable increase in the fine structure of the fluorescence spectra of polycyclic aromatic hydrocarbons occurs. Such spectra can be used for their identification by comparing them with the spectra of standard compounds. On-line coupling of a liquid chromatograph and a mass spectrometer has also been used to identify separated polycyclic aromatic hydrocarbons[37].

12.3. REFERENCES

1 N.A. Parris, *J. Chromatogr. Sci.*, 17 (1979) 541.
2 R.J. Crowley, S. Siggia and P.C. Uden, *Anal. Chem.*, 52 (1980) 1224.
3 D.E. Hirsch, R.L. Hopkins, H.J. Coleman, F.O. Cotton and C.J. Thompson, *Anal. Chem.*, 44 (1972) 915.
4 L.R. Snyder and H.D. Warren, *J. Chromatogr.*, 15 (1964) 344.
5 M. Popl, J. Mostecký and Z. Havel, *J. Chromatogr.*, 53 (1970) 233.
6 M. Popl, V. Dolanský and J. Mostecký, *Anal. Chem.*, 44 (1972) 2082.
7 M. Popl, M. Stejskal and J. Mostecký, *Anal. Chem.*, 47 (1975) 1947.
8 D.M. Jewell, R.G. Ruberto and B.E. Davis, *Anal. Chem.*, 44 (1972) 2318.
9 S. Marsh and C. Grandjean, *J. Chromatogr.*, 147 (1978) 411.
10 A.D. Thruston, *J. Chromatogr. Sci.*, 16 (1978) 254.
11 W.C. Eisenberg, *J. Chromatogr. Sci.*, 16 (1978) 145.
12 A.M. Siouffi, J.C. Traynard and G. Guiochon, *J. Chromatogr. Sci.*, 15 (1977) 469.
13 A.F. Haeberger, M.E. Snook and O.T. Chortyk, *Anal. Chim. Acta*, 80 (1975) 303.
14 *Anal. Chem.*, 46 (1974) 1200A.

15 G.P. Blümer and M. Zander, *Z. Anal. Chem.*, 288 (1977) 277.
16 D.M. Ordemann and H.F. Walton, *Anal. Chem.*, 48 (1976) 1728.
17 J.R. Jadamec, W.A. Saner and Y. Talmi, *Anal. Chem.*, 49 (1977) 1316.
18 R. Yost, J. Stoveken, and W. MacLean, *J. Chromatogr.*, 134 (1977) 73.
19 W.A. Saner, J.R. Jadamec and R.W. Sager, *Anal. Chem.*, 50 (1978) 749.
20 H.S. Hertz, J.M. Brown, S.N. Chesler, F.R. Guenther and W.E. May, *Anal. Chem.*, 52 (1980) 1650.
21 H. Wehmer and M. Teschner, *J. Chromatogr.*, 204 (1981) 481.
22 T.H. Monrey, S. Siggia, P.C. Uden and R.J. Crowley, *Anal. Chem.*, 52 (1980) 885.
23 I. Berthold, *Erdoel Kohle Erdgas Petrochem. Brennst. Chem.*, 28 (1975) 93.
24 C.G. Vaughan, B.B. Wheals and M.J. Whitehouse, *J. Chromatogr.*, 78 (1973) 203.
25 J.J. Kirkland, *Chromatographia*, 8 (1975) 661.
26 R. Thoms and M. Zander, *Z. Anal. Chem.*, 282 (1976) 443.
27 A.M. Krstulovic, D.M. Rosie and P.R. Brown, *Anal. Chem.*, 48 (1976) 1383.
28 R. Majors, *Varian Application Bull.*, 11 No. 1 (1977) 6.
29 W. Slavin, A.T. Rhys-Williams and R.F. Adams, *J. Chromatogr.*, 134 (1977) 121.
30 L.S. Ettre, *J. Chromatogr. Sci.*, 16 (1978) 396.
31 A.L. Colmsjö and C.E. Östman, *Anal. Chem.*, 52 (1980) 2093.
32 R.E. Majors, *J. Chromatogr. Sci.*, 18 (1980) 488.
33 D. Ishii, K. Asai, K. Hibi, T. Jonokuchi and M. Nagaya, *J. Chromatogr.*, 144 (1977) 157.
34 N.A. Parris, *J. Chromatogr.*, 157 (1978) 161.
35 J.A. Schmit, R.A. Henry, R.C. Williams and J.F. Dieckman, *J. Chromatogr. Sci.*, 9 (1971) 645.
36 J.K. Selkirk, *Spectra-Physics Chromatogr. Rev.*, 2 No. 1 (1975) 1.
37 P.R. Jones and S.K. Yang, *Anal. Chem.*, 47 (1975) 1000.
38 M. Dong, D.C. Locke and E. Ferrand, *Anal. Chem.*, 48 (1976) 368.
39 J.K. Selkirk, *Nature (London)*, 270 (1977) 603.
40 A.R. Oyler, D.L. Bodenner, K.J. Welch, R.J. Liukkonen, R.M. Carlson, H.L. Kopperman and R. Caple, *Anal. Chem.*, 50 (1978) 837.
41 A. Colmsjö and U. Stenberg, *Anal. Chem.*, 51 (1979) 145.
42 K. Ogan, E. Katz and W. Slavin, *Anal. Chem.*, 51 (1979) 1315.
43 G.R. Sirota, J.F. Uthe, C.J. Musial and V. Zitko, *J. Chromatogr.*, 202 (1980) 294.
44 B.P. Dunn and R.J. Aymorer, *Anal. Chem.*, 52 (1980) 2027.
45 J.M. Colin, G. Vion, M. Lamotte and J. Joussot-Dubien, *J. Chromatogr.*, 204 (1981) 135.
46 K. Ogan, E. Katz and T.J. Porro, *J. Chromatogr. Sci.*, 17 (1979) 597.
47 M. Schönmann, *Varian Bull.*, LC 113 (1980).
48 J.S. Warner, R.M. Riggin and T.M. Engel, in L. Petrakis and F.T. Weiss (Editors), *Petroleum in the Marine Environment*, American Chemical Society, Washington, 1980, p. 87.
49 E. Johnson, A. Abu-Shumays and S.R. Abbott, *J. Chromatogr.*, 134 (1977) 107.
50 J. Chmielowiec, *J. Chromatogr. Sci.*, 19 (1981) 296.
51 J.L. Di-Cesare, M.W. Dong and L.S. Ettre, *Chromatographia*, 14 (1981) 257.
52 J. Müller and E. Rohbock, *Talanta*, 27 (1980) 673.
53 W.E. May, S.N. Chesler, S.P. Cram, B.H. Gump, H.S. Hertz, D.P. Enagonio and S.M. Dyszel, *J. Chromatogr. Sci.*, 13 (1975) 535.

Chapter 13

ALCOHOLS AND PHENOLS, ETHERS

13.1. ALIPHATIC AND AROMATIC ALCOHOLS

Lower monohydric alcohols are best analysed by gas chromatography, while liquid column chromatography is suitable for the separation of higher free alcohols, their derivatives and phenols. The application of gradient elution to the separation of aliphatic alcohols is subject to detection problems, because these compounds are neither UV-absorbing nor fluorescent. Therefore, it is necessary to prepare suitable derivatives that would increase the UV absorptivity. It is possible to form, for example, 3,5-dinitrobenzoates prior to liquid chromatographic separation by heating the sample containing alcohols for 1 h with 3,5-dinitrobenzoyl chloride[1]. Another possibility is offered by the reaction of alcohols with phenyl isocyanate to form alkylphenylurethanes. The reaction can be performed at ambient temperature without the addition of catalysts. The reaction mixture should be shaken and allowed to stand for 15 min, then the excess of the reagent is removed by reaction with an excess of an alcohol that is not of interest in the analysed mixture. The most sensitive detection is provided at 230 nm, which is the absorption maximum of alkylphenylurethanes. Reversed-phase chromatography on an octadecylsilica column (SIL 60-D 10 C_{18}, 250 x 4 mm I.D.) was applied to the chromatography of alkylphenylurethanes of lower aliphatic alcohols using a multilinear gradient of acetonitrile in water (from 25 to 75% in 18 min). This procedure allows the resolution of all of the homologues and some of the isomeric derivatives. The method has been applied to the analysis of residual 2-ethylhexanol in polymer samples[2,24].

Aromatic alcohols and phenols can be easily detected by UV absorption. Separation of aromatic alcohols was achieved, for example, by using a linear concentration gradient of methylene chloride-isopropanol in *n*-hexane on a 500 mm column packed with silica (Micro Pak Si-60)[3].

13.2. HALOGENO-, NITRO-, ALKYL- AND OTHER TECHNICALLY IMPORTANT PHENOLS

Much more frequent and important are the applications of gradient elution liquid chromatography to the analysis of phenolic compounds. Some phenolics, in particular halogenated phenols and nitrophenols, are toxic pollutants in

References on p. 299.

industrial waste waters. UV absorption at a wavelength in the range 210-280 nm (most frequently 254 nm) has often been used in the gradient elution chromatography of phenols. The sensitivity of detection of halogenophenols was increased using a post-column reactor, where the phenols reduced cerium(IV) sulphate in 11.5 mol/l sulphuric acid, being added continuously to the column effluent at 25^{0}C to produce fluorescent Ce^{3+} ions, which were monitored by fluorimetric detector using an excitation wavelength of 260 nm and an emission wavelength of 350 nm[13].

An electrochemical detector using a tubular graphite anode made of carbon black/polyethylene (50:50) (Dow Chemical, Findlay, OH, U.S.A.) also proved very sensitive for the detection of mono- to pentachlorophenols. The detection was superior to UV detection at 210 nm and was applied to the trace determination of phenolic compounds in water[18].

13.2.1. Chromatography on ion-exchange and unsubstituted resins

In most earlier applications, anion-exchange columns were used for gradient elution separations of phenolic compounds. Ion-exchange resins in the acetate form were mainly used for this purpose, with gradients of acetic acid in aqueous methanolic solutions. Using continuous linear or stepwise gradients, mixtures containing 4-chlorophenol, 2-chlorophenol, 2,4-dichlorophenol, 2,6-dichlorophenol and 2,4,6-trichlorophenol or those containing 2,3,4,5-tetrachlorophenol, 2,3,4,6-tetrachlorophenol, 2,3,5,6-tetrachlorophenol and pentachlorophenol or those containing bisphenol A, 2-bromobisphenol A, 2,2'-dibromobisphenol A, 2,6-dibromobisphenol A, 2,2',6-tribromobisphenol A and 2,2',6,6'-tetrabromobisphenol A were separated[4,5].

Isomeric nitrophenols, cresols, xylenols or di- and monochlorophenols were separated on a De-Acidite anion-exchange column in the acetate form using a gradient of triethylamine-acetate buffer in aqueous methanol[6].

Phenol, 2,6-dichlorophenol, 2,5-dichlorophenol and 3,5-dichlorophenol were separated on Amberlite XAD-2 macroporous polystyrene-divinylbenzene resin using elution with 0.01 mol/l phosphate buffer of pH 12.2 in a solution with a gradient of ethanol from 15 to 30%.

A common feature of these earlier separations of phenolic compounds was a poor efficiency and long analysis times.

13.2.2. Chromatography on silica gel

Highly efficient separations of phenolic compounds have been achieved in both normal-phase and reversed-phase systems. Phenolic compounds can be separated on microparticulate silica columns using a linear gradient from 20 to 60% of methylene chloride in *n*-hexane[7]. A direct on-line sample enrichment technique was reported which makes use of a silica column (Spherisorb, 10 μm, 250 x 4.6 mm I.D.) modified with alkalis to contain alkali metal silicates. On such a column, acidic compounds are strongly retained from relatively weakly polar solvents, such as cyclohexane-methylene chloride (1 : 1). A linear gradient of methanol concentration in this solvent would elute phenolic compounds. Substituted alkyl-, halogeno- and nitrophenols could be separated using this technique[8].

13.2.3. Reversed-phase chromatography

Reversed-phase chromatography on octylsilica[9] and, more frequently, octadecylsilica has been very successfully applied to the separation of phenolic compounds using elution with linear gradients of methanol or acetonitrile in water[10-17]. This method was used for the separation of alkylphenols[10,11]. Mixtures containing mono-, di-, tri- and tetramethylphenols, ethylphenols and propylphenols, including a number of positional isomers, could be successfully separated in 60 min on a PAH 10 - C18 column using a linear gradient from 10 to 40% of acetonitrile in water. Isomeric *m*- and *p*-cresol and three xylenols could not be separated. The method was applied to the analysis of alkylphenols in extracts of a coal liquefaction products[11]. Benzo[*a*]pyrene phenols were chromatographed using a gradient of methanol in water[12]. Other highly toxic phenols, such as nitro- and halogenophenols were separated on octadecylsilica columns using linear gradients of acetonitrile in water[13,14]. The method was applied to the analysis of these solutes as pollutants in water samples. Linear gradients of acetonitrile in water and octadecylsilica columns were also used for the analysis of phenol, guaiacol, *o*-vanillin and eugenol in pulp mill effluents[13] and for the separation of bisphenol A and impurities in the technical product used as a raw material for polymer production[15]. If certain combinations of phenols are present in sample mixtures, neither a binary gradient of methanol in water nor a binary gradient of acetonitrile in water is successful and the application of a ternary water-methanol-acetonitrile gradient is necessary[16].

The dissociation of phenolic compounds may be suppressed by working at an acidic pH, which may sometimes lead to improved resolution and peak symmetry. Phosphoric acid[18] or a phosphate buffer[19] added to the starting eluent at a concentration of 0.01 or 0.02 mol/l was found to yield good separations of chlorinated phenols in chromatography on octadecylsilica columns using linear gradients of acetonitrile or methanol in the buffered mobile phase. An example

References on p. 299.

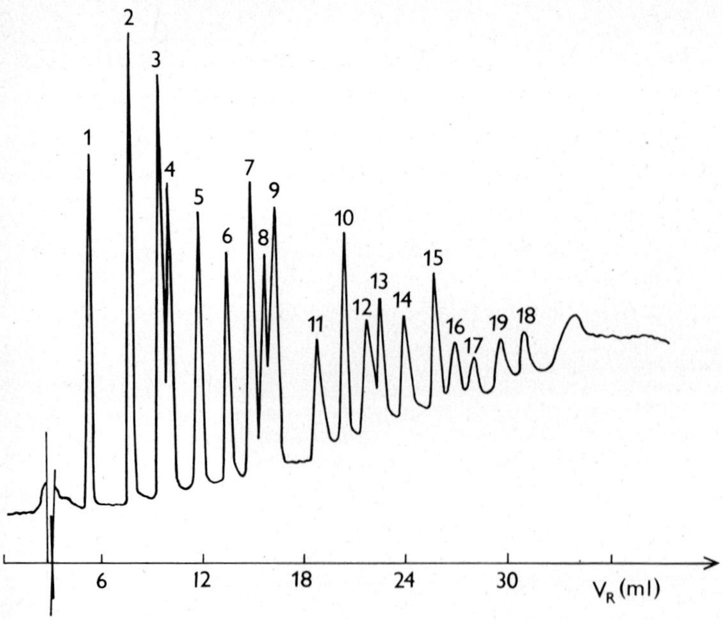

Fig. 13.1. Separation of a mixture of chlorinated phenols and phenol by reversed-phase gradient elution liquid chromatography. Column: Hypersil ODS, 5 μm, 230 x 4.6 mm I.D. Gradient: linear, 3.3% B/min. Solvents: A, 0.02 mol/l KH_2PO_4, pH 4.0 in 56% (v/v) methanol in water; B, 0.02 mol/l KH_2PO_4, pH 4.0, in 80% (v/v) methanol in water. Flow-rate: 1 ml/min. Detection: UV, 280 nm. Peaks: 1 = phenol; 2 = *o*-chlorophenol; 3 = *p*-chlorophenol; 4 = *m*-chlorophenol; 5 = 2,6-dichlorophenol; 6 = 2,3-dichlorophenol; 7 = 2,5-dichlorophenol; 8 = 2,4-dichlorophenol; 9 = 3,4-dichlorophenol; 10 = 3,5-dichlorophenol; 11 = 2,3,6-trichlorophenol; 12 = 2,3,4-trichlorophenol; 13 = 2,4,6-trichlorophenol; 14 = 2,4,5-trichlorophenol; 15 = 3,4,5-trichlorophenol; 16 = 2,3,5,6-tetrachlorophenol; 17 = 2,3,4,6-tetrachlorophenol; 18 = 2,3,4,5-tetrachlorophenol; 19 = pentachlorophenol. (Reproduced from ref. 19 with permission.)

of the separation of a mixture of 19 chlorophenols using these conditions is shown in Fig. 13.1[19].

Ion-pair chromatography on an octylsilica column employing elution with a linear gradient of methanol in an aqueous mobile phase containing a constant concentration ($5 \cdot 10^{-3}$ mol/l) of tetrabutylammonium phosphate at pH 7.5 (Waters PIC A reagent) and $35^{o}C$ successfully separated and determined polynitrophenols (3-hydroxy-2,4-dinitrophenol, 3-hydroxy-2,4,6-trinitrophenol, 2-amino-4,6-dinitrophenol, 3-hydroxy-4,6-dinitrophenol, 2,6-dinitrophenol, 2,4-dinitrophenol, 2-methyl-4,6-dinitrophenol, 2,4,6-trinitrophenol and 3-methyl-2,4,6-trinitrophenol) in waste water samples. Good precision of determination was reported[9].

13.3. NATURALLY OCCURRING PHENOLICS AND RELATED COMPOUNDS

Reversed-phase gradient elution chromatography on octadecylsilica has been utilized for the separation and determination of various naturally occurring phenolic compounds, such as phenolic compounds in tobacco leaves[20], phenolic lipids in cashewnut-shell liquid[21], the flavonoids vitexin and isovitexin in *Passiflora incarnata*[22] and toxic gossypol in plants[23].

Glucuronide and sulphate conjugates of *p*-nitrophenol as metabolites in human urine have been analysed by chromatography on a pellicular anion-exchange resin using elution with a linear gradient of potassium chloride in 1.0 mol/l formic acid at $80^{o}C$[25].

Tannin and related polyphenols in various natural materials[26,27], plants[27], tea extract[28], cider and wine[29] have been determined by liquid chromatography on porous-layer silica gel (Corasil II) using a linear gradient of tetrahydrofuran-methanol-acetic acid (200 : 800 : 8)[26] and, more frequently, by reversed-phase liquid chromatography. In the latter, elution with linear gradients of simultaneously increasing concentrations of ethanol and ethyl acetate in 0.5 mol/l potassium dihydrogen phosphate was used for chromatography on porous-layer Zipax-hydrocarbon phase[27], a linear gradient of acetone in 0.5% acetic acid was employed on microparticulate octadecylsilica[28] and a linear gradient of methanol in water served for elution on microparticulate octyl- or hexylsilica chemically bonded phases[29].

13.4. ETHERS

Aromatic ethers and related compounds (safrole, *trans*-isosafrole, anethole, myristicin, *trans*-isomyristicin, dillapiole, parsleyapiole, methyleugenol, *trans*-methylisoeugenol, *cis*-isoelemicin, elemicin, *trans*-isoelemicin and asarone) have been separated by chromatography on a microparticulate bonded nitrile phase using a linear gradient of tetrahydrofuran in *n*-heptane with UV detection at 280 and 260 nm[30].

Lacinilene C_7-methyl ether in cotton powder (a suspected cause of byssinosis, a respiratory disorder) has been determined using normal-phase chromatography on microparticulate silica with a linear gradient of tetrahydrofuran in *n*-hexane containing 1% of acetic acid and UV detection at 340 nm[31].

13.5. REFERENCES

1 J.F. Lawrence and R.W. Frei, *Chemical Derivatization in Liquid Chromatography*, Elsevier, Amsterdam, 1976, p. 151.
2 B. Björkoqvist and H. Toivonen, *J. Chromatogr.*, 153 (1978) 265.

3 *Varian Application Bull.*, 6, No. 4 (1972) 14.
4 N.E. Skelly, *Anal. Chem.*, 33 (1961) 271.
5 N.E. Skelly and W.B. Crummett, *J. Chromatogr.*, 55 (1971) 309.
6 D.E. Thomas and J.D.R. Thomas, *Analyst (London)*, 94 (1969) 1099.
7 *Micro Pak Column Brochure*, Varian Assoc., Palo Alto, CA.
8 D.L. Stalling, J.D. Petty and L.M. Smith, *J. Chromatogr. Sci.*, 19 (1981) 18.
9 J.C. Hoffsommer, D.J. Glover and C.Y. Hazzard, *J. Chromatogr.*, 195 (1980) 435.
10 P.J. Schoenmakers, H.A.H. Billiet and L. de Galan, *J. Chromatogr.*, 205 (1981) 13.
11 K. Ogan and E. Katz, *Anal. Chem.*, 53 (1981) 160.
12 J.K. Selkirk, R.C. Croy and M.V. Gelboin, *Cancer Res.*, 36 (1976) 922.
13 A.W. Wolkoff and L.H. Larose, *J. Chromatogr.*, 99 (1974) 731.
14 P.A. Realini, *J. Chromatogr. Sci.*, 19 (1981) 124.
15 P. Szap, I. Kesse and J. Klapp, *J. Liq. Chromatogr.*, 1 (1978) 89.
16 P. Jandera, J. Churáček and H. Colin, *J. Chromatogr.*, 214 (1981) 35.
17 M.H. Simatupang, *Chromatographia*, 12 (1979) 661.
18 D.N. Armentrout, J.D. McLean and M.V. Long, *Anal. Chem.*, 51 (1979) 1039.
19 K. Ugland, E. Lundanes, T. Greibrokk and A. Bjorseth, *J. Chromatogr.*, 213 (1981) 83.
20 W.A. Court, *J. Chromatogr.*, 130 (1977) 287.
21 J.H.P. Tyman, V. Tychopoulos and B.A. Colenutt, *J. Chromatogr.*, 213 (1981) 287.
22 V. Quercia, L. Turchetto, N. Pierini, V. Cuozzo and G. Percaccio, *J. Chromatogr.*, 161 (1978) 396.
23 S.A. Abou-Donia, J.M. Lasker and M.B. Abou-Donia, *J. Chromatogr.*, 206 (1981) 606.
24 T. Jupille, *J. Chromatogr. Sci.*, 17 (1979) 160.
25 M.W. Anders and J.P. Lattore, *J. Chromatogr.*, 55 (1971) 409.
26 T.H. Beasley, H.W. Ziegler and A.D. Bell, *Anal. Chem.*, 49 (1977) 238.
27 T. Okuda, K. Mori, K. Seno and T. Hatano, *J. Chromatogr.*, 171 (1979) 313.
28 *Modular Liquid Chromatograph LC 750 Brochure*, Applied Chromatography Systems, Luton, Great Britain.
29 A.G.H. Lea, *J. Chromatogr.*, 194 (1980) 62.
30 L.W. Wulf, C.W. Nagel and A.L. Branen, *J. Chromatogr.*, 161 (1978) 271.
31 J.H. Wall, L.L. Muller and R.S. Berni, *J. Liq. Chromatogr.*, 3 (1980) 561.

Chapter 14

CARBONYL COMPOUNDS: ALDEHYDES, KETONES, QUINONES

14.1. ALDEHYDES AND KETONES

Aliphatic aldehydes and ketones are not frequently chromatographed in their free forms using modern liquid column chromatography because of the relatively low sensitivity of detection. These compounds absorb in the UV region; however, the absorption band corresponds to the $n \rightarrow \pi^*$ transition and has a relatively low molar absorptivity. Therefore, derivatives with stronger UV absorption are often prepared in order to enhance the sensitivity and improve the detection limit.

14.1.1. Chromatography of free aldehydes and ketones on ion-exchange resins

In earlier applications, aldehydes and ketones in their free forms were chromatographed on ion-exchange columns, packed with either cation or anion exchangers. Two chromatographic mechanisms were utilized for the separation of aliphatic carbonyl compounds. In "salting-out chromatography", the retention of non-ionic compounds was enhanced by adding an electrolyte (salt) to the mobile phase. During the separation, a stepwise gradient of a decreasing concentration of a suitable salt, such as ammonium sulphate, was applied to decrease gradually the retention of later eluted compounds and to speed-up the analysis[1]. In so-called "solubilization chromatography", a stepwise gradient of an increasing concentration of the organic solvent (methanol, ethanol or acetic acid) was used for the same purpose[2]. (In fact, the mechanism of reversed-phase chromatography on ion-exchange resins was utilized in these applications.) Using either of the two methods, it was possible to achieve the separation of aliphatic aldehydes and ketones containing 1-4 carbon atoms in salting-out chromatography and 1-10 carbon atoms in solubilization chromatography. Post-column reaction with bichromate ions was used to detect the carbonyl compounds in the fractions of the eluate by monitoring the absorbance of the Cr(III) ions formed. This detection could only be used in salting-out chromatography[1], while in solubilization chromatography, where reducing alcohols were used as the components of the mobile phase, reaction of carbonyl compounds with hydroxylamine in the fractions of eluate was employed for quantitation[2]. The efficiency

References on p. 305.

of these separations was very low and analyses of mixtures containing about seven components took from several hours to two or three days.

14.1.2. Normal- and reversed-phase chromatography of aldehydes, ketones and their phenylhydrazones

In more recent applications of gradient elution liquid column chromatography to the separation of aliphatic aldehydes and ketones, the high specificity of the reaction between these compounds and 2,4-dinitrophenylhydrazine has been most frequently utilized. The 2,4-dinitrophenylhydrazones may be formed by precipitation after adding the reagent to aqueous samples or by extraction of carbonyl compounds into a non-aqueous medium containing 2,4-dinitrophenylhydrazine.

Gas chromatography is often used for the chromatography of 2,4-dinitrophenylhydrazones because of the high selectivity of the derivatization reaction. Liquid column chromatography offers potential advantages over gas chromatography for the separation of these derivatives. In gas chromatography, thermal decomposition of the derivatives may occur at the temperatures necessary for separation; a non-linear detector response and eventual deterioration of the columns may often result. These problems are eliminated in liquid chromatography. 2,4-Dinitrophenylhydrazones are detected in the UV region at 340-360 or 254 nm.

Normal-phase chromatography on silica (LiChrospher Si 100) was used to separate the 2,4-dinitrophenylhydrazones of glyoxal, glycolaldehyde, glyceraldehyde and several monosaccharides using a linear gradient of methanol and chloroform in isooctane[3].

The use of a polar (ethereal) bonded phase (Permaphase ETH) and a linear concentration gradient of chloroform in *n*-hexane made it possible to accomplish the separation of 2,4-dinitrophenylhydrazones of lower aliphatic aldehydes (hexanal, *n*-pentanal, butanal, crotonaldehyde, acetaldehyde, formaldehyde, benzaldehyde, salicylaldehyde, glyoxal and glutaraldehyde) in 30 min at 40°C[4]. Formaldehyde in tobacco smoke was converted into its dinitrophenylhydrazone prior to chromatographic separation on a coupled column system of a nitrile column (μBondapak CN) and a silica gel column (μPorasil) connected in series. A concave ternary gradient of dichloromethane and acetonitrile in isooctane was employed for elution. Using this method, a number of other carbonyl compounds potentially present in cigarette smoke, such as acetaldehyde, furfural, 5-hydroxymethylfurfural, acrolein and crotonaldehyde, could be simultaneously analysed[5].

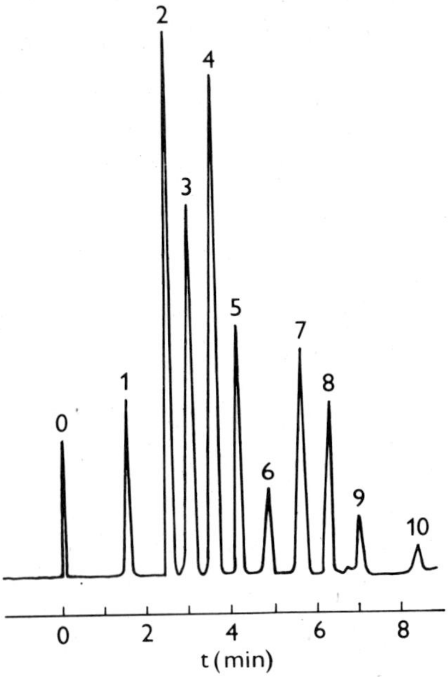

Fig. 14.1. Separation of 2,4-dinitrophenylhydrazones of aliphatic aldehydes by reversed-phase gradient elution liquid chromatography. Column: μBondapak C_{18}, 300 x 3.9 mm I.D. Gradient: linear, 5% B/min. Solvents: A, acetonitrile-water (70 : 30); B, acetonitrile. Flow-rate: 3 ml/min. Detection: UV, 340 nm. Peaks: 1 = 2,4-dinitrophenylhydrazine (reagent); 2-10 = derivatives of propionaldehyde (2); butyraldehyde (3); pentanal (4); hexanal (5); heptanal (6); octanal (7); nonanal (8); decanal (9); and dodecanal (10). (Reproduced from ref. 8 with permission.)

Reversed-phase chromatography on octadecylsilica columns with linear or convex gradients of methanol[6,7] or acetonitrile[8] in water has been applied to the separation of 2,4-dinitrophenylhydrazones of homologous aldehydes or ketones containing 1-12 carbon atoms. Fig. 14.1 shows the separation of homologous series of these derivatives of C_3-C_{10} aliphatic aldehydes using a linear gradient of acetonitrile in water[8]. This method was applied to the analysis of trace amounts of aldehydes in *n*-octanol[8], the separation and determination of carbonyl compounds in cheese[6] and the analysis of acetaldehyde formed during chemical degradation of carcinostatic nitrosoureas in a study of their alkylating activities in vitro[7].

Aromatic and a number of heterocyclic carbonyl compounds possess strong UV absorption and it is therefore not necessary to form dinitrophenylhydrazone derivatives. Acetophenone, propiophenone, *n*-butyrophenone and caprophenone were separated by reversed-phase chromatography on a μBondapak C_{18} (300 x 3.9 mm I.D.)

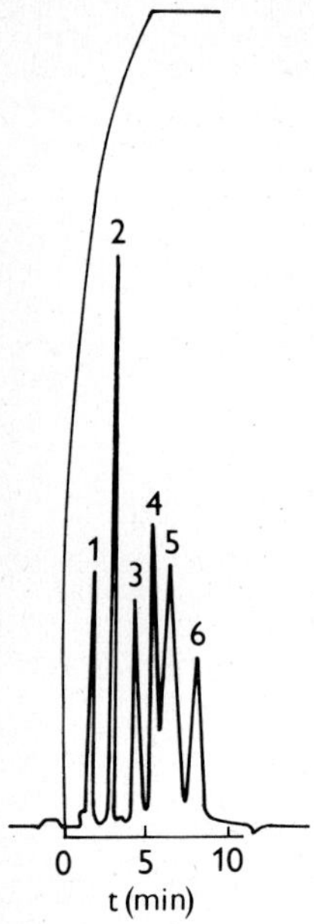

Fig. 14.2. Separation of quinones by reversed-phase gradient elution chromatography. Column: Bondapak C_{18}/Corasil, 1220 x 2.3 mm I.D. Gradient: convex 0-100% B in 5 min. After the end of the gradient, elution with pure solvent B followed. Solvents: A, methanol-water (30 : 70); B, methanol-water (70 : 30). Detection: UV, 254 nm. Peaks: 1 = benzoquinone; 2 = 1,4-naphthoquinone; 3 = anthraquinone; 4 = 2-methylanthraquinone; 5 = 2-ethylanthraquinone; 6 = 2-*tert.*-butylanthraquinone. (Reproduced from ref. 12 with permission.)

column using a linear gradient from 50 to 80% of methanol in water at 30°C. The method was applied in the analysis of the degradation products of the pesticide carbofuran[9].

The naturally occurring glycosides, which modify the auxin-receptor interaction in corn, were analysed for their active constituents, benzoxazolinones, after enzymatic hydrolysis and heating of the samples, by normal-phase chromatography on silica using a gradient of chloroform in *n*-hexane[10].

14.2. QUINONES

Quinones have been analysed using reversed-phase chromatography on porous octylsilica[11] (Zorbax C-8) or on superficially porous octadecylsilica (Permaphase ODS, Bondapak C_{18}/Corasil)[12-14] using gradients of methanol in water, with addition of phosphoric acid to the starting eluent (water)[13]. Linear gradients are usually employed, but a convex gradient profile may speed up the analysis, as shown in the example in Fig. 14.2, where *p*-benzoquinone, 1,4-naphthoquinone, anthraquinone, 2-methylanthraquinone, 2-ethylanthraquinone and 2-*tert.*-butyl-anthraquinone were separated in less than 10 min[12]. The method can be applied to the analysis of quinones and alkylquinones in industrial samples of dye intermediates and of photographic developers[11], or of naturally occuring quinones, such as anthraquinones in rhubarb[13] or ubiquinones (redox carriers in the respiratory chain; individual ubiquinones differ in the number of isoprenoid side-chains) in various animal and human tissues[14]. UV detection at 254 nm was used in these examples.

Normal-phase chromatography on superficially porous silica (Corasil II, 1000 x 3.2 mm I.D.) with an exponential (convex) gradient of ethyl acetate in cyclohexane was used to separate and determine pharmaceutically important anthraquinones that occur naturally in plants (chrysophanol, physcion, emodin and aloe emodin). The separation took 20 min and the compounds were detected by UV absorption at 280 nm[15].

Bis-substituted aminoalkylaminoanthraquinones with different substituents, used as cancer chemotherapeutic agents, were chromatographed on chemically bonded medium-polarity phases in reversed-phase systems. A convex gradient of methanol in 0.05 mol/l aqueous phosphate buffer (pH 8) was used for elution with a column packed with a nitrile bonded phase, μBondapak CN, while elution with a concave gradient of acetonitrile-methanol (1 : 1) in 0.05 mol/l acetate buffer (pH 8) was employed in chromatography on an amino-bonded phase, μBondapak NH_2[16].

14.3. REFERENCES

1 A. Breyer and W. Rieman, *Anal. Chim. Acta*, 18 (1958) 204.
2 J. Sherma and W. Rieman, *Anal. Chim. Acta*, 19 (1958) 134.
3 S. Honda and K. Kakehi, *J. Chromatogr.*, 152 (1978) 405.
4 L.J. Papa and L.P. Turner, *J. Chromatogr. Sci.*, 10 (1972) 747.
5 C.T. Mansfield, B.T. Hodge, R.B. Hege, Jr. and W.C. Hamlin, *J. Chromatogr. Sci.*, 15 (1977) 301.
6 L. Piergiovanni and G. Volonterio, *Riv. Ital. Sostanze Grasse*, 53 (1976) 99.
7 K.I. Nakamura, M. Asami, S. Orita and K. Kawada, *J. Chromatogr.*, 168 (1979) 221.
8 P.R. Demko, *J. Chromatogr.*, 179 (1979) 361.
9 E.J. Kikta, A.E. Stange and S. Lam, *J. Chromatogr.*, 138 (1977) 321.
10 A. Pessi and D. Scalorbi, *J. Chromatogr.*, 177 (1979) 162.
11 *LC Column Report – Zorbax C-8*, DuPont, Wilmington, DE.
12 *Solvent Delivery Systems for LC*, Waters Assoc., Milford, MA, 1975.
13 V. Castagnolo, G. Pettinari and G.A. De Vries, *Bull. Chim. Farm.*, 115 (1976) 376.
14 S. Imabayashi, T. Nakamura, Y. Sawa, J. Hasegawa, K. Sakaguchi, T. Fujita, Y. Mori and K. Kawabe, *Anal. Chem.*, 51 (1979) 534.
15 P.P. Rai, T.D. Turner and S.A. Matlin, *J. Chromatogr.*, 110 (1975) 401.
16 R.F. Taylor and L.A. Gaudio, *J. Chromatogr.*, 187 (1980) 212.

Chapter 15

SUGARS AND THEIR DERIVATIVES, POLYSACCHARIDES

Monosaccharides, oligosaccharides, polysaccharides and a number of sugar derivatives such as alditols, sugar acids, amino sugars and glycosides can be conveniently chromatographed by liquid chromatography, which offers distinct advantages for the analysis of carbohydrates over gas chromatography, where methyl-, trimethylsilyl- or trifluoroacetyl derivatives must be formed prior to the chromatographic separation. During the derivatization step, several isomers are formed from one saccharide, which may make the interpretation of the chromatogram difficult.

The application of liquid chromatography is subject to detection problems because carbohydrates show neither fluorescence nor significant absorption at wavelengths higher than 210-220 nm. Refractometric detection, sometimes used under isocratic conditions, has a low sensitivity and cannot be applied in gradient elution chromatography. Therefore, post-column reaction with colour-forming reagents has frequently been used in the gradient elution chromatography of carbohydrates, especially in earlier applications. The coloured reaction products are monitored by their absorption in the visible region. The most frequently used reactions are with orcinol in concentrated sulphuric acid (monitored at 420 nm) or with anthrone reagent in sulphuric acid (monitored at 640 nm), but other reactions can also be employed, e.g., periodate or bichromate oxidation, or reaction with phenol in sulphuric acid. A survey of post-column reaction systems for detection of carbohydrates is given in ref. 1. Such reaction detectors may be used as the integrated part of carbohydrate analysers, together with the gradient elution device and an ion-exchange column.

15.1. ION-EXCHANGE CHROMATOGRAPHY OF SUGARS, SUGAR PHOSPHATES AND AMINO SUGARS

In earlier work, chromatography on ion-exchange columns yielded more successful separations than chromatography on silica. The best results were achieved using anion-exchange chromatography with borate ions. The stabilities of the borate complexes of the individual sugars differ from one another and so do the strengths of their interactions with anion-exchange resins.

During the chromatography of carbohydrates, the concentration of the borate buffer and the pH are increased simultaneously in a linear or multilinear manner.

References on p. 313.

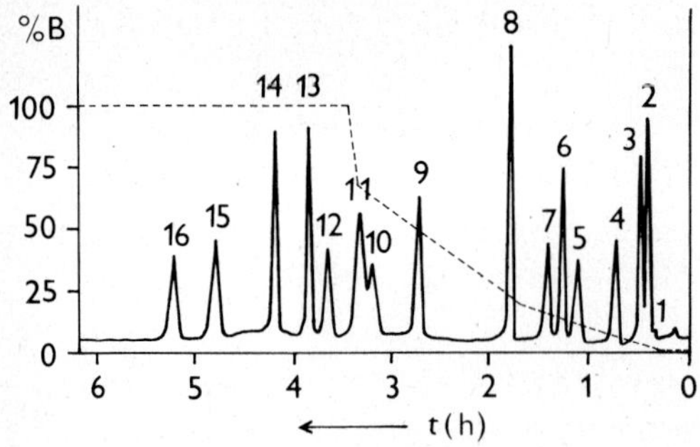

Fig. 15.1. Separation of a sugar mixture by gradient elution chromatography on an ion-exchange column. Column: anion-exchange resin Durrum DA-X4, 20 μm, 280 x 6 mm I.D. Gradient: as indicated by the broken line. Solvents: A, 0.1 mol/l borate buffer (pH 8.0); B, 0.5 mol/l borate buffer (pH 10.0). Flow-rate: 1 ml/min. Temperature: 50°C. Detection: photometric at 418 nm in post-column derivatization system based on reaction of carbohydrates with orcinol-sulphuric acid reagent, continuously added to the column effluent. Peaks: 1 = 2-deoxyribose; 2 = sucrose; 3 = trehalose; 4 = cellobiose; 5 = maltose; 6 = rhamnose; 7 = lactose; 8 = ribose; 9 = mannose; 10 = fructose; 11 = arabinose; 12 = galactose, 13 = xylose; 14 = glucose; 15 = gentiobiose; 16 = melibiose. (Reproduced from ref. 3 with permission.)

The original method of Kesler[2] was modified and improved depending on the character of the sugars to be separated and of the anion exchanger used[18,19,22]. The concentration of the borate ion usually changes from 0.05-0.12 mol/l (pH 7-8) to 0.5-0.6 mol/l (pH 10-10.5) and the column temperature is kept at 50-55°C during the chromatographic run. Small-particle polystyrene-divinylbenzene anion-exchange resins (particle diameter ca. 10-20 μm), such as Aminex A-28, Durrum DA-X4 or DA-X4F or Jeol LC-R-3 have been used in more recent applications for separations of mono- and oligosaccharides[3-6]. For example, the complete separation of sixteen mono- and disaccharides could be accomplished in 6 h using a multilinear borate ion concentration gradient (Fig. 15.1).

When organic polyhydric compounds are added to the mobile phase, they compete with sample carbohydrates for complex-forming borate anions. This effect may sometimes improve the separation and minimize the alkaline rearrangement of saccharides in the column. Mixtures of oligosaccharides have been separated on a column (65 x 0.6 cm I.D.) packed with Chromobeads 5 anion-exchange resin using a gradient formed from boric acid, glycerol, sodium chloride and sodium tetraborate solutions in a nine-chamber Autograd apparatus[20]. 2,3-Butanediol, which forms less strong complexes with borate ions, may be used instead of glycerol and a stepwise gradient instead of the continuous one[21].

A rapid anion-exchange separation of twelve mono- and disaccharides was accomplished in ca. 15 min using elution with a gradient of acetate ions in acetate buffer (from 0.15 to 6 mol/l, at pH 4.4) and a post-column reaction with cerium(IV) sulphate with continuous fluorimetric monitoring of the concentration of the cerium(III) ions produced[39].

Sugar phosphates formed as glycolysis intermediates have been separated on a column packed with AG 2-X4 (Cl^-) resin (200-400 mesh, 500 x 10 mm I.D.) using a concave gradient of ammonium chloride in the mobile phase containing a sufficient concentration of alkaline borate to ensure complexing of the sugar phosphates[17].

A chromatographic method has been proposed that makes it possible to analyse simultaneously amino acids, amino sugars and neutral sugars in mixtures in samples of protein hydrolysates using a single-run operation on two columns connected in series, the first packed with a sulphonated cation-exchange resin and the second with a strong anion-exchange resin. This method employs detection based on post-column derivatization of sugars by the orcinol method and of amino sugars and amino acids by the ninhydrin method[38]. Further details of this method are given in Chapter 20.

Sugar acids have been separated using chromatography on anion-exchange columns, where the application of mobile phase gradients sometimes proved useful. The application of a linear gradient of formic acid or a linear or stepwise gradient of sodium formate made possible the chromatographic separation of mixtures of oligogalacturonic acids (mono- to octa-) in chromatography on weak Dowex 3 or on strong Dowex 1-X8 anion-exchange columns in the formate form[24,25]. This method was also used for the separation of epimeric pairs of 3-deoxy-D-hexulosonic acids[26].

Cation-exchange chromatography on a column packed with Chromobeads B was applied to the separation of amino sugars and linear gradients of increasing pH (from 2.8 to 3.85) in a mobile phase consisting of pyridine acetate buffers allowed the separation of some hexosamines and their acid derivatives[27] and the separation of acidic and neutral amino acids from muramic acid and hexosaminuronic acids, as well as glucosamine, galactosamine, quinovosamine and fucosamine[28]. A mixture containing D-glucosamine, D-galactosamine and various amino acids was separated on a column packed with the cation-exchange resin Chromobeads C-2, using elution with a gradient formed in a nine-compartment Autograd system from buffers of various pH (2.75-6.10), containing sodium citrate, hydrochloric acid, sodium chloride, methanol and thiodiglycol in various concentrations[29].

Anion exchangers based on matrices other than polystyrene-divinylbenzene may appear more advantageous for certain applications. The glycol methacrylate anion exchanger DEAE-Spheron allows the distribution constants of the sugar-borate complexes to be decreased without a significant decrease in efficiency in comparison with polystyrene-divinylbenzene resins and the analysis to be

References on p. 313.

speeded up. Mixtures of mono- and oligosaccharides were separated in about 2-4 h using elution with a multi-linear gradient of borate ion at pH 8.5[7] or stepwise elution with borate buffers of increasing ionic strength and pH[40].

15.2. CHROMATOGRAPHY OF OLIGOSACCHARIDES ON CELLULOSE COLUMNS

Oligosaccharides are usually too strongly retained on ion exchangers in borate buffers to allow their successful separation. Such separations can be achieved on columns packed with cellulose, but at the cost of long analysis times. Thus, mixtures containing up to fifteen homologous glucose oligomers in wort, beer and brewing syrups could be separated and determined using chromatography on a column (450 x 15 mm I.D.) packed with Whatman No. CF12 cellulose powder employing a linear concentration gradient of *n*-butanol in an aqueous ethanolic mobile phase (starting eluent water-ethanol-*n*-butanol, 24.5 : 23 : 52.5; terminating eluent water-ethanol-*n*-butanol, 42 : 25 : 33). However, the separation took a whole day[8].

15.3. CHROMATOGRAPHY OF SUGARS ON SILICA GEL COLUMNS

Silica has also been used in some recent applications of carbohydrate analysis by gradient elution liquid chromatography. For example, a method for the determination of sugars commonly found in milk oligosaccharides was developed that employs reduction of mannose and fucose with sodium borohydride, followed by perbenzoylation of the reduction products, fucitol and mannitol, at 37°C for 2 h in pyridine containing 10% benzoic anhydride and 5% 4-dimethylamino-pyridine. After removal of excess of reagent on a Sep-pak cartridge and concentration of the sample solution by evaporation, the benzoylated alditols are separated by chromatography on a Zipax column (500 x 2.1 mm I.D.) using a linear gradient of dioxane in *n*-heptane and detected at 230 nm. Galactitol and glucitol perbenzoates cannot be separated under these conditions and, if they are present in the reduced sample, liquid chromatography should be combined with borate electrophoresis[9].

Perbenzoylated carbohydrates (xylose, mannose, glucose, galactose, maltose and maltotriose) were separated on a column (2000 x 2 mm I.D.) packed with Corasil II by gradient elution with a linear concentration change of diethyl ether in *n*-hexane (0-100% in 110 min) using UV detection at 254 nm[36].

15.4. CHROMATOGRAPHY OF SUGARS ON BONDED AMINO PHASES

Reversed-phase chromatography on octadecylsilica or octylsilica columns cannot be used successfully for the analysis of free sugars, which are very polar and are not significantly retained on these columns. However, stationary phases with chemically bonded polar amino- or aminoalkyl functional groups enter into preferential interactions with the hydroxyl groups of the sugars and make possible separations with good chromatographic resolution. Mobile phases containing an organic solvent, most often acetonitrile, in water are used. Owing to the polar (hydrogen bonding) interactions between the bonded amino groups and sugars, the chromatographic system behaves as a normal-phase system, where the mobile phase competes for interaction sites with sugar molecules. In addition, bonded amino phases have some properties of weak anion exchangers. Therefore, an increase in water content in the mobile phase leads to decreased retention of sugars. In gradient elution liquid chromatography of sugars on bonded amino phases, such as Spherisorb S5 NH_2 or Micro Pak NH_2, the concentration of water in the mobile phase increases with time. The carbohydrates are eluted in the order monosaccharides, disaccharides, trisaccharides, etc., as in the anion-exchange chromatography of sugar-borate complexes. Mixtures containing 9-12 sugars may be separated in 10-30 min using linear or multilinear gradients of water in acetonitrile (usually from 10-15% to 40-80% water)[10-13].

Mixtures of photosynthesis intermediates containing carbohydrates, nucleotides and sugar phosphates were chromatographed on a bonded amino phase using elution with a gradient of potassium dihydrogen phosphate in water (0.005-0.4 mol/l)[41].

In these applications, UV detection at 192 nm was employed[10,12,13], but extreme care should be taken with regard to the purity of the water and organic solvents used. Only acetonitrile can be employed as the organic solvent at such low wavelengths. The use of a mass detector has been described, where the solvent is evaporated after nebulization in a heated column, producing finely divided particles of non-volatile solutes that pass through a light beam. Light scattered from the particles is detected by a photomultiplier and its signal is amplified and recorded. The detection limit of this detector corresponds to ca. 2-3 μg of the sugar. Fig. 15.2 shows the separation of nine sugar standards on Spherisorb S5 NH_2 using gradient elution (acetonitrile-water) with this type of detector. The method has been applied to plant extracts and to glucose syrup to determine the oligosaccharide composition in these samples[11].

References on p. 313.

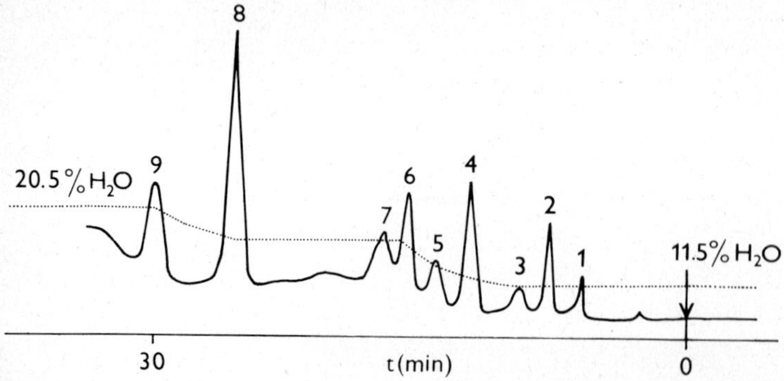

Fig. 15.2. Separation of a sugar mixture by gradient elution chromatography on a column packed with a bonded amino phase. Column: Spherisorb S5 NH_2, 250 x 5 mm I.D. Gradient: as indicated by the broken line. Solvents: A, 88.5% (v/v) acetonitrile in water; B, 79.5% (v/v) acetonitrile in water. Flow-rate: 2 ml/min. Temperature: 90°C. Detection: light-scattering mass detector. Peaks: 1 = arabinose; 2 = xylose; 3 = ribose; 4 = fructose; 5 = mannose; 6 = glucose; 7 = galactose; 8 = sucrose; 9 = maltose. (Reproduced from ref. 11 with permission.)

15.5. CHROMATOGRAPHY OF MUCOPOLYSACCHARIDES

Glycosaminoglycans (mucopolysaccharides) have been separated on anion-exchange resins (Dowex AG 1-X2) using gradient elution with increasing concentrations of sodium and magnesium chlorides[14] or on a short column packed with DEAE-cellulose (DE-52, 5 x 1 cm I.D.) using a gradient of lithium chloride (from 0.2 to 1.2 mol/l) in 0.05 mol/l sodium acetate solution (pH 4) at 60°C, where resolution mainly according to degree of sulphation, but also according to molecular weight, was achieved[15]. A complete recovery of acidic glycosaminoglycans was achieved on the anion-exchange cellulose, in contrast to chromatography on columns packed with ion-exchange resins. The method was applied to the analysis of papain digests of mammalian connective tissue specimens.

Heparin was fractionated on a Phenyl-Sepharose Cl-4B column (6 x 0.6 cm I.D.) using a gradient of ammonium sulphate in 0.01 *M* hydrochloric acid[16].

Spectrophotometric analysis of fractions of the eluate was usually employed for the detection and quantitation of mucopolysaccharide components.

15.6. CHROMATOGRAPHY OF GLYCOSIDES

The presence of aglycones, which are less polar than carbohydrates, in the molecules of glycosides makes it possible to use reversed-phase chromatography on chemically bonded non-polar phases, and also usually considerably improves the possibility of using UV detection. Digitalis glycosides were chromatographed on an octylsilica column (250 x 3 mm I.D.) using a linear gradient of acetonitrile in water (from 12.5 to 21%) and the separated compounds were detected at 225 nm[30]. Addition of an acid or salt to water as the starting eluent may improve the separation in certain instances. The flavonoid glycoside hesperidin was chromatographed on octadecylsilica using a linear gradient of methanol in 0.03 mol/l potassium dihydrogen phosphate[31]. Glycosides from grapefruit peel were separated on octadecylsilica using gradient elution chromatography with a linear concentration change of methanol in 0.6% aqueous perchloric acid with UV detection at 280 nm[32]. 3-Glucosides, 3,5-diglucosides, 3-(6-O-*p*-coumaryl)-glucosides and 3-(6-O-*p*-coumarylglucoside)-5-glucosides of anthocyanidine were separated by chromatography on octadecylsilica using a convex gradient of methanol (0-20%) in 15% acetic acid in water[33]. Glycosidic flavonoids were chromatographed on a bonded alkylphenyl phase using a convex gradient of ethanol (12.5-40%) in water containing 0.1% acetic acid and were detected at 254 nm[34]. Glycoflavones were also separated on a bonded amino phase using a linear gradient of acetonitrile in water (10-90% in 15 min) and the separated compounds, isoorientin-3'-O-glucoside and 4'-O-glucoside, were detected at 254 nm. In this example, the affinity of the carbohydrate part of the glycoside molecule to the polar amino phase is obviously utilized as the basis of the chromatographic mechanism. Esters of indole-3-acetic acid and myoinositol or myoinositol glucosides in corn extracts were purified on a partially sulphonated cation-exchange column using a linear gradient of acetone in water[37].

15.7. REFERENCES

1 K. Čapek and J. Staněk, Jr., in Z. Deyl, K. Macek and J. Janák (Editors), *Liquid Column Chromatography*, Elsevier, Amsterdam, 1975, p. 465.
2 R.B. Kesler, *Anal. Chem.*, 39 (1967) 1416.
3 W. Voelter and H. Bauer, *J. Chromatogr.*, 126 (1976) 693.
4 H. Bauer and W. Voelter, *Chromatographia*, 9 (1976) 433.
5 J.F. Kennedy and J.E. Fox, *Carbohydr. Res.*, 54 (1977) 13.
6 W.H. Morrison, M.F. Lou and P.G. Hamilton, *Anal. Biochem.*, 71 (1976) 415.
7 Z. Chytilová, O. Mikeš, J. Farkaš, P. Štrop and P. Vrátný, *J. Chromatogr.*, 153 (1978) 37.
8 G.E. Otter, J.A. Popplewell and L. Taylor, *J. Chromatogr.*, 49 (1970) 462.
9 P.F. Daniel, *J. Chromatogr.*, 176 (1979) 260.
10 R.E. Majors, *J. Chromatogr. Sci.*, 15 (1977) 334.
11 R. Macrae and J. Dick, *J. Chromatogr.*, 210 (1981) 138.
12 J. Hettinger and R.E. Majors, *Varian LC Application Bull.*, 10 (1976) 6.
13 *Applied Science Division Catalog*, Applied Science Labs., State College, PA, 1980.
14 B. Radhakrishnamurthy, E. Dalferes, H. Ruiz and G.S. Berenson, *Anal. Biochem.*, 82 (1977) 445.
15 A. Hallén, *J. Chromatogr.*, 71 (1972) 83.
16 A. Ogamo, K. Matsuzaki, H. Uchiyama and K. Nagasawa, *J. Chromatogr.*, 213 (1981) 439.

17 G. Bedetti, G.D'Agnolo and F. Pocchiari, *J. Chromatogr.*, 49 (1970) 53.
18 Y.C. Lee, J.F. McKelvy and D. Lang, *Anal. Biochem.*, 27 (1969) 567.
19 R.L. Jolley and M.L. Freeman, *Clin. Chem.*, 14 (1968) 538.
20 P. Auriol, *C. R. Acad. Sci., Ser. D*, 269 (1969) 1183.
21 E.F. Walborg and L.E. Kondo, *Anal. Biochem.*, 29 (1969) 433.
22 D.J. MacLaurin and J.W. Green, *Can. J. Chem.*, 47 (1969) 3947, 3957.
23 J.I. Ohms, J. Zec, J.V. Benson and J.A. Patterson, *Anal. Biochem.*, 20 (1967) 51.
24 R. Derungs and H. Deuel, *Helv. Chim. Acta*, 37 (1954) 657.
25 C.W. Nagel and T.M. Wilson, *J. Chromatogr.*, 41 (1969) 410.
26 O. Portsmouth, *Carbohydr. Res.*, 8 (1969) 193.
27 K. Brendel, N.O. Roszel, R.W. Wheat and E.A. Davidson, *Anal. Biochem.*, 18 (1967) 147.
28 K. Brendel, R.S. Steele, R.W. Wheat and E.A. Davison, *Anal. Biochem.*, 18 (1967) 161.
29 M. Monsigny, *Bull. Soc. Chim. Biol.*, 50 (1968) 2188.
30 *Application of LiChrosorb RP-8*, Merck, Darmstadt, G.F.R.
31 C.T. Seitz and R.E. Wingard, *J. Agr. Food Chem.*, 26 (1978) 278.
32 N.W. Preston and C.F. Timberlake, *J. Chromatogr.*, 214 (1981) 222.
33 M. Williams, G. Hrazdina, M.M. Wilkinson, J.G. Sweeny and G.A. Iacobucci, *J. Chromatogr.*, 155 (1978) 389.
34 M. Vanhaelen and R. Vanhaelen-Fastré, *J. Chromatogr.*, 187 (1980) 255.
35 H. Becker, G. Wilking and K. Hostettmann, *J. Chromatogr.*, 136 (1977) 174.
36 J. Lehrfeld, *J. Chromatogr.*, 120 (1976) 141.
37 A. Ehmann and R.S. Bandurski, *J. Chromatogr.*, 72 (1972) 61.
38 M.M. Tikhomirov, A.Ya. Khorlin, W. Voelter and H. Bauer, *J. Chromatogr.*, 167 (1978) 197.
39 W.W. Pitt, Jr., C.D. Scott, W.F. Johnson and G. Jones, *Clin. Chem.*, 16 (1970) 657.
40 P. Vrátný, O. Mikeš, J. Farkaš, P. Štrop, J. Čopíková and K. Nejepinská, *J. Chromatogr.*, 180 (1979) 39.
41 C. Giersch, *J. Chromatogr.*, 172 (1979) 153.

Chapter 16

LOWER CARBOXYLIC ACIDS, ESTERS AND ORGANIC PEROXIDES

16.1. CARBOXYLIC ACIDS

Aliphatic carboxylic acids do not absorb strongly in the UV region. Therefore, UV detection can be used only at wavelengths near 200 nm and fluorimetric detection of free acids cannot be used at all. To enhance the possibility of detection, UV-absorbing or fluorescent derivatives are often formed from these acids prior to the chromatographic separation. If the acids are chromatographed in their free forms, post-column derivatization may be also used, which was very common in earlier applications. Aromatic carboxylic acids can be conveniently detected using UV detectors.

16.1.1. Anion-exchange chromatography

In earlier applications, anion-exchange chromatography with gradient elution was frequently used for separations of carboxylic acids. Most often, anion-exchange columns were employed in the acetate or formate form and concentration gradients of acetic acid, sodium acetate or formic acid were applied. It was possible to separate benzoic and the three isomeric hydroxybenzoic acids using elution with an increasing concentration of acetic acid in methanolic solution on a column packed with Dowex 2-X8[1]. Pyruvic, glutaric, citric, 2-ketoglutaric and *trans*-aconitic acids were separated by chromatography on a column packed with Dowex 1-X8 using linear gradient of sodium acetate in water (0-1.2 mol/l) at 67°C. Post-column reaction of the acids in the eluate with dichromate and monitoring of the decrease in dichromate concentration at 424 nm was used for detection[2]. A number of aliphatic and aromatic carboxylic acids may be separated on a column packed with Dowex 1-X10 using a gradient of formic acid (from 0 to 25 mol/l)[3].

Katz and Burtis[4] reported the elution volumes of 114 standard compounds in an anion-exchange system based on a column (31.6 x 0.45 cm I.D.) packed with Dowex 1-X8, 10 μm) and a gradient of acetate buffer (pH 4.4) from 0.015 to 6 mol/l as the eluent. Among these compounds, a number of aromatic carboxylic acids could be separated[4].

References on p. 322.

However, anion-exchange chromatographic separations of carboxylic acids on columns packed with ion-exchange resins were very time consuming and took from several hours to several days. The use of columns packed with porous-layer anion-exchange resins (such as Zipax SAX) allowed the separation of seventeen aromatic carboxylic acids, including eleven benzenecarboxylic acids (mono- to hexa-) in ca. 2 h with a linear gradient (from 0.01 to 0.1 mol/l) of sodium nitrate in 0.01 mol/l borate buffer (pH 9.1)[5].

16.1.2. Chromatography on silica gel columns

Chromatography of carboxylic acids on "naked" silica gel with mobile phases composed of neutral organic solvents usually yields asymmetric peak shapes owing to strong adsorption of the acids on the active centres of silica. Therefore, an acid has to be added to the mobile phase to saturate the active centres of the adsorbents prior to the chromatographic run.

Chromatography on silica gel coated with 0.05-0.25 mol/l sulphuric acid was used as the basis of an automated method for the analysis of organic acids in an instrument resembling an amino acid analyser. A continuous gradient of *tert*.-amyl alcohol in chloroform was applied using a Varigrad gradient apparatus. The separated acids in the eluate were subject to reaction with an indicator (*o*-nitrophenol in absolute methanol) in a post-column reactor and the colour developed was monitored at 350 nm[6-8]. The method was applied to separation of a number of physiologically important acids, such as Krebs cycle intermediates. However, separations of mixtures containing acetic, pyruvic, fumaric, glutaric, lactic, succinic, α-ketoglutaric, malic and citric acids took 3-5 h.

The chromatographic control of an antiflammatory agent, tolmetin (a substituted indoleacetic acid), may be mentioned as another example of the gradient elution chromatography of carboxylic acids on a silica gel column.

Tolmetin and nine related acidic, esteric and amidic compounds, which may be present as impurities in this product, were separated in 25 min with a concave gradient of isopropanol in *n*-hexane containing 0.25% of acetic acid on a LiChrosorb Si 60 column (10 μm), 250 x 2.2 mm I.D., using UV detection at 254 nm[44].

16.1.3. Reversed-phase chromatography of free aliphatic carboxylic acids

In recent applications of gradient elution liquid column chromatography for the separation of carboxylic acids, reversed-phase chromatography on chemically bonded 5 or 10 μm octadecyl- or octylsilica packing materials has been by far the most frequently used method. In contrast to conventional time-consuming methods employing ion-exchange resins, the separation even of complex samples may be achieved in 30 min or less. This method has been applied to aliphatic carboxylic acids, α-keto acids, aromatic acids, substituted indoleacetic acids, UV-absorbing acidic compounds in urine, etc.

Gradient separation of butyric, octanoic, dodecanoic and hexadecanoic acids was accomplished using a linear gradient of methanol in water (from 60 to 100%) on an octadecylsilica column (μBondapak C_{18}, 10 μm, 300 x 3.9 mm I.D.) using conductimetric detection. The column was equilibrated with a sodium salt and the remaining free silanol groups were converted into $Si\text{-}O^{-}Na^{+}$ groups prior to the chromatographic run. During the chromatography, sodium ions exchange for H^{+} ions of the acids and thus all of the chromatographed acids are converted into their sodium salts, which are detected in the eluate with high sensitivity by conductivity monitoring[45].

However, this method of detection cannot be used if the mobile phase contains salts (buffers), whose presence has been often reported to improve separations because of the suppression of the dissociation of the acids or owing to salting-out effects. Electrochemical detection may be used for readily oxidizable or reducible compounds, such as uric acid and some related compounds that were chromatographed on a μBondapak C_{18} (10 μm) column (300 x 4.6 mm I.D.) with a linear gradient of acetonitrile (from 0 to 60%) in 0.1 mol/l aqueous potassium dihydrogen phosphate (pH 2.5) and detected by direct electrochemical oxidation at +0.8 V[9].

16.1.4. Reversed-phase chromatography of carboxylic acids derivatives

Pre-column derivatization has become popular in the reversed-phase liquid column chromatography of non-UV-absorbing carboxylic acids, not only because of the enhancement of the detection sensitivity. The formation of a derivative of an acid usually results in a marked increase in its hydrophobicity, which usually improves the separation in reversed-phase systems. Phenacyl and *p*-bromophenacyl esters are the most frequently used derivatives for this purpose. The acids are subject to reaction with phenacyl or *p*-bromophenacyl bromide with addition of triethylamine or another catalyst (the use of potassium fluoride has been reported) prior to injection on to the column. Linear gradients of acetonitrile or methanol in water or in water with addition of phosphoric acid in connection with octadecyl-[10] or octylsilica[11] columns are used for separation. The acid derivatives in the eluate are detected using UV detection at 254 nm. The method has been applied, for example, to the determination of quinic and other acids in plant extracts[11] and the analysis of gibberellins (plant growth hormones)[43].

References on p. 322.

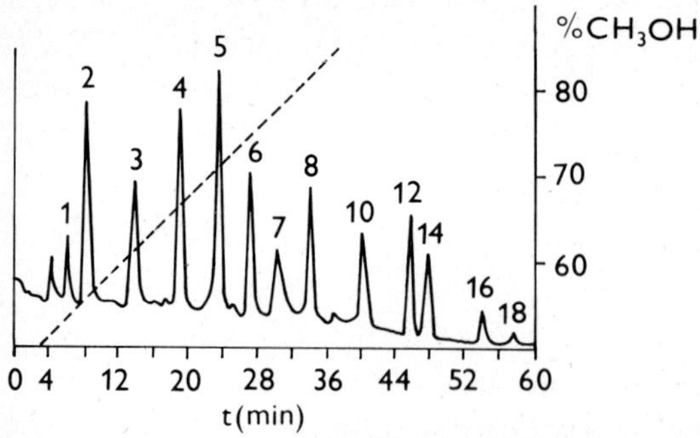

Fig. 16.1. Separation of 4-methyl-7-methoxycoumarin esters of aliphatic saturated carboxylic acids by reversed-phase gradient elution chromatography. Column: Nucleosil 10 C_{18}, 250 x 2 mm I.D. Gradient: elution with solvent A for 3 min, then 1% B/min. Solvents: A, 50% (v/v) methanol in water; B, methanol. Flow-rate: 1 ml/min. Detection: fluorimetric, λ_{ex} = 340 nm; λ_{em} = 420 nm. Peaks: derivatives of: 1 = formic (C_1); 2 = acetic (C_2); 3 = propionic (C_3); 4 = *n*-butyric (C_4); 5 = *n*-valeric (C_5); 6 = caproic (C_6); 8 = caprylic (C_8); 10 = capric (C_{10}); 12 = lauric (C_{12}); 14 = myristic (C_{14}); 16 = palmitic (C_{16}); 18 = stearic (C_{18}) acids. (Reproduced from ref. 12 with permission.)

2-Naphthacyl esters of carboxylic acids are also suitable for use in reversed-phase gradient elution liquid chromatography. The derivatives are formed by heating with 2-bromoacetonaphthone and dimethylformamide in dry acetonitrile. Crown ethers can also be used as catalysts instead of dimethylformamide. 2-Naphthacyl esters of lactic acid and of lower saturated unbranched carboxylic acids (from formic to valeric) were separated on an octadecylsilica column (250 x 3 mm I.D.) using a convex gradient of acetonitrile (from 38 to 75%) in water at 35°C. Either UV detection at 245 nm or fluorescence detection using an emission wavelength of 450 nm and an excitation wavelength of 290 nm can be used, because the derivatives show significant fluorescence with a maximum at 450 nm[41].

The formation of fluorescent esters of 4-bromomethyl-7-methoxycoumarin with monocarboxylic acids and a variety of other acidic compounds of biomedical interest offers the possibility of using fluorimetric detection in the reversed-phase chromatography of these derivatives. If the excitation wavelength is selected at 340 nm and the emission wavelength at 420 nm, the sensitivity of fluorimetric detection is about one order of magnitude greater than that of UV detection at 340 nm (absorption maximum). The fluorescent derivatives of saturated unbranched aliphatic fatty acids between formic and stearic acids could be completely separated within 60 min using chromatography on a Nucleosil 10 C-18 (10 μm) column (250 x 2 mm I.D.) with a linear gradient of methanol in water (50-100%)[12] (see Fig. 16.1).

Derivatization of carboxylic acids with 1-chloromethylisatin and a crown ether or triethylamine as a catalyst in dry dimethylformamide at 50°C has been proposed. Isatinyl methyl esters of C_1-C_{10}, C_{12}, C_{14} and C_{16} saturated unbranched carboxylic acids were separated in ca. 50 min on an octylsilica column (250 x 4 mm I.D.) using a gradient of methanol in water from 50 to 100% and UV detection at 240 nm[40].

16.1.5. Reversed-phase chromatography of α-keto acids

The sensitivity of detection of α-keto acids in the UV region may be enhanced by using derivatization with *o*-phenylenediamine in 2 mol/l hydrochloric acid at 80°C to form 2-quinoxalinol derivatives. These derivatives were chromatographed on a LiChrosorb RP-8 (5 μm) column (250 x 2.1 mm I.D.) using two-step elution with the first isocratic step followed by the second, gradient step, where the concentration of acetonitrile in water changed in a linear manner from 20 to 50% in 20 min. This method was used for the separation and determination of phenylpyruvic, α-ketobutyric, *p*-hydroxyphenylpyruvic, α-ketovaleric, α-ketoisovaleric, α-ketoisocaproic and α-keto-β-methylvaleric acids in urine for the purpose of diagnostic indications of phenylketonuria and other metabolism deficiencies[13]. The free α-keto acids in urine were also chromatographed on an octadecylsilica column using elution with a linear gradient of methanol in aqueous ammonium acetate[14].

16.1.6. Reversed-phase chromatography of aromatic carboxylic acids on chemically bonded alkyl phases

Substituted benzoic and cinnamic acids have often been chromatographed on octadecyl-[15-17] and octylsilica columns[18,19] with UV detection at 254, 280 or 312 nm. Binary gradients of methanol in water were reported to give no or poor separations of mono-, di- and trimethoxycinnamic acids. Therefore, a ternary gradient of decreasing methanol concentration (from 30 to 15%) and simultaneously increasing tetrahydrofuran concentration (from 0 to 15%) in water was employed and yielded a satisfactory separation of eight hydroxy- and methoxycinnamic acid metabolites[19]. The acids are often chromatographed at a low pH to suppress their dissociation and to improve the separation possibilities[16,17]. A 0.1 mol/l potassium dihydrogen phosphate solution (pH 2.1-2.5) was used as the starting eluent and the concentration of acetonitrile or methanol in the mobile phase was increased with time in the chromatography of acidic constituents in urine samples on octadecylsilica columns, using UV detection at 254 and 280 nm, or dual detection (UV at 280 nm and fluorimetric with excitation at 238 nm and emission at 340 nm). This dual detection proved very useful for the identification of aromatic acids. This method gave excellent results; it was possible to

References on p. 322.

separate over 100 acidic compounds in less than 30 min (Fig. 4.16)[16]. The method was applied to the determination of some clinically significant aromatic acids, such as vanillylmandelic acid, in urine for the diagnosis of neural and adrenal medullary tumours[16,17].

The dissociation of carboxylic acids is also suppressed if a gradient of methanol (linear or convex) in water containing 0.6-2.5% of acetic acid is employed[18,20]. Ten free hydroxycinnamic acids and their esters with glucose and quinic acid were separated in 25 min under these conditions, using UV detection at 312 nm. These acids are widespread in higher plants and the method has been applied to their analysis in plant tissues[18]. It was also possible to separate several substituted indole acids and their esters in immature seeds and other plant tissues in less than 20 min using this technique, but detection was affected at 254 nm[20].

Ion-pair chromatography of carboxylic acids on octadecylsilica columns (μBondapak C_{18}, 300 x 4 mm I.D.) using linear or multi-linear concentration gradients of methanol or acetonitrile in water, both solvents containing 0.005 mol/l tetrabutylammonium phosphate (Waters reagent PIC A), was employed for the separation of phthalic acid isomers[21] (UV detection at 254 nm) and as a rapid routine method for the quantitative analysis of porphyrin carboxylic acids in urine (fluorimetric detection)[22].

16.1.7. *Reversed-phase chromatography of aromatic carboxylic acids on organic gels and on cation exchangers*

Gels or cation exchangers based on organic matrices have occasionally been used for the gradient elution liquid column chromatography of carboxylic acids. It was possible to separate thirteen phenolic acids naturally occurring in plants on a polyvinylpyrrolidone column (Polyclar AT, 60 μm) using an exponential gradient of ethanol in water[23]. Phenolic acids in tobacco leaves were separated by chromatography on Sephadex LH-20 using a linear gradient of methanol in chloroform[24]. Indoleacetic acids and esters (glycosides) were purified on a styrene-divinylbenzene cation-exchange resin, Dowex 50W-X2, using elution with a linear gradient of acetone in water[25]. Twelve acidic compounds commonly utilized as components of analgesic drugs were separated in 2 h on a column (250 x 4 mm I.D.) packed with a styrene-divinylbenzene copolymer (5-10 μm) using a gradient of pH from 2.5 to 8.5 in 0.2 mol/l phosphate buffer in 25% methanol (an alcohol concentration gradient could not be used because it would have damaged the column bed)[26]. Humic substances (a complex mixture of organic acids extracted from soils, sediments and natural waters) can be rapidly and conveniently chromatographed on a methyl methacrylate resin, Amberlite XAD-8, using a pH gradient (from 3 to 11) in phosphate or phosphate-borate buffer with UV detection at 254 nm. A fingerprint chromatogram of the humic acids is obtained and the compounds are separated into two classes, containing phenolic and carboxylic substances, prior to further characterization[27,28].

16.2. ESTERS OF LOWER CARBOXYLIC ACIDS

Reversed-phase chromatography on octadecylsilica columns was used for separations of esters of carboxylic acids like for those of free acids. Parabens (esters of lower alcohols with *p*-hydroxybenzoic acid, frequently used as preservatives in food or skin cream formulations) were chromatographed on these columns using linear gradients of methanol[29], acetonitrile[30] or ethanol[31] in water or using a linear ternary gradient of methanol and acetonitrile in water[32]. The compounds were detected in the UV region at 254 nm.

Phthalate esters were chromatographed on octadecylsilica columns using convex or linear gradients of methanol in water[33,34], a concave gradient of methanol in water containing 1% acetic acid[35] or a ternary gradient of acetonitrile and tetrahydrofuran in water[36]. The phthalate esters were detected by UV absorption at 254 nm. The method was applied to the analysis of phthalate plasticizers in aqueous extracts from plastic materials and in environmental water samples. In the latter applications, a hold-up period was inserted into the early stage of the gradient to allow the elution of naturally occurring organic compounds in water samples sufficiently separated from the phthalate esters[34].

Wyerone ester derivatives (antifungal compounds produced by plants in response to microbial infection) from plant extracts were separated on a C18 column using a ternary linear gradient of methanol and acetonitrile in water containing 5% formic acid[37].

Phorbol esters (carcinogenic esters of terpenic alcohols) in urine were determined by liquid chromatography on a silica gel column (5 μm Spherisorb) using a linear gradient of methanol in 2,2,4-trimethylpentane-isopropanol (1 : 1)[42].

Several naturally occurring esters of phenolic acids, including chlorogenic acid isomers, were separated on an octylsilica column using a linear gradient of methanol in 0.01 mol/l phosphoric acid and on a diol chemically bonded phase using a linear gradient of *n*-propanol in methylene chloride containing 0.5-1% of acetic acid. The compounds were monitored with UV detector at 280 nm[38].

16.3. ORGANIC PEROXIDES

Several organic peroxides (di-*tert.*-butyl peroxide, nonanoyl peroxide; *tert.*-butyl peroctoate, *tert.*-butylperbenzoate, cumene hydroperoxide, *tert.*-butyl

References on p. 322.

hydroperoxide and di-*n*-butyl peroxidicarbonate) were separated on a silica gel column using a linear gradient of isopropanol in *n*-hexane (from 0.1 to 2%) and UV detection at 220 nm[39].

16.4. REFERENCES

1 N.E. Skelly and W.B. Crummett, *Anal. Chem.*, 35 (1963) 1680.
2 H. Veening and R.C. Zerfing, *Anal. Chem.*, 38 (1966) 1312.
3 G.L. Lawson and J.W. Purdie, *Mikrochim. Acta*, (1961) 415.
4 S. Katz and C.A. Burtis, *J. Chromatogr.*, 40 (1969) 270.
5 J. Aurenge, *J. Chromatogr.*, 84 (1973) 285.
6 L. Kesner and E. Muntwyler, *ACS Winter Meeting, Phoenix, AZ, January 17-21, 1966.*
7 *JEOL Chromatography*, No. 9, 1973.
8 *Automatic Organic Acid Analyzer - Technical Bulletin*, Waters Assoc., Milford, MA.
9 A.M. Krstulovic, L.M. Bertani-Dziedzic, S.E. Gitlow and K. Lohse, *J. Chromatogr.*, 164 (1979) 363.
10 J.M. Miller, I.D. Brindle, S.D. Cater and J.H. Clark, *Anal. Chem.*, 14 (1980) 2430.
11 L. Nagels, C. Debeuf and E. Esmans, *J. Chromatogr.*, 190 (1980) 411.
12 W. Dünges and N. Seiler, *J. Chromatogr.*, 145 (1978) 483.
13 T. Hayashi, H. Todoriki and H. Naruse, *J. Chromatogr.*, 224 (1981) 197.
14 J.C. Liao, N.E. Hoffman, J.J. Barboriak and D.A. Roth, *Clin. Chem.*, 22 (1977) 802.
15 J.B. Murphy and C.A. Stutte, *Anal. Biochem.*, 86 (1978) 220.
16 I. Molnár, Cs. Horváth and P. Jatlow, *Chromatographia*, 5 (1978) 260.
17 L.M. Bertani-Dziedzic, A.M. Krstulovic, S. Ciriello and S.E. Gitlow, *J. Chromatogr.*, 164 (1979) 345.
18 J. Krause and D. Strack, *J. Chromatogr.*, 176 (1979) 465.
19 E. Roggendorf and R. Spatz, *J. Chromatogr.*, 204 (1981) 263.
20 V. Sjut, *J. Chromatogr.*, 209 (1981) 107.
21 *Paired-ion Chromatography*, Technical Bulletin, Waters Assoc., Milford, MA, 1976.
22 H.D. Mayer, K. Jaocb, W.Vogt and M. Knedel, *J. Chromatogr.*, 199 (1980) 339.
23 G. Alibert, *J. Chromatogr.*, 80 (1973) 173.
24 M.E. Snook, P.J. Fortson and O.T. Chortyk, *Anal. Chem.*, 53 (1981) 374.
25 A. Ehmann and R.S. Bandurski, *J. Chromatogr.*, 72 (1972) 61.
26 T. Hanai, H.F. Walton, J.D. Navratil and D. Warren, *J. Chromatogr.*, 155 (1978) 261.
27 P. McCarthy, M.J. Peterson, R.L. Malcolm and E.M. Thurman, *Anal. Chem.*, 51 (1979) 2041.
28 M.A. Curtis, A.F. Witt, S.B. Schram and L.B. Rogers, *Anal. Chem.*, 53 (1981) 1195.
29 S.R. Bakalyar and R.A. Henry, *J. Chromatogr.*, 126 (1976) 327.
30 T. Alfredson and L. Tallman, *Varian Liquid Chromatogr. Bull.*, LC 115.
31 *Micromeritics Application Notes*, No. 44, Micromeritics, Norcross, GA.
32 *Spectra-Physics 8000 HPLC Brochure.*
33 *Waters Application Bull.*, Dec. 1973.
34 A. Otsuki, *J. Chromatogr.*, 133 (1977) 402.
35 S.C. Amundson, *J. Chromatogr. Sci.*, 16 (1978) 170.
36 T. Alfredson and L. Tallman, *Varian Liquid Chromatogr. Bull.*, LC 116.
37 A.E.A. Portez, R.V. Smallman and J.W. Mansfield, *J. Chromatogr.*, 172 (1979) 498.
38 L. Nagels, W. Van Dongen, J. De Brucker and H. De Pooter, *J. Chromatogr.*, 187 (1980) 181.

39 L.A. Cornish, R. Ferrie and J.E. Paterson, *J. Chromatogr. Sci.*, 19 (1981) 85.
40 G. Gübitz, *J. Chromatogr.*, 187 (1980) 208.
41 W. Distler, *J. Chromatogr.*, 192 (1980) 240.
42 D.L. Berry, *Spectra-Physics Chromatogr. Rev.*, 3, No. 2 (1977) 5.
43 R.O. Morris and J.B. Zaerr, *Anal. Lett.*, A11 (1978) 73.
44 R.K. Gilpin and C.A. Janicki, *J. Chromatogr.*, 147 (1978) 501.
45 E.M. Thurman, *J. Chromatogr.*, 185 (1979) 625.

Chapter 17

FATTY ACIDS AND LIPIDS

Column liquid chromatographic methods may offer certain advantages for the purification, fractionation and analysis of non-volatile or thermolabile compounds such as fatty acids and lipids. The gradient elution technique is particularly useful in the liquid chromatography of lipids, where the compounds usually differ very significantly in their retention under isocratic conditions. The most serious limitations in the use of modern column liquid chromatography for lipid analysis are due to the detection possibilities of these usually non-fluorescent and non-UV-absorbing solutes. Therefore, the need for a special detection technique was felt more strongly in the application of liquid chromatography to lipid analysis than in other applications. Mass-transport (moving-wire flame-ionization detectors) and, more recently, infrared detectors have been most frequently employed for the HPLC of lipids. Chemical derivatization is also often used.

17.1. CHROMATOGRAPHY OF FATTY ACID DERIVATIVES

Like the lower carboxylic acids (Chapter 16), the individual fatty acids are most frequently separated using chromatography of their appropriate UV-absorbing or fluorescent derivatives, which not only makes it possible to use common detectors, but also improves significantly the possibilities of chromatographic separation. The possibilities of derivatization and the choice of chromatographic systems are similar for lower and higher carboxylic acids. The most frequently used derivatives in the chromatography of fatty acids are phenacyl-, naphthacyl- and benzyl esters, anilides and derivatives with substituted coumarin, but other derivatization reactions can also be used[1].

Reversed-phase gradient elution liquid chromatography on non-polar chemically bonded phases is most frequently used for separations of fatty acid derivatives.

17.1.1. Separation of phenacyl esters of fatty acids

p-Bromophenacyl esters of higher carboxylic acids can be conveniently chromatographed on octadecylsilica columns using linear or stepwise gradients of methanol or acetonitrile in water and UV detection at 254 nm. Using this system, homologous derivatives of saturated unbranched acids from C_1 to C_{24} can easily be separated and quantified[2-4]. Acetonitrile-water gradients seem to offer better resolution of carboxylic acid derivatives with more than fourteen carbon atoms than methanol-water gradients and are therefore usually preferred[2,41]. The range of the compositional change during the gradient depends on the number of carbon atoms in the acids in sample mixtures, but usually linear gradients from 40-70% to 100% acetonitrile are adequate for separations of fatty acid derivatives with 6-24 carbon atoms on columns 300-900 mm long[1-4]. The selection of the operating conditions is more critical if the unsaturated fatty acid derivatives are to be separated from each other and from the phenacyl esters of saturated fatty acids. The introduction of one or more double bonds into the acid molecule decreases the retention on non-polar stationary phases. This effect increases with increasing number of double bonds in the acid molecule and may lead to overlapping of the peaks of acids with different combinations of carbon number and double bond number. The selectivity for such pairs or groups of acids depends on the type of bonded phase. Octylsilica columns have been reported to yield better separations than octadecylsilica columns of the derivatives of oleic and elaidic acids (C_{18} with a single double bond) from palmitic acid (C_{16}, saturated) and of linoleic acid (C_{18} with two double bonds) from myristic acid (C_{14}, saturated), and the *cis*- and *trans*-isomeric unsaturated acids derivatives were also better separated from one another on an octylsilica column[2].

Using columns packed with a stationary phase containing bonded alkyl groups longer than C-18 may also significantly improve the selectivity of the separation of certain pairs of fatty acid phenacyl esters that are very difficult to separate on an octadecylsilica column. On C_{30} bonded phase (two 300 x 3.9 mm I.D. columns in series), it was possible to accomplish the complete separation of *p*-bromophenacyl esters of 40 saturated and unsaturated fatty acids containing 3-24 carbon atoms in 3.5 h using a convex gradient from 40 to 100% acetonitrile in water in the first step followed by a linear gradient from 0 to 100% dioxane in acetonitrile in the second step (Fig. 17.1). This two-step combination of aqueous and non-aqueous reversed-phase gradient elution chromatography allowed the difficult separations of the pairs arachidonic-erucic, behenic-nervonic, elaidic-vaccinic and other acid derivatives to be achieved[5]. Reversed-phase chromatography of *p*-bromophenacyl derivatives has been applied to the analysis of fatty acids in butter (after previous hydrolysis of fatty acid esters)[2], biological tissues[5] and river water[4].

Methoxyphenacyl esters of fatty acids can be separated on octadecylsilica columns using elution with a convex gradient of acetonitrile in water[1,6].

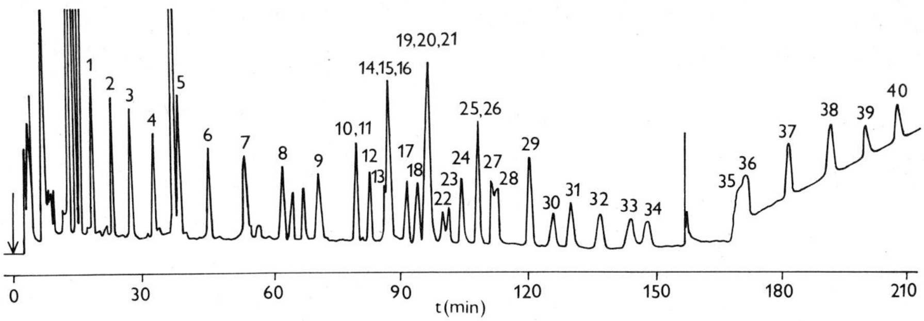

Fig. 17.1. Separation of *p*-bromophenacyl esters of fatty acids by reversed-phase gradient elution chromatography. Column: C_{30}/μPorasil, 20 μm; two columns, 300 x 3.9 mm I.D., in series. Gradient: convex, from 40% (v/v) acetonitrile in water to 100% (v/v) acetonitrile in 150 min, followed by a linear gradient from acetonitrile to *p*-dioxane in 150 min. Flow-rate: 1 ml/min. Detection: UV at 254 nm. Peaks: *p*-bromophenacyl esters of the following fatty acids: 1 = propionic (C_3); 2 = butyric (C_4); 3 = valeric (C_5); 4 = caproic (C_6); 5 = heptanoic (C_7); 6 = caprylic (C_8); 7 = nonanoic (C_9); 8 = capric (C_{10}); 9 = undecanoic (C_{11}); 10 = lauric (C_{12}); 11 = myristoleic ($C_{14:1}$); 12 = 5,8,11,14,17-eicosapentaenoic ($C_{20:5}$); 13 = 4,7,10,13,16,19-docosahexaenoic ($C_{22:6}$); 14 = α-linolenic ($C_{18:3}$); 15 = γ-linolenic ($C_{18:3}$); 16 = tridecanoic (C_{13}); 17 = arachidonic ($C_{20:4}$); 18 = palmitoleic ($C_{16:1}$); 19 = myristic (C_{14}); 20 = linoleic ($C_{18:2}$); 21 = palmitelaidic ($C_{16:1}$); 22 = linolelaidic ($C_{18:2}$); 23 = 11,14,17-eicosatrienoic ($C_{20:3}$); 24 = pentadecanoic (C_{15}); 25 = oleic ($C_{18:1}$); 26 = vaccinic ($C_{18:1}$); 27 = elaidic ($C_{18:1}$); 28 = palmitic (C_{16}); 29 = heptadecanoic (C_{17}); 30 = *cis*-5-eicosenoic ($C_{20:1}$); 31 = stearic (C_{18}); 32 = erucic ($C_{22:1}$); 33 = nonadecanoic (C_{19}); 34 = brassidic ($C_{22:1}$); 35 = arachidic (C_{20}); 36 = nervonic ($C_{24:1}$); 37 = heneicosanoic (C_{21}); 38 = behenic (C_{22}); 39 = tricosanoic (C_{23}); 40 = lignoceric (C_{24}). (Reprinted from ref. 5 by courtesy of Marcel Dekker, Inc.)

17.1.2. Chromatography of other fatty acid derivatives

Fatty acids may be converted first to acyl chlorides by reaction with triphenylphosphine (or with polystyryldiphenylphosphine) and carbon tetrachloride at 80°C and then to *p*-methoxyanilides by reaction of the acyl chlorides with *p*-methoxyaniline in ethyl acetate at 80°C. *p*-Methoxyanilides of saturated and of some unsaturated fatty acids (C_6-C_{24}) were separated on a C_{18} column using a convex gradient of acetonitrile in water and UV detection at 254 nm[7].

Fluorescent 4-methyl-7-methoxycoumarin derivatives (see Chapter 16) of fatty acids, including capric, caproic, oleic and lauric acids, were separated on a silica gel column, 250 x 2 mm I.D., using a ternary multi-linear gradient of acetonitrile and dichloromethane in isooctane with fluorimetric detection[8].

A mixture of these derivatives of C_4-C_{18} fatty acids could be separated in ca. 40 min on an octadecylsilica column using two-step elution with the first, isocratic step employing elution with the mixture acetonitrile-tetrahydrofuran-water (1 : 1 : 3) followed with the linear gradient step of increasing concentration of the mixture acetonitrile-tetrahydrofuran (1 : 1) in the starting eluent. Another separation example is shown in Fig. 16.1. Here, as with phenacyl esters of fatty acids, the resolution of saturated acids from certain unsaturated acids (such as the palmitic acid derivative from the oleic acid derivative) is difficult, but can be improved by addition of silver nitrate to the mobile phase, which enhances the interactions of the mobile phase with compounds containing double bonds[39].

Fatty acids condense with 9,10-diaminophenanthrene in methyl phosphate-chloroform solution to form 2-substituted phenanthrimidazoles, which can be chromatographed on an octadecylsilica column using fluorimetric detection with an excitation wavelength of 255 nm and an emission wavelength of 382 nm. The detection limits in chromatography using this pre-column derivatization method were reported to correspond to picogram amounts of the acids and the peak areas in the homologous series are more constant than those of 4-methyl-7-methoxycoumarin derivatives. A linear gradient of methanol in water (1-30%) made it possible to separate even-carbon-number carboxylic acid derivatives (from C_2 to C_{20}) in 40 min (Fig. 17.2)[9].

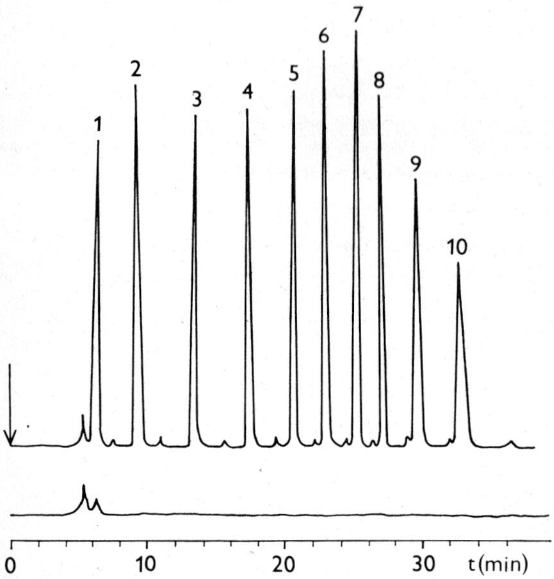

Fig. 17.2. Separation of 2-substituted phenanthrimidazoles derived from fatty acids. Column: ODS-Hypersil (250 x 4 mm I.D.). Gradient: linear, from 70 to 99% (v/v) methanol in water, 1 %/m. Flow-rate: 0.8 ml/min. Detection: fluorimetric, λ_{ex} = 255 nm; λ_{em} = 382 nm. Lower trace: reagent blank. Peaks in upper trace: derivatives of the following fatty acids: 1 = acetic (C_2); 2 = butyric (C_4); 3 = caproic (C_6); 4 = caprylic (C_8); 5 = capric (C_{10}); 6 = lauric (C_{12}); 7 = myristic (C_{14}); 8 = palmitic (C_{16}); 9 = stearic (C_{18}); 10 = arachidic (C_{20}). (Reproduced from ref. 9 with permission.)

17.2. LIPIDS

Modern liquid chromatography has been successfully applied to rapid separations of lipids into classes. The application of modern liquid chromatography to lipid analysis was reviewed by Aitzetmüller[10,11]. The gradient elution technique is very useful for the separation of lipid classes with significant differences in polarity.

Chromatography on silica gel columns can be successfully used for the separation of lipids into classes in order of increasing polarity: hydrocarbons; cholesterol esters and wax esters; methyl esters of fatty acids; triglycerides; 1,3-diglycerides; 1,2-diglycerides; free cholesterol (and other sterols); free fatty acids; monoglycerides and other polar lipids. A column packed with 5 or 10 μm silica gel (LiChrosorb Si 60) and a sequence of two gradients from solvent A (tetrachloromethane-isooctane, 34 : 66) to solvent B (chloroform-dioxane-*n*-hexane, 40 : 11 : 49) and then to solvent C (chloroform-methanol-diisopropyl ether, 34 : 36 : 30) made possible this separation in less than 15 min[12,13]. Various other solvent gradients, such as linear gradient of diethyl ether in light petroleum[14] a linear gradient of acetone in *n*-hexane[43], a sequence of two concave gradients, from *n*-heptane to *n*-heptane-diisopropyl ether (1 : 1) and then from this solvent to diisopropyl ether-ethanol (95 : 5)[11], a multi-linear gradient using toluene-*n*-hexane (1 : 1) as solvent A and toluene-ethyl acetate (3 : 1) with 1.2% formic acid as solvent B[15] and a stepwise gradient using diethyl ether, ethyl acetate, acetic acid and water[16] have also been applied for the fractionation of lipids on silica columns. In these analyses, instrumentation using a moving wire as the system for the transport of the solutes in the eluate to the flame ionization detector (mass-transport detector, see Part III) was employed. In this detector, only a small portion of the sample can be utilized to produce the response signal, which depends on the flow-rate of the mobile phase and on a number of other factors. The method has been applied to the analysis of lipids in sebum, serum and other tissues, skin fat, frying fat, food components, technical emulsifiers, food emulsifiers and polymer additives.

For example, Erdahl et al.[42] analysed commercial soya bean lecithin on a silica gel column using a multi-solvent gradient elution system with a transport flame-ionization detector and resolved the product into 19 fractions.

A similar separation of free acids, sterols and mono-, di- and triglycerides in soya bean lecithin was achieved in 15 min, using a silica gel column (Micro Pak SI-10, 10 μm, 500 x 2.1 mm I.D.), elution with a linear gradient of ethanol in *n*-hexane-chloroform (9 : 1) and a moving-wire flame-ionization detector[19].

References on p. 332.

Partial glycerides and sorbitol fatty acid esters in emulsifiers were separated on a μPorasil (10 μm) column (300 x 4 mm I.D.) using elution with a gradient of ethanol-water (90 : 10) in isooctane-isopropanol (99 : 1) and UV detection at 220 nm[44].

Glycerol and polyglycerol esters in food emulsifiers could be separated on a LiChrosorb-Diol (10 μm) column (250 x 4.6 mm I.D.) using gradient elution with an increasing concentration of isopropanol in *n*-hexane and UV detection at 220 nm. Using this method, it was possible to resolve in 1 h not only glycerol mono-, 1,2,-di-, 1,3,-di- and triesters of fatty acids and mono-, di-, tri- and polyesters of polyglycerols, but also to differentiate some of numerous polyglycerol ester positional isomers. The Diol column does not separate esters differing in the chain length of the polyol[45].

A mixture of mono-, di- and triesters of glycerol with stearic (or palmitic) acid were separated into these classes in 48 min using gradient elution chromatography with a linear gradient of benzene in isooctane on a nitrile phase (Durapak OPN-Porasil C column, 36-75 μm, 350 x 6 mm I.D.) using a moving-wire flame-ionization detector[17].

Rapid separation of lipids into the classes fatty acid methyl esters, triglycerides, 1,3-diglycerides, 1,2-diglycerides and monoglycerides was achieved on an amino- + cyano-bonded column (Partisil CN + Partisil NH_2, 250 x 4.6 mm I.D.) in 20 min using a linear gradient of acetonitrile-*n*-hexane-chloroform (35 : 25 : 65) in *n*-hexane-chloroform (60 : 65). Glycerides were detected in the eluate using an IR detector at 5.72 μm (absorption maximum of the carbonyl band). Careful balancing of the amounts of *n*-hexane, acetonitrile and chloroform in the two mixed solvents was necessary to produce a baseline with a relatively little drift. The method has been applied to the analysis of soya bean oil and commercial shortenings[18].

The individual saturated triglycerides (from triacetin to tristearin) were separated using non-aqueous reversed-phase gradient elution chromatography on an octadecylsilica column with a linear gradient of methylene chloride-tetrahydrofuran (47.75 : 52.25) in acetonitrile and IR detection at 5.75 μm (the composition of the components of the gradient was balanced to achieve equal absorption at this wavelength)[20].

On-line coupling of a mass spectrometer with a liquid chromatograph has proved useful in the characterization of lipids. Non-aqueous reversed-phase chromatography on an octadecylsilica column with a linear gradient of chloroform in acetonitrile-acetone (2 : 1) was applied to the separation of wax esters in eye wax samples. The eluate from the chromatographic column was introduced into the ion source of the mass spectrometer using a moving-belt LC-MS interface, where electron-impact ionization was used and the total ion current was monitored to register the chromatogram. In addition, by selecting a fragment mass characteristic of a specific acid (such as M = 243 of a C-15 acid ion $C_{14}H_{29}\ COOH_2^+$ and M = 264 of oleic acid ion), the distribution of various alcohols in esters of this acid is obtained in the resulting chromatogram, where each peak corresponds to the elution of a particular ester of the acid[21].

17.2.1. *Chromatography of phospholipids and sphingolipids*

Phospholipids in various biological tissues can be analysed using gradient elution liquid chromatography on silica gel columns using linear, stepwise or multi-linear gradients of increasing solvent polarity, such as *n*-hexane-isopropanol-water[22,23], dichloromethane-methanol-ammonia[24] or chloroform-*n*propanol-acetic acid-water[25], and the following sample components could be separated: phosphatidic acid, diphosphatidylglycerol, phosphatidylethanolamine, phosphatidylglycerol, phosphatidylinositol, phosphatidylserine, phosphatidylcholine, sphingomyelin, diphosphatidylinositol, lysophosphatidylcholine and lysophosphatidylethanolamine. In addition to separation of phospholipids into different classes, separation of molecular species could be achieved for certain phospholipids, such as phosphatidylcholines and sphingomyelins[21,23]. In addition to the mass-transport detector, end absorption of phosphates at 206 nm was used for their UV detection[22,23] or labelled compounds containing ^{32}P were detected using radioactivity monitoring of the eluate[25]. Derivatization of lipid extracts with dansyl chloride made it possible to detect selectively dansyl derivatives of amino group-containing phospholipids using fluorimetric detection[24].

Chemically bonded polar phases can also be used for the separation of lipids in biological tissues, such as in gradient elution operation on a bonded amino phase using a multi-linear gradient of methanol-water (1 : 1.5) in the solvent *n*-hexane-isopropanol (5.5 : 8)[26] or in chromatography on a Diol column using linear gradient of acetonitrile in water (from 80 to 100% in 15 min)[27].

Chromatography on anion-exchange resins using linear ionic strength gradients in borate buffers at pH 8.5 or 10 has also been used for the separation of phospholipids, but with a lower efficiency than in above-mentioned systems[28,29].

Sphingolipids from human plasma and other biological tissues have been chromatographed on superficially porous silica (Zipax) columns, usually after previous derivatization using a benzoylation reaction, with gradients of 1,4-dioxane, methanol or ethyl acetate in *n*-hexane and UV detection at 230-280 nm. Good separation of mono-, di-, tri- and tetrahexosylceramides was reported and the method was used for diagnostic purposes[30-33].

References on p. 332.

17.2.2. Prostaglandins

Prostaglandins in physiological fluids have been analysed by liquid chromatography on silica gel columns using elution with a gradient of methanol and acetic acid in chloroform[34,40] or a gradient of acetonitrile in chloroform-*n*-hexane (2 : 1)[35]. A gradient of methanol in chloroform was used for elution in chromatography on a column packed with Carbowax 400 on Porasil C[36].

^{14}C-Labelled prostaglandins were separated into fractions on a μBondapak Fatty Acid Analysis column (a bonded amino phase, the exact composition of which is proprietary information of Waters Assoc.) by elution with a linear gradient of methanol with addition of acetic acid in chloroform. This step was followed by separation of the individual compounds in fractions using reversed-phase liquid chromatography or gas chromatography[37].

Ion-pair reversed-phase gradient elution chromatography has been applied to the separation and quantitation of ^{14}C-labelled prostaglandin E_1 from lung effluent metabolites. An octadecylsilica column and elution with a gradient of methanol (from 56-62% to 100%) in an aqueous methanolic mobile phase containing 3% acetic acid and addition of tetrabutylammonium phosphate were employed[38].

17.3. REFERENCES

1 T. Jupille, *J. Chromatogr. Sci.*, 17 (1979) 160.
2 H. Engelhardt and H. Elgass, *J. Chromatogr.*, 158 (1978) 249.
3 R.F. Borch, *Anal. Chem.*, 47 (1975) 2437.
4 D.A. Hullett and S.J. Eisenreich, *Anal. Chem.*, 51 (1979) 1953.
5 K. Takayama, H.C. Jordi and F. Benson, *J. Liq. Chromatogr.*, 3 (1980) 61.
6 R.A. Miller, N.E. Bussell and C. Ricketts, *J. Liq. Chromatogr.*, 1 (1978) 291.
7 N.E. Hoffman and J.C. Liao, *Anal. Chem.*, 48 (1976) 1104.
8 S.G. Zelenski and J.W. Huber, *Chromatographia*, 11 (1978) 645.
9 J.B.F. Lloyd, *J. Chromatogr.*, 189 (1980) 359.
10 K. Aitzetmüller, *J. Chromatogr.*, 113 (1975) 231.
11 K. Aitzetmüller, *J. Chromatogr. Sci.*, 13 (1975) 454.
12 K. Aitzetmüller, *J. Chromatogr.*, 139 (1977) 61.
13 K. Aitzetmüller and J. Koch, *J. Chromatogr.*, 145 (1978) 195.
14 G. Cavina, G. Moretti, R. Alimenti and B. Callinella, *Riv. Ital. Sostanze Grasse*, 54 (1977) 341.
15 E.W. Hammond, *J. Chromatogr.*, 203 (1981) 397.
16 D. Vandamme, B. Declercq, A. Denco, V. Blaton and H. Peeters, *Anal. Biochem.*, 69 (1975) 29.
17 R. Fischer, *Chromatographia*, 4 (1974) 207.
18 K. Payne-Wahl, G.F. Spencer, R.D. Plattner and R.O. Butterfield, *J. Chromatogr.*, 209 (1981) 61.
19 K. Kiuchi, T. Ohta and H. Ebine, *J. Chromatogr. Sci.*, 13 (1975) 461.
20 N.A. Parris, *J. Chromatogr. Sci.*, 17 (1979) 541.
21 W.H. McFadden, *J. Chromatogr. Sci.*, 17 (1979) 2.
22 W.M.A. Hax and W.S.M. Geurts van Kessel, *J. Chromatogr.*, 142 (1977) 735.
23 W.S.M. Geurts van Kessel, W.M.A. Hax, R.A. Demel and J. DeGier, *Biochim. Biophys. Acta*, 486 (1977) 524.
24 S. Shi-Hua Chen, A.Y. Kow and Hsin-Hsin Y. Chen, *J. Chromatogr.*, 208 (1981) 339.

25 C.P. Blom, F.A. Deierkauf and J.C. Riemersma, *J. Chromatogr.*, 171 (1979) 331.
26 V.L. Hanson, J.Y. Park, T.W. Osborn and R.M. Kiral, *J. Chromatogr.*, 205 (1981) 393.
27 R.L. Briand, S. Harold and K.G. Blass, *J. Chromatogr.*, 223 (1981) 277.
28 D. Lairon, J. Amic, H. Lafont, G. Nalbone, N. Domingo and J. Hauton, *J. Chromatogr.*, 88 (1974) 183.
29 J.K. Kaitaranta, P.J. Geiger and S.P. Bessman, *J. Chromatogr.*, 206 (1981) 327.
30 J.E. Evans and R.H. McCluer, *Biochim. Biophys. Acta*, 270 (1972) 565.
31 M.D. Ullman and R.M. McCluer, *J. Lipid Res.*, 18 (1977) 371.
32 F.B. Jungalwala, L. Hayes and R.M. McCluer, *J. Lipid Res.*, 18 (1977) 285.
33 M.D. Ullman, R.E. Pyeritz, H.W. Mosser and D.A. Wenger, *Clin. Chem.*, 26 (1980) 1499.
34 J.M. Tusell and E. Gelpi, *J. Chromatogr.*, 181 (1980) 295.
35 M.V. Merritt and G.E. Bronson, *Anal. Chem.*, 48 (1976) 1851.
36 W.L. Hubbard and T.J. Watson, *Prostaglandins*, 12 (1976) 21.
37 A.R. Whorton, K. Carr, M. Smigel, L. Walker, K. Ellis and J.A. Oates, *J. Chromatogr.*, 163 (1979) 64.
38 M. Ghias-ud-Elin, E.B. Olson and J. Rankin, *J. Chromatogr.*, 192 (1980) 463.
39 E.L. Johnson, *LC at Work*, No. 49, Varian, Palo Alto, CA, 1977.
40 S.R. Abbott, *J. Chromatogr. Sci.*, 18 (1980) 540.
41 H.C. Jordi, *J. Liq. Chromatogr.*, 1 (1978) 215.
42 W.L. Erdahl, A. Stolyhwo and O.S. Privett, *J. Amer. Oil Chem. Soc.*, 50 (1973) 513.
43 K. Nakamura and I. Matsumoto, *Nippon Kagaku Kaishi*, (1976) 104.
44 H. Brueschweiler, *Mitt. Geb. Lebensmittelunters. Hyg.*, 68 (1977) 46.
45 N. Garti and A. Aserin, *J. Liq. Chromatogr.*, 4 (1981) 1173.

Chapter 18

STEROIDS

It would be neither possible nor reasonable to review here the enormous number of papers describing applications of HPLC for the purification and separation of steroid compounds. Several reviews of these applications have been published already[1-4]. Applications in which liquid column chromatography, often using the gradient elution technique, is utilized for the purification and isolation of small amounts of steroids from a natural material prior to their characterization and quantitation by gas chromatography, radioimmunoassay, mass spectrometry or another suitable method have frequently been described, but the use of HPLC for the separation and quantitation of steroids has become increasingly important recently.

The steroids, which contain a system of four condensed rings as the basic structure, include a wide variety of compounds that occur in various natural materials and differ in structure, such as sterols, androgens, estrogens, gestagens, corticosteroids, bile acids and other steroid acids, steroidal glycosides and steroidal insect hormones (polyhydroxylated steroids). These compounds are usually classified as lipids, but the individual steroid sub-groups differ considerably in polarity. Therefore, it is possible to use either polar adsorbents and polar chemically bonded phases or non-polar chemically bonded phases and gels and even ion exchangers for the liquid column chromatography of different steroids.

UV detection is convenient for monitoring a number of steroids in column effluents. A combination with another detection system, such as the on-line combination of liquid chromatography with mass spectrometry, may be valuable for the identification of steroids in biological samples[29]. A subsequent radioimmunoassay of fractions from HPLC columns is useful for the precise quantitation of steroids, as it eliminates possible errors caused by non-steroidal UV-absorbing contaminants from biological samples[22,23]. With some steroids that do not absorb significantly in the near-UV region, a moving-wire flame-ionization detector has been used for monitoring column effluents[9-11].

A number of earlier publications (some of them reviewed in refs. 1-4) described the separation of steroids on polar adsorbents into fractions prior to further quantitation. Thus, for example, unsaponifiable lipids such as terpenoid hydrocarbons, carotenoids and sterols extracted from biological materials could be fractionated on a column packed with acid-washed alumina, deactivated with water, using gradient elution with an increasing concentration of diethyl ether in light petroleum. In this way the separation of squalene, phytoene, β-carotene, lanosterol, cholesterol and ergosterol could be achieved[5]. Using elution with a linear gradient of diethyl ether in *n*-hexane on four columns in series packed with silica (Adsorbosil, 160-200 mesh) saturated with 25% silver nitrate, specific interactions of π-electrons with Ag^+ ions allowed saturated, mono-unsaturated and di-unsaturated desmethyl sterols to be fractionated prior to further analysis by TLC and GLC. Cholesterol, cholestanol, sitosterol, camposterol, stigmasterol, ergostanol, ergosterol and some other sterols could be isolated and characterized in samples of citrus juice lipids[6].

18.1. CHROMATOGRAPHY OF STEROID HORMONES

Most published applications concern separations of steroid hormones such as corticosteroids, androsterones and estrogens and their metabolites in biological materials. A linear gradient of dichloromethane in light petroleum was used for the fractionation of adrenocortical steroids and ketosteroids in biological mixtures on a silica gel column prior to their determination in the fractions[7]. Mestranol, norethynodral, norethindrone, lynestrenol, ethylestrenol, allylestrenol, estradiol benzoate and some other steroids occurring in formulations with estrogenic and progestational activity were fractionated on a silica gel column (600 x 6 mm I.D., Bio-Rad silicic acid, 325 mesh) using a concave gradient of diethyl ether in light petroleum[8]. This packing material in columns 300-1000 mm long was applied to the separation of corticosteroid mixtures using elution with linear or concave gradients of methanol in chloroform and continuous monitoring of the steroids in the eluate with a UV or mass-transport flame-ionization detector[9-11].

Microparticular silica allowed far more rapid and efficient separations of hormonal steroids to be accomplished than in the above examples. Rapid purification of milligram amounts of impure progesterone was achieved using a gradient of ethanol in dichloromethane on a Zorbax SIL column and large-scale preparative purification of impure cholesterol phenylacetate was achieved on two 500 x 23 mm I.D. columns of Spherosil XOA-400 using stepwise gradient elution with dichloromethane in *n*-hexane[12]. Elution with two subsequent steps of linear gradients of isopropanol in *n*-hexane on a Micro Pak Si-60 column gave a rapid separation of estrone, estradiol and estriol in 10 min[13]. A mixture of eight fluorinated corticosteroids was separated on a 5 μm silica gel column using elution with a multi-linear gradient of methanol in 2,2,4-trimethylpentane at a constant concentration of diisopropyl ether, ethyl acetate, acetonitrile and acetic acid

(37.5 : 7.5 : 5 : 0.2) in the mobile phase[14]. The principal natural corticosteroids 11-deoxycorticosterone, 11-dehydrocorticosterone, 11-deoxycortisol, corticosterone, cortisone, aldosterone, cortisol and prednisolone) in adrenocortical extracts, rat adrenal tissue extracts and human plasma extracts were separated in 30 min on 5 µm silica columns using linear or multi-linear gradients of methanol in chloroform[15,16]. In all of these applications, a UV detector at 254 (or 242) nm was employed to monitor the steroids in the column effluents.

Chromatography on chemically bonded polar phases has also been applied successfully to the separation of steroid hormones. Corticosteroids were separated on an ethereal bonded phase (Permaphase ETH) using elution with a gradient of chloroform in *n*-hexane[17] and on a nitrile bonded phase using a linear ternary gradient of methanol and dichloromethane (20 : 80) in isooctane[18]. Adrenal and gonadal steroids (progesterone, androstenedione, 17α-hydroxyprogesterone, adrenosterone, Δ^4-pregnen-20β-olone, corticosterone, aldosterone, estriol, estradiol, testosterone, etc.) were separated on columns packed with a bonded nitrile phase using elution with a gradient of isopropanol in *n*-hexane[19] or with a ternary gradient of isopropanol and dichloromethane in *n*-hexane[20], on columns with a bonded nitro phase using elution with a multi-linear gradient of ethanol in *n*-hexane[19], on columns containing a bonded amino phase with a multi-linear ternary gradient of isopropanol and dichloromethane in *n*-hexane[20,21] and on columns packed with a Diol bonded phase using elution with a linear gradient of isopropanol in *n*-hexane[22-24] or a ternary gradient of dichloromethane and dioxane in *n*-hexane[22]. A Diol column with a dioxane gradient has been reported to provide the best separations. However, to achieve complete separation of complex mixtures of hormonal steroids in biological samples, it is often necessary to combine separations on two or more columns, such as Diol and octadecylsilica columns[22].

Testosterones and androsterones have been separated on porous-layer octadecylsilica bonded phases using linear or exponential concentration gradients of methanol or acetonitrile in water[25-29]. Chromatography on a microparticulate octadecylsilica bonded phase using elution with a gradient of methanol in 0.01 mol/l ammonium acetate (pH 6.9) may be used for the fractionation of estradiol-17β and its various metabolites in biological fluids and tissues into six groups, which can be further separated in isocratic runs on columns packed with chemically bonded octadecyl and Diol phases. Approximately 25 conjugated and non-conjugated standards could be separated in this way[24]. A combination of gradient elution chromatography on octadecylsilica and Diol columns has also been applied for this purpose[22].

References on p. 340.

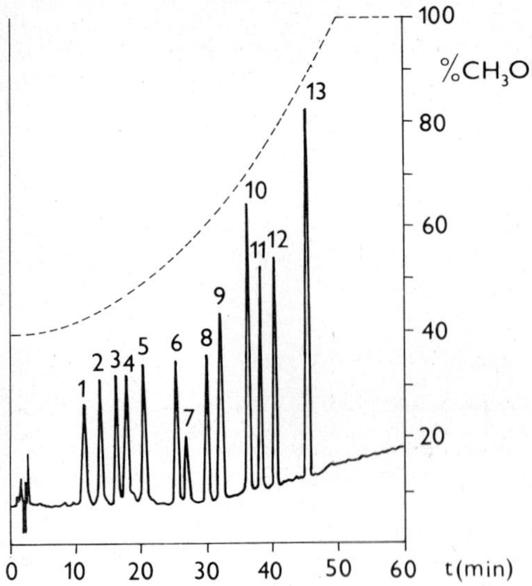

Fig. 18.1. Separation of a mixture of steroid hormones and related compounds by reversed-phase gradient elution chromatography. Column: Hypersil ODS, 5 µm, 150 x 4.6 mm I.D. Gradient: as indicated by the broken line. Solvents: A, water; B, methanol. Flow-rate: 1 ml/min. Temperature: 45°C. Detection: UV at 240 nm. Peaks: 1 = 18-hydroxy-11-deoxycorticosterone; 2 = aldosterone; 3 = 18-hydroxy-corticosterone; 4 = cortisone; 5 = cortisol; 6 = 11β-hydroxyandrostenedione; 7 = 18-hydroxy-11-deoxycorticosterone; 8 = corticosterone; 9 = 11-deoxycorti-costerol; 10 = androsterone-3,17-dione; 11 = 11-deoxycorticosterone; 12 = 17α-hydroxyprogesterone; 13 = progesterone. (Reproduced from ref. 33 with permission).

Testosterone propionate, phenylpropionate, isocaproate and decanoate were rapidly separated on an octylsilica (10 µm) column with a linear gradient of acetonitrile concentration in water[30].

Chromatography on 5 µm octadecylsilica columns using a combination of three different runs with concave concentration gradients of methanol in water, aceto-nitrile in water or dioxane in water allowed good separations of at least 43 naturally occurring adrenal steroids and testisteroids[31]. The best results were achieved with columns where the residual silanol groups had been carefully removed[32,33]. An example of the separation of thirteen adrenal steroid standards on an octadecylsilica column using a concave gradient of methanol concentration in water is shown in Fig. 18.1.

A concave gradient of methanol in water was employed for the separation of anabolic drugs of abuse, such as methandienone in the presence of conjugated and non-conjugated natural steroids, on an octadecylsilica column. The method was used in tests for drug abuse in athletics and in cattle feed[34].

Sephadex gel (G-10) was used for the purification of estrogens by gradient elution using an exponential pH gradient[35]. On DEAE-Sephadex, which is a weak anion exchanger, it was possible to separate free steroids from their glycosidic and sulphate conjugates using a linear gradient of sodium chloride (from 0 to 0.2 mol/l)[36].

18.2. CHROMATOGRAPHY OF SAPONINS

Saponins (glycosidic steroid conjugates in plants) in soya beans could be separated into twelve fractions by anion-exchange chromatography on Dowex 1-X4 (-400 mesh, 700 x 9 mm I.D.) column using a linear gradient from 1.5 to 3 mol/l of propionic acid in water. Individual compounds in fractions of the eluate were further analysed by TLC[37].

18.3. CHROMATOGRAPHY OF BILE ACIDS

Bile salts and free and conjugated bile acids have been chromatographed on pellicular silica with a ternary gradient of ethyl acetate and ethanol in chloroform as the eluent (dihydroxy bile salts, cholate, taurine conjugated dihydroxy bile salts, taurocholate, glycochenodeoxycholate and glycocholate could be separated)[38] and on a microparticular bonded amino phase using elution with a gradient of dichloromethane in 2,2,4-trimethylpentane[39].

18.4. CHROMATOGRAPHY OF ECDYSTEROIDS (STEROIDAL INSECT HORMONES)

Gradient elution liquid column chromatography has been applied to the separation of ecdysteroids, which are an important family of steroids mainly found in invertebrates and in some plants. A combination of normal-phase isocratic elution chromatography and chromatography on a chemically bonded phenyl phase (Bondapak phenyl/Corasil, 37-50 μm, 1000 x 3 mm I.D.) using elution with a linear gradient from 0 to 20% ethanol in water allowed the separation of ecdysterone, inokosterone, makisterone A, α-ecdysterone and ponasterone A in insect samples[40]. These steroids could be separated in a single gradient run on a microparticulate octylsilica column (Zorbax C-8, 6 μm, 250 x 4.6 mm I.D.) using a linear gradient from 10 to 30% of acetonitrile in water[41].

Polar ecdysteroids and some of their possible biosynthetic intermediates of lower polarity (20-hydroxyecdysone, ecdysone, 2β,3β-dihydroxy-5β-cholest-7-en-6-one and 3β-hydroxy-5α-cholest-7-en-6-one) could be separated in 25 min using elution with a convex gradient of methanol in water on a C_{18} or C_{22} bonded phase column, but the removal of free silanol groups from the bonded phase, for example

References on p. 340.

by reaction with trimethylchlorosilane, is necessary for good results[42]. 20-Hydroxyecdysone, ecdysone and accompanying 3-dehydro and 3-epi derivatives could be separated using a combination of chromatography on a 5 μm octadecylsilica column with a linear gradient of methanol in water (from 25 to 50%) and of isocratic elution chromatography on an aminopropyl bonded phase[43]. UV detection at 240-254 nm is suitable for monitoring of ecdysteroids in the eluate from HPLC columns.

18.5. REFERENCES

1 Ž. Procházka, in Z. Deyl, K. Macek and J. Janák (Editors), *Liquid Column Chromatography*, Elsevier, Amsterdam, 1975, Ch. 28.
2 F.A. Fitzpatrick, *Advan. Chromatogr.*, 16 (1978) 37.
3 E. Heftmann and I.R. Hunter, *J. Chromatogr.*, 165 (1979) 283.
4 M. Kautsky (Editor), *Steroid Analysis by HPLC: Recent Applications*, Marcel Dekker, New York, 1980.
5 B.H. Davies and E.I. Mercer, *J. Chromatogr.*, 46 (1970) 161.
6 H.E. Nordby and S. Nagy, *J. Chromatogr.*, 79 (1973) 147.
7 D.F. Johnson, N.S. Lamontagne, G.C. Riggle and F.O. Anderson, *Anal. Chem.*, 43 (1971) 1712.
8 G. Cavina, G. Moretti, A. Mollica and R. Antonini, *J. Chromatogr.*, 60 (1971) 179.
9 G. Cavina, G. Moretti and A. Cantafora, *Proc. VII Int. Symp. Chromatogr. Electrophoresis, Brussels, 1972*, p. 148.
10 G. Cavina, G. Moretti and A. Cantafora, *J. Chromatogr.*, 80 (1973) 89.
11 A. Cantafora, G. Cavina, G. Moretti and B. Gallinella, *Farmaco, Ed. Prat.*, 29 (1974) 351.
12 D.R. Baker, R.A. Henry, R.C. Williams, D.R. Hudson and N.A. Parris, *J. Chromatogr.*, 83 (1973) 233.
13 *Varian Aerograph Series 4100/4200 LC Brochure*, p. 9.
14 J.P. Thomas, A. Brun and J.P. Bounine, *J. Chromatogr.*, 139 (1977) 21.
15 G. Cavina, G. Moretti, R. Alimenti and B. Gallinella, *J. Chromatogr.*, 175 (1979) 125.
16 R. Ballerini, M. Chinal and M. Ghelardoni, *J. Chromatogr.*, 193 (1980) 413.
17 Z. Saito, T. Hashiba, M. Miyamoto and R. Takeda, *Nippon Naibumpi Gakkai Zasshi*, 54 (1977) 775.
18 *Spherisorb-Spherical Materials and Columns for HPLC*, Phase Separations, Queensferry, Great Britain, 1976.
19 M. Schöneshöfer and H.J. Dulce, *J. Chromatogr.*, 164 (1979) 17.
20 R.E. Majors and M.J. Hopper, *J. Chromatogr. Sci.*, 12 (1974) 767.
21 R.E. Majors, *VIA*, 7, No. 2 (1973) 8, and 7, No. 4, (1973) 6, Varian, Palo Alto, CA.
22 M. Schöneshöfer and H.J. Dulce, *J. Chromatogr.*, 164 (1979) 17.
23 M. Schöneshöfer, J. Maxeiner and A. Fenner, *J. Chromatogr.*, 224 (1981) 229.
24 W. Slikker, G.W. Lipe and G.O. Newport, *J. Chromatogr.*, 224 (1981) 205.
25 D. Hertzog, *Inf. Chim.*, 122 (1973) 133.
26 R.A. Henry, J.A. Schmit and J.F. Dieckman, *J. Chromatogr. Sci.*, 9 (1971) 513.
27 A. Pryde, *J. Chromatogr. Sci.*, 12 (1974) 486.
28 C.D. Carr and K. Loeffer, *LC at Work*, 21, Varian, Palo Alto, CA.
29 P.J. Arpino, B.G. Dawkins and F.W. McLafferty, *J. Chromatogr. Sci.*, 12 (1974) 574.
30 *LC 750 Liquid Chromatograph*, Applied Chromatography Systems, Luton, Great Britain.
31 M.J. O'Hare, E.C. Nice, R. Magee-Brown and H. Bullman, *J. Chromatogr.*, 125 (1976) 357.

32 E.C. Nice and M.J. O'Hare, *J. Chromatogr.*, 166 (1978) 263.
33 M.J. O'Hare, E.C. Nice and M. Capp, *J. Chromatogr.*, 198 (1980) 23.
34 C.G.B. Frischkorn and H.E. Frischkorn, *J. Chromatogr.*, 151 (1978) 331.
35 H.J. Horst, *J. Chromatogr.*, 58 (1971) 227.
36 R. Hobkirk and S. Davidson, *J. Chromatogr.*, 54 (1971) 431.
37 W.J. Wolf and B.W. Thomas, *J. Chromatogr.*, 56 (1971) 281.
38 R.W.R. Baker, J. Ferrett and G.M. Murphy, *J. Chromatogr.*, 146 (1978) 137.
39 S. Okuyma, D. Uemura and Y. Hirata, *Chem. Lett.*, 7 (1976) 679.
40 M.W. Gilgan, *J. Chromatogr.*, 129 (1976) 447.
41 R. Lafont, G. Martin-Sommé and J.-C. Chambert, *J. Chromatogr.*, 170 (1979) 185.
42 I.D. Wilson, C.R. Bielby, E.D. Morgan and A.E.M. McLean, *J. Chromatogr.*, 194 (1980) 343.
43 L.N. Dinan, P.L. Donnathey, H.H. Rees and T.W. Goodwin, *J. Chromatogr.*, 205 (1981) 139.

Chapter 19

AMINES, NITROGEN BASES AND OTHER RELATED NITROGEN-CONTAINING COMPOUNDS

Liquid column chromatography has been applied to the analysis of aliphatic mono-, di- and polyamines, aromatic amines and nitrogen-containing heterocycles. Of these compounds, aliphatic di- and polyamines and catecholamines are of special biochemical and physiological interest, while certain aromatic amines and aza-arenes are interesting as industrial products and intermediates, components of raw materials and environmental pollutants with potential carcinogenic properties. Bases of nucleic acids are treated in the chapter on nucleic acid constituents and substituted ureas in chapter on pesticides.

19.1. CHROMATOGRAPHY OF ALIPHATIC AMINES AND POLYAMINES ON ION EXCHANGERS

Columns packed with strong sulphonated cation exchangers, such as Dowex 50W, Amberlite CG-120 and Aminex A-4 and A-5, or with weak carboxylic cation exchangers, such as Amberlite IRP-64, Amberlite CG-50 and Zeo-Karb 226, were used mainly in earlier applications of gradient elution liquid chromatography to the separation of aliphatic amines. Gradients of increasing pH or salt concentration in solutions of phosphate, acetate, citrate, borate or salicylate buffers were employed to achieve the separation of amines. In these systems, the separations were lengthy and took several hours, even if fine ion-exchange resins with particle diameter of 10-20 μm were used[1-4]. For example, the separation of a mixture of sixteen amines (methylamine, ethylamine, allylamine, *n*-propylamine, isobutylamine, *n*-butylamine, 1,2-propanediamine, histamine, isoamylamine, *n*-amylamine, tyramine, putrescine, phenylethylamine, cadaverine, serotonin and hexamethylenediamine) on an Aminex A-4 column (17 μm, 120 x 6 mm I.D.) was achieved in 8 h using a combined stepwise-gradient elution with citrate, borate and salicylate buffers (pH 5.28-12.50)[4].

In these techniques, the amines in the column effluents were quantitated using post-column derivatization with ninhydrin reagent and continuous monitoring of the colour developed by this reaction at 570 and 440 nm, as usual in classical automated amino acid analysis systems. Because the same columns and similar buffers could be used as in amino acid analysers, these instruments could be employed for the chromatography of aliphatic amines[4-7]. Using a modification of

References on p. 350.

the buffer system, basic amino acids and primary aliphatic amines could be simultaneously analysed in a single chromatographic run. The modified buffer gradient system allowed the analysis time for amine mixtures such as in the above example to be shortened to approximately 3 h[5].

The influence of the particle size of the cation-exchange resin (sulphonic type) and of the careful selection of buffers for gradient elution may be illustrated by a simpler chromatographic example, where only the separation of the polyamines putrescine, spermine and spermidine is of interest. This separation took 3 h on a 90 x 5 mm I.D. column packed with Dowex 50W-X4 (200-400 mesh) with a gradient of potassium ion (from 0.5 to 3.0 mol/l) in 0.02 mol/l phosphate buffer at pH 8[8], whereas it could be achieved in 2 h on an Aminex A-5 (13-20 μm) column (150 x 4.6 mm I.D.) using a convex pH gradient from 9.2 to 11.8[9], and in only 22 min using a 45 x 4 mm I.D. column packed with Technicon C-2 cation-exchange resin (8% cross-linked, 8-12 μm particle diameter) and a gradient of sodium ion from 2 to 3.5 mol/l in citrate buffer of pH 5.25[7].

These three polyamines and their N-acetyl derivatives (N-monoacetylputrescine and two isomeric N-monoacetylated spermidines) could be separated in 40 min by chromatography on a cation-exchange column (250 x 3.5 mm I.D.), packed with the chemically bonded cation exchanger Partisil PXS 10-25 SCX, using an acetate buffer (pH 4.6) with a gradient of sodium salt (added as nitrate) from 0.01 to 1.0 mol/l as the mobile phase[10].

Free homologous aliphatic primary amines (C_1-C_7) were separated by chromatography on a column packed with anion-exchange resin [Hitachi 2632 (OH^-), 18 μm, 550 x 9 mm I.D.] using elution with a linear gradient of acetone in water and a continuous flow coulometric detection in the eluate. The separation is based on a reversed-phase mechanism. The coulometric detector employs measurement of the current resulting from the electrolytic oxidation of hydroquinone in a basic medium provided by the presence of an amine in the eluate[25].

The importance of the chromatographic analysis of putrescine, spermidine, cadaverine and spermine increased after the discovery on increased levels of these polyamines in biological fluids in situations such as heart hypertrophy and growth. This means that their determination in human urine or blood may be used for the early diagnosis of malignant tumour growth in humans and a rapid and sensitive analytical method was needed for this purpose. This was the reason for a search for chromatographic methods other than ion-exchange techniques, which are generally lengthy (with a few exceptions, such as of the techniques described in refs. 7 and 10).

19.2. CHROMATOGRAPHY OF AMINE DERIVATIVES ON POLAR CHEMICALLY BONDED PHASES

Polar chemically bonded phases have appeared useful for the rapid chromatography of urinary aliphatic di- and polyamines in the form of UV-absorbing or fluorescent derivatives. A column (1000 x 2.1 mm I.D.) packed with an ethereal porous-layer bonded phase (Zipax Permaphase ETH, 30-50 μm) was used to separate tosylated (*p*-toluenesulphonylated) derivatives of putrescine, spermine and spermidine in 25 min using elution with an exponential gradient of acetonitrile in water (from 25 to 75%) at 35°C. The separated derivatives were detected by UV absorption at 254 nm[11]. Dansyl (5-dimethylaminonaphthalene-1-sulphonyl) derivatives (fluorescent) of putrescine, spermidine and spermine were separated in ca. 30 min[12] and dansyl derivatives of monoacetylputrescine, N^1-acetylspermidine and N^8-acetylspermidine were resolved in ca. 15 min[13] on a bonded nitrile phase (Micro Pak CN-10, 250 x 2.5 mm I.D.) using concave or linear ternary gradients of dichloromethane and isopropanol in *n*-hexane with fluorimetric (or UV) detection.

19.3. REVERSED-PHASE CHROMATOGRAPHY OF AMINE AND ISOCYANATE DERIVATIVES AND OF FREE ALIPHATIC AMINES

Reversed-phase chromatography on microparticular octadecyl-[14-19] or octylsilica[20] columns (5-10 μm) has been applied for rapid separations of fluorescent dansyl[14-16,20,43] and fluorescamine[17-19] derivatives of diamines and polyamines formed by pre-column derivatization reactions, using fluorimetric detection ($\lambda_{ex} \approx$ 360-390 nm; $\lambda_{em} \approx$ 450-510 nm). Linear[14] or multi-linear[20] concentration gradients of methanol[20] or acetonitrile[14,43] in water (see separation in Fig. 19.1)[20] or linear concentration gradients of methanol in aqueous borate[17] or acetate[19] buffers were employed for elution. Ion-pair reversed-phase gradient elution chromatography has been applied successfully for rapid separations (ca. 30 min) of dansyl-[15] and fluorescamine[16,18] derivatives of di- and polyamines, using elution with a gradient of methanol[16,18] or acetonitrile[15] in mobile phases containing 0.02 mol/l heptanesulphonic acid[15] or 0.035 mol/l benzenesulphonic acid[16,18] as the ion-pairing reagent.

Reversed-phase chromatography of the fluorescent derivatives of di- and polyamines has been applied successfully to analyses of urine and blood and samples of other biological materials. The detection limits reported are generally in the picomole range.

An alternative to the chromatography of fluorescent derivatives is offered by the use of post-column derivatization reactions with fluorescamine reagent

References on p. 350.

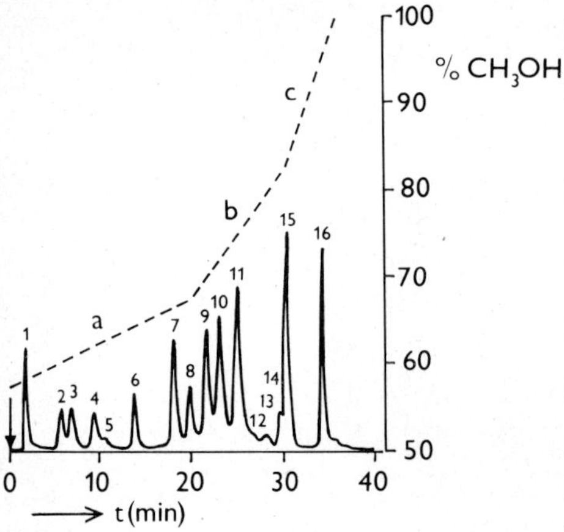

Fig. 19.1. Separation of a mixture of dansyl amine derivatives by reversed-phase gradient elution chromatography. Column: Hibar pre-packed LiChrosorb RP-8, 7 μm, 250 x 3 mm I.D. Gradient: multi-linear, as indicated by the broken line; (a) 20 min 0.5% B/min; (b) 10 min 1.5% B/min; (c) 6 min 3% B/min. Solvents: A, water; B, methanol. Flow-rate: 1 ml/min. Detection: fluorimetric; λ_{ex} = 360 nm; λ_{em} = 510 nm. Peaks: 1 = 5-dimethylaminonaphthalene-1-sulphonic acid (hydrolysed reagent); 2-16 = dansyl derivatives of: (2) ammonia; (3) ethanolamine; (4) methylamine; (5) 2-oxopyrrolidine (reaction product of 4-aminobutyric acid); (6) dimethylamine; (7) bis-derivative of agmatine; (8) 2-phenylethylamine; (9) bis-derivative of putrescine; (10) bis-derivative of cadaverine; (11) bis-derivative of 1,6-diaminohexane; (12) bis-derivative of histamine; (13) O,N-bisderivative of serotonin; (14) O,N-bis-derivative of *p*-tyramine; (15) tris-derivative of spermidine; (16) tetrakis-derivative of spermine. (Reproduced from ref. 20 with permission.)

or *o*-phthalaldehyde - 2-mercaptoethanol reagent and monitoring the fluorescence in the eluate. This method has been applied in the cation-exchange chromatography of polyamines[8-10] and also in the single-run reversed-phase chromatography of amino acids and di- and polyamines in biological samples on a μBondapak C_{18} column (300 x 3.9 mm I.D.), using elution with a linear gradient of acetonitrile in the mobile phase containing aqueous sodium acetate[21].

Reversed-phase chromatography on chemically bonded alkylsilica column is suitable for the separation of fluorescent derivatives of homologous primary alkylamines. Dansyl derivatives are appropriate for the purpose[22], but the sensitivity can be improved by using pre-column derivatization with 1,2-naphthoylenebenzimidazole 6-sulphochloride (1,2-NBI-6-SO_2Cl). Fig. 19.2 shows the chromatographic separation of homologous derivatives of C_1-C_{10} primary aliphatic amines on an octadecylsilica column with a linear gradient of methanol in water (70-100%) as the mobile phase[23].

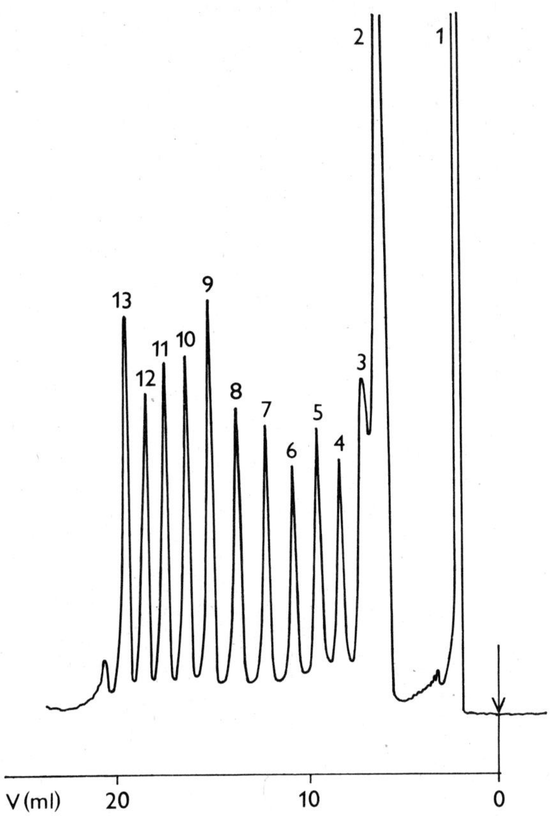

Fig. 19.2. Separation of a homologous series of 1,2-naphthoylenebenzimidazole-6-sulphonyl derivatives of saturated aliphatic *n*-amines by reversed-phase gradient elution liquid chromatography. Column: Silasorb C_{18}, 10 μm, 300 x 4.2 mm I.D. Gradient: linear, 70-100% B in 18 min (1.67% B/min). Solvents: A, water; B, methanol. Flow-rate: 1 ml/min. Detection: fluorimetric; λ_{ex} = 365 nm; λ_{em} > 400 nm. Peaks: 1 = 1,2-naphthoylenebenzimidazol-6-sulphonic acid (hydrolysed reagent); 2 and 3 = amide and imide derivative of ammonia; 4-13 = derivatives of amines: (4) methyl; (5) ethyl; (6) *n*-propyl; (7) *n*-butyl; (8) *n*-amyl; (9) *n*-hexyl; (10) *n*-heptyl; (11) *n*-octyl; (12) *n*-nonyl; (13) *n*-decyl.

Primary aliphatic amines have also been separated on an octadecylsilica column as N,N'-disubstituted ureas formed by pre-column derivatization with phenyl isocyanate. Elution with a linear gradient of acetonitrile in water and UV detection at 240 nm were employed, with a reported detection limit in nanogram range[24].

This derivatization reaction has also been used for the chromatography of aliphatic and aromatic diisocyanates, which were converted into ureas by reaction with N-4-nitrobenzyl-N-*n*-propylamine in heptane solution and then subjected to chromatography on a silica column (Partisil 5, 50 x 4.5 mm I.D.) using elution with a linear gradient of isopropanol in dichloromethane (1-9%) with UV detection at 254 nm. The method can be applied to the analysis of isocyanates used in the urethane foam industry (these compounds cause skin and respiratory tract irritations)[26].

19.4. CHROMATOGRAPHY OF CATECHOLAMINES

Biogenic catecholamines have been separated using gradient elution chromatography on cation-exchange columns. Elution with various gradients using borate, carbonate or phosphate buffers in the pH range 10-12 allowed the separation of phenolic amines on columns packed with fine-particle sulphonated polystyrene cation-exchange resins to be achieved[27]. Concentration gradients of phosphate buffers (0.03-0.5 mol/l NaH_2PO_4, at pH 3-4.5) were used for the chromatography of catecholamines on superficially porous cation-exchange columns (Zipax SCX)[28,29].

Linear gradients of methanol[30,44] or acetonitrile[31] in potassium dihydrogen phosphate buffer make it possible to accomplish rapid separations of catecholamines and their metabolites on microparticulate octadecylsilica columns. For example, the separation of norepinephrine, kynurenine, epinephrine, dopa, dopamine, methyldopa, deoxyepinephrine, tyramine, isoproterenol, 5-hydroxytryptophan, serotonin, tryptophan, 5-hydroxyindole-3-acetic acid, anthranilic acid, indoleacetamide, indole-3-lactic acid, indole-3-acetic acid and indole-3-propionic acid could be achieved in about 30 min using this technique.

UV detection at 254 or 280 nm[28,30,31] or fluorimetric detection utilizing UV absorption or the native fluorescence[30,44] of some of these aromatic and heterocyclic compounds may be used for the detection of catecholamines and related compounds in the eluate; the sensitivity of the method can be improved and the detection limits decreased to ca. 20-30 pg of the individual compounds using post-column derivatization by reaction with *o*-phthalaldehyde - 2-mercaptothiazole reagent or by reaction with trihydroxyindole and continuous monitoring of the fluorescence formed in the reaction with a fluorimetric detector[29]. An electrochemical (voltammetric) detection method using a three-electrode detector with a glassy carbon working electrode has also been applied successfully to the sensitive detection of these oxidizable compounds[31].

These techniques have been applied to the determination of catecholamines and their metabolites in urine and various biological tissues[29-31].

19.5. CHROMATOGRAPHY OF OTHER AROMATIC AMINES, AMIDES AND NITRO COMPOUNDS

Various aromatic amino, amido and nitro compounds, important as technical products and as potential carcinogenic environmental pollutants, have been analysed by gradient elution liquid column chromatography with UV detection. Aniline and six substituted halogeno anilines were separated in 15 min on a 150 mm long octadecylsilica column (Zorbax C_{18}) using a linear concentration gradient from 20 to 60% of acetonitrile in 0.15 mol/l phosphate buffer[32].

The acaricide amitraz and its metabolites 2,4-dimethylaniline and 2,4-dimethylformanilide were chromatographed on a nitro-bonded phase (Nucleosil NO_2, 5 µm, 200 x 4 mm I.D.) using a linear gradient of diethyl ether (from 10 to 100%) in *n*-hexane containing 0.3% triethylamine[33].

A silica gel column (Partisil, 10 µm, 250 x 4.6 mm I.D.) was employed for the separation of eight carcinogenic nitro-, amino- and acetylamino-anthracenes and -anthraquinones using a linear gradient of isopropanol in *n*-hexane (1-10%)[34].

Substituted amino-, nitro-, dinitro-, acetoxy- and acetamidobiphenyls were chromatographed on silica gel or on bonded nitrile phase columns using a linear gradient from 0 to 100% of dichloromethane in isooctane[35] or a linear ternary gradient of isopropanol-dichloromethane (30 : 70) in *n*-hexane[36], or on octadecylsilica columns using linear gradients of methanol or acetonitrile in water, with addition of 0.1% acetic acid[36].

Using similar operating conditions, substituted amino-, amido- and nitrofluorenes could be separated on a silica gel column using elution with a gradient of propanol and dichloromethane (30 : 70) in *n*-hexane[37] or on chemically bonded octadecylsilica phases using linear gradients of methanol or acetonitrile in water containing 0.1% acetic acid[37,38].

19.6. CHROMATOGRAPHY OF NITROGEN-CONTAINING HETEROCYCLES

Various nitrogen-containing heterocycles (aza-arenes) with two to five rings could be rapidly separated and determined using UV detection, either on microparticulate silica gel columns using elution with linear gradients of isopropanol in chloroform[39] or methanol in dichloromethane[40] or on an octadecylsilica bonded phase by elution with linear gradients of acetonitrile[41] or isopropanol[40] in water. The method has been employed for the analysis of air pollution by aza-arenes on solid particles suspended in air[39,41] and for the study of the metabolism of aza-arenes in biological tissues[40].

5-Chloro-8-hydroxy-7-isoquinoline conjugates were analysed by anion-exchange chromatography on Zipax SAX (500 x 2 mm I.D.) columns, using a linear gradient of sodium perchlorate (0.1-0.5 mol/l) in a borate buffer at pH 9.5, with UV detection at 254 nm[42]. Carcinogenic 4-nitroquinoline-1-oxide and its 4-hydroxylamino and 4-amino metabolites in blood plasma were determined by reversed-phase chromatography on an octadecylsilica column using elution with an increasing concentration gradient of methanol in water[45].

References on p. 350.

Plant hormones of the indole type in biological samples were separated and determined by reversed-phase chromatography on a microparticulate octadecylsilica column using a linear gradient of acetic acid in ethanol for elution[46].

19.7. REFERENCES

1 S. Holder and H.J. Bremer, *J. Chromatogr.*, 25 (1966) 48.
2 T.L. Perry and W.A. Schroeder, *J. Chromatogr.*, 12 (1963) 358.
3 M. Yoshioka, A. Ohara, H. Kondo and H. Kanazawa, *Chem. Pharm. Bull.*, 17 (1969) 1276.
4 H. Hatano, K. Sumizu, S. Rokushika and F. Murakami, *Anal. Biochem.*, 35 (1970) 377.
5 S. Rokushika, S. Funakoshi, F. Murakami and H. Hatano, *J. Chromatogr.*, 56 (1971) 137.
6 L.J. Marton and P.L.Y. Lee, *Clin. Chem.*, 21 (1975) 1721.
7 H. Adler, M. Margoshes, L.R. Snyder and C. Spitzer, *J. Chromatogr.*, 143 (1977) 125.
8 B. Kneip, M. Raymondjean, D. Bogdanovsky, L. Bachner and G. Shapira, *J. Chromatogr.*, 162 (1979) 547.
9 H. Veening, W.W. Pitt and G. Jones, *J. Chromatogr.*, 90 (1974) 129.
10 P.K. Bondy and Z.N. Canellakis, *J. Chromatogr.*, 224 (1981) 371.
11 T. Hayashi, T. Sugiura, S. Kawai and T. Ohno, *J. Chromatogr.*, 145 (1978) 141.
12 N.E. Newton, K. Ohno and M.M. Abdel-Monem, *J. Chromatogr.*, 124 (1976) 277.
13 M.M. Abdel-Monem and J.L. Merdink, *J. Chromatogr.*, 222 (1981) 363.
14 Y. Saeki, N. Uehara and S. Shirakawa, *J. Chromatogr.*, 145 (1978) 221.
15 N.D. Brown, R.B. Sweet, J.A. Kintzios, H.O. Cox and B.P. Doctor, *J. Chromatogr.*, 164 (1979) 35.
16 K. Haraguchi, M. Kai, K. Kohashi and Y. Ohkura, *J. Chromatogr.*, 202 (1980) 107.
17 K. Samejima, *J. Chromatogr.*, 96 (1974) 250.
18 M. Kai, T. Ogata, K. Haraguchi and Y. Ohkura, *J. Chromatogr.*, 163 (1979) 151.
19 A. Licht, R.L. Bowman and S. Stein, *J. Liq. Chromatogr.*, 4 (1981) 825.
20 N. Seiler, B. Knödgen and F. Eisenbeiss, *J. Chromatogr.*, 145 (1978) 29.
21 N. Seiler and B. Knödgen, *J. Chromatogr.*, 221 (1980) 227.
22 P. Jandera, V. Říha and J. Churáček, *Chem. Listy*, 74 (1980) 1209.
23 P. Jandera, H. Pechová, J. Královský, D. Tocksteinová and J. Churáček, *Chromatographia*, 16 (1982) 275.
24 B. Björkqvist, *J. Chromatogr.*, 204 (1981) 109.
25 K. Tanaka, T. Ishizuka and H. Sunahara, *J. Chromatogr.*, 172 (1979) 484.
26 C.R. Hastings Vogt, Chan Yan Ko and T.R. Ryan, *J. Chromatogr.*, 134 (1977) 451.
27 T.A. Wheaton and I. Stewart, *Anal. Biochem.*, 12 (1965) 585.
28 A. Kojima-Suda, *Ind. Hlth. Jap.*, 13 (1975) 69.
29 K.I. Okamoto, Y. Ishida and K. Asai, *J. Chromatogr.*, 167 (1978) 205.
30 A.M. Krstulovic and A.M. Powell, *J. Chromatogr.*, 171 (1979) 345.
31 A.M. Krstulovic, M. Zakaria, K. Lohse and L. Bertani-Dziedzic, *J. Chromatogr.*, 186 (1979) 733.
32 E.M. Lores, D.W. Bristol and R.F. Moseman, *J. Chromatogr. Sci.*, 16 (1978) 358.
33 F.E. Rickett, *J. Chromatogr.*, 142 (1977) 705.
34 B. Shaikh, M.R. Hallmark, R.K. Hallmark, W.B. Manning, A. Pinnock and J.C. Kawalek, *J. Chromatogr.*, 195 (1980) 392.
35 *Spectra-Physics Chromatogr. Rev.*, 3, No. 1 (1977) 1.
36 H.J. Johnson, S.F. Cernosek and R.M. Gutierez-Cernosek, *J. Chromatogr.*, 161 (1978) 259.
37 H.J. Johnson, S.F. Cernosek and R.M. Gutierez-Cernosek, *J. Chromatogr.*, 177 (1979) 297.
38 S.S. Thorgeirsson and W.L. Nelson, *Anal. Biochem.*, 75 (1976) 122.

39 M.W. Dong, D.C. Locke and D. Hoffmann, *Environ. Sci. Technol.*, 11 (1977) 612.
40 C. Stubley, J.G.P. Stell and D.W. Mathieson, *J. Chromatogr.*, 177 (1979) 313.
41 M. Dong, D.C. Locke and D. Hoffmann, *J. Chromatogr. Sci.*, 15 (1977) 32.
42 C.T. Chen, K. Hayakawa, T. Imanari and Z. Tamura, *Chem. Pharm. Bull.*, 23 (1975) 2173.
43 E. Johnson, A. Abu-Shumays and S.R. Abbott, *J. Chromatogr.*, 134 (1977) 107.
44 A.M. Krstulovic and C. Matzura, *J. Chromatogr.*, 163 (1979) 72.
45 H. Lund, D.J. Speelman, A. Culling, J.S. Kittredge and T. Albrecht, *J. Liq. Chromatogr.*, 4 (1981) 299.
46 M. Wurst, Z. Přikryl and V. Vančura, *J. Chromatogr.*, 191 (1980) 129.

Chapter 20

AMINO ACIDS

The number of papers dealing with the chromatography of amino acids is enormous, mainly owing to the importance of the quantitative evaluation of the amino acid content of physiological fluids and tissues and protein and peptide hydrolysates. In these hydrolysates, 18-20 amino acids are present that must be resolved. In physiological fluids, some other amino acids, amides of acidic amino acids (asparagine and glutamine) and certain peptides are present in addition to the amino acids common in proteins.

Analyses of amino acids have been performed mainly by ion-exchange chromatography and special instruments (amino acid analysers) were constructed for this purpose that allow automated analytical runs. The quantitation of amino acids in these instruments is provided by post-column reaction photometric detectors, where ninhydrin as the reagent is added continuously to the eluate from the analytical column and the red products of the reaction of ninhydrin with amino acids are monitored at 570 nm (and at 440 nm for the yellow reaction products of ninhydrin with proline or hydroxyproline). Columns packed with sulphonated polystyrene-divinylbenzene cation-exchange resins and elution with citrate buffers are generally employed.

Other liquid chromatographic techniques, in particular reversed-phase chromatography of free or derivatized amino acids and ion-pair chromatography, have been reported to provide rapid separations with excellent efficiency and resolution of the individual amino acids, and these techniques are likely to compete successfully even with ion-exchange chromatography on microparticulate resins or on efficient pellicular or microparticulate chemically bonded exchangers on a silica gel support.

Most free amino acids can be detected in the UV region only at wavelengths lower than 220 nm. In order to improve the possibilities of detection of most amino acids and to echance the detection sensitivity, post-column derivatization reactions other than reaction with ninhydrin were introduced (using fluorescamine, *o*-phthalaldehyde, etc.) or the amino acids were chromatographed in the from of suitable UV-absorbing or fluorescent derivatives (pre-column derivatization techniques). Phenylthiohydantoins and other substituted hydantoins, derivatives with dansyl chloride, *o*-phthalaldehyde and substituted nitro- and dinitrophenyl

References on p. 361.

derivatives of amino acids have been most frequently used for this purpose[24]. In this way, an improvement in detection limits of several orders of magnitude over the ninhydrin detection was obtained[8].

The various sensitive methods of amino acid analysis have been applied not only to the analysis of protein and peptide hydrolysates, but also to the analysis of free amino acids in physiological fluids[21,22,62,67,68,72,74,77,80-82], where the content of certain amino acids may be important for the diagnosis of diseases caused by aberrations in amino acid metabolism (phenylketonuria, tyrosinosis, etc.) and in the analysis of food and beverages[63,76], sea water[74], etc.

20.1. ION-EXCHANGE CHROMATOGRAPHY OF FREE AMINO ACIDS

Two different methods are used as the basis of amino acid analysis by ion-exchange chromatography. In the system of Spackman et al.[1], two cation-exchange columns are used, where one sample aliquot is analysed for neutral and acidic amino acids on the longer column using two-step elution with buffers of pH 3.25 and 4.25, and the other aliquot is chromatographed on the shorter column using a more concentrated buffer of pH 5.28 (separation of basic amino acids). Piez and Morris[2] improved the system for amino acid analysis by using a single column and elution with a continuous gradient of concentration and of pH using citrate buffers in a Varigrad gradient mixer. The amino acids are eluted roughly in order of increasing basicity, but the presence of aromatic rings contributes to the retention. Using these systems, all of the most important amino acids found in protein and peptide hydrolysates (except glutamine and asparagine) could be separated. One analysis run originally took more than 20 h.

Numerous modifications of the basic amino acid analysis systems have been reported aimed at improving the reproducibility and decreasing the analysis time. These amendements concerned mainly the geometry of columns and the use of spherical resins with smaller particle diameters and narrower size distributions (see, for example refs. 3-7). In addition, different types and modifications of commercial amino acid analysers were introduced, some of which make possible simultaneous analyses of three samples on three parallel ion-exchange columns[6]. The original gradient buffer system of Piez and Morris[2] was also modified[86,87]. Stepwise single-column elution with citrate buffers, introduced by Hamilton and co-workers[8,9], was modified and compared as an alternative to continuous gradient elution[10]. Further details of amino acid analysers can be found, for example, in ref. 23.

The efficiency of separation was found to be improved when using lithium citrate buffers[10,11,87], or mixed lithium-sodium citrate buffers[12] instead of sodium citrate buffers. The lithium buffers allow good resolution of glutamine and asparagine. Citrate-phosphate and citrate-borate buffers were used to improve the separation further[13-15]. The reproducibility and flexibility of the gradient formation were improved by using precise pumps for mixing two or three buffers of high and low pH according to a pre-set programme instead of the Varigrad mixing device[5,13,16].

A chromatographic method was proposed that makes it possible to perform the analysis of amino acids, amino sugars and neutral sugars in mixtures in samples of protein hydrolysates using a single-run operation on two columns connected in series, the first packed with a sulphonated cation-exchange resin and the second with a strong anion-exchange resin (each packing of 8 μm particle diameter) using elution with a pH gradient (from 3.25 to 9.4) in citrate-borate buffers. In this combination of an amino acid and carbohydrate analyser, a post-column ninhydrin reaction was used for amino acid detection and a post-column reaction with orcinol-sulphuric acid was used for sugar detection with a photometric detector in the visible region[15].

The various improvements in column efficiency and the buffer systems for gradient elution allowed the analysis time to be decreased to ca. 4 h[4,5], and even separations in less than 1 h have been reported[13,89].

Amino acid analysers can also be utilized for the analysis of certain less common amino acids or of amino acids of certain types, such as sulphur-containing, branched-chain or aromatic amino acids, where modification of the components used for gradient formation is often necessary[17,18].

Amino acid analysers make possible the successful analysis of amino acid mixtures; however, these are expensive special-purpose instruments that can be used efficiently for routine amino acid analysis only. In many laboratories, there is a need for occasional amino acid analyses among analyses of various other types of samples. Therefore, a rapid separation method for amino acids, compatible with a general-purpose liquid chromatograph and a sensitive detection, have been sought.

Gradient elution chromatography on cation-exchange columns using increasing pH gradients in aqueous citrate buffers for elution has been attempted for the analysis of free amino acids with post-column derivatization based on the reaction with fluorescamine[25] or o-phthalaldehyde[26] and monitoring by a flow-through reaction fluorimetric detector.

References on p. 361.

20.2. REVERSED-PHASE CHROMATOGRAPHY OF FREE AMINO ACIDS ON CHEMICALLY BONDED PHASES

Reversed-phase chromatography with UV detection at 200-220 nm has been applied for the separation of free amino acids on a bonded amino column packing material using a multi-linear gradient of acetonitrile in phosphate buffer (pH 4.3)[19], for the separation of non-polar amino acids and small peptides on octadecyl and octyl bonded phases with a gradient of acetonitrile in 0.5 mol/l perchloric acid[20] and for separations of thyroidal iodoamino acids on columns packed with a bonded C_{18} phase by elution with gradients of methanol in phosphate buffers[21,22].

Certain aromatic acids (such as tyrosine and tryptophan) and their metabolites can be detected either using their UV absorption at 254 or 280 nm or their native fluorescence. These acids were chromatographed on octadecylsilica columns using linear gradients of acetonitrile[80] or methanol[81,82] in potassium dihydrogen phosphate buffers (pH 2.5-5.5). Electrochemical detection has also been employed for compounds of this type with a sensitivity comparable to that of fluorescence monitoring of amino acid derivatives[82].

Post-column derivatization reaction with *o*-phthalaldehyde was employed in the ion-pair chromatography of free amino acids (Asp, Ser, Glu, Gly, Thr, Ala, Cys, Val, Met, Tyr, Ile, Leu, Phe, His, Lys, Trp, Arg) on an octadecylsilica column using elution with a combined gradient of increasing concentrations of heptane-sulphonate, acetonitrile and methanol in an aqueous mobile phase containing a constant concentration of acetic acid and sodium decylsulphate. The separation took about 40 min[29]. Comparable results had previously been achieved on an octylsilica column using ion-pair gradient elution chromatography with a multi-linear gradient of *n*-propanol concentration in a mobile phase containing an aqueous solution of 0.017 mol/l sodium citrate and 0.5% dodecyl sulphate, employing post-column ninhydrin reaction detection[30]. Ion-pair reversed-phase chromatography of free amino acids provides better results than chromatography on chemically bonded ion-exchange phases.

20.3. CHROMATOGRAPHY OF DERIVATIVES

20.3.1. Thiohydantoin derivatives

Phenyl- and methylthiohydantoin-amino acids are formed by reaction of the amino acids with phenyl isothiocyanate or with methyl isothiocyanate in pyridine medium. Phenylthiohydantoin(PTH)-amino acids have been separated by chromatography on pellicular[31] and microparticulate totally porous octadecylsilica columns using elution with linear gradients of acetonitrile in water[32,33].

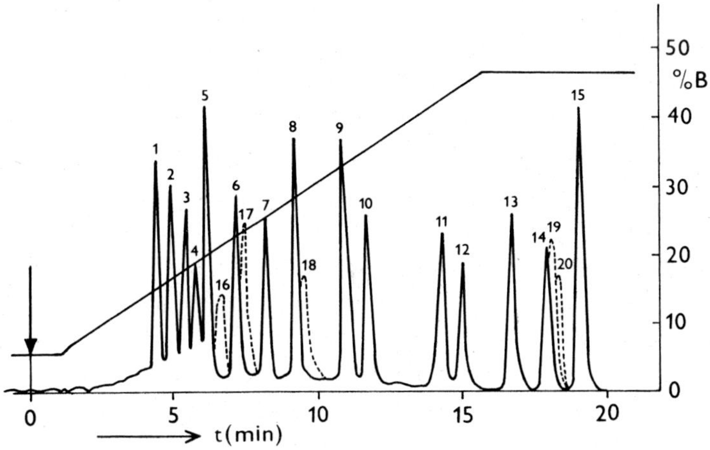

Fig. 20.1. Separation of a mixture of phenylthiohydantoin derivatives of amino acids by reversed-phase gradient elution liquid chromatography. Column: μBondapak C_{18}, 300 x 4 mm I.D. Gradient: linear, 5-45% B in 15 min (2.67% B/min). Solvents: A, water-methanol-acetic acid-acetone (900 : 100 : 2.5 : 0.05), pH adjusted to 4.1 with NaOH; B, water-methanol-acetic acid (100 : 900 : 0.25). Flow-rate 2.5 ml/min. Detection: UV at 254 nm. Peaks of PTH-amino acids: 1 = Asp; 2 = Asn; 3 = Ser; 4 = Thr; 5 = Gly; 6 = Gln; 7 = Glu; 8 = Ala; 9 = Hy-Pro; 10 = Tyr; 11 = Pro; 12 = Val/Met; 13 = Trp; 14 = Ileu; 15 = Lys; 16 = His; 17 = Cm-Cys; 18 = Arg; 19 = Phe; 20 = Leu. Peaks drawn with broken lines indicate the elution profiles of the remaining important amino acids, which were not present in the sample mixture. (Reproduced from ref. 49 with permission.)

However, linear or concave gradients of acetonitrile (from 5-15% to 50-90%)[34-40] and methanol (from 15-20% to 50-90%)[41-43,47-49] and linear ternary gradients with increasing concentrations of acetonitrile and tetrahydrofuran[44] in 0.01 or 0.005 mol/l aqueous sodium acetate (pH 4-5) have been most frequently used for elution and allowed rapid separations of twenty common phenylthiohydantoin-amino acids in about 30 min or less (see the example in Fig. 20.1). A linear gradient of ethanol (from 10 to 50%) in 0.01 mol/l sodium acetate[45], a multi-linear gradient of acetonitrile in lithium acetate solution[46] and a concave gradient of acetonitrile (from 5 to 60%) in aqueous diethylenetriamine-trichloroacetic acid (pH 4.2)[50] were also tested as eluting agents for these separations. Ion-pair reversed-phase chromatography on an octadecylsilica column with a linear gradient of acetonitrile (from 10 to 30%) in 0.005 mol/l heptanesulphonic acid (pH 3.5) as the eluent has also been applied for the separation of phenylthiohydantoin-amino acids[51].

Porous polystyrene-divinylbenzene copolymers (XAD) with small particle diameters were successfully used as column packing materials for the reversed-phase chromatography of phenylthiohydantoin, dinitrophenyl and dansyl derivatives of amino acids using linear gradients of acetonitrile in water (with addition of a salt to the mobile phase)[52,53].

Phenylthiohydantoin-amino acids have also been separated by normal-phase gradient elution liquid chromatography on microparticulate silica using a ternary concave gradient of methanol and *n*-propanol in *n*-hexane, which allowed the separation of all the common amino acids except arginine and histidine in ca. 40 min[54].

Polar nitrile (and aminonitrile) bonded phases have also been employed for separations of phenylthiohydantoin-amino acids with linear gradients of isopropanol and dichloromethane in *n*-hexane (in ca. 30-40 min)[55-57]. Successful separations were also achieved on a bonded nitrile phase in a reversed-phase system using a multi-linear gradient of methanol and acetonitrile in 0.02 mol/l ammonium acetate[58].

Phenylthiohydantoin[27] and methylthiohydantoin-amino acids[28] can be also separated by gradient elution cation-exchange chromatography with UV detection.

Methylthiohydantoin[59] and diphenylindenonylthiohydantoin-amino acids[60] were successfully separated in ca. 30 min on octadecylsilica columns using linear or concave gradients of acetonitrile in 0.01 mol/l aqueous sodium acetate (pH 4-5), i.e., under the commonest conditions for the separation of phenylthiohydantoin-amino acids.

UV detection, usually at 254 nm, was used for monitoring thiohydantoin derivatives of amino acids in column effluents with detection limits in the picomole range.

20.3.2. Dansyl derivatives

Fluorescent dansyl (dns) derivatives of amino acids have most frequently been separated by reversed-phase chromatography on octadecyl-[61-66] and octylsilica[67-69] chemically bonded phases. Linear (less frequently convex or concave) gradients of acetonitrile in water[61], aqueous acetic [62,63] or phosphoric acid[67], aqueous sodium acetate[64,68] or sodium phosphate solution[65] allowed rapid separations (in ca. 30-60 min) of common amino acid derivatives with detection limits in the picomole or femtomole range using fluorimetric detection.

Karger et al.[69] employed mobile phases containing chelates of metal cations [zinc(II) complexes with 4-dodecyldiethylenetriamine] that yield high selectivity for dns-amino acid separations owing to the possibilities of the complexation of these derivatives in the outer coordination sphere of the metal-chelate system. Thus, it was possible to separate fifteen common amino acids (Arg, Glu, Thr, Pro, Ser, Ala, Val, Cy-SO_3H, Asp, Gly, Met, Ileu, Leu, Phe, Trp) in 10 min using elution either with a linear gradient of increasing ionic strength (concentration of ammonium trifluoroacetate, 0-0.52 mol/l) in 60% aqueous acetonitrile or with

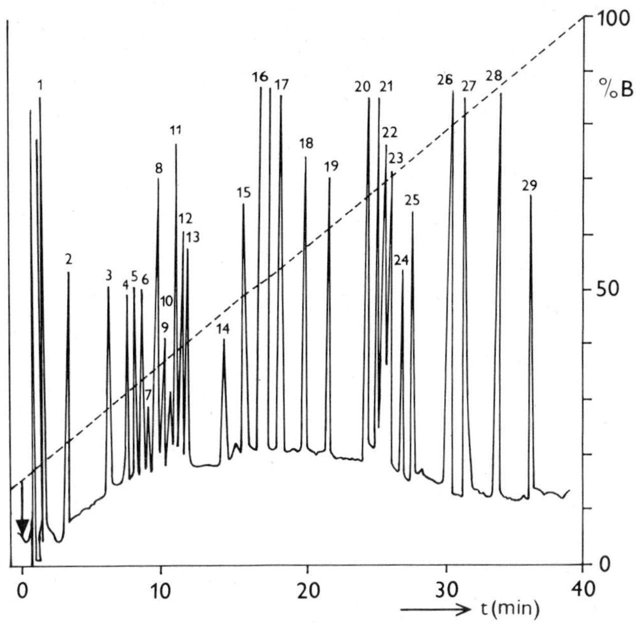

Fig. 20.2. Separation of a mixture of dns derivatives of amino acids by reversed-phase gradient elution liquid chromatography. Column: C_{18} Hypersil, 5 µm, 150 x 4.6 mm I.D. Gradient: linear, 14-100% B in 40 min (2.15% B/min). Solvents: A, methanol-tetrahydrofuran-acetic acid-triethylamine (5 : 3 : 0.57 : 0.088); B, methanol-tetrahydrofuran-acetic acid-triethylamine (70 : 3 : 0.57 : 0.088). Flow-rate: 2 ml/min. Temperature: 50°C. Detection: UV at 254 nm (fluorimetric detection would provide improved sensitivity). Peaks: 1 = unknown, probably dansylation by-products; 2 = $CySO_3H$; 3 = Asn; 4 = Gln; 5 = $MetSO_2$; 6 = Ser; 7 = Asp; 8 = Arg; 9 = Glu; 10 = dns-Lys; 11 = Gly; 12 = dns-NH_2; 13 = Thr; 14 = Ala; 15 = unknown; 16 = dns-NHMe; 17 = Pro; 18 = Met; 19 = Val; 20 = Trp; 21 = Phe; 22 = Ileu; 23 = Leu; 24 = unknown; 25 = $(Cys)_2$; 26 = Orn; 27 = Lys; 28 = His; 29 = Tyr. (Reproduced from ref. 66 with permission.)

a linear gradient of acetonitrile concentration (from 40 to 60%) at a constant ionic strength (0.15 mol/l ammonium acetate). The mobile phase always contained 10^{-3} mol/l zinc chelate. Later, Tapuhi et al.[66] used mobile phase containing chiral chelate of nickel(II) (complex with L-prolyl-*n*-dodecylamide) for the successful separation of amino acid enantiomers. The separation system was based on a two-column arrangement with a switching possibility. On the first (octadecyl-silica) column (150 x 4.6 mm I.D., C_{18} Hypersil, 5 µm), it was possible to separate all of the common dns-amino acid derivatives in about 40 min using a linear gradient of methanol in the mobile phase containing tetrahydrofuran and acetic acid-triethanolamine buffer at a constant concentration (Fig. 20.2).

By appropriate column switching, it was possible to transfer the separated amino acid from the first column to the second column (60 x 4.6 mm I.D., C_8 Hypersil, 5 µm) to achieve chiral separation with a mobile phase containing

$4 \cdot 10^{-3}$ mol/l L-prolyl-*n*-octylamide, $8 \cdot 10^{-3}$ mol/l Ni^{2+} and $8.75 \cdot 10^{-2}$ or $1.75 \cdot 10^{-1}$ mol/l ammonium acetate (pH 9.2) in methanol-water (60 : 40). Here, D- and L-enantiomers of each acid selected could be clearly separated and their ratio determined from the chromatogram.

Dns derivatives of the common amino acids (Ileu, Leu, Val, Pro, Phe, Met, Ala, Ly, Tyr, Gly, Trp, Glu, Thr, Ser, Asp, Cys, His) were separated in ca. 30 min on two columns packed with microparticulate silica at 65°C using combined gradient and isocratic elution in the mobile phase benzene-pyridine-acetic acid [the concentration of a mixed solvent pyridine-acetic acid (10 : 1) increased linearly in the gradient step][70]. Dns derivatives of amino acids could also be separated on a bonded amino phase column using elution with a multi-linear gradient of acetonitrile and acetic acid in dichloromethane[71].

20.3.3. *o-Phthalaldehyde derivatives*

Amino acid derivatives formed by pre-column reaction with *o*-phthalaldehyde and 2-mercaptoethanol or ethanethiol (highly fluorescent thio-substituted isoindoles) were rapidly separated (in 20-60 min) on octadecylsilica columns using elution with a gradient of acetonitrile in aqueous phosphate[72,84,85] or aqueous triethylamine acetate[73] buffers, with a gradient of methanol in citrate-phosphate buffer[74] or with a ternary gradient of methanol and tetrahydrofuran in 0.05 mol/l aqueous sodium acetate[75]. A complex five-component gradient (methanol-dimethylfuran-phosphoric acid-acetic acid-water) has also been employed for this purpose[76]. Fluorimetric detection allowed determinations with detection limits at the picomole level.

20.3.4. *Other derivatives*

The separation of 2,4-dinitrophenyl derivatives of the common amino acids could be achieved in about 50 min on an octadecylsilica column using elution with a linear gradient of acetonitrile in an aqueous isopropanol solution of pyridine and formic acid with UV detection at 360 nm[77].

Enantiomeric amino acids could be separated after derivatization to N-*d*-10-camphorsulphonyl *p*-nitrobenzoates, derivatives with a chromophore and a centre of optical asymmetry. A column packed with a microparticulate bonded amino phase and a linear ternary gradient of isopropanol and dichloromethane in isooctane allowed the separation of diastereoisomers of several amino acids in just over 1 h with UV detection at 254 nm[78,79].

20.4. SEPARATION OF POLYAMINOPOLYCARBOXYLIC ACIDS

Carrier ampholytes used in isoelectric focusing techniques (mixtures of polyaminopolycarboxylic acids) have been analysed on an Aminex A-6 ion-exchange column using elution with a pH gradient (from pH 3.8 to 12) at 60°C[83].

20.5. REFERENCES

1 D.H. Spackman, S. Moore and W.H. Stein, *Anal. Chem.*, 30 (1958) 1190.
2 K.A. Piez and L. Morris, *Anal. Biochem.*, 1 (1960) 187.
3 E.J. Miller and K.A. Piez, *Anal. Biochem.*, 16 (1966) 320.
4 B.C. Starcher, L.Y. Wenger and L.D. Johnson, *J. Chromatogr.*, 54 (1971) 425.
5 H. Tschesche, C. Frank and H. Ebert, *J. Chromatogr.*, 85 (1973) 35.
6 J.P. Ellis and J.B. Garcia, *J. Chromatogr.*, 87 (1973) 419.
7 G.W. Robinson, *J. Chromatogr.*, 111 (1975) 416.
8 P.B. Hamilton, D.C. Bogue and R.A. Anderson, *Anal. Chem.*, 32 (1960) 1782.
9 P.B. Hamilton, *Anal. Chem.*, 35 (1963) 2055.
10 J.L. Young and M. Yamamoto, *J. Chromatogr.*, 78 (1973) 221.
11 Y. Houpert, P. Tarallo and G. Siest, *J. Chromatogr.*, 115 (1975) 33.
12 J.L. Young and M. Yamamoto, *J. Chromatogr.*, 78 (1973) 349.
13 C. Murren, D. Stelling and G. Felstead, *J. Chromatogr.*, 115 (1975) 236.
14 K. Ohtsuki and H. Hatano, *J. Chromatogr.*, 148 (1978) 536.
15 M.M. Tikhomirov, A.Ya. Khorlin, W. Voelter and H. Bauer, *J. Chromatogr.*, 167 (1978) 197.
16 A.J. Thomas, *Proc. Nutr. Soc.*, 29 (1970) 108.
17 S.E. Moeller, *Anal. Biochem.*, 79 (1977) 590.
18 T. Ubuka, M. Kinuta, R. Akagi and S. Kiguchi, *J. Chromatogr.*, 188 (1980) 442.
19 R. Schuster, *Anal. Chem.*, 52 (1980) 617.
20 I. Molnár and Cs. Horváth, *J. Chromatogr.*, 142 (1977) 623.
21 M.T.W. Hearn, W.S. Hancock and C.A. Bishop, *J. Chromatogr.*, 157 (1978) 337.
22 M.T.W. Hearn and W.S. Hancock, *J. Liq. Chromatogr.*, 2 (1979) 217.
23 Z.J. Zmrhal, J.C. Heathcote and R.J. Washington, in Z. Deyl, K. Macek and J. Janák (Editors), *Liquid Column Chromatography*, Elsevier, Amsterdam, 1975, Ch. 32.
24 Z. Deyl and M. Juřicová, in Z. Deyl, K. Macek and J. Janák (Editors), *Liquid Column Chromatography*, Elsevier, Amsterdam, 1975, Ch. 33.
25 W. Voelter and K. Zech, *J. Chromatogr.*, 112 (1975) 643.
26 J.R. Cronin and P.E. Hare, *Anal. Biochem.*, 81 (1977) 151.
27 S. Ganno, K. Ojima, K. Fujita and S. Takeuchi, *Hitachi Hyoron*, 56 (1974) 1043; *C.A.*, 82 (1975) 69970.
28 V.M. Stepanov, S.P. Katrukha, L.A. Baratova, L.P. Belyanova and V.P. Korzhenko, *Anal. Biochem.*, 43 (1971) 209.
29 M.K. Radjai and R.T. Hatch, *J. Chromatogr.*, 196 (1980) 319.
30 J.C. Kraak, K.M. Jonker and J.F.K. Huber, *J. Chromatogr.*, 142 (1977) 671.
31 J.X. Devries, R. Frank and C. Birr, *FEBS Lett.*, 55 (1975) 65.
32 *Applied Science Catalogue*, Applied Science Labs., State College, PA, 1980.
33 *Columns and Packings for HPLC*, DuPont Wilmington, DE.
34 R.V. Vivilecchia, R.L. Cotter, R.J. Limpert, N.Z. Thimot and J.N. Little, *J. Chromatogr.*, 99 (1974) 407.
35 M.R. Downing and K.G. Mann, *Anal. Biochem.*, 74 (1976) 298.
36 C.L. Zimmerman, E. Appella and J.J. Pisano, *Anal. Biochem.*, 77 (1977) 569.
37 M.N. Margolies and A. Brauer, *J. Chromatogr.*, 148 (1978) 429.
38 *RP High Efficiency LC Columns*, *μBondapak C_{18}*, Waters Assoc., Milford, MA, 1974.
39 *Micro Sil HPLC Columns*, Micromeritics, Norcross, GA.

40 T.F. Bronzert and H.B. Bronzert, *LC Applications Brief*, DuPont, Wilmington, DE, 1976.
41 J. Elion, M. Downing and K. Mann, *J. Chromatogr.*, 155 (1978) 436.
42 W.F. Moo-Penn, D.L. Jue, K.C. Bechtel, M.H. Johnson, R.M. Schmidt, P.R. McCurdy, J. Fox, J. Bonaventura, J.B. Sullivan and C. Bonaventura, *J. Biol. Chem.*, 251 (1976) 7557.
43 W.F. Moo-Penn, M.H. Johnson, K.C. Bechtel and D.L. Jue, *J. Chromatogr.*, 172 (1979) 476.
44 N.H.C. Cooke and K. Olsen, *J. Chromatogr. Sci.*, 18 (1980) 512.
45 T. Greibrokk, E. Jensen, G. Ostvald, *J. Liq. Chromatogr.*, 3 (1980) 1277.
46 P.W. Moser and E.E. Rickli, *J. Chromatogr.*, 176 (1979) 451.
47 M. Holman and G. Hawk, *Proteins, Edman Degradation and LC*, Waters LC Application Literature, Waters Assoc., Milford, MA, 1974.
48 J. Bridgen, *Sci. Tools*, 24 (1977) 1.
49 A.S. Bhown, J.E. Mole, A. Weissinger and J.C. Bennett, *J. Chromatogr.*, 148 (1978) 532.
50 W.D. Annan, *J. Chromatogr.*, 173 (1979) 194.
51 *PTH Amino Acids, Micromeritics Application Notes No. 57*, Micromeritics, Norcross, GA.
52 E.P. Kroeff and D.J. Pietrzyk, *Anal. Chem.*, 50 (1978) 502.
53 Z. Iskandarani and D.J. Pietrzyk, *Anal. Chem.*, 53 (1981) 489.
54 E.W. Matthews, P.G.H. Byfield and I. MacIntyre, *J. Chromatogr.*, 110 (1975) 369.
55 C.D. Carr, K.C. Judah and M.N. Munk, *VIA*, 8, No. 2, 1974, 3, Varian, Palo Alto, CA.
56 *Perkin-Elmer Instrument News*, 25, No. 1 (1975) 12.
57 C. Bollet and M. Caude, *J. Chromatogr.*, 121 (1976) 323.
58 S.J. Di Mari, J.P. Robinson and J.H. Hash, *J. Chromatogr.*, 213 (1981) 91.
59 M.J. Horn, P.A. Hargrave and J.K. Wang, *J. Chromatogr.*, 180 (1979) 111.
60 I.N. Mancheva, R.N. Nikolov and J. Pfletschinger, *J. Chromatogr.*, 213 (1981) 99.
61 *The Series 3B Solvent Delivery Systems*, Perkin-Elmer, Norwalk, CT, 1980.
62 R. Bongiovanni, K.D. Burman, R.K. Garis and T. Boehm, *J. Liq. Chromatogr.*, 4 (1981) 813.
63 J.R. Helbert, W.C. Herwig, R.E. Wagener, M.C. Cieslak and E. Chicoye, cited in G. Charalambous (Editor), *Liquid Chromatographic Analysis of Food and Beverages*, Vol. 2, Academic Press, London, 1979, p. 237.
64 S. Weiner and A. Tishbee, *J. Chromatogr.*, 213 (1981) 501.
65 J.M. Wilkinson, *J. Chromatogr. Sci.*, 16 (1978) 547.
66 Y. Tapuhi, N. Miller and B.L. Karger, *J. Chromatogr.*, 205 (1981) 325.
67 G.J. Schmidt, D.C. Olson and W. Slavin, *J. Chromatogr.*, 164 (1979) 355.
68 G. Powis and M.M. Ames, *J. Chromatogr.*, 170 (1979) 195.
69 B.L. Karger, W.S. Wong, R.L. Viavattene, J.N. Lepage and G. Davies, *J. Chromatogr.*, 167 (1978) 253.
70 E. Bayer, E. Grom, B. Kaltenegger and R. Uhmann, *Anal. Chem.*, 48 (1976) 1106.
71 E. Johnson, A. Abu-Shumays and S.R. Abbott, *J. Chromatogr.*, 134 (1977) 107.
72 D.W. Hill, F.H. Walters, T.D. Wilson and J.D. Stuart, *Anal. Chem.*, 51 (1979) 1338.
73 B.R. Larsen and F.G. West, *J. Chromatogr. Sci.*, 19 (1981) 259.
74 P. Lindroth and K. Mopper, *Anal. Chem.*, 51 (1979) 1667.
75 B.N. Jones, S. Pääbo and S. Stein, *J. Liq. Chromatogr.*, 4 (1981) 565.
76 W.J. Hurst and R.A. Martin, *J. Agr. Food Chem.*, 28 (1980) 1039.
77 E.W. Bachmann, J. Frei and M. Muehlemann, *Chromatographia*, 12 (1979) 345.
78 H. Furukawa, Y. Mori, Y. Takeuchi and K. Ito, *J. Chromatogr.*, 136 (1977) 428.
79 S.R. Abbott, *J. Chromatogr. Sci.*, 18 (1980) 540.
80 A.M. Krstulovic, S. Ciriello, L. Bertani-Dziedzic and S.E. Gitlow, *J. Chromatogr.*, 164 (1979) 495.
81 A.M. Krstulovic, P.R. Brown, D.M. Rosie and P.B. Champlin, *Clin. Chem.*, 23 (1977) 1984.

82 D.A. Richards, *J. Chromatogr.*, 175 (1979) 293.
83 R.K. Brown, J.M. Lull, S. Lowenkron and J.C. Bagshaw, *Anal. Biochem.*, 71 (1976) 325.
84 J.D. Stuart, T.D. Wilson, D.W. Hill, F.H. Walters and S.Y. Feng, *J. Liq. Chromatogr.*, 2 (1979) 809.
85 G.H.T. Wheler and J.T. Russell, *J. Liq. Chromatogr.*, 4 (1981) 1281.
86 J.V. Benson and J.A. Patterson, *Anal. Biochem.*, 13 (1965) 265.
87 J.V. Benson, M.J. Gordon and J.A. Paterson, *Anal. Biochem.*, 18 (1967) 228.
88 S. Stein, P. Böhlen, J. Stone, W. Dairman and S. Udenfriend, *Arch. Biochem. Biophys.*, 155 (1973) 203.
89 R.J. Benson, *Amer. Lab.*, (1972) 53.

Chapter 21

PEPTIDES AND PROTEINS

Proteins, the essential components of living matter, are composed of polypeptide chains. Enzymes and a number of hormones are structurally based on proteins. A knowledge of the amino acid sequence in peptides and in proteins is very important in biochemical research. The number of published applications of liquid chromatography to the separation of peptides and proteins is enormous and perhaps outnumbers the sum of liquid chromatography applications to the separation of all the other compounds. It would be impossible here to give a review of all separation methods in which gradient elution has been applied to the chromatography of peptides and proteins. Therefore, a brief survey of some generally applicable methods employing the gradient elution technique is given and illustrated by only a few examples, to give a picture of what gradient elution liquid column chromatography means in studies of these compounds. A systematic, comprehensive survey of liquid chromatography applications in this field may be found elsewhere (see, for example, refs. 1-3).

Applications of liquid chromatography to the preparative purification and fractionation of proteins are important. High-performance liquid chromatography (HPLC) has brought new dimensions to the analysis of proteins, peptides and amino acids with respect to speed of analysis, separation efficiency and sensitivity, which makes it possible to achieve micro-scale separations of complex mixtures of peptides in small amounts of samples in less than 1 h, in contrast to earlier elaborate and time-consuming methods giving low yields of material[133].

To determine the amino acid composition of proteins and peptides, it is necessary first to subject these compounds to total or sequential hydrolysis. In total hydrolysates the ratio of the individual amino acids can be determined, but sequential hydrolysis allows the determination of the sequence of amino acids in peptide chains. Edman degradation is a non-destructive, subtractive method of protein analysis using repeated degradation of peptides or proteins in three-step cycles. In the first step, the peptide or protein reacts with phenylisothiocyanate to give the phenylthiocarbamyl (PTC) derivative. In the second step, the derivatized N-terminal amino acid is hydrolysed from the remainder of the protein or peptide and its 2-anilino-5-thiazoline derivative is formed and converted into the stable 3-phenyl-2-thiohydantoin (PTH) derivative

References on p. 376.

in the third step. Because only the N-terminal amino acid is cleaved in this procedure and the remainder of the peptide or protein remains intact, the whole procedure can be repeated as many times as necessary to find the amino acid sequence.

Liquid chromatography is very suitable for combination with Edman degradation as a rapid and reliable analytical technique that makes it possible to purify and to check the purity of original proteins or peptides and of PTC derivatives formed in the first step of Edman degradation, to separate, identify and quantitate PTH-amino acid derivatives formed in each three-step Edman degradation cycle. In all of these liquid chromatography applications gradient elution, either continuous or stepwise, may be used.

In the subsequent determination of the amino acid sequence in protein and peptide hydrolysates, the methods for amino acid analysis are applied; the applications of the gradient elution technique for this purpose were treated in Chapter 20.

Prior to the determination of the amino acid ratio and/or sequence or for other biochemical research, it is necessary to purify and separate proteins and peptides in mixtures, and liquid column chromatography is uniquely suited for this purpose. However, the specific fragile character of the biopolymers should be borne in mind when selecting the chromatographic conditions, in order to avoid the risk of possible denaturation of proteins. This risk is particularly important with enzymes, where their specific topography, which gives rise to their particular activities, may become lost by structural changes when the enzyme is brought into contact with an unsuitable column packing material[3].

In numerous earlier applications of liquid chromatography to the purification and fractionation of peptides and proteins, fractions of the eluate were collected and the amount of compounds eluted was determined using spectrophotometric methods, measurement of the radioactivity of labelled compounds, measurement of enzymatic activity, etc. In more recent applications, mainly those using efficient columns, the eluates are monitored continuously.

The detection of aromatic peptides employs UV absorption at 280 nm. UV detection at 254 nm has often been utilized in the chromatography of proteins, hormones and enzymes based on proteins. Non-aromatic peptides were detected in the UV region between 200 and 210 nm. Enzymes in the eluate can be detected by continuous measurement of their enzymatic activity and labelled compounds by continuous monitoring of radioactivity. Post-column derivatization is often used for the detection of proteins and peptides. Derivatives with *o*-phthalaldehyde or other reagents are detected in the column effluents employing flow-through reactors connected with fluorimetric detectors or with photometric detectors operating in the UV or visible region.

21.1. ION-EXCHANGE CHROMATOGRAPHY

Ion-exchange chromatography has been most frequently used in the chromatography of peptides and proteins. The separation of peptides and proteins by ion-exchange chromatography is based on electrostatic interactions between the charged exchange groups of the exchanger and the opposite charge sites of the peptide or protein. The amphoteric character of these compounds allows both cation and anion exchangers to be used. At the isoelectric point, the positive and negative charges of the solute are in equilibrium. At a pH lower than that corresponding to the isoelectric point, the protein or peptide is retained on a cation exchanger owing to the electrostatic forces, whereas it is unretained on an anion exchanger. The opposite holds true at a pH above that corresponding to the isoelectric point. However, the use of anion exchangers is often preferred owing to the better stability of proteins in their anionic forms.

To improve the possibilities of penetration of large protein molecules into the interior of the particles, exchangers with wide pores are selected. Because the rigidity of exchangers with an organic matrix decreases and their swelling increases with increasing pore sizes, which is obviously disadvantageous from the point of view of column bed stability and maximum tolerable flow-rates, narrow-pore ion exchangers are sometimes preferred for the chromatography of very large proteins.

Polyions such as peptides and proteins are normally very firmly attached to ion exchangers by many ionic groups in their molecules. To achieve their elution, it is necessary to employ a buffer with pH about 1 unit above (in connection with cation exchangers) or below (with anion exchangers) the isoelectric point of the protein or peptide. Therefore, optimum conditions of elution should be established for each compound and it is generally necessary to change the composition of the mobile phase in the course of elution. Stepwise elution can be used for this purpose; however, all substances released by a change of buffer composition are eluted together, unfractionated, usually as peaks with sharp fronts and pronounced tailing. False peaks may occur if the buffer composition is changed before the complete elution of the tailing peak has been achieved. This drawback is avoided when using continuous gradients for elution, which yields symmetrical peaks, better resolved from each other than in stepwise elution.

During gradient elution, the solutes are eluted in order of their isoelectric points. In a gradient of increasing ionic strength at a pH close to the isoelectric points of solutes, the salt in the eluent competes with the solutes for the exchange groups of the exchanger and this competing effect increases with increasing ionic strength of the mobile phase.

References on p. 376.

The ionic strength can be changed by changing the concentration of the buffer or, more frequently, by increasing the concentration of a neutral salt. Ideally, the ionic strength of the starting buffer should be low. However, buffers with very low ionic strengths have low buffering capacities and large changes in ionic strength may cause large changes in the volumes of swollen exchangers of certain types and, therefore, buffers with very low ionic strengths should be avoided.

If a pH gradient is applied for chromatography of proteins or peptides on a cation-exchange column, the original pH is selected below that corresponding to the lowest isoelectric point of the compound in the sample mixture and then it is increased with time to achieve the elution and separation. In contrast, the original pH above the isoelectric points of solutes is selected in anion-exchange chromatography and it is decreased during the elution.

Cation- or anion-exchange polystyrene-divinylbenzene resins[4-8,11], ion exchangers based on modified cellulose[9-23], dextran (Sephadex)[24-32,130], agarose (Sepharose)[32,33] or glycol methacrylate (Spheron) gels[34-39] and inorganic ion exchangers, such as hydroxyapatite with anion-exchange properties[40-42], have been widely applied for the purification and fractionation of peptides and proteins by liquid chromatography. Most classical ion exchangers do not withstand elevated pressures and therefore rapid analyses at elevated flow-rates of the mobile phase are not possible. Spheron exchangers and some exchangers based on styrene-divinylbenzene copolymers are exceptions from this rule. However, proteins may become denatured by virtue of hydrophobic interactions with the polystyrene matrix. Really rapid analyses of proteins and peptides have become possible only with ion exchangers based on microparticulate inorganic supports with chemically bonded ion-exchange functional groups, which are sufficiently rigid and form stable column beds at high pressures. An organic surface layer bonded on the inorganic support is useful for protecting proteins from direct contact with the inorganic surface and contributes to the chromatographic partitioning process in a similar manner to ion-exchange resins or modified gels. An important advance was the introduction of column packing materials prepared by chemical bonding of glycerylpropylsilane on the surface of controlled porosity glass. The product retains its terminal epoxy group, which can be subjected to reaction with a nucleophilic reagent to give the required chemically modified material. These materials have been marketed under the name Glycophase/CPG. Ion exchangers of this type (DEAE-, QAE-, SP- or SM-Glycophase/CPG)[43-51] or ion exchangers chemically bonded on a silica gel support[52-55] made possible high-speed separations of proteins and peptides. Because ion-exchange chromatography can resolve proteins with small charge differences, it is possible to separate isoenzymes that have nearly identical amino acid sequences in these highly selective and efficient chromatographic systems.

Concentration (ionic strength) or pH gradients in mobile phases are usually formed from solvents containing different concentrations of phosphate, acetate or pyridine-acetate buffers in water.

Just to obtain an insight into the variety of applications possible, we can mention a few examples, such as chromatography of serum proteins on an anion-exchange resin using a concentration gradient of acetate buffer[8]; the determination of reduced and oxidized glutathione in wheat flours and doughs on a cation-exchange resin with a gradient of hydrochloric acid concentration[7]; the separation of proteins in human and animal sera by chromatography on DEAE-cellulose anion exchanger using concentration[14] or pH[12] gradients in phosphate buffers, concentration gradients in Tris buffers[16,22,23,131] or gradients of increasing concentration of sodium chloride in water[21]; the fractionation of tyrosine-rich proteins from wool on a QAE-cellulose anion exchanger using a gradient of ionic strength in alkaline β-alanine buffer[10]; and the separation and purification of the enzymes β-glucan hydrolase[15] and β-N-acetyl-D-hexosaminidase[19] on DEAE-cellulose using an ionic strength (concentration of alkali metal chloride) gradient in a phosphate buffer. We can further mention the chromatography of insulin and its derivatives on a DEAE-Sephadex anion exchanger using a concentration gradient of sodium chloride in a Tris-HCl buffer containing 7 mol/l urea[26]; the chromatography of thymopoietins on QAE-Sephadex using a concentration gradient of potassium chloride[132]; the fractionation of whey proteins and casein in cows' milk on the same material using a concentration gradient of sodium chloride in a phosphate buffer[24]; the chromatography of enzymes ribonucleases on a CM-Sephadex cation exchanger using combined pH and ionic strength gradients in a phosphate buffer[31]; the chromatography of α-bungarotoxin and other components of snake venom on the same column packing with a linear concentration gradient of ammonium acetate in water (0.05-1 mol/l)[130]; and the fractionation of human serum proteins by anion-exchange chromatography on DEAE-Sepharose Cl-6B using a linear gradient of increasing sodium chloride concentration in 0.05 mol/l Tris-HCl buffer at pH 8.6[33]. On Spheron ion exchangers, similar types of compounds can be chromatographed as on cellulose or polydextran exchangers, but Spheron materials have the advantages of better mechanical rigidity and stability at higher pressures, which allows more rapid separations to be achieved at higher flow-rates of the mobile phase. An example of gradient elution chromatography of human plasma on DEAE-Spheron column is shown in Fig. 21.1. The elution profile here is similar to that obtained on DEAE-cellulose[34]. On hydroxyapatite columns, for example, fractionation of myosin[40,41] and histones[42] was achieved using ionic strength gradients in phosphate buffers. By chromatography on DEAE-Glycophase/CPG using linear concentration

References on p. 376.

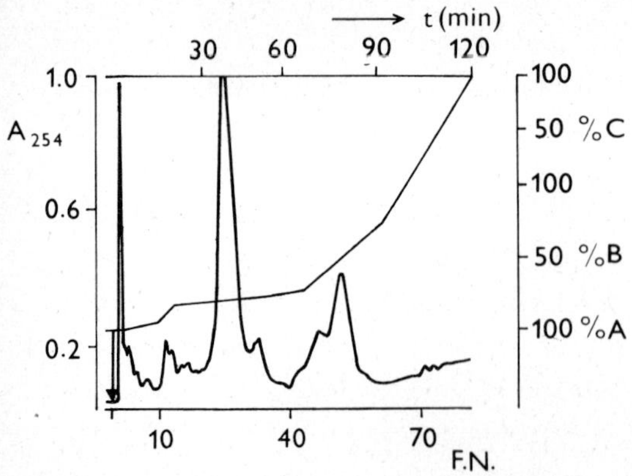

Fig. 21.1. Protein mapping of human plasma by ion-exchange liquid chromatography on a modified diethylaminoethyl-Spheron anion exchanger. Column: DEAE-Spheron, 25-32 µm, 190 x 8 mm I.D. Gradient: multi-linear, as indicated by the thin full line, from 0 to 100% B in A and then from 0 to 100% C in B. Buffers: Solvent A, 0.025 mol/l phosphoric acid-Tris (pH 8.5); solvent B, 0.5 mol/l phosphoric acid-Tris (pH 3.5); solvent C, solvent B with 1.0 mol/l KCl. Flow-rate: 3.75 ml/min. Fractions collected at 1.5-min intervals. F.N. = fraction number; A_{254} = absorbance at 254 nm. Sample: 2 ml of human plasma, centrifuged before application. (Reproduced from ref. 34 with permission.)

gradients of sodium chloride in acetate or Tris-HCl buffers (pH 8), it was possible to separate isoenzymes of lactate dehydrogenase, hexokinase, etc., in 25-40 min[45-50]. Human serum proteins and rat kidney and rat liver isoenzymes could be rapidly separated by anion-exchange chromatography on a PEI 6-LiChrospher Si 500-column (a pellicular anion exchanger on a silica gel support) using linear gradients of sodium acetate concentration in 0.02 mol/l Tris-acetate buffer (pH 8)[53].

21.2. CHROMATOFOCUSING

pH Gradients may be generated in an ion-exchange column itself, when it is originally equilibrated with a buffer with a pH different to that of the buffer used for elution. Using a special blend of buffers with even buffering capacities over a wide pH range and special ion exchangers, a smooth continuous pH gradient may be generated on the column. This technique, so-called "chromatofocusing"[56-59], is treated in more detail in Part III. Very similar in principle is so-called "ampholyte-displacement chromatography"[60]. An example of the separation of soluble proteins extracted from elk muscle using the chromatofocusing technique on agarose-derived ion exchangers (Polybuffer exchangers) is shown in Fig. 21.2[59].

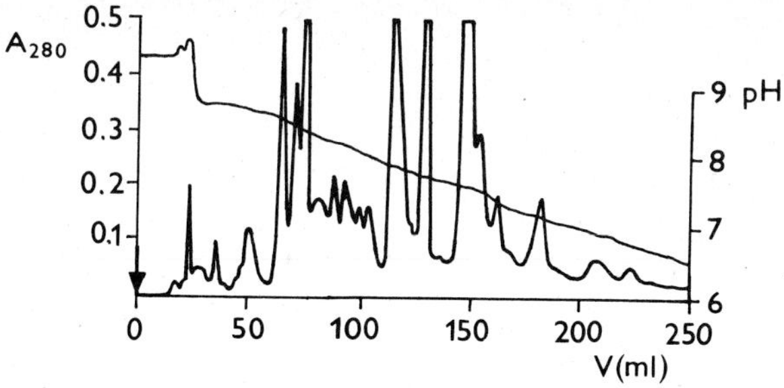

Fig. 21.2. Fractionation of soluble proteins extracted from elk muscle by chromatofocusing. Column: Pharmacia C 10/40; packed with polybuffer exchanger, PBE 94, bed height = 45 mm. Column equilibration buffer, 0.025 mol/l ethanol-amine•HCl (pH 9.4); elution buffer, 0.0075 mmol/pH unit/ml Polybuffer 96 (pH 6). Column self-generated pH gradient is indicated by the full line. Flow-rate: 20 cm/h. Detection: UV at 280 nm. A_{280} = absorbance at 280 nm; V = elution volume. Sample: 5 ml of supernatant from elk meat homogenate. (Reproduced from ref. 59 with permission.)

21.3. CHROMATOGRAPHY ON POLYAMIDE GELS

Polyamide gels with amide groups can behave as weak anion exchangers and bond peptides and proteins by electrostatic forces. Therefore, these compounds can be chromatographed on polyamide gels, such as Bio-Gel P, using elution with pH or ionic strength gradients in acetate, phosphate, Tris-HCl or other buffers. This technique was applied, for example, to separations of methionyl peptides[61], the chromatography of iodinated L-tyrosyl-L-tyrosine peptides[62] and the resolution of venom proteins[63].

21.4. CHROMATOGRAPHY ON POLAR ADSORBENTS AND ON POLAR BONDED PHASES

Chromatography on polar adsorbents is relatively rarely used for separations of proteins and peptides. UV-absorbing derivatives of peptides (phenylthiohydantoin and benzoyl derivatives) were chromatographed on silica gel using elution with a continuous gradient of ethanol in chloroform[64], propanol, methanol and dichloromethane in *n*-hexane[65] or stepwise gradients of anhydrous ethanol and acetic acid in dichloromethane[66]. UV detection at 254 nm is suitable with these systems.

Peptides of human foetal globin and of murine 1a antigens were separated by chromatography on a nitrile phase bonded on silica (Zorbax CN) using a continuous gradient of organic solvent in aqueous acidic medium (acetone in phosphate buffer, pH 2.1[67], or propanol in 0.1% trifluoroacetic acid[68]) for elution.

References on p. 376.

21.5. HYDROPHOBIC INTERACTION CHROMATOGRAPHY

It has been observed that in chromatography on gels in aqueous media, proteins and peptides are eluted in disagreement with their expected elution order according to their molecular sizes and that the retention is influenced by the mobile phase composition. Therefore, it is obvious that, in addition to a possible steric exclusion mechanism, other interactions control the chromatographic process. It has been established that these interactions are hydrophobic, i.e., that they originate in attraction forces between water molecules, by virtue of which the molecules of biopolymers are excluded from the mobile phase on to the gel surface. This effect is of a similar nature as in reversed-phase chromatography on chemically bonded non-polar alkyl or aryl phases. Peptides, proteins and other biopolymers could be separated on non-polar or slightly polar organic gels (or on materials with a chemically bonded organic layer on an inorganic support, such as Glycophase-G/CPG)[71] using this separation mechanism and the corresponding chromatographic technique has been called "hydrophobic interaction chromatography"[69,70]. As in chromatography on chemically bonded non-polar phases, retention increases with increasing concentration of a salt and decreases with increasing concentration of an organic solvent in the mobile phase. Gels containing ethereal bonds and hydroxy or ester groups, such as unmodified Spheron gel[72], can be used for the hydrophobic interaction chromatography of proteins and peptides, but agarose gels with coupled hydrophobic groups, such as octyl, phenyl or benzyl, have been specially developed for this purpose. The organic solvent for the mobile phase gradient should be selected with care and the solubility of proteins in the mobile phase should be borne in mind. Most often, gradients of increasing concentration of ethylene glycol (0-50%) in acetate[73], phosphate[74] or Tris-HCl[75] buffers or of decreasing concentration of a salt (phosphate or sulphate) in water[76] and combinations of these gradients[74,75] are used, but a linear gradient of ethanol in water[77] and even of urea in a phosphate buffer[78] have also been employed. Various peptides, proteins and enzymes have been chromatographed using this technique. An example of the separation of β-amylase from barley meal is shown in Fig. 21.3.

21.6. REVERSED-PHASE CHROMATOGRAPHY OF PEPTIDES AND POLYPEPTIDES

Recently, octadecyl and octyl stationary phases chemically bonded on a microparticulate (5 or 10 μm) silica gel support have been used as column packing materials for reversed-phase chromatography. As in hydrophobic interaction chromatography on organic gels, the separation is based on hydrophobic interactions and the peptides are separated on the basis of differences in their

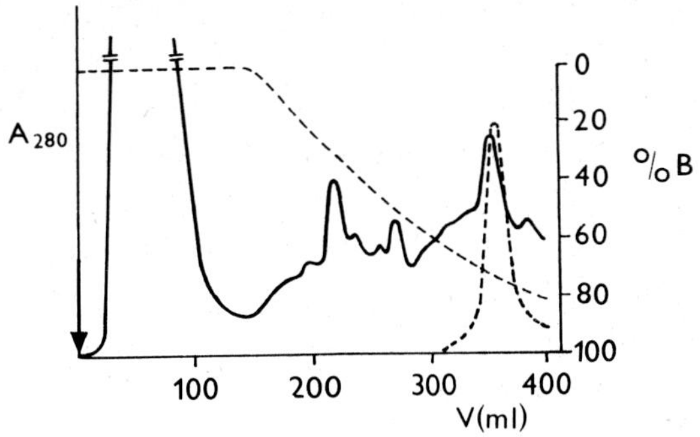

Fig. 21.3. Fractionation of β-amylase from barley meal by hydrophobic interaction chromatography with gradient elution. Column: Phenyl-Sepharose CL-4B, column K 16/20, bed volume 30 ml. Gradient: as indicated by broken line. Solvents: A, 0.01 mol/l sodium phosphate buffer (pH 6.8), 25% saturated with ammonium sulphate; B, 50% (v/v) ethylene glycol in water. Flow-rate: 0.417 ml/min. Detection: UV at 280 nm. A_{280} = absorbance at 280 nm; V = elution volume. Sample: barley meal. β-Amylase enzymatic activity is shown by the broken-line peak. (Reproduced from ref. 74 with permission.)

hydrophobicities. This technique has been applied in the high-speed chromatography of a variety of peptides and smaller polypeptides. Gradients with an increasing content of the organic solvent in an aqueous mobile phase are usually used for elution.

A linear gradient of acetonitrile in water used for the chromatography of collagen peptides[79] is an exception, because acidic buffers or aqueous solutions of acids are usually necessary for good separations. Gradients of methanol[80], acetonitrile[30,81-87] or propanol[68,102] in a 0.1% aqueous solution of phosphoric[30,81-89] acetic[80], formic[102] or trifluoroacetic acid[68,87-89], a linear gradient of acetonitrile in acidified 0.155 mol/l sodium chloride solution (pH 2.1)[90,91], gradients of acetonitrile[92-99], methanol[100] or isopropanol-2-methoxyethanol[101] in phosphate buffers (usually at pH 2-2.5), gradients of acetonitrile[103-105] or methanol-acetic acid[106,107] in 0.01 mol/l ammonium acetate buffer (pH 4-6), a linear gradient of methanol in Tris buffer[108], a linear gradient of n-propanol in sodium citrate (pH 4.5)[109] and a linear gradient of n-propanol concentration in pyridinium formate buffer[110] have been used for elution of peptides.

The selectivity of reversed-phase separations of peptides can sometimes be improved by using ion-pair formation in the mobile phase. Both cationic (such as tetrabutylammonium phosphate[112], triethylammonium phosphate[113-116] and dodecylamine[117]) and anionic (such as hexylsulphonate, decylsulphonate, pentafluoropropanoic acid, heptafluorobutyric acid and undecafluorocaproic acid[118]) ion-pairing reagents have been used for this purpose. The ion-pairing agent was usually contained in the mobile phase at a constant concentration, while the content of the organic solvent (acetonitrile[113-118], methanol[112] or isopropanol[116]) changed linearly during gradient elution.

Reversed-phase gradient elution chromatography on octyl- and octadecylsilica columns is suitable for the efficient and rapid analytical and preparative mapping of complex peptide mixtures and usually yields better separations in a shorter time than other liquid chromatography techniques. However, it would be difficult to achieve complete resolution of all peptides in very complex mixtures, such as in tryptic digests of larger proteins, where it could be necessary to re-chromatograph a zone of partially resolved peptides in a second, independent isocratic run.

To give a few examples of applications of the reversed-phase chromatography of peptides and small proteins, we can mention chromatography of proinsulin[30], peptides of human foetal globin[68], neuropeptides[87], hypothalamic oligopeptides[114], enkephalin peptides[80,92,108], argotensins[107], parathyroid[91] and adrenocorticotropic[92,113] hormones, pteroyl-oligo-γ-L-glutamines[95], growth hormones[82], 19-S thyloglobulin[85], opioid peptides[88], oxytocin, vasopressin, samostatin[96], tuftsin[97], calcitonin[112], apoliproteins[115], peptides of rat casein[82,86], tryptic digest of haemoglobin[84,103,104], of egg white lysozome[105], of sheep 19-5 thyroglobulin and of bovine thyrotrophin[117]. A high-efficiency separation of four polypeptides is shown in Fig. 21.4[119].

21.7. AFFINITY CHROMATOGRAPHY OF PROTEINS

Affinity chromatography is a highly selective separation method that allows the purification of proteins, protein-based enzymes and hormones and other biopolymers on the basis of their biological functions or individual chemical structures. This technique makes it possible to remove large amounts of contaminating substances from the solutes of interest in biological material or to separate the active form from the denatured forms of the same substance. Therefore, the merits of affinity chromatography for biological research are considerable, and this method has found wide acceptance since its first introduction about 15 years ago[120].

In principle, the method is based on the interactions between a biospecific ligand covalently bonded to a suitable support and the substance of interest, the specific structure of which "fits" the structure of bonded (immobilized) ligand. For example, an enzyme bonded as the ligand can be used for the purification of a substrate analogue, inhibitor or cofactor, or an antibody bonded as the ligand may be suitable for the purification of an antigen or of a virus, a nucleic acid as the ligand for purification of nucleic acid polymerase or of a binding protein, etc.

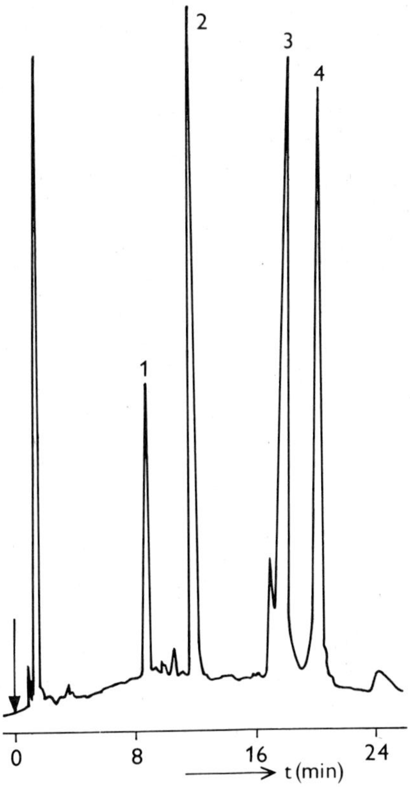

Fig. 21.4. Separation of four polypeptides by reversed-phase gradient-elution chromatography. Column: Micro Pak MCH, 150 x 5.4 mm I.D. Gradient: linear, 18-60% B in 42 min (1% B/min). Solvents: A, triethylammonium phosphate buffer (pH 2.2); B, acetonitrile. Flow-rate: 1 ml/min. Detection: UV at 210 nm. Peaks: 1 = bombesin; 2 = somatostatin; 3 = insulin; 4 = glucagon. (Reproduced from ref. 119 with permission.)

The use of the method consists in several consecutive stages:

(a) A suitable biospecific ligand is covalently bonded to a chromatographic bed material (matrix), such as an agarose gel (Sepharose 4B, Sepharose CL), previously activated for easier covalent attachment of the ligand. It is important that the attached, "immobilized", ligand retains its specific binding activity for the substance of interest.

(b) The gel with the "immobilized" ligand is packed into a column of suitable dimensions, the sample is applied, from which substances capable of biospecific interactions with the immobilized ligand become adsorbed on the column, while the non-sorbed impurities are washed away from the column with a starting buffer.

References on p. 376.

(c) Elution of the sorbed substances from the column using eluents that decrease the affinity forces between the substances and the immobilized ligand. This is usually achieved by either a stepwise or a continuous change in properties of the eluent and selective or non-selective eluents can be employed, depending on the nature of the binding between the ligand and solutes. Non-selective elution methods involve pH or ionic strength gradients. A change (usually a decrease) in pH during gradient elution alters the degree of ionization of charged groups at the binding sites and causes desorption. A gradient of increasing ionic strength (e.g., of sodium chloride concentration) also leads to desorption of sorbed solutes at a concentration of about 1 mol/l. Urea and guanidine hydrochloride at high concentrations are useful eluents, but they should be removed from the purified proteins immediately after the elution, otherwise they could cause denaturation of proteins.

Selective (affinity) elution makes use of an increasing concentration gradient of an eluting agent that either competes for binding to the adsorbed substance or for binding to the ligand.

To give a few rather arbitrarily chosen examples from the wealth of published information, we can mention chromatography of phosphofructokinase on immobilized Cibacron Blue F 36-A using a linear gradient (0-1.5 mol/l) of ammonium sulphate[121], purification of dipeptidyl peptidase IV on the immobilized peptide Gly-Pro-NH-$(CH_2)_6$-NH using a gradient (0-0.2 mol/l) of sodium chloride[122], fractionation of proteolytic enzymes trypsin and chymotrypsin on immobilized aminocaproylproflavin using a gradient (0-0.08 mol/l) of sodium chloride[123], fractionation of *Escherichia coli* B aminoacyl-tRNA synthetases on immobilized hexylamine using a gradient (0-0.4 mol/l) of potassium chloride[124], purification of alkaline phosphatase on immobilized triazine dyes using a gradient (0-1 mol/l) of potassium chloride[125], fractionation of a complex mixture of dehydrogenases on immobilized adenosine 5'-monophosphate using a pH gradient (pH 6-10)[126], purification of trypsin and chymotrypsin on immobilized low-molecular-weight soybean protease inhibitors using a pH gradient (pH 8-4)[127] and purification of indolethylamine N-methyltransferase on immobilized S-adenosylhomocysteine[128]. In most of these applications, agarose gels were used as substrates.

Comprehensive information about the various techniques and applications of affinity chromatography can be found in a monograph by Turková[129].

21.8. REFERENCES

1 I. Kluh, in Z. Deyl, K. Macek and J. Janák (Editors), *Liquid Column Chromatography*, Elsevier, Amsterdam, 1975, Ch. 34.
2 Z. Prusík, Ref. 1, Ch. 35.
3 O. Mikeš, Ref. 1, Ch. 36.

4 A. Castel, J.M. de Graaf and B. Kraal, *J. Chromatogr.*, 174 (1979) 474.
5 P. Johnson, *J. Chromatogr. Sci.*, 17 (1979) 406.
6 R.A. Bradshaw, O.J. Bates and J.R. Benson, *J. Chromatogr.*, 187 (1980) 27.
7 F. Weber and W. Grosch, *Z. Lebensm.-Unters.-Forsch.*, 167 (1978) 87.
8 T.D. Schlabach and S.R. Abbott, *Clin. Chem.*, 26 (1980) 1504.
9 E.A. Peterson and H.A. Sober, *J. Amer. Chem. Soc.*, 78 (1956) 751.
10 J.M. Gillespie, *J. Chromatogr.*, 72 (1972) 319.
11 H. Determann, N. Mayer and T. Wieland, *Nature (London)*, 223 (1969) 499.
12 J. Wagner and H.J. Duck, *J. Chromatogr.*, 66 (1972) 67.
13 K. James and D.R. Stanworth, *J. Chromatogr.*, 15 (1964) 324.
14 Z. Hrkal, Z. Vodrážka and J. Rejnková, *J. Chromatogr.*, 72 (1972) 198.
15 J.J. Marshall, *J. Chromatogr.*, 71 (1972) 367.
16 D.B. Ramsden and L.N. Louis, *J. Chromatogr.*, 84 (1973) 87.
17 R.L. Munier, A.M. Drapier and B. Faivre, *Chromatographia*, 6 (1973) 305.
18 J.J. Marshall, *J. Chromatogr.*, 76 (1973) 257.
19 R.B. Ellis, J.U. Ikonne and P.K. Massan, *Anal. Biochem.*, 63 (1975) 5.
20 *DEAE-Sephacel*, Pharmacia, Uppsala, Sweden, 1977.
21 W.G. Edwards and C.A. Wellington, *J. Chromatogr.*, 135 (1977) 463.
22 *Separation News*, 6 (1979), Pharmacia, Uppsala, Sweden.
23 D. Berger and B. Gilliard, *J. Chromatogr.*, 210 (1981) 33.
24 J. Hladík and J. Káš, *J. Chromatogr.*, 75 (1973) 117.
25 *Separation News*, 3 (1973), Pharmacia, Uppsala, Sweden.
26 *Separation News*, 1 (1976), Pharmacia, Uppsala, Sweden.
27 M. Pace and J.E. Dixon, *J. Chromatogr.*, 147 (1978) 521.
28 P. Sörup, *J. Chromatogr.*, 160 (1978) 221.
29 P. Cammarata and C. Ceccarini, *J. Chromatogr.*, 168 (1979) 395.
30 J.G.R. Hurrell, R.J. Fleming and M.T.W. Hearn, *J. Liq. Chromatogr.*, 3 (1980) 473.
31 S. Moore, *J. Chromatogr.*, 159 (1978) 3.
32 *Separation News*, 1 (1977), Pharmacia, Uppsala, Sweden.
33 *DEAE-Sepharose CL-6B, CM-Sepharose CL-6B for Ion Exchange Chromatography*, Pharmacia, Uppsala, Sweden, 1976.
34 O. Mikeš, P. Štrop, J. Zbrožek and J. Čoupek, *J. Chromatogr.*, 119 (1976) 339.
35 O. Mikeš, P. Štrop and J. Sedláčková, *J. Chromatogr.*, 148 (1978) 237.
36 O. Mikeš, *Int. J. Peptide Protein Res.*, 14 (1979) 393.
37 O. Mikeš, P. Štrop, J. Zbrožek and J. Čoupek, *J. Chromatogr.*, 180 (1979) 17.
38 O. Mikeš, P. Štrop, M. Smrž and J. Čoupek, *J. Chromatogr.*, 192 (1980) 159.
39 O. Mikeš, J. Sedláčková, L. Rexová-Benková and J. Omelková, *J. Chromatogr.*, 207 (1981) 99.
40 J.M. Mozersky, J.D. Pettinati and S.D. Kolman, *J. Chromatogr.*, 65 (1972) 387.
41 P. Appel, E.G. Richards and H.S. Olcott, *J. Chromatogr.*, 76 (1973) 351.
42 H. Blüthmann, *J. Chromatogr.*, 137 (1977) 222.
43 S.-H. Chang, R. Noel and F.E. Regnier, *Anal. Chem.*, 48 (1976) 1839.
44 S.-H. Chang, K.M. Gooding and F.E. Regnier, *J. Chromatogr.*, 125 (1976) 103.
45 P.J. Kudirka, R.R. Schroeder, T.E. Hewitt and E.C. Toren, *Clin. Chem.*, 22 (1976) 471.
46 R.R. Schroeder, P.J. Kudirka and E.C. Toren, *J. Chromatogr.*, 134 (1977) 83.
47 T.D. Schlabach, S.-H. Chang, K.M. Gooding and F.E. Regnier, *J. Chromatogr.*, 134 (1977) 91.
48 T.D. Schlabach and F.E. Regnier, *J. Chromatogr.*, 158 (1978) 349.
49 F.H. Bissett, *J. Chromatogr.*, 178 (1979) 515.
50 J.A. Fulton, T.D. Schlabach, J.E. Kerl and E.C. Toren, *J. Chromatogr.*, 175 (1979) 283.
51 T.D. Schlabach, J.A. Fulton, P.B. Mockridge and E.C. Toren, *Anal. Chem.*, 52 (1980) 729.
52 R.W. Stout, A.R. Cashmore, J.K. Coward, C.G. Horvath and J.R. Bertino, *Anal. Biochem.*, 71 (1976) 119.
53 A.J. Alpert and F.E. Regnier, *J. Chromatogr.*, 185 (1979) 375.
54 M. Dizdaroglu and M.G. Simic, *J. Chromatogr.*, 195 (1980) 119.

55 R. Wood, L. Cummings and T. Jupille, *J. Chromatogr. Sci.*, 18 (1980) 551.
56 L.A.A. Sluyterman and O. Elgersma, *J. Chromatogr.*, 150 (1978) 17.
57 L.A.A. Sluyterman and J. Wijdenes, *J. Chromatogr.*, 150 (1978) 31.
58 L.A.A. Sluyterman and J. Wijdenes, *J. Chromatogr.*, 206 (1981) 429 and 441.
59 *Chromatofocusing*, Pharmacia, Uppsala, Sweden, 1981.
60 J.P. Emond and M. Pagé, *J. Chromatogr.*, 200 (1980) 57.
61 L. Bachner, J.-P. Boissel and H. Wajcman, *J. Chromatogr.*, 193 (1980) 491.
62 J. Michelot, D. Godeneche, J.C. Maurizis and G. Meyniel, *J. Chromatogr.*, 188 (1980) 431.
63 T.P.A. Kruck and D.M. Logan, *J. Chromatogr.*, 206 (1981) 368.
64 T.F. Gabriel, J. Michalewsky and J. Meienhofer, *J. Chromatogr.*, 129 (1976) 287.
65 Z. Deyl, *J. Chromatogr.*, 127 (1976) 91.
66 W. Kullmann, *J. Liq. Chromatogr.*, 4 (1981) 1121.
67 N. McMillan, J. Cecka, L. Hood, D.B. Murphy and H. McDewitt, *Nature (London)*, 277 (1979) 663.
68 J.D. Pearson, W.C. Mahoney, M.A. Hermodson and F.E. Regnier, *J. Chromatogr.*, 207 (1981) 325.
69 S. Hjertén, *Advan. Chromatogr.*, 19 (1981) 111.
70 S. Hjertén, J. Rosengren and S. Påhlman, *J. Chromatogr.*, 101 (1974) 281.
71 L.J. Fischer, R.L. Thies and D. Charkowski, *Anal. Chem.*, 50 (1978) 2143.
72 P. Štrop, F. Mikeš and Z. Chytilová, *J. Chromatogr.*, 156 (1978) 239.
73 S. Allenmark and B. Servenius, *J. Chromatogr.*, 153 (1978) 239.
74 *Octyl-Sepharose CL-4B; Phenyl-Sepharose CL-4B for Hydrophobic Interaction Chromatography*, Pharmacia, Uppsala, Sweden, 1976.
75 T. Laas, *J. Chromatogr.*, 111 (1975) 373.
76 J. Hardwick and P. Hechtman, *J. Chromatogr.*, 190 (1980) 385.
77 W.J. Gelsema, C.L. de Ligny, W.M. Blanken, R.J. Hamer, A.M.P. Roozen and J.A. Bakker, *J. Chromatogr.*, 196 (1980) 51.
78 L.K. Creamer and A.R. Matheson, *J. Chromatogr.*, 210 (1981) 105.
79 C. Balck, D.M. Douglas and M.L. Tanzer, *J. Chromatogr.*, 190 (1980) 393.
80 M. Knight, Y. Ito and T.N. Chase, *J. Chromatogr.*, 212 (1981) 356.
81 W.S. Hancock, C.A. Bishop and R.L. Prestidge, *Anal. Biochem.*, 89 (1978) 203.
82 M.T.W. Hearn, *J. Liq. Chromatogr.*, 3 (1980) 1255.
83 M.C. Kemp, W.L. Hollaway, R.L. Prestidge, J.C. Bennett and R.W. Compans, *J. Liq. Chromatogr.*, 4 (1981) 587.
84 C.A. Bishop, W.S. Hancock, S.O. Brennan, R.W. Carrell and M.T.W. Hearn, *J. Liq. Chromatogr.*, 4 (1981) 599.
85 M.T.W. Hearn and B. Grego, *J. Chromatogr.*, 203 (1981) 349.
86 A.A. Hobbs, B. Grego, M.G. Smith and M.T.W. Hearn, *J. Liq. Chromatogr.*, 4 (1981) 651.
87 J.R. McDermott, A.I. Smith, J.A. Biggins, M. Chyad Al-Noaemi and J.A. Edwardson, *J. Chromatogr.*, 222 (1981) 371.
88 C.E. Dunlap, S. Gentleman and L.I. Lowney, *J. Chromatogr.*, 160 (1978) 191.
89 M. Schöneshöfer and A. Fenner, *J. Chromatogr.*, 224 (1981) 472.
90 E.C. Nice, M. Capp and M.J. O'Hare, *J. Chromatogr.*, 185 (1979) 413.
91 J.M. Zanelli, M.J. O'Hare, E.C. Nice and P.H. Corran, *J. Chromatogr.*, 223 (1981) 59.
92 E.C. Nice and M.J. O'Hare, *J. Chromatogr.*, 162 (1979) 401.
93 M.J. O'Hare and E.C. Nice, *J. Chromatogr.*, 171 (1979) 209.
94 F. Erni, K. Krummen and A. Pellet, *Chromatographia*, 12 (1979) 399.
95 B.T. Bush, J.H. Frenz, W.R. Melander, Cs. Horváth, A.R. Cashmore, R.N. Dryer, J.O. Knipe, J.S. Coward and J.R. Bertino, *J. Chromatogr.*, 168 (1979) 343.
96 K. Krummen, *J. Liq. Chromatogr.*, 3 (1980) 1243.
97 A.A. Amoscato, G.F. Babcock and K. Nishioka, *J. Chromatogr.*, 205 (1981) 179.
98 J.L. Meek and Z.L. Rossetti, *J. Chromatogr.*, 211 (1981) 15.
99 A. Harmar, J.G. Schofield and P. Keen, *Nature (London)*, 284 (1980) 267.
100 W. Mönch and W. Dehnen, *J. Chromatogr.*, 140 (1977) 260.
101 W. Mönch and W. Dehnen, *J. Chromatogr.*, 147 (1978) 415.
102 B.N. Jones, R.V. Lewis, S. Pääbo, K. Kojima, S. Kimura and S. Stein, *J. Liq. Chromatogr.*, 3 (1980) 1373.

103 J.B. Wilson, H. Lam, P. Pravatmuang and T.H.J. Huisman, *J. Chromatogr.*, 179 (1979) 271.
104 W.A. Schroeder, J.B. Shelton and J.R. Shelton, *J. Chromatogr.*, 174 (1979) 385.
105 L. Haeffner-Gormley, N.H. Poludniak and D.B. Wetlaufer, *J. Chromatogr.*, 214 (1981) 185.
106 J.P.H. Burbach, J.G. Loeber, J. Verhoef, V.M. Wiegant and E. Dekloet, *Nature (London)*, 283 (1980) 465.
107 J.A.D.M. Tonnaer, J. Verhoef, V.M. Wiegant and W. De Jong, *J. Chromatogr.*, 183 (1980) 303.
108 K.M. Wu, J.W. Sloan and W.R. Martin, *J. Chromatogr.*, 202 (1980) 500.
109 J.F.M. Kinkel, G. Heuver and J.C. Kraak, *Chromatographia*, 13 (1980) 145.
110 P. Böhlen and G. Kleeman, *J. Chromatogr.*, 205 (1981) 65.
111 E.H. Creaser and G.J. Hughes, *J. Chromatogr.*, 144 (1977) 69.
112 P.W. Lambert and B.A. Roos, *J. Chromatogr.*, 198 (1980) 293.
113 N.G. Seidah, R. Routhier, S. Benjannet, N. Larivière, F. Gossard and M. Chrétien, *J. Chromatogr.*, 193 (1980) 291.
114 D.M. Desiderio, J.L. Stein, M.O. Cunningham and J.Z. Sabbatini, *J. Chromatogr.*, 195 (1980) 369.
115 W.S. Hancock, J.D. Capra, W.A. Bradley and J.T. Sparrow, *J. Chromatogr.*, 206 (1981) 59.
116 W.S. Hancock and J.T. Sparrow, *J. Chromatogr.*, 206 (1981) 71.
117 M.T.W. Hearn, B. Grego and W.S. Hancock, *J. Chromatogr.*, 185 (1979) 429.
118 H.P.J. Bennett, C.A. Browne and S. Solomon, *J. Liq. Chromatogr.*, 3 (1980) 1353.
119 F.E. Regnier, *VIA*, 14, No. 2 (1980) 12, Varian, Palo Alto, CA.
120 R. Axén, J. Porath and S. Ernback, *Nature (London)*, 214 (1967) 1302.
121 H.J. Böhme, G. Kopperschläger, J. Schulz and E. Hofmann, *J. Chromatogr.*, 69 (1972) 209.
122 K. Kojima, T. Hama, T. Kato and T. Nagatsu, *J. Chromatogr.*, 189 (1980) 233.
123 J.H. Brantner, R.G. Medicus and R.A. McRorie, *J. Chromatogr.*, 129 (1976) 97.
124 H. Jakubowski, *J. Chromatogr.*, 139 (1977) 331.
125 V. Bouriotis and P.D.G. Dean, *J. Chromatogr.*, 206 (1981) 521.
126 C.R. Lowe, M.J. Harvey and P.D.G. Dean, *Eur. J. Biochem.*, 41 (1974) 347.
127 K.D. Lin, D.L. Hwang and D.E. Foard, *J. Chromatogr.*, 195 (1980) 385.
128 J.P.G. Mack and M.B. Slaytor, *J. Chromatogr.*, 157 (1978) 153.
129 J. Turková, *Affinity Chromatography*, Elsevier, Amsterdam, 1978.
130 C.Y. Lee, S.L. Chang, S.T. Kau and Shing-Hui Luh, *J. Chromatogr.*, 72 (1972) 71.
131 T. Gotoh, T. Ochiai and K. Shikama, *J. Chromatogr.*, 60 (1971) 260.
132 N. Kelemen, F. Lasmoles and G. Milhaud, *J. Chromatogr.*, 177 (1979) 271.
133 F. Lottspeich, A. Henschen and K.-P. Hupe (Editors), *High Performance Liquid Chromatography in Protein and Peptide Chemistry*, De Gruyter, Berlin, 1981.

Chapter 22

NUCLEIC ACIDS AND THEIR CONSTITUENTS

Nucleic acids, the fundamental genetic material of living organisms, are structurally based on several simple heterocyclic compounds of two main types: (a) pyrimidines and (b) purines. The main individual bases, (a) cytosine, uracil and thymine and (b) adenine, guanine, hypoxanthine and xanthine, differ from one another in the substitution of the pyrimidine or purine rings. These bases and some their alkyl derivatives are bound with sugars (ribose or deoxyribose) in molecules of nucleosides. Phosphorylated nucleoside derivatives (nucleotides) with one, two or three phosphoric acid groups are combined in a given sequence in polymeric nucleic acids. The individual nucleic acids differ in the number and arrangement of nucleotides in their molecules. Deoxyribonucleic acids (DNAs) consist essentially of a long, non-branching deoxyribosophosphate chain with four fundamental nitrogenous bases (adenine, guanine, cytosine and thymine) while ribonucleic acids (RNA) contain ribose instead of deoxyribose and the individual RNAs differ more distinctly in the type and arrangement of nitrogenous bases and in molecular size than DNAs (rRNA with molecular weight between $0.5 \cdot 10^6$ and $2 \cdot 10^6$ daltons, tRNA with molecular weight ca. $2.5 \cdot 10^4$ daltons and mRNA with mean molecular weight 10^5 daltons). Nucleic acids, oligonucleotides and nucleotides behave as weak acids in aqueous solutions because of the presence of sugar-phosphate moieties in the molecules, while the bases and nucleosides have a very weakly basic character.

The problems connected with the analysis of nucleic acids and their constituents are similar to those involved in the analysis of proteins, peptides and amino acids. Liquid chromatographic methods used for the purification and separation of the individual nucleic acids are similar to those used for the purification and separation of individual proteins. However, the techniques used for the separation of individual bases, nucleosides and nucleotides by liquid chromatography differ from the methods used for liquid chromatographic separations of amino acids. The quantitation of nucleic acids constituents is very important for the determination of the composition of and sequence in the individual nucleic acids.

22.1. NUCLEIC ACIDS

Liquid chromatographic methods for the fractionation and purification of nucleic acids using gradient elution can be classified according to the column

References on p. 390.

packing materials and separation mechanism. Chromatography on hydroxyapatite columns, columns packed with diatomaceous earth impregnated with different natural proteins and synthetic polyamino acids, columns packed with organic exchangers and with various modified and non-modified gels, and affinity and reversed-phase chromatography were tested and utilized for this purpose in connection with various salt and buffer gradients. These individual separation techniques have been reviewed in detail[1-3].

22.1.1. Chromatography on hydroxyapatite columns

In chromatography on hydroxyapatite columns, the retention of nucleic acids is attributed to the interactions between the negatively charged phosphate groups of the nucleic acids and the positively charged Ca^{2+} ions from the surface of the hydroxyapatite crystal. The strength of retention depends both on the molecular size and on the three-dimensional macromolecular conformation of nucleic acids or their fragments, which allows their purification and fractionation. The elution is performed with buffer solutions in the pH range 5.4-7.8, usually at pH 6.8. The elution is based on the competition between buffer anions and the phsophate groups of nucleic acids for Ca^{2+} ions of hydroxyapatite, and therefore an increase in the molarity of the buffers decreases the retention. The eluting buffer should not change significantly the characteristics of the nucleic acid molecules and should have an acceptable buffering capacity. Phosphate buffers are most suitable in this respect and continuous or stepwise gradients of increasing buffer molarities are most frequently used for elution[2-7]. Sometimes the resolution is improved by applying a simultaneous temperature gradient with a gradient of salt concentration.

22.1.2. Chromatography on diatomaceous earth impregnated with polyamino acids

Diatomaceous earth impregnated with polyamino acids has often been used as a column packing material for the fractionation of nucleic acids. The separation mechanism is based on the particular affinity of certain parts of nucleic acids for polyamino acids. Polylysine[8-12], polyarginine[11,12], polyornithine[11,12], methylalbumin[13], histone[14], etc., have been used for impregnation and elution has been performed with a linear gradient of increasing concentration of sodium chloride, usually in phosphate, citrate or Tris-HCl buffers at pH 7 ± 0.3.

22.1.3. Chromatography on cellulose, dextran and chemically bonded anion exchangers

Chromatography on hydroxyapatite and polyamino acid-diatomaceous earth columns is the best method for separating DNA from RNA and higher oligonucleotides, for removing proteins and for separating native and denatured DNA. Chromatography on anion-exchange columns results in fractionation into many peaks, usually not all of which are fully characterized. DEAE-cellulose, DEAE-Sephadex and their derivatives (benzoyldiethylaminoethylcellulose) are more suitable than resinous exchangers for fractionation of nucleic acids because of the fragile character of biopolymers. Stepwise[15] or continuous[16,18,19] gradients of sodium chloride in a suitable buffer (containing magnesium chloride and combined with a pH or an organic solvent concentration gradient) are used for elution. A gradient of urea can be also used[19] (in concentrated aqueous solutions of urea, other effects are suppressed and the separation is controlled essentially only by the number of charge-carrying groups in the molecules of solutes).

More recently, anion exchangers chemically bonded on a silica gel support (10 μm) have been tested with similar elution systems to speed up the analysis[17].

22.1.4. Chromatography on unmodified and modified dextran, agarose and polyamide gels

Unmodified dextran, polyamide and composite dextran-polyamide gels posses affinities for certain nucleic acids. The chromatographic mechanism has not yet been fully explained, but possibly hydrophobic effects are combined with specific interactions between parts of nucleic acids and some functional groups (such as hydroxy groups) of the gels, as is indicated by the fact that a specific structure of the gel is required for successful separations (for example, Sepharose 2B and 6B and Ultragel AcA 44 are much better suited than Sepharose 2B and Cl-4B for the separation of mixed yeast tRNAs). On the other hand, steric exclusion effects appear to be of only minor importance. As the retention of nucleic acids increases with increasing salt concentration in the mobile phase (hydrophobic effect), linear gradients of decreasing salt concentration in an appropriate buffer are suitable for elution, such as a gradient from 2 to 0 mol/l ammonium sulphate in acetate buffer (pH 4.25)[20,21]. However, linear gradients of increasing salt (sodium chloride) concentration have also been applied for the fractionation of nucleic acids[22]. To enhance the influence of the hydrophobic effect on separation, hydrophobic derivatives of agarose gels can be employed; for example, naphthoyl-Sepharose Cl-4B in connection with a gradient of decreasing ammonium sulphate

References on p. 390.

concentration (form 0.75 to 0 mol/l) in 0.01 mol/l phosphate buffer at pH 6.8 was used for the fractionation of tRNA from *Escherichia coli*[23].

22.1.5. *Affinity chromatography*

Affinity chromatography on agarose gels as supports with immobilized lysine[24], spermine[25,26], hexylamine[27], heparin[28] or DNA[29] using linear gradients of increasing concentrations of sodium, potassium or magnesium chloride (usually in Tris-HCl buffer at pH 7.5-8) for elution has been employed for the fractionation of different species of rRNA and tRNA.

22.1.6. *High-speed reversed-phase chromatography*

Modern column packing materials for reversed-phase chromatography allow rapid separations of various DNA and RNA species, usually in a final step after a previous separation using some of the above-described chromatographic methods. Such separations were accomplished on columns packed with RPC-5 resin (small particle diameter inert polychlorotrifluoroethylene beads) using elution with a linear gradient of increasing sodium chloride concentration in 0.01 mol/l acetate buffer (pH 4.5) containing 0.01 mol/l magnesium chloride and 0.001 mol/l β-mercaptoethanol[18,30]. Rapid separations have also been achieved on octadecyl-silica (5-10 μm) columns using linear gradients of methanol in water[7] or aqueous ammonium formate[31].

22.2. NUCLEIC ACID CONSTITUENTS

To determine the content and sequence of constituents in nucleic acids, the acids should first be subjected to hydrolysis or enzymic degradation, which would yield, according to the reaction conditions, oligonucleotides, nucleotides, nucleosides or bases. These constituents, either from nucleic acid hydrolysates or from reaction mixtures of organic synthesis, have frequently been chromatographed on ion-exchange columns. Cation exchangers are preferred for chromatography of nucleosides and bases, whereas nucleotides are usually chromatographed on anion exchangers. Detailed reviews of these methods may be found elsewhere[32,33].

22.2.1. *Ion-exchange chromatography of nucleosides and bases*

Nucleosides and bases have been separated on columns packed with styrene-divinylbenzene sulphonated cation-exchange resins and the analysis time was decreased in certain applications by using continuous or stepwise gradient elution with ionic strength or pH gradients in buffered mobile phases, such as using a linear gradient of increasing concentration (from 0.3 to 1 mol/l) of citrate buffer at pH 5 for the chromatography of purines and pyrimidines in urine on a column packed with Chromobeads A (600 mesh)[34]. Chromatography on strongly basic fine-particle styrene-divinylbenzene anion-exchange resins employing gradient elution with a changing concentration (from 0.15 to 3 mol/l) of acetate buffer at pH 4.4 has also been tested for the separation of bases and nucleosides[35,36].

22.2.2. *Chromatography of nucleotides on anion-exchange resins*

Because nucleotide mixtures are usually more complex than samples containing only bases and nucleosides, the application of gradient elution is more useful here than in the chromatography of nucleosides and bases. Ionic strength gradients are generally more popular than pH gradients, because the retention of nucleotides on anion exchangers is more influenced by changes in ionic strength than by an increase or decrease in pH[32].

In the chromatography of nucleotides and complex mixtures of bases, nucleosides and nucleotides on anion-exchange styrene-divinylbenzene resins, stepwise or continuous gradients of increasing concentrations of hydrochloric acid[37,38] or sodium chloride in buffered solutions[39], gradients of increasing concentrations of formic acid-sodium formate[40-45], acetate[46-48], citrate[49,50] or borate[51] buffers, and gradients of increasing ammonium carbonate concentration[52] have been employed for elution.

22.2.3. *Chromatography of oligonucleotides on cellulose and dextran anion exchangers*

Oligonucleotide mixtures have been successfully separated on DEAE-cellulose columns. The elution order was in agreement with increasing lengths of the oligonucleotide chains and the elution employed an increasing concentration gradient of hydrogen carbonate-carbonate buffer[53], linear gradients of ammonium acetate[54] and ammonium formate[54], salt gradients in the presence of 7 *M* urea[54-56] (where principally only the number of charged groups in the solutes controls the retention) and a gradient of triethylammonium acetate in an aqueous solution of methanol and isopropanol[57]. Gradient elution with an increasing concentration of sodium chloride in acetate buffer was used for chromatography of oligonucleotides on DEAE-Spheron 300[58]. Oligonucleotides were also separated on DEAE-Sephadex[59,61-63] and QAE-Sephadex[60,64] columns using elution with a linear gradient of sodium chloride in a Tris-HCl buffer in either the presence[59,62,63] or absence[60,64] of 7 mol/l urea or with a linear gradient of ammonium hydrogen carbonate[61].

Chromatography on QAE-Sephadex[65-67] and DEAE-Sephadex[68-72] columns with gradients of increasing sodium chloride concentration in Tris-HCl[65,66] or in sodium carbonate-hydrogen carbonate[67] buffers or gradients of hydrogen carbonate in water[68-72] have also been employed for the separation of mononucleotides[65-70], cyclic nucleotides[71] and pharmacologically important thiopurine nucleotides[72].

Separations of nucleotides on classical ion-exchange resins and on cellulose and dextran ion exchangers are usually time consuming (with the exception of the use of short columns packed with very fine particles).

22.2.4. *Chromatography of nucleotides on pellicular anion exchangers*

The introduction of pellicular anion exchangers brought major improvements in both the resolution and the speed of ion-exchange chromatography of nucleotides[73]. Chromatography on columns packed with strong pellicular anion exchangers in connection with linear gradients of potassium dihydrogen phosphate buffers in water allowed relatively high-speed separations of nucleotides to be achieved, such as of common 2'- and 3'-ribonucleotides[73], mono-, di- and triphosphates of ribonucleosides[74-85], 5'-ribonucleotides, 5'-deoxynucleotides[76], dinucleotides[76,79] and cyclic nucleotides[82] in various biological samples. Continuous UV detection at 254 nm allowed analyses to be performed at picomole levels. The success of this method led to the construction of special-purpose gradient elution liquid chromatographs (nucleotide analysers), such as the Varian LCS 1000. In some applications, a gradient of phosphate buffer was combined with a gradient of potassium chloride, because of the stronger affinity of Cl^- ions for the anion exchanger[75,78,83-85]. Gradients of formate buffer[73] yield changed selectivities for certain nucleotide pairs, but these buffers may corrode stainless-steel parts of the instruments. Acetate buffer gradients were applied in the chromatography of 3'- and 5'-cyclic ribonucelotides[86] and a sulphate gradient was employed for the separation of oligonucleotides[87]. Operation at elevated temperatures (50-80°C) is common, as in the chromatography of nucleotides on classical ion-exchange resins. For more detailed reviews of the chromatography of nucleotides on pellicular anion exchangers, see for example refs. 32, 76 and 81.

The gradients used in work with pellicular anion exchangers are also employed in the chromatography of nucleotides and oligonucleotides on porous-layer chemically bonded anion exchangers, such as Permaphase AAX, e.g., gradients of potassium dihydrogen phosphate buffers[88,89] or of potassium chloride in phosphate buffers[90].

22.2.5. *Chromatography of nucleotides on microparticulate chemically bonded anion exchangers*

Linear[91,93], concave[94] and multi-linear[95] gradients of phosphate buffer concentration or linear gradients of potassium chloride concentration in phosphate buffers[96-98] have also been employed for the chromatography of nucleotides[91-94, 96,98], oligonucleotides[97] and nucleotide-nucleoside mixtures[95] on totally porous chemically bonded anion exchangers on a silica gel (5-10 μm) support (Partisil-10 SAX, LiChrosorb AN, Whatman SAX-10). An example of the separation of a mixture of nucleoside 5'-mono-, di- and triphosphates by gradient elution chromatography on a chemically bonded anion-exchange column is shown in Fig. 22.1[93]. Phosphate buffer gradients in mobile phases containing magnesium ions were tested for the chromatography of nucleosides and nucleotides to study the influence of the formation nucleotide-magnesium(II) complexes on chromatographic separations on chemically bonded totally porous microparticulate anion exchangers[99]. A combination of gradients of phosphate buffer and acetonitrile was also employed for the chromatographic separation of nucleotides, nucleosides and bases[93,100] and gradients of triethylammonium acetate were used for the separation of oligonucleotide mixtures[101-103] on this type of anion-exchange column.

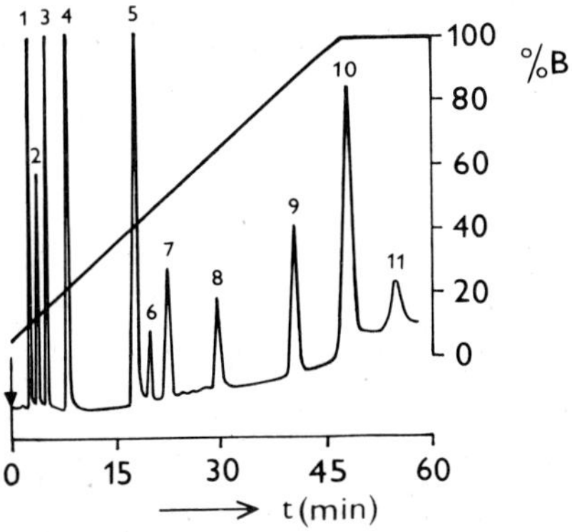

Fig. 22.1. Separation of nucleoside 5'-mono-, di- and triphosphates by gradient elution chromatography on a column packed with a chemically bonded anion exchanger. Column: Micro Pak AX-10, 10 μm (weak anion-exchange bonded phase on LiChrosorb Si 60 silica support), 300x4 mm I.D. Gradient: linear, 5-100% B in 47.5 min (2% B/min). Solvents: A, 0.01 mol/l KH_2PO_4 (pH 2.85); B, 0.75 mol/l KH_2PO_4 (pH 4.4). Flow-rate: 2 ml/min. Detection: UV at 260 nm; 0.5 and 0.2 a.u.f.s. Peaks: 1 = cytidine monophosphate; 2 = adenosine monophosphate; 3 = uridine monophosphate; 4 = guanosine monophosphate; 5 = cytidine diphosphate; 6 = uridine diphosphate; 7 = adenosine diphosphate; 8 = guanosine diphosphate; 9 = cytidine triphosphate; 10 = adenosine triphosphate; 11 = guanosine triphosphate. (Reproduced from ref. 93 with permission.)

References on p. 390.

22.2.6. Reversed-phase chromatography of nucleotides, nucleosides and bases on chemically bonded phases and on organic resins

The chromatography of nucleotides, nucleosides and bases on chemically bonded non-polar alkylsilica phases has become very popular in recent years and can compete successfully in resolution and speed of separation with chromatography on pellicular and microparticulate chemically bonded anion exchangers. Octadecylsilica[97,104-114,116-119] and, less frequently, octylsilica[97,115] columns have been used for this purpose. Elution is usually performed with a gradient of organic solvent concentration in an aqueous buffer solution. Here again, phosphate buffers (pH 5-7) are most frequently used as the starting eluent[97,104-117] in connection with linear (less frequently concave[97,116] or multi-linear[117]) gradients of increasing methanol concentration for the chromatography of nucleosides[104-108,110,111,113], bases[104-108,111-113,115], alkylated bases[114], mixtures of nucleotides and nucleosides[104-108,113,116], cyclic nucleotides[109] and oligonucleotides[97,117]. A linear gradient of methanol concentration in formate buffer was applied to the separation of bases, nucleosides and their methylated derivatives[118] and a linear methanol gradient in acetate buffer for the chromatography of some nucleosides[119]. Less frequently, gradients of methanol, acetonitrile or ethanol in water without the addition of a buffer have been used for the chromatography of bases and their alkyl derivatives[120], nucleosides[123], carcinogenic nucleoside conjugates with polyaromatic hydrocarbons[121,122] or zeatin and related cytokinins[134,135] in cellular extracts and tRNA hydrolysates.

A typical example of the reversed-phase gradient elution chromatographic separation of nucleotides, nucleosides and bases in a mixture is shown in Fig. 22.2[108].

Ion-pair chromatography on octadecylsilica columns has also been successfully applied to the chromatography of purine and pyrimidine bases[124], nucleosides[124], nucleotides[124,126] and oligonucleotides[126].

Linear gradients of methanol[124,125] or acetonitrile[126] in solutions containing tetrabutylammonium phosphate[124,126] or sulphate[125] as ion-pairing reagents in aqueous medium, sometimes with addition of a salt (ammonium chloride)[125] or acetic acid[124], have been used for elution. A linear gradient of methanol in 0.1 mol/l triethylammonium acetate was used as the eluent in the chromatography of oligonucleotides on octadecylsilica column[127].

Elution with linear gradients of triethylammonium hydrogen carbonate was employed for the separation of mixtures containing bases, nucleosides and nucleotides[128] and for the resolution of 3':5'- and 2':3'-cyclic ribonucleotides from their related mononucleotides[129] by reversed-phase chromatography on column packed with Amberlite XAD-4 (a styrene-divinylbenzene copolymer).

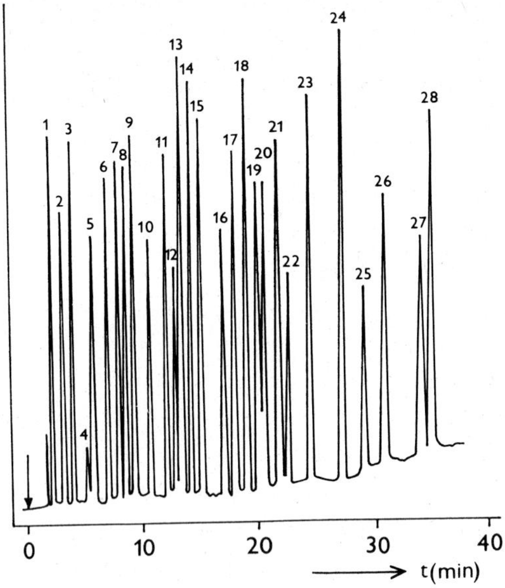

Fig. 22.2. Separation of 0.1-0.5 nmol of nucleosides, bases, nucleotides, aromatic amino acids and metabolites by reversed-phase gradient elution chromatography. Column: C18 on 10 μm totally porous silica, 300x4.6 mm I.D. Gradient: linear, 0-100% B in 87 min (1.15% B/min). Solvents: A, 0.02 mol/l KH_2PO_4 (pH 5.6); B, 60% (v/v) methanol in water. Flow-rate 1.5 ml/min. Detection: UV at 254 nm. Peaks: 1 = cytosine; 2 = orotidine; 3 = uracil; 4 = *l*-tyrosine; 5 = cytidine; 6 = hypoxanthine; 7 = uridine; 8 = 5-aminoimidazolocarboxamide riboside; 9 = 7-methylinosine; 10 = 7-methylxanthosine; 11 = 7-methylguanosine; 12 = β-nicotinamide adenine dinucleotide; 13 = inosine; 14 = guanosine; 15 = 2'-deoxyinosine; 16 = 2'-deoxythimidine; 17 = 1-methylinosine; 18 = N_1-methylguanosine; 19 = N_2-methylguanosine; 20 = kynurenic acid; 21 = adenosine; 22 = theobromine; 23 = N_1,N_2-dimethylguanosine; 24 = theophylline; 25 = dyphilline; 26 = 6-methyladenosine; 27 = indole-3-propionic acid; 28 = caffeine. (Reproduced from ref. 108 with permission.)

A column packed with a bonded amino phase was employed for the chromatography of nucleotides and oligonucleotides using elution with a linear gradient of potassium dihydrogen phosphate in water[130].

UV detection at 254 nm is well suited for the chromatography of nucleotides, nucleosides and bases on modern chemically bonded phases and is commonly employed. However, simultaneous monitoring of UV absorption and radioactivity is useful in investigating the metabolic pathways of labelled nucleic acid constituents[113]. Simultaneous monitoring of UV absorption at two wavelengths (254 and 280 nm) or of UV absorption and of fluorescence may help in identifying some nucleosides and bases, in particular if the stopped-flow spectral scanning technique is employed[105,115].

The so-called enzymatic peak-shift technique is also helpful in the identification (and in achieving better resolution and quantitation of poorly resolved

References on p. 390.

peaks) of nucleotides. Here, the specificity of enzyme reactions with a nucleotide or with a class of nucleotides is utilized. Chromatograms of nucleotide mixtures before and after reaction with a specific enzyme (such as hexokinase for reaction of adenosine triphosphate to adenosine diphosphate) are compared and the peaks of the respective nucleotide (or peaks of a group of nucleotides) is identified by the change in its relative position on the chromatograms. For more details, see ref. 76.

22.2.7. *Normal-phase chromatography of nucleic acid constituents*

Normal-phase chromatography is rarely used in the separation of nucleic acid constituents, but the separation of phosphoinositide nucleotides has been reported, using a column packed with silica gel impregnated with oxalate solution as the stationary phase and a linear gradient of ammonia in chloroform-methanol-water[131].

22.2.8. *Chromatography of oligonucleotides in other systems*

In the chromatography of oligonucleotides, analogous methods to those for nucleic acid separations may be used; for example, chromatography on hydroxyapatite columns using phosphate buffer concentration gradients[132] or affinity chromatography using elution with buffer concentration gradients[133].

22.3. REFERENCES

1 S. Zadražil, in Z. Deyl, K. Macek and J. Janák (Editors), *Liquid Column Chromatography, A Survey of Modern Techniques and Applications*, Elsevier, Amsterdam, 1975, Ch. 38.
2 R.M. Kothari and V. Schankar, *Chromatogr. Rev.*, 98 (1974) 449.
3 G. Bernardi, *Methods Enzymol.*, 21 (1971) 95.
4 A.L. Mazin and G.E. Sulimova, *Biokhimia*, 40 (1975) 115.
5 M. Spencer, *J. Chromatogr.*, 166 (1978) 435.
6 M. Spencer, E.J. Neave and N.L. Webb, *J. Chromatogr.*, 166 (1978) 447.
7 F.A. Beland, K.L. Dooley and D.A. Casciano, *J. Chromatogr.*, 174 (1979) 177.
8 S.R. Ayad and J. Blamire, *J. Chromatogr.*, 48 (1970) 456.
9 S.R. Ayad, P. Premachandra and R.P.F. Gregory, *J. Chromatogr.*, 58 (1971) 235.
10 S.R. Ayad and J.G. Delinassions, *J. Chromatogr.*, 85 (1973) 162; 130 (1977) 396.
11 C.W. Helleiner, *J. Chromatogr.*, 72 (1972) 400.
12 D. Jarvis, R. Loeser, P. Herrlich and R. Röschenthaler, *J. Chromatogr.*, 52 (1970) 158.
13 J.P. Garel, G. Nullans and P. Mandel, *J. Chromatogr.*, 56 (1971) 154.
14 S.R. Ayad and A.E. Wilkinson, *J. Chromatogr.*, 66 (1972) 277.
15 F. Millard and B.W. Fox, *J. Chromatogr.*, 107 (1975) 125.
16 S.J. Morris, *J. Chromatogr.*, 82 (1973) 349.
17 C.T. Wehr and S.R. Abbott, *J. Chromatogr.*, 185 (1979) 453.
18 R. Patnaik, R.M. Kothari and M.W. Taylor, *J. Chromatogr.*, 97 (1974) 193.
19 J.D. Cherayil and H.M. Bock, *Biochemistry*, 4 (1965) 1174.

20 C.J.O.R. Morris, *J. Chromatogr.*, 159 (1978) 33.
21 G.W. Hatfield, *Methods Enzymol.*, 59 (1979) 215.
22 S. Zadražil, J. Šatava and Z. Šormová, *J. Chromatogr.*, 91 (1974) 451.
23 S. Hjertén, *Advan. Chromatogr.*, 19 (1981) Ch. 4, p.111.
24 D.S. Jones, H.K. Lundgren and F.T. Jay, *Nucleic Acids Res.*, 3 (1976) 1569.
25 G. Seibert and R.K. Zahn, *J. Chromatogr.*, 161 (1978) 324.
26 A. Joachimiak, M. Barciszewska, J. Barciszewski and M. Wiewidrowski, *J. Chromatogr.*, 180 (1979) 157.
27 T. Dzieglielewski and H. Jakubowski, *J. Chromatogr.*, 103 (1975) 364.
28 A. Joachimiak, T. Zwierzyński, D. Radocki, M. Barciszewska and J. Barciszewski, *J. Chromatogr.*, 206 (1981) 600.
29 C. Nüsslein and B. Heyden, *Biochem. Biophys. Res. Commun.*, 47 (1972) 282.
30 R. Dion and R.J. Cedergren, *J. Chromatogr.*, 152 (1978) 131.
31 R.F. Newbold, W. Warren, A.S.C. Medcalf and J. Amos, *Nature (London)*, 283 (1980) 596.
32 P.R. Brown, *High Pressure Liquid Chromatography - Biochemical and Biomedical Applications*, Academic Press, New York, 1973.
33 S. Zadražil, in Z. Deyl, K. Macek and J. Janák (Editors), *Liquid Column Chromatography, A Survey of Modern Techniques and Applications*, Elsevier, Amsterdam, 1975, Ch. 37.
34 P.V. Byrne and I.V. Chapman, *J. Chromatogr.*, 88 (1974) 190.
35 N.G. Anderson, J.G. Green, M.L. Barber and F.C. Ladd, *Anal. Biochem.*, 6 (1963) 153.
36 M. Hori and E. Konishi, *J. Biochem.*, 56 (1964) 375.
37 T. Shimizu, F. Uda, H. Matsumiya, K. Miyauchi and S. Ganne, *Bunseki Kagaku (Jap. Anal.)*, 18 (1969) 632.
38 Dar-San Hsu and S.S. Chen, *J. Chromatogr.*, 192 (1980) 193.
39 A. Floridi, C.A. Palmerini and C. Fini, *J. Chromatogr.*, 138 (1977) 203.
40 J. Lerner and A.I. Schepartz, *J. Chromatogr.*, 35 (1968) 37.
41 P. Virkola, *J. Chromatogr.*, 51 (1970) 195.
42 H. Schmitz, *Biochem. Z.*, 325 (1954) 554.
43 S. Zadražil and Z. Šormová, *Collect. Czech. Chem. Commun.*, 26 (1961) 2651.
44 M.R.V. Murthy and P.K. Ranjekar, *J. Chromatogr.*, 76 (1973) 337.
45 C. Bernofsky and W.J. Gallagher, *Anal. Biochem.*, 67 (1975) 611.
46 M. Hori, *Methods Enzymol.*, 12A (1967) 381.
47 P.R. Brown, S. Bobick and F.L. Hanley, *J. Chromatogr.*, 99 (1974) 587.
48 K. Seta, M. Washitake, T. Anmo, N. Takai and T. Okuyama, *J. Liq. Chromatogr.*, 4 (1981) 129.
49 J.X. Khym, *J. Chromatogr.*, 97 (1974) 277.
50 J. Maybaum, F.K. Klein and W. Sadee, *J. Chromatogr.*, 188 (1980) 149.
51 B. Bakay, E. Nissinen and L. Sweetman, *Anal. Biochem.*, 86 (1978) 65.
52 W.E. Cohn and F.J. Bollum, *Biochim. Biophys. Acta*, 48 (1961) 588.
53 M. Staehlin, E.A. Peterson and H.A. Sober, *Arch. Biochem. Biophys.*, 85 (1959) 289.
54 M. Uziel and H. Gassen, *Biochemistry*, 8 (1969) 1643.
55 G.W. Rushizky, J.H. Mozejko, P.C. Woodford and H.A. Sober, *Anal. Biochem.*, 46 (1972) 443.
56 R.V. Tomlinson and G.M. Tener, *Biochemistry*, 2 (1963) 697, 703.
57 H. Köster and R. Frank, *Chromatographia*, 10 (1976) 497.
58 J. Šatava, O. Mikeš and P. Štrop, *J. Chromatogr.*, 180 (1979) 31.
59 G.M. Tener, *Methods Enzymol.*, 12A (1967) 398.
60 H. Schott, *J. Chromatogr.*, 172 (1979) 179.
61 Y. Lapidot and I. Barzilay, *J. Chromatogr.*, 71 (1972) 275.
62 H.R. Matthews, *Eur. J. Biochem.*, 7 (1968) 96.
63 D.A. Lloyd and S. Mandeles, *Biochemistry*, 9 (1970) 932.
64 H. Schott and H. Watzlawick, *J. Chromatogr.*, 196 (1980) 435.
65 H. Schott and H. Schwarz, *J. Chromatogr.*, 157 (1978) 197.
66 H. Schott, *J. Chromatogr.*, 187 (1980) 119.
67 D.G. Kaufman and J.W. Grisham, *J. Chromatogr.*, 57 (1971) 275.

68 W. Loesche, R. Bublitz, A. Horn, W. Koehler, H. Petermann and U. Till, *J. Chromatogr.*, 92 (1974) 166.
69 H.-D. Hunger and H. Reinbothe, *J. Chromatogr.*, 97 (1974) 273.
70 J. Arendes, R.K. Zahn and W.E.G. Müller, *J. Chromatogr.*, 140 (1977) 118.
71 T. Rosett, J.G. Smith, I. Matsuo, P.A. Bailey, D.B. Smith and S. Surakiat, *J. Chromatogr.*, 49 (1970) 308.
72 D.J. Nelson, C.J.L. Bugge, H.C. Krasny and T.P. Zimmerman, *J. Chromatogr.*, 77 (1973) 181.
73 C.G. Horváth, B.A. Preiss and S.R. Lipsky, *Anal. Chem.*, 39 (1967) 1422.
74 H.W. Shmukler, *J. Chromatogr. Sci.*, 8 (1970) 653.
75 P.R. Brown, *J. Chromatogr.*, 52 (1970) 257.
76 C.A. Burtis and D. Gere, *Nucleic Acid Constituents by Liquid Chromatography*, Varian, Palo Alto, CA, 1970.
77 C.A. Burtis, M.N. Munk and F.R. McDonald, *Clin. Chem.*, 16 (1970) 667.
78 P.R. Brown, *J. Chromatogr.*, 57 (1971) 383.
79 W.P. Kennedy and J.C. Lee, *J. Chromatogr.*, 51 (1970) 203.
80 G. Brooker, *Anal. Chem.*, 42 (1970) 1108.
81 C. Horváth in J.A. Marinsky and Y. Marcus (Editors), *Ion Exchange and Solvent Extraction*, Vol. 5, Marcel Dekker, New York, 1973, Ch. 3.
82 W.H. Schmukler, *J. Chromatogr. Sci.*, 10 (1972) 137.
83 D.P. Brenton, K.H. Astrin, M.K. Cruishank and J.E. Seegmiller, *Biochem. Med.*, 17 (1977) 231.
84 J.A. Nelson and L.M. Rose, *J. Chromatogr.*, 106 (1975) 188.
85 D. Perret, *J. Chromatogr.*, 124 (1976) 187.
86 S.P. Dutta, A. Mittelman and G.B. Chheda, *Anal. Biochem.*, 75 (1976) 649.
87 T.F. Gabriel and J.E. Michalewski, *J. Chromatogr.*, 80 (1973) 263.
88 R.A. Henry, J.A. Schmit and R.C. Williams, *J. Chromatogr. Sci.*, 11 (1973) 358.
89 D.R. Baker, R.C. Williams and J.C. Steichen, *J. Chromatogr. Sci.*, 12 (1974) 499.
90 J.H. van Boom and J.F.M. de Rooy, *J. Chromatogr.*, 131 (1977) 169.
91 L.M. Rose and R.W. Brockman, *J. Chromatogr.*, 133 (1977) 335.
92 S.-C. Chen, P.R. Brown and D.M. Rosie, *J. Chromatogr. Sci.*, 15 (1977) 218.
93 E.H. Edelson, J.G. Lawless, C.T. Wehr and S.R. Abbott, *J. Chromatogr.*, 174 (1979) 409.
94 W. Plunkett, S. Chubb and B. Barlogie, *J. Chromatogr.*, 221 (1980) 425.
95 R.T. Pon and K.K. Ogilvie, *J. Chromatogr.*, 205 (1981) 202.
96 R.A. Hartwick and P.R. Brown, *J. Chromatogr.*, 112 (1975) 651.
97 J.R. Miksic and P.R. Brown, *J. Chromatogr.*, 142 (1977) 641.
98 T.L. Riss, N.L. Zorich, M.D. Williams and A. Richardson, *J. Liq. Chromatogr.*, 3 (1980) 133.
99 E. Grushka and F.K. Chow, *J. Chromatogr.*, 199 (1980) 283.
100 C.T. Wehr, *VIA*, 13, No. 1 (1979) 18, Varian, Palo Alto, CA.
101 M. Dizdaroglu, W. Hermes, C. von Sonntag and H. Schott, *J. Chromatogr.*, 169 (1979) 429.
102 M. Dizdaroglu and W. Hermes, *J. Chromatogr.*, 171 (1979) 321.
103 M. Dizdaroglu, M.G. Simic and H. Schott, *J. Chromatogr.*, 188 (1980) 273.
104 R.A. Hartwick and P.R. Brown, *J. Chromatogr.*, 126 (1976) 679.
105 A.M. Krstulovic, P.R. Brown and D.M. Rosie, *Anal. Chem.*, 49 (1977) 2237.
106 R.A. Hartwick, C.M. Grill and P.R. Brown, *Anal. Chem.*, 51 (1979) 34.
107 R.A. Hartwick, A.M. Krstulovic and P.R. Brown, *J. Chromatogr.*, 186 (1979) 659.
108 R.A. Hartwick, S.P. Assenza and P.R. Brown, *J. Chromatogr.*, 186 (1979) 647.
109 A.M. Krstulovic, R.A. Hartwick and P.R. Brown, *Clin. Chem.*, 25 (1979) 235.
110 H. Ratech, G.J. Thorbecke and R. Hirshhorn, *J. Chromatogr.*, 183 (1980) 499.
111 S.P. Assenza and P.R. Brown, *J. Liq. Chromatogr.*, 3 (1980) 41.
112 P.F. Agris, J.G. Tompson, C.W. Gehrke, K.C. Kuo and R.H. Rice, *J. Chromatogr.*, 194 (1980) 205.
113 H.K. Webster and J.M. Whaum, *J. Chromatogr.*, 209 (1981) 283.

114 A.H. Csárnyi, M. Vajda and J. Sági, *J. Chromatogr.*, 204 (1981) 213.
115 N.M. Ram and J.C. Morris, *J. Liq. Chromatogr.*, 4 (1981) 791.
116 P.D. Schweinsberg and Ti Li Loo, *J. Chromatogr.*, 181 (1980) 103.
117 R.W. Tyson and E. Wickstrom, *J. Chromatogr.*, 192 (1980) 485.
118 C.R. Blakley, J.J. Carmody and M.L. Vestal, *Anal. Chem.*, 52 (1980) 1656.
119 M.G. Carnes, M.L. Brenner and C.R. Andersen, *J. Chromatogr.*, 108 (1975) 95.
120 P. Jandera, J. Churáček and L. Svoboda, *J. Chromatogr.*, 174 (1979) 35.
121 A.M. Jeffrey, S.H. Blobstein, I.B. Weinstein and R.G. Harvey, *Anal. Biochem.*, 73 (1976) 378.
122 A. Panthananickal and L.J. Marnett, *J. Chromatogr.*, 206 (1981) 253.
123 R.W. Chapman, R.O. Morris and J.B. Zaerr, *Nature (London)*, 262 (5564) (1976) 153.
124 A.A. Qureshi, N. Prentice and W.C. Burger, *J. Chromatogr.*, 170 (1979) 343.
125 N.E. Hoffman and J.C. Liao, *Anal. Chem.*, 49 (1977) 2231.
126 D.W. Nierenberg, A.L. Pogolotti and D.V. Santi, *J. Chromatogr.*, 190 (1980) 212.
127 D. Molko, R. Derbyshire, A. Guy, A. Roget, R. Teoule and A. Boucherle, *J. Chromatogr.*, 206 (1981) 493.
128 T. Uematsu and R.J. Suhadolnik, *J. Chromatogr.*, 123 (1976) 347.
129 T. Uematsu and T. Sasaki, *J. Chromatogr.*, 179 (1979) 229.
130 M.F. Aukaty, S.N. Bubenschikova, V.K. Kagrammanova and L.A. Baratova, *J. Chromatogr.*, 137 (1977) 351.
131 P.H. Cooper and J.N. Hawthorne, *J. Chromatogr.*, 87 (1973) 267.
132 T. Kawasaki, *J. Chromatogr.*, 151 (1978) 95.
133 H. Schott, *J. Chromatogr.*, 96 (1974) 79.
134 R.A. Andersen and T.R. Kemp, *J. Chromatogr.*, 172 (1979) 509.
135 R. Horgan and M.R. Kramers, *J. Chromatogr.*, 173 (1979) 263.

Chapter 23

ALKALOIDS, DRUGS, PHARMACEUTICAL PREPARATIONS

In this chapter, some applications of gradient elution liquid chromatography for the analysis of pharmaceuticals (excluding vitamins and antibiotics, which are treated in separate chapters), drugs of abuse and their metabolites and compounds of interest in forensic medicine and toxicology are described. Such compounds necessarily differ widely in their structures, and so do the chromatographic techniques applied in their analysis. It is possible to classify some of these compounds further into several subclasses on the basis of their structural similarities, such as alkaloids, barbiturates and sulpha drugs, but it would be neither possible nor useful to try to classify consistently all of these compounds on a structural basis.

23.1. ALKALOIDS

Alkaloids are a heterogeneous group of organic compounds with a tertiary, quaternary or, less frequently, secondary amino group in the molecule, formed in plants by transformations of amino acids. Some of them are obtained from plant material, others are partially or completely synthesized. A number are strongly toxic, others are used in pharmaceutical products (analgesics, narcotics, etc.) and certain alkaloids belong to hallucinogenic and narcotic drugs of abuse. The necessity for suitable rapid analytical methods for these compounds in therefore obvious. These compounds are analysed in plant samples, synthetic mixtures, commercial pharmaceutical formulations and, together with their metabolites, in physiological fluids and other biological materials. A survey of some examples of the applications of gradient elution liquid chromatography in analytical and preparative separations of alkaloids and their metabolites is given in Table 23.1.

It can be seen that it is often necessary to add diethylamine to the mobile phase in normal-phase chromatography on silica gel columns[3,5,7] or to use silica gel columns impregnated with silver ions[3,7]. Normal-phase chromatography on bonded amino phases using elution with a gradient of ethanol in diethyl ether may be successfully applied for the analysis of certain alkaloids[8,9]; however, as with other separation problems, reversed-phase chromatography on chemically bonded alkyl phases has become increasingly popular. Here, methanol or acetonitrile gra-

References on p. 405.

TABLE 23.1

EXAMPLES OF SEPARATION OF ALKALOIDS, ILLICIT DRUGS AND THEIR METABOLITES BY GRADIENT ELUTION LIQUID CHROMATOGRAPHY

Compounds	Material	Column	Solvent A	Solvent B	Gradient	Detection	Ref.
Methylxanthines: theophylline, caffeine, theobromine	Biological fluids	Durapak OPN, 2000 x 4.2 mm	*n*-Hexane	Isopropanol	Linear, 14-22% B, 1% B/min	UV, 270 nm	1
Theophylline and its metabolites	Urine	Ultrasphere ODS, 5 μm, 250 x 4.6 mm	0.01 *M* CH_3COONa + 0.005 *M* TBAHS*	50% A in CH_3OH	Multi-linear, 9-99% B	UV, 280 nm	2
Caffeine, thebaine, LSD, procaine, morphine	Synthetic mixture	Silica gel coated with Ag^+, 110 x 4.5 mm	Chloroform-*n*-hexane (1:1)	Chloroform-methanol-diethylamine (90:10:0.1)	Multi-linear, 0-60% B	UV, 254 nm	3
Caffeine, morphine, monoacetylmorphine, strychnine, diamorphine, quinine, cocaine, barbitone	Illicit diamorphine preparations	Zipax SCX (pellicular cation exchanger), 1200 x 2.1 mm	0.2 *M* H_3BO_3 + 40% NaOH (pH 9.3)	0.2 *M* H_3BO_3-acetonitrile-propanol (86:12:2) + 40% NaOH (pH = 9.8)	Linear, 0-100% B, 16.7% B/min	UV, 254, 270, 235 nm	4
Morphine, codeine, thebaine, papaverine, cryptopine, narcotine	Opium	Corasil II, 1000 x 2.8 mm	*n*-Hexane	Chloroform-methanol-diethylamine (100:300:1)		UV, 254 nm	5
Nicotine, nornicotine, anabasine, anatobine, cotinine, myosmine, dipyridyl	Tobbaco smoke	LiChrosorb RP-18	0.07 *M* triethylamine + H_3PO_4 (pH 7.56)	Acetonitrile	Linear, 0-25% B, 0.28% B/min	UV, 254 nm	6
Atropine, scopolamine, caffeine, ergotamine, barbiturates	Pharmaceuticals	LiChrosorb Si100, 5 μm, coated with Ag^+	Chloroform-*n*-hexane (1:1)	Chloroform-ethanol-diethylamine (90:10:0.5)	Linear, 16-92% B, 15% B/min	UV, 254 nm	7

Clavine alkaloids, lysergic acid derivatives	*Claviceps purpurae*	Micro Pak NH_2 10 µm, 250 x 2 mm	Diethyl ether-ethanol (79:21)	Diethyl ether-ethanol (21:79)	Linear, 0-100% B, 4% B/min	UV, 225, 240, 254 nm	8
Ergot alkaloids		LiChrosorb NH_2	Diethyl ether	Ethanol	Linear, 0.3% B/min	UV, 310 nm	9
Ergot alkaloids	Ergotamine preparations	µBondapak C_{18}, 10 µm, 300 x 4 mm	0.01 *M* $(NH_4)_2CO_3$-acetonitrile (3.7:0.3)	0.01 *M* $(NH_4)_2CO_3$-acetonitrile (1:1)	Linear, 0-100% B, 6.7% B/min	UV, 320 nm	10
Ergot alkaloids	Pharma-ceuticals	Nucleosil C_{18}, 5 µm, 150 x 3 mm	0.01 *M* $(NH_4)_2CO_3$ in water	Acetonitrile	Stepwise, 4.25-50% B and 15-60% B	UV, 254 nm	18
Quinidine, quinine and their metabo-lites	Rat urine**	µBondapak C_{18}, 300 x 7.8 mm	Water-acetic acid (99:1)	Water-aceto-nitrile-acetic acid (40:59:1)	Concave, 10-85% B	UV, 254 nm	11
Cinchonine,cincho-nidine, quinine, quinidine, quino-line	Rat urine**	RS II C 18-HL, 20 µm, 500 x 8 mm	Water	Methanol	Concave, 50-100% B	UV, 280 nm	12
Tubocurarine	Curare	µBondapak C_{18}, 10 µm, 300 x 4.6 mm	0.025 *M* TMAH*** -water (90:10 and 75:25)	0.025 *M* TMAH*** -water (10:90 and 55:45)	Linear, 0-100% and 10-85% B, 2.2 and 2.5% B/min	UV, 280 nm	13
Colchicine and colchicoside	Colchicum seeds	LiChrosorb C_{18}, 30 µm, 500 x 3 mm	Water	Acetonitrile	Multi-linear, 0-30% B	UV, 254 nm	14
Digoxin, digi-toxin and metabo-lites	Synthetic mixture	µBondapak C_{18}, 10 µm, 300 x 4 mm	Water	Acetonitrile	Linear, 25-40% B, 5% B/min	UV, 220 nm	15
Digoxin, digi-toxin and metabo-lites	Blood plasma	µBondapak C_{18} 300 x 3.9 mm	Water	Methanol	Concave, 53-62% B	UV, 254 nm	16
Tomatidine, sol-amidine, solaso-dine, rubijervine, veratramine, jervi-ne	*Solanum chacoense***	Porasil A, 37-75 µm, 2400 x 10 mm	Acetone-*n*-hexane (2:1)	Acetone-water (97:3)	Concave, 0-100% B	Analysis in frac-tions	17

(Continued on p. 398)

References on p. 405.

Table 23.1 (continued)

Compounds	Material	Column	Solvent A	Solvent B	Gradient	Detection	Ref.
Vinca rosea alkaloids	*Vinca rosea*	Alumina, 18-30 μm, 6000 x 7.9 mm	Methylene chloride	Ethanol	Linear, 0-5% B	UV, 280 nm	19
Vinca rosea alkaloids	*Vinca rosea*	R Sil-C18-HL-D, 10 μm, 250 x 4.6 mm	Water	Methanol	Linear, 50-85% B, 1% B/min	UV, 290 nm	20
Pyrrolizidine alkaloids	*Senecio vulgaris*	Bondapak CN, 10 μm, 300 x 4 mm	0.01 *M* $(NH_4)_2CO_3$ (pH 7.8)	Tetrahydrofuran	Linear, 13-26% B, 0.42% B/min	UV, 235 nm	21
Pyrrolizidine alkaloids	*Radix consolidae*	XAD-2[§], 50-100 μm, 260 x 13 mm	0.5 *M* KCl in methanol-water	0.5 *M* KCl + HCl in methanol-water	Linear, not specified	UV, 254 nm	22
Pyrrolizidine alkaloids	*Senecio vulgaris*, *Senecio jacobaea*	PRP-1[§], 150 x 4.1 mm	0.1 *M* NH_4OH in water	Acetonitrile	Linear, 10-30% B, 1% B/min	UV, 220 nm	23
Ellipticine and related compounds		μBondapak C_{18}, 10 μm, 300 x 4 mm	0.005 *M* Na heptanesulphonate + 0.032 *M* acetic acid in methanol-water (1:1)	0.005 *M* Na heptanesulphonate + 0.032 *M* acetic acid in methanol-water (9:1)	Concave, 20-60% B in 12 min	Fluorimetric, UV, 313 nm	48
Heroin, LSD, cocaine, phenobarbital	Illicit drugs	Corasil II, 37-50 μm, 1000 x 2.3 mm	0.5% Ethanol + 0.25% dioxane + 0.13% cyclohexylamine in hexane	10% Ethanol + 20% dioxane + 1.4% cyclohexylamine in hexane	Linear, 0-100% B	UV, 254 nm	24
LSD	Urine	RP-8 7 μm, 100 x 3.2 mm	Water	Acetonitrile	Linear, 10-90% B, 14% B/min	Mass spectrometry	25

Benzoylecgonine (cocaine metabolite)	Urine	µBondapak C_{18}	Water	Methanol	Multi-linear, 5-100% B	UV, 254 nm	26
Δ^9-Tetrahydrocannabinol (from marijuana)	Blood plasma	Silica gel	*n*-Heptane	Methylene chloride		UV, 273 nm	27

*TBAHS = tetrabutylammonium hydrogen sulphate (ion-pair technique).
**Preparative separation.
***TMAH = tetramethylammonium hydroxide (ion-pair technique).
§Styrene-divinylbenzene resin.

References on p. 405.

TABLE 23.2

EXAMPLES OF SEPARATION OF DRUGS, PHARMACEUTICALS, DRUGS OF ABUSE AND THEIR METABOLITES BY GRADIENT ELUTION LIQUID CHROMATOGRAPHY

Compounds	Material	Column	Solvent A	Solvent B	Gradient	Detection	Ref.
Barbiturates:							
Barbiturates and hydroxylated metabolites		Pellicular anion exchanger, 3000 x 1 mm	0.1 *M* NaCl in water	1.0 *M* NaCl in water	Linear, 0-100% B, 2.5% B/min	UV, 254 nm	28
Free barbiturates		LiChrosorb Si 100-C18, 10 μm, 300 x 4.2 mm	Water	Methanol	Linear, 10-100% B, 1.67 % B/min	UV, 254 nm	29
Dansyl derivatives of barbiturates	Blood serum	C18/pellicular silica, 30 μm, 500 x 2.5 mm	Water	Methanol	Delayed linear, 0-65% B, 3,25% B/min	Fluorimetric	30
Sulphonamides:							
7 sulpha drugs		Micro Pak CH-10 250 x 2.1 mm	Water	Methanol	Concave, 2-47% B in 40 min	UV, 254 nm	31
Sulpiride	Body fluids	Nucleosil C_{18}, 5 μm, 150 x 4 mm	Water-acetic acid (99:1)	Water-acetic acid-acetonitrile (49:1:50)	Linear, 12-60% B, 4.8% B/min	Fluorimetric	32
Analgesics:							
Caffeine, acetaminophen, aspirin		Zipax SAX	Water, buffered at pH 9.2	A + 0.1 *M* $NaClO_4$	Linear, 0-100% B, 5% B/min	UV, 254 nm	33
Caffeine, phenacetin,aspirin,salicylic acid, salicylamide, etc.		Styrene-divinylbenzene copolymer, 5-10 μm, 250 x 4 mm	0.2 *M* H_3PO_4 in 25% (v/v) aqueous methanol	0.2 *M* Na_2HPO_4 in 25% (v/v) aqueous methanol	Convex, 0-100% B in 120 min	UV, 254 nm	34

References on p. 405.

Caffeine, phenacetin, propylphenazone, aspirin, paracetamol, N-acetylsalicylamide, diallylbarbituric acid, *o*-epoxybenzamide, karbromal	Tablets	LiChrosorb RP-8 10 μm, 250 x 4.6 mm	Water	Acetonitrile	Various linear gradients	UV, 254 nm	35
Phenacetin and its metabolites	Urine	μBondapak C_{18}, 600 x 4 mm	0.01 *M* $(NH_4)_2HPO_4$ in water	Acetonitrile	Concave, 5-40% B in 30 min	UV, 254 nm	36
Aspirin, salicylic acid	Tablets	Poly(methyl methacrylate) resin	*n*-Hexane-diethyl ether-acetic acid	*n*-Hexane-ethanol	Linear, 0-100% B	UV, 280 nm	37
Zomepirac and its derivatives		LiChrosorb Si 60, 10 μm, 250 x 2.1 mm	0.25% acetic acid in *n*-hexane	0.25% acetic acid + 10% isopropanol in *n*-hexane	Concave, 3-90% B in 20 min	UV, 254 nm	38
Other pharmaceuticals and drugs:							
Chlordiazepoxide and its N-desmethyl metabolite	Blood serum	μBondapak C_{18}, 300 x 4 mm	1 m*M* phosphate buffer (pH 8) in water	Methanol	Linear, 40-90% B, 17% B/min	UV, 254 nm	39
Benzodiazepines		Spherisorb ODS, 10 μm, 250 x 3 mm	Water	Methanol-acetonitrile (1:1)	Linear, 50-99% B, 32.7% B/min	UV, 254 nm	40
Carbamazepine metabolites	Rat urine	μBondapak C_{18} 300 x 3.9 mm	Methanol-water-acetic acid (20:80:0.1)	Methanol-water-acetic acid (50:50:0.1)	Linear, 10-90% B, 2% B/min	UV, 254 nm	41
Methadone hydrochloride	Oral solution	LiChrosorb Si 100-C18	HCOOH + $HCOONH_4$ in water	Acetonitrile	Linear	UV, 254 and 280 nm	42
Phenylketazones and metabolites	Blood plasma, urine	μBondapak C_{18} 10 μm, 300 x 4 mm	0.01 *M* CH_3COONa in methanol-water (1:1) (pH 4)	Methanol	Linear 0-100% B, 10% B/min	UV, 254 nm	43

(Continued on p. 402)

Table 23.2 (continued)

Compounds	Material	Column	Solvent A	Solvent B	Gradient	Detection	Ref.
Diftalone and its metabolites	Blood plasma	Permaphase ODS, 1000 x 2.2 mm	Water	Acetonitrile-water (1:4)	Linear, 0-20% B, 6% B/min	UV, 254 nm	44
Lignocaine, hydrocortisone acetate, butazolidine	Pharmaceuticals	Spherosil, 5 µm, 150 x 4.8 mm	2% Ethylamine in cyclohexane	2% Ethylamine in ethanol	Delayed linear, 15-40% B, 6.2% B/min	UV, 254 nm	45
Chlorambucil, its metabolite and prednisolone conjugate	Blood plasma	µBondapak C_{18}	0.175 *M* CH_3COOH in methanol-water (60:40)	Methanol	Linear, 0-100% B, 10% B/min	UV, 254 nm	46
Trithiozine and its metabolites	Blood plasma, urine	LiChrosorb RP-8, 10 µm, 250 x 4.6 mm	Water	Methanol	Delayed linear, 30-50% B, 6.7% B/min	UV, 254 nm	47
Mecinazone		Permaphase ETH	0.1% ammonia in water	Methanol	Linear, 10-100% B	UV, 254 nm	49
Azosemide	Urine	Zorbax ODS, 5 µm, 250 x 4.6 mm	Acetate buffer in water (pH 4.05)	Acetonitrile	Linear, 10-40% A, 3% B/min	UV, 239 nm	50
Melphalan	Urine	HCODS/SIL X-C18, 10 µm, 250 x 2.6 mm	0.0175 *M* acetic acid in water	Acetonitrile	Concave, 12-80% B in 14 min	UV, 263 nm	51
Diethylsilboestrol and its derivatives (carcinogenic)	Physiological fluids	LiChrosorb RP-18, 5 µm, 250 x 10 mm	Water	Methanol	Convex, 10-100% B in 50 min	UV, 254 nm	52
Amitriptyline	Urine	µBondapak C_{18}	0.4% phosphate buffer in water (pH 3.5)	Methanol		UV, 254 nm	53

Disulphiram and metabolites	Blood plasma	μBondapak alkylphenyl, 10 μm, 300 x 3.9 mm	Water	Acetonitrile	Linear, 51-75% B, 2.4% B/min	UV, 280 nm	54
Cyclosporin A	Blood plasma	LiChrosorb RP-8, 5 μm, 125 x 3 mm	Water-aceto-nitrile-methanol (600:200:200)	Water-aceto-nitrile-methanol (50:750:200)	Delayed linear, 0-48% B, 9.6% B/min	UV, 210 nm	55
Mebendazole and metabolites	Blood plasma	LiChrosorb RP-8, 10 μm, 250 x 4 mm	Methanol-water (55:45)	0.05 *M* ammonium phosphate (pH 5.5) in solvent A	Delayed concave, 0-45% B in 14 min	UV, 254 nm	56
Clanobutine and metabolites	Blood plasma, urine	LiChrosorb RP-8, 10 μm, 250 x 4 mm	Water + H_2SO_4 (pH 2.8)	Acetonitrile	Linear, 25-40% B 0.75% B/min	UV, 230 nm	57

References on p. 405.

dients in water[12,14-16,20,23] or in buffered aqueous salt solutions[6,10,11,18,21] are used or ion-pair reversed-phase chromatography with concentration gradients of organic solvents is employed[2,13,48]. Styrene-divinylbenzene resins[22,23] and bonded nitrile phases[21] have been employed in reversed-phase systems and ion-exchange chromatography on pellicular cation exchangers[4] has also proved useful.

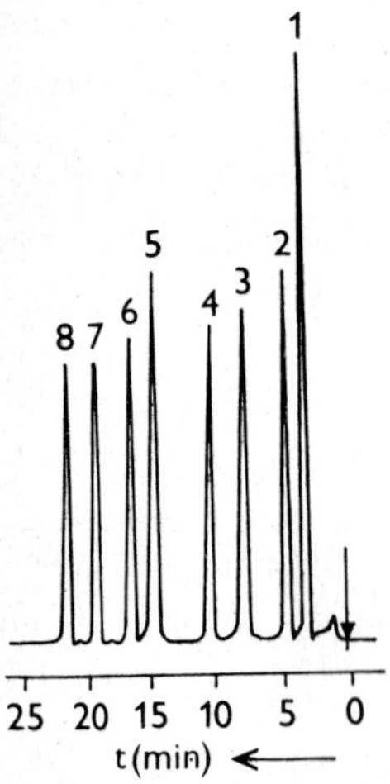

Fig. 23.1. Separation of digoxin, digitoxin and their metabolites by reversed-phase gradient elution chromatography. Column: μBondapak C_{18}, 10 μm, 300 x 4 mm I.D. Gradient: linear, from 25 to 40% B in A, 5% B/min. Solvents: A, water; B, acetonitrile. Flow-rate: 2.2 ml/min. Detection: UV at 220 nm. Peaks: 1 = digoxigenin; 2 = digoxigenin monodigitoxoside; 3 = digoxigenin bisdigitoxoside; 4 = digoxin; 5 = digitoxigenin; 6 = digitoxigenin monodigitoxoside; 7 = digitoxigenin bisdigitoxoside; 8 = digitoxin. (Reproduced from ref. 15 with permission.)

Fig. 23.1 shows an example of the gradient elution reversed-phase separation of cardiac glycosidic alkaloids[15].

23.2. PHARMACEUTICALS AND DRUGS

Liquid chromatography has been widely applied in the analysis of pharmaceuticals and drugs, both in commercial preparations and in physiological fluids, where it is often applied to study the metabolism of applied medicaments. A survey of some examples of gradient elution liquid chromatography in the analysis of barbiturates, sulphonamides, analgesics and other pharmaceuticals and drugs is given in Table 23.2. It can be seen that chromatography on silica gel[38,45] and ion-exchange chromatography[28,33] are still used for the analysis of certain medicaments, but reversed-phase chromatography on chemically bonded phases has become an increasingly used technique. Here, buffered systems[36,39,42,43,50,53,56] or mobile phases with addition of acetic acid[32,41,46,51] or sulphuric acid[57] are frequently used, but gradients of methanol or acetonitrile in water[29-31,35,40,44,47,52,54,55] are also often applied. Organic copolymer resins[34,37] and chemically bonded medium-polarity phases[49] may be employed in certain applications.

UV detection at 254 or at 280 nm is suitable for most applications, because many pharmaceuticals absorb in the UV region. Fluorimetric detection can be used in certain applications, mainly if pre-column or post-column derivatization (such as dansylation of barbiturates[30]) is employed to improve the sensitivity of determination of trace amounts of drugs in physiological fluids.

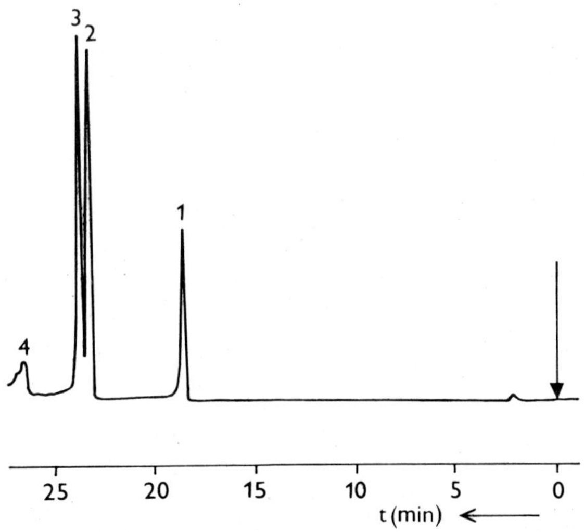

Fig. 23.2. Separation of a commercial analgesic product by reversed-phase gradient elution chromatography. Column: Silasorb C18, 10 μm, 300 x 4.2 mm I.D. Gradient: concave, from 0 to 100% B in A in 30 min. Solvents: A, water; B, methanol. Flow-rate: 1 ml/min. Detection: UV at 254 nm. Peaks (methanolic extract of an analgesic dosage): 1 = caffeine; 2 = aminophenazine; 3 = phenacetin; 4 = impurity.

Fig. 23.2 shows an example of the gradient elution reversed-phase chromatographic separation of components in a commercial analgesic tablet.

23.2. REFERENCES

1 C.V. Manion, D.W. Shoeman and D.L. Azarnoff, *J. Chromatogr.*, 101 (1974) 169.
2 K.T. Muir, J.H.G. Jonkman, D.-S. Tang, M. Kunitani and S. Riegelman, *J. Chromatogr.*, 221 (1980) 85.

3 C.R. Vogt, J.S. Baxter and T.R. Ryan, *J. Chromatogr.*, 150 (1978) 93.
4 P.J. Twitchett, *J. Chromatogr.*, 104 (1975) 205.
5 H.W. Ziegler, T.M. Beasley and D.W. Smith, *J. Ass. Offic. Anal. Chem.*, 58 (1975) 888.
6 J.J. Piade and D. Hoffmann, *J. Liq. Chromatogr.*, 3 (1980) 1505.
7 R. Aigner, H. Spitzy and R.W. Frei, *J. Chromatogr. Sci.*, 14 (1976) 381.
8 M. Wurst, M. Flieger and Z. Řeháček, *J. Chromatogr.*, 150 (1978) 477.
9 M. Wurst, M. Flieger and Z. Řeháček, *J. Chromatogr.*, 174 (1979) 401.
10 H. Bethke, B. Delz and K. Stich, *J. Chromatogr.*, 123 (1976) 193.
11 S.E. Barrow, A.A. Taylor, E.C. Horning and M.G. Horning, *J. Chromatogr.*, 181 (1980) 219.
12 M. Verzele and E. Geeraert, *J. Chromatogr. Sci.*, 18 (1980) 559.
13 F.P.B. van der Maeden, P.T. van Rens, F.A. Buytenhuys and E. Buurman, *J. Chromatogr.*, 142 (1977) 715.
14 G. Forni and G. Massarani, *J. Chromatogr.*, 131 (1977) 444.
15 M.C. Castle, *J. Chromatogr.*, 115 (1975) 437.
16 H.A. Nelson, S.V. Lucas and T.P. Gibson, *J. Chromatogr.*, 163 (1979) 169.
17 I.R. Hunter, M.K. Walden, J.R. Wagner and E. Heftmann, *J. Chromatogr.*, 119 (1976) 223.
18 F. Erni, R.W. Frei and W. Lindner, *J. Chromatogr.*, 125 (1976) 265.
19 S. Taylor, W.E. Jones, D.E. Dorman, E.E. Logsdon and G.H. Svoboda, *J. Pharm. Sci.*, 64 (1975) 1953.
20 M. Verzele, L. de Tacye, J. van Dyck, G. de Decker and C. de Pauw, *J. Chromatogr.*, 214 (1981) 95.
21 C.W. Qualls and H.J. Segall, *J. Chromatogr.*, 150 (1978) 202.
22 H.J. Huizing and T.M. Malingré, *J. Chromatogr.*, 176 (1979) 274.
23 H.S. Ramsdell and D.R. Buhler, *J. Chromatogr.*, 210 (1981) 154.
24 M.L. Chan, C. Whetsell and J.D. McChesney, *J. Chromatogr. Sci.*, 12 (1974) 512.
25 W.H. McFadden, *J. Chromatogr. Sci.*, 18 (1980) 97.
26 A.P. Graffeo, D.C.K. Lin and R.L. Foltz, *J. Chromatogr.*, 126 (1976) 717.
27 J.L. Valentine, P.J. Bryant, P.L. Gutshall, O.M. Gan and E.R. Thompson, *J. Pharm. Sci.*, 66 (1977) 1263.
28 M.W. Anders and J.P. Latorre, *Anal. Chem.*, 42 (1970) 1430.
29 P. Jandera, J. Churáček and L. Svoboda, *J. Chromatogr.*, 174 (1979) 35.
30 W. Dünges, G. Naundorf and N. Seiler, *J. Chromatogr. Sci.*, 12 (1974) 655.
31 *Sulfa Drugs, Micromeritics Application Notes No. 8*, Micromeritics, Norcross, GA.
32 G. Alfredsson, G. Sedvall and F.-A. Wiesel, *J. Chromatogr.*, 164 (1979) 187.
33 *Pharmaceutical Applications of HPLC*, DuPont, Wilmington, MA.
34 T. Hanay, H.F. Walton, J.D. Navratil and D. Warren, *J. Chromatogr.*, 155 (1978) 261.
35 R. Schuster and M. Riedmann, *HPLC of Pharmaceutical Products*, Hewlett-Packard, Avondale, PA, April 1977.
36 *Analysis of Pharmaceutical Products*, Waters Assoc., Milford, MA, 1976.
37 R.G. Baum and F.F. Cantwell, *J. Pharm. Sci.*, 67 (1978) 1066.
38 C.Y. Ko and C.A. Janicki, *J. Chromatogr.*, 190 (1980) 429.
39 H.B. Greizerstein and C. Wojtowicz, *Anal. Chem.*, 49 (1977) 2235.
40 *Spectra-Physics Chromatogr. Rev.*, 3, No. 1 (1977).
41 M.G. Horning and K. Lertratanangkoon, *J. Chromatogr.*, 181 (1980) 59.
42 T.H. Beasley and M.W. Ziegler, *J. Pharm. Sci.*, 66 (1977) 1749.
43 T. Marunaka, T. Shibata, Y. Minami and Y. Umeno, *J. Chromatogr.*, 183 (1980) 331.
44 S. Roggia, G. Grossoni, G. Pelizza, B. Ratti and G.G. Gallo, *J. Chromatogr.*, 124 (1976) 169.
45 M. Caude and Le Xuan Phan, *Chromatographia*, 9 (1976) 20.
46 D.R. Newell, L.I. Hart and K.R. Harrap, *J. Chromatogr.*, 164 (1979) 114.
47 T. Crolla, M. Borsa and G. Pifferi, *J. Chromatogr.*, 222 (1981) 257.
48 G. Muzard and J.-B. Le Pecq, *J. Chromatogr.*, 169 (1979) 446.
49 R. Gimet and J. Rabiant, *Farmaco, Ed. Prat.*, 32 (1977) 214.
50 R. Seiwell and C. Brater, *J. Chromatogr.*, 182 (1980) 257.

51 A.E. Ahmed and Ting-Feng Hsu, *J. Chromatogr.*, 222 (1981) 453.
52 G.D. Newport, S.K. Headley, J.P. Freeman and W. Slikker, *J. Liq. Chromatogr.*, 3 (1980) 1053.
53 D.R. Hawkins and I. Midgley, *J. Pharm. Pharmacol.*, 30 (1978) 547.
54 P.D. Masso and P.A. Kramer, *J. Chromatogr.*, 224 (1981) 457.
55 W. Niederberger, P. Schaub and T. Beveridge, *J. Chromatogr.*, 182 (1980) 454.
56 R.J. Allan, H.T. Goodman and T.R. Watson, *J. Chromatogr.*, 183 (1980) 311.
57 K. Zech and G. Ludwig, *Chromatographia*, 9 (1976) 638.

Chapter 24

VITAMINS AND CAROTENOIDS

Vitamins are organic compounds with various structures, formed in plants, which are necessary as biocatalysts in human and animal organisms. They are usually classified into two groups: fat-soluble vitamins (such as vitamins A, D, E and K) and water-soluble vitamins (such as vitamin C and B group vitamins). Liquid column chromatography is very suitable for the analysis of these compounds, because of the mild conditions used.

24.1. FAT-SOLUBLE VITAMINS

Fat-soluble vitamins are less polar than water-soluble vitamins. Vitamins A, E, D_2 and D_3 and their analogues and metabolites have been analysed by chromatography on silica gel columns using linear or exponential gradients of dichloromethane[1] or chloroform[2] in *n*-hexane, gradients of methanol in dichloromethane[3] and ternary gradients of methanol and dichloromethane[4] or of isopropanol and chloroform[5] in *n*-hexane.

Fat-soluble vitamins have also been analysed by chromatography on superficially[6-8] and totally porous octadecylsilica columns[9,10], on totally porous octylsilica[11] or octadecyl- and phenylsilica[12] columns, using linear[6-11] or multilinear[12] gradients of methanol[6,8,9,12] or acetonitrile[11] in water or linear gradients of methanol in 1% aqueous ammonium carbonate[7,10].

Non-aqueous reversed-phase chromatography on octadecylsilica columns has recently been successfully applied to the determination of vitamins A, D_2, D_3, E and K_1 using delayed linear gradients of acetonitrile in methanol-ethyl acetate (86:14)[13] and to the separation of K vitamins and related compounds by elution with a linear gradient of tetrahydrofuran or dichloromethane in acetonitrile[14].

24.2. CAROTENOIDS

Carotenoid compounds (a group including coloured organic pigments closely related to vitamin A) have been chromatographed on columns packed with microparticulate silica gel by elution with a linear[15] or concave[16] gradient of acetone in *n*-hexane. This gradient has also been applied to the chromatography of carotenoids on a column packed with diatomaceous earth[17]. UV-absorbing acetone can be used for the formation of gradients because the detection of these coloured compounds

References on p. 413.

is performed in the visible region (between 400 and 490 nm). Carotenoids have also been chromatographed on an alumina column by elution with a concave gradient of diethyl ether in light petroleum[18].

In reversed-phase systems, rapid separations or various carotenoid mixtures were achieved on octadecylsilica columns by elution with convex gradients of acetone or acetonitrile in water[19,20]. This method can be applied to the identification and determination of carotenoids in various plants, fruits and vegetables and to the analysis of synthetic carotenoid additives in food[19].

24.3. WATER-SOLUBLE VITAMINS

24.3.1. Chromatography on ion exchangers and organic gels

Water-soluble vitamins have been analysed by ion-exchange chromatography. Small-diameter particle cation-exchange resins (Aminex A-5) were used as the column packing material and B vitamins and related compounds (pyridoxamine, pyridoxol, pyridoxal and nicotinamide) were separated using combined ionic strength and pH gradients in mobile phases containing aqueous ammonium formate[21] or phosphate[22] buffer.

Pellicular and porous-layer chemically bonded ion exchangers have also been applied in high-speed separations of water-soluble vitamins. Mixtures of various B_2 vitamins (riboflavin, folic acid and nicotinic acid) were chromatographed on a column packed with Permaphase AAX anion exchanger using elution with a concave gradient of sodium dihydrogen phosphate (pH 4.4)[23]. Folic acid and related compounds (derivatives and metabolites) in food[24] and blood serum[25] were subjected to analysis on anion-exchange columns (Pellionex SAX[24] or Permaphase AAX[25]) using elution with a gradient of potassium chloride in phosphate buffer[24] or with a linear gradient of potassium perchlorate in water[25]. Thiamine (vitamin B_1), thiamine antagonists (oxythiamine, pyrothiamine) and their phosphate esters in rat livers were determined by liquid chromatography on a Vydac SAX anion-exchange column. A linear gradient of potassium dihydrogen phosphate in water (pH 6) was employed for elution[26].

Some water-soluble vitamins are retained on dextran gel, which made possible the separation of thiamine and its derivatives (thiochrome, thiamine phosphate, diphosphate and disulphide) on a Sephadex G-10 column by elution with a linear gradient of hydrochloric acid in water[27].

24.3.2. Reversed-phase chromatography on chemically bonded phases

Because of the relatively highly polar character of water-soluble vitamins, the addition of buffers, salts or acids to the aqueous-organic mobile phase is usually necessary in order to improve separations in reversed-phase systems. For example, mixtures of vitamins B_1, B_2, B_3 and B_6 were chromatographed on an octadecylsilica column by elution with a linear gradient of methanol in 1% aqueous ammonium carbonate[10]. A concave gradient of methanol in 0.05 mol/l acetate buffer (pH 4) was employed for the separation of cobalamines (vitamin B_{12})[28] and a linear gradient of methanol in aqueous acetic acid (0.4 mol/l) served as the eluent in the chromatography of lipoic acid and analogues[29] on octadecylsilica columns.

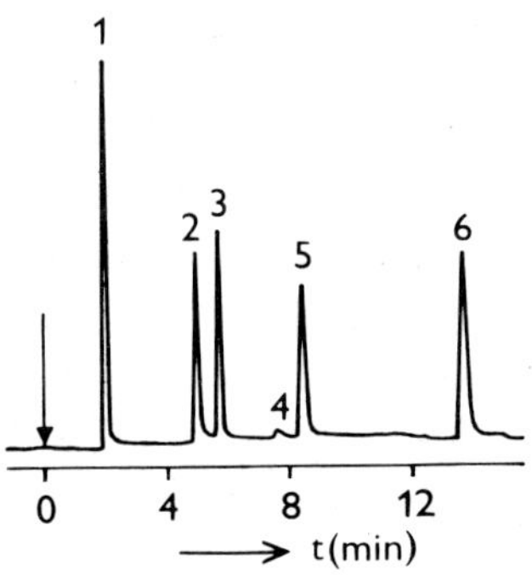

Fig. 24.1. Separation of a mixture of water-soluble vitamins by reversed-phase gradient elution ion-pair chromatography. Column: Micro Pak MCH-10. Gradient: linear, from 0 to 30% B in A, 2% B/min. Solvents: A, 0.01 mol/l KH_2PO_4 + 0.01 mol/l pentanesulphonic acid + 0.005 mol/l trimethylamine in water (pH 2.2); B, 0.01 mol/l pentanesulphonic acid + 0.005 mol/l trimethylamine in 90% (v/v) acetonitrile in water. Flow-rate: 2 ml/min. Detection: UV at 254 nm. Peaks: 1 = ascorbic acid; 2 = niacin; 3 = niacinamide; 4 = pyridoxine; 5 = thiamine; 6 = riboflavin. (Adapted from ref. 30.)

Ion-pair reversed-phase chromatography has become popular for the separation of vitamins and elution with concentration gradients of organic solvents in the mobile phase is useful for shortening the analysis time. Because most water-soluble vitamins contain both acidic (such as carboxylic) and basic (amino- or heterocyclic nitrogen) groups, they can be chromatographed with the addition of either cationic or anionic ion-pairing reagents to the mobile phase. Pharmaceutical multi-vitamin preparations[30,31] and food vitamin additives[32] were analysed for the content of vitamin C (ascorbic acid) and B vitamins (niacin, niacinamide, thiamine, pyridoxine, riboflavin, folic acid) by chromatography on octadecyl-

References on p. 413.

silica[30,32] and octylsilica[31] columns using elution with linear[30,32] or multi-linear[31] gradients of methanol[31] or acetonitrile[30,32] in an aqueous mobile phase (buffered at pH 2.2-4 with phosphate buffers) containing 0.01 mol/l pentane-sulphonic acid[30] or heptanesulphonic acid[31,32]. Fig. 24.1 shows an example of a separation obtained under these conditions[30].

Thiamine, its derivatives and phosphate esters were chromatographed on an octadecylsilica column using elution with a linear gradient of methanol in a 0.05 mol/l aqueous solution of tetrabutylammonium ions[26]. Folic acid and its clinically important derivatives were separated under similar conditions, but a multi-linear gradient of methanol was used[33].

Chemically bonded amino phases may be used for the chromatography of water-soluble vitamins, but it is important to adjust the pH appropriately in order to achieve reproducible results. For example, mixtures containing niacinamide, pyridoxine, riboflavin, thiamine, niacin, pantothenic acid, ascorbic acid and cobalamin were successfully separated in less than 10 min on a LiChrosorb NH_2 (250 x 4.6 mm I.D.) column by elution with a linear gradient from 63 to 86.5% acetonitrile in 0.005 mol/l KH_2PO_4[34].

TABLE 24.1

OPTIMUM DETECTION WAVELENGTHS FOR SOME IMPORTANT VITAMINS

Type	Vitamin	Wavelength (nm)
Fat-soluble	D (calciferol)	265
	E (tocopherol)	288
	A (retinol)	330
Water-soluble	B_1 (thiamine)	234
	B_2 (riboflavine)	272
	B_2 [niacin (nicotinamide, niacinamide)]	218
	B_2 (pantothenic acid)	200
	B_6 (pyridoxine)	286
	B_{12} (cobalamin)	212
	C (ascorbic acid)	268
	B_2 [folic (pteroylglutamic) acid]	278

Continuous UV detection is suitable with both fat-soluble and water-soluble vitamins and can be performed at 254 or 280 nm. However, optimum wavelengths for the detection of different vitamins may differ with respect to their absorption maxima and possible interferences, as shown in Table 24.1[34].

24.4. REFERENCES

1 M.K. Head and E. Gibbs, *J. Food Sci.*, 42 (1977) 395.
2 J. Vermont, M. Deleuil, A.J. de Vries and C.L. Guillemin, *Anal. Chem.*, 47 (1975) 1329.
3 N. Ikekawa and N. Koizumi, *J. Chromatogr.*, 119 (1976) 227.
4 *Application Sheet E 771,* Laboratory Data Control, Riviera Beach, Fl.
5 K.J. Bombaugh, *J. Chromatogr.*, 107 (1975) 201.
6 R.C. Williams, J.A. Schmit and R.A. Henry, *J. Chromatogr. Sci.*, 10 (1972) 494.
7 J.N. Little and G.J. Fallick, *J. Chromatogr.*, 112 (1975) 389.
8 E.W. Matthews, P.G.H. Byfield, K.W. Colston and I.M. Evans, *FEBS Lett.*, 48 (1974) 122.
9 Pye Unicam LC-XP Liquid Chromatography Series prospectus, Pye Unicam, Cambridge, Great Britain, 1978.
10 E.C. Conrad, in G. Charalambous (Editor), *Liquid Chromatographic Analysis of Food and Beverages,* Academic Press, London, 1979, p. 237.
11 *LC 750 Brochure,* Applied Chromatography Systems, Luton, Great Britain.
12 S.A. Barnett and L.W. Frick, *Anal. Chem.*, 51 (1979) 641.
13 S.A. Barnett, L.W. Frick and H.M. Baine, *Anal. Chem.*, 52 (1980) 610.
14 Y. Harroon, M.J. Shearer and P. Barkhan, *J. Chromatogr.*, 206 (1981) 333.
15 A. Fiksdahl, J.T. Mortensen and S. Liaaen-Jensen, *J. Chromatogr.*, 157 (1978) 111.
16 S.K. Hajibrahim, P.J.C. Tibbetts, C.D. Watts, J.R. Maxwell, G. Eglinton, H. Colin and G. Guiochon, *Anal. Chem.*, 50 (1978) 549.
17 I. Stewart, *J. Ass. Offic. Anal. Chem.*, 60 (1977) 132.
18 A.H. Goldie and R.E. Subden, *J. Chromatogr.*, 84 (1973) 192.
19 K. Lange, *Dissertation,* Nürnberg, 1976.
20 R. Ohmacht, *Chromatographia*, 12 (1979) 565.
21 A.K. Williams and P.D. Cole, *J. Agr. Food Chem.*, 23 (1975) 915.
22 A. Floridi, C. Fini, C.A. Palmerini and A. Rossi, *Riv. Sci. Tecnol. Alimenti Untr. Um.*, 6 (1976) 197.
23 R.C. Williams, D.R. Baker and J.A. Schmit, *J. Chromatogr. Sci.*, 11 (1973) 618.
24 C.K. Clifford and A.J. Clifford, *J. Ass. Offic. Anal. Chem.*, 60 (1977) 1248.
25 S.K. Chapman, B.C. Greene and R.R. Streiff, *J. Chromatogr.*, 145 (1978) 302.
26 B.C. Hemming and C.J. Gubler, *J. Liq. Chromatogr.*, 3 (1980) 1697.
27 J. Davídek, F. Pudil and J. Seifert, *J. Chromatogr.*, 140 (1977) 316.
28 E.P. Frenkel, R.L. Kitchens and R. Prough, *J. Chromatogr.*, 174 (1979) 393.
29 S.C. Howard and D.B. McCormick, *J. Chromatogr.*, 208 (1981) 129.
30 T. Wehr and L. Correia, *Liquid Chromatography LC 114*, Varian, Palo Alto, CA.
31 J.A. Apffel, T.W. Alfredson and R.E. Majors, *J. Chromatogr.*, 206 (1981) 43.
32 F.L. Vandemark and G.J. Schmidt, *J. Liq. Chromatogr.*, 4 (1981) 1157.
33 B.A. Allen and R.A. Newman, *J. Chromatogr.*, 190 (1980) 241.
34 R. Schuster, *Application Note AN 232-6,* Hewlett-Packard, Avondale, PA, 1978.

Chapter 25

ANTIBIOTICS

Antibiotics are compounds produced by certain microorganisms and fungi, with common antimicrobial effects. Antibiotics belong to very different groups of organic compounds and therefore different chromatographic systems are suitable for individual subclasses of antibiotics. Liquid chromatography is suitable for the separation of antibiotics and gradient elution is often used to shorten the analysis time. The method has been applied to the separation, purification, determination and identification of individual antibiotics, their derivatives, metabolites and degradation products. Generally, continuous UV detection is used in the high-performance liquid column chromatography of antibiotics.

25.1. PENICILLINS

Penicillins can be separated using chromatography on ion-exchange resins. For example, benzylpenicillin and its degradation products were analysed by chromatography on Chromobeads type B cation-exchange resin using elution with a gradient of lithium citrate buffer[1].

Reversed-phase chromatography on chemically bonded octadecyl-[2-4] and octylsilica[5] has been applied to the analysis of penicillins. Rapid separation of penicillins V and G was achieved by elution with a linear gradient of methanol in water (Fig. 25.1)[2]. Ampicillin and its degradation products in pharmaceutical preparations were separated, identified and quantitated by reversed-phase chromatography using a convex gradient of acetonitrile in 0.01 mol/l aqueous phosphate buffer[5]. A gradient of acetonitrile in aqueous ammonium acetate was employed for the separation of monobasic penicillins and cefazolin in biological samples[4]. Chromatography of the phenacyl esters of some natural penicillins on an octylsilica column with a linear gradient of acetonitrile in water as the eluent was tested with the aim of improving both the separation and detection sensitivity[3].

25.2. CEPHALOSPORINS

Reversed-phase chromatography on an octadecylsilica column with a gradient of methanol in 0.01 mol/l phosphate buffer as the eluent was employed for the separation of diastereomeric cephalosporins[6].

References on p. 418.

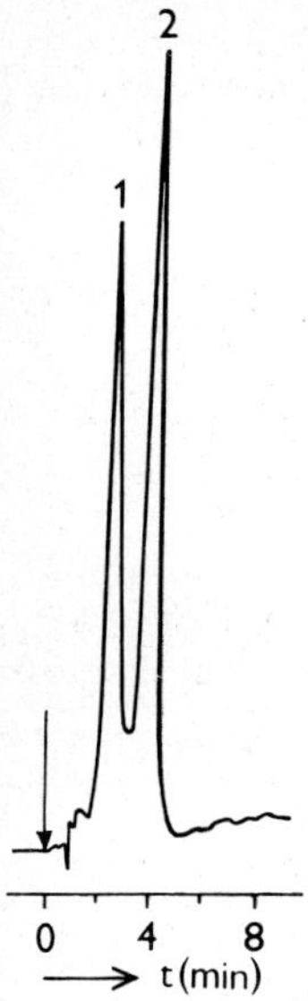

Fig. 25.1. Separation of penicillins by reversed-phase gradient elution liquid chromatography. Column: Micro Pak CH, 250 x 2 mm I.D. Gradient: linear, from 10 to 25% B in A, 1% B/min. Solvents: A, water; B, methanol. Flow-rate: 0.83 ml/min. Detection: UV at 254 nm, 0.16 a.u.f.s. Peaks: 1 = penicillin G; 2 = penicillin V. (Reproduced from ref. 2 with permission.)

25.3. TETRACYCLINES

Antibiotics of the tetracycline type have been analysed by reversed-phase chromatography on chemically bonded octadecylsilica[7-9], octylsilica[10,11] and phenylsilica[12] phases, using elution with a pH gradient[10] and linear[7] and multilinear[8,11] concentration gradients of methanol in water[11], acetonitrile[7,12] and methanol[8] in aqueous phosphate buffers (0.02 mol/l, with or without addition of ethylenediaminetetraacetic acid[8]). A concave gradient of isopropanol in an aqueous solution of ethylenediaminetetraacetic acid and diethanolamine has also been employed as the eluent[9]. These techniques were employed for rapid separations and the analysis of individual antibiotics in pharmaceutical preparations[7] and for the determination of their analogues and degradation products[9,10] (for example, epitetracycline, chlortetracycline, epianhydrotetracycline, anhydrotetracycline, penimocycline, oxytetracycline, methacycline, epidoxycycline, deoxycycline, epipenimocycline).

25.4. PEPTIDE- AND RELATED ANTIBIOTICS

Reversed-phase chromatography on chemically bonded octadecyl- and octylsilica columns and on columns packed with organic gels can be successfully applied for the analysis of peptide antibiotics.

Bacitracin, polymyxin, cizinlin, colistin, etc., were chromatographed on columns packed with totally[13] or superficially[14] porous octadecylsilica using elution with a binary convex gradient of acetonitrile in aqueous phosphate buffer[13] or with a ternary convex gradient of methanol and acetonitrile in aqueous phosphate buffer[14].

Polymyxin antibiotics (colistin A, B and C) were chromatographed on non-polar Amberlite XAD-2 styrene-divinylbenzene resin using elution with a linear gradient of hydrochloric acid in aqueous methanol (1:1) containing 0.2 mol/l sodium chloride[15].

Nikkomycins in biological fluids were assayed by reversed-phase ion-pair chromatography on a LiChrosorb RP-8 column using elution with a multi-linear gradient of methanol in 0.001 mol/l aqueous heptanesulphonic acid buffered at pH 3.75 with ammonium formate[16].

Linear gradients of ammonium formate or triethylammonium hydrogen carbonate were used as the eluents in the preparative purification of phleomycins[17] and of bleomycins[18,19] by chromatography on carboxymethyl-Sephadex cation-exchange columns. Reversed-phase chromatography on chemically bonded octadecylsilica columns using elution with multi-linear gradients of methanol in aqueous ammonium formate has also been employed for the separation of bleomycins[20].

25.5. CARBOHYDRATE ANTIBIOTICS

Gentamicin C 1, C 1A and C 2 were separated on a column packed with cellulose phosphate P-11 using elution with a convex gradient of sodium chloride (from 2 to 2.2 mol/l) in water[21].

Pre-column derivatization with 1-fluoro-2,4-dinitrobenzene yielded UV-absorbing derivatives of neomycin sulphates, which could be separated on silica gel columns (LiChrosorb Si 100, 5 μm) by gradient elution chromatography with ternary chloroform-tetrahydrofuran-water gradients (from 680:314:6 to 542:450:8)[22].

25.6. OTHER ANTIBIOTICS

Reversed-phase chromatography on a porous-layer chemically bonded phenyl phase (Phenyl/Corasil) using elution with a gradient of acetonitrile in 1% aqueous ammonium formate (pH 4) was employed for the separation and determination of adriamycin and its metabolites[23].

References on p. 418.

6-Dimethylgriseofulvin, a phenolic antibiotic, was determined in urine using normal-phase gradient elution liquid chromatography with a linear ternary gradient of methanol and isopropanol (1:1) in *n*-heptane[24].

Vancomycin was separated into three biologically active fractions by chromatography on the dextran cation exchanger CM-Sephadex G-50 using elution with a gradient from 0 to 0.13 mol/l of ammonium hydrogen carbonate in water[25].

25.7. REFERENCES

1 P. Adriaens, B. Meesschaert, H. Eyssen, E. Roets and H. Vanderhaeghe, *J. Chromatogr.*, 144 (1977) 93.
2 T. Ottake and M. Yaguchi, *LC at Work*, 23, Varian, Palo Alto, CA.
3 S. Lam and E. Grushka, *J. Liq. Chromatogr.*, 1 (1978) 33.
4 A. Tsuji, E. Miyamoto, O. Kubo and T. Yamana, *J. Pharm. Sci.*, 68 (1979) 812.
5 C. Larsen and H. Bundgaard, *J. Chromatogr.*, 147 (1978) 143.
6 H. Breuer, U.O. Treuner, M.J. Schneider, M.G. Young and H.I. Basch, *J. Antibiot.*, 31 (1978) 546.
7 K. Tsuji and J.H. Robertson, *J. Pharm. Sci.*, 65 (1976) 400.
8 G. Chevalier, C. Bollet, P. Rohrbach, C. Risse, M. Caude and R. Rosset, *J. Chromatogr.*, 124 (1976) 343.
9 G.D. Mack and R.B. Ashworth, *J. Chromatogr. Sci.*, 16 (1978) 93.
10 K. Gstrein and F. Nachtmann, *Acta Pharm. Technol.*, 26 (1980) 247.
11 C. Bollet, P. Oliva and M. Caude, *J. Chromatogr.*, 149 (1978) 625.
12 N. Muhammad and J.A. Bodnar, *J. Pharm. Sci.*, 69 (1980) 928.
13 K. Tsuji and J.H. Robertson, *J. Chromatogr.*, 112 (1975) 663.
14 K. Tsuji, J.H. Robertson and J.A. Bach, *J. Chromatogr.*, 99 (1974) 597.
15 Y. Kimura, H. Kitamura, T. Araki, K. Noguchi, M. Baba and M. Hori, *J. Chromatogr.*, 206 (1981) 563.
16 H.P. Fiedler, *J. Chromatogr.*, 204 (1981) 313.
17 T. Ikekawa, F. Iwami, H. Hiranaka and H. Umezawa, *J. Antibiot., Ser. A*, 17 (1964) 194.
18 H. Umezawa, Y. Suhara, T. Takita and K. Maeda, *J. Antibiot.*, 19 (1966) 210.
19 T.T. Sakai and J.M. Riordan, *J. Chromatogr.*, 178 (1979) 302.
20 W.J. Rzeszotarski, W.C. Eckelman and R.C. Reba, *J. Chromatogr.*, 124 (1976) 88.
21 A.H. Thomas and S.D. Tappin, *J. Chromatogr.*, 97 (1974) 280.
22 K. Tsuji, J.F. Goetz, W. van Meter and K.A. Gusciora, *J. Chromatogr.*, 175 (1979) 141.
23 J.J. Langone, H. van Vunakis and V.R. Bachner, *Biochem. Med.*, 12 (1975) 283.
24 E. Papp, K. Magyar and H.J. Schwarz, *J. Pharm. Sci.*, 65 (1976) 441.
25 G.K. Best, N.M. Best and N.N. Durham, *Antimicrob. Ag. Chemother.*, 1968 (1969) 115.

Chapter 26

SULPHONIC ACIDS, SYNTHETIC DYES, DYE INTERMEDIATES

Aromatic sulphonic acids derived from benzene, naphthalene, anthracene and anthraquinone and their amino- and hydroxysubstituted derivatives are important intermediates in production of synthetic dyes. Alkyl- and alkylarylsulphonic acids are widely used as anionic surface-active agents. Therefore, there is a need for rapid quantitative analytical methods for such an industrially important class of compounds.

Most synthetic water-soluble dyes contain one or more sulphonic (or, less frequently, carboxylic) groups that improve their solubility in water, and can be analysed using similar methods to those for other sulphonic acids. For convenience, water-insoluble dyes that do not contain sulphonic groups are also included in this chapter.

Because most compounds treated here possess aromatic rings, their detection does not represent any serious problem and can be performed conveniently using conventional UV detectors at 254 or 280 nm.

26.1. AROMATIC SULPHONIC ACIDS; ACID DYES

Sulphonic acids are very strong organic acids with an acidity comparable to that of strong mineral acids. Therefore, HPLC is potentially superior to other methods for the analysis of sulphonic acids, such as GC. However, earlier attempts to separate sulphonic acids by liquid column chromatography suffered from difficult problems and the use of ion-exchange chromatography was only partially successful, because sulphonic acids were usually too strongly retained on strong anion-exchange resins and separations by salting-out chromatography on cation-exchange columns were lengthy[1].

In one of the first relatively rapid ion-exchange chromatography applications for the separation of aromatic sulphonic acids, a 500 x 2 mm I.D. column packed with an intermediate strength anion exchanger, Bio-rex 5, was used. Elution was achieved using various linear, convex or concave gradients of lithium chloride (from 0 to 5 mol/l) in water-acetonitrile-methanol (1:1:1). Using this method, it was possible to separate a mixture of 4-biphenyl sulphonate, 4-hydroxy-4'-biphenyl sulphonate and 4,4'-biphenyl disulphonate in ca. 50 min, a mixture containing *o*- and *p*-aminobenzenesulphonic acids in ca. 40 min and a mixture of

References on p. 423.

orthanilic and dimethylorthanilic acids in ca. 40 min. The acids were detected continuously at 250 nm[2].

The introduction of porous-layer anion exchangers on inorganic supports seemed promising for the ion-exchange chromatography of sulphonic acids and, indeed, a number of rapid separations of some simple mixtures of aromatic sulphonic acids were achieved, many of them with the aid of gradient elution. For example, partial separation of a mixture of five aromatic sulphonic acids (benzene-, *p*-toluene-, 2,5-dimethylbenzene-, *p*-chlorobenzene- and naphthalene-2-sulphonic acids) was accomplished in ca. 35 min on a Zipax SAX column, using stepwise elution with 0.0025 and 0.005 mol/l aqueous perchloric acid[3]. The same column and elution with gradients of sodium borate or sodium perchlorate in water were used for the separation of a mixture containing R-salt, G-salt and 2-naphthol-3,6,8-trisulphonic acid as impurities in 6-hydroxy-2-naphthalenesulphonic acid (Schaeffer's salt)[4], for the separation of naphthionic acid, 2-naphthol-6,8-disulphonic, 2-naphthol-3,6-disulphonic and 2-naphthol-3,6,8-trisulphonic acids as impurities (reaction intermediates) in F.D. & C. Red No.2[5], for the separation of phenylhydrazine-*p*-sulphonic acid, sulphanilic acid, pyrazolone and tartrazine[6], for the separation of 4,4'-(diazoamino)bis(5-methoxy-2-methylbenzenesulphonic) acid and other impurities (intermediates) in F.D. & C. Red No.40[7-9], for the chromatography of 4,4'-(diazoamino)benzenedisulphonic acid and other intermediates as impurities in F.D. & C. Yellow No.6[10-12] and for the separation and determination of intermediates and subsidiary colours in Orange B[13,14], F.D. & C. Red No.2[15], Yellow No.5[15] and Blue No.2[16].

Successful separations of sulphonic acids have been achieved by ion-pair chromatography. Reversed-phase ion-pair systems are more convenient and simpler to operate than those in normal-phase ion-pair chromatography. A mixture of carboxylic and sulphonic acids (4-amino-, 3-amino-, 4-hydroxy- and 3-hydroxybenzenecarboxylic, benzenesulphonic, benzoic, toluene-4-sulphonic, 2,4-dimethylbenzenesulphonic and 2-hydroxybenzenesulphonic acids) was separated in ca. 25 min by chromatography on a column of silica gel (LiChrosorb Si 60) coated with 1-pentanol using elution with a linear gradient of decreasing concentration of tetrabutylammonium ion (from 0.1 to 0. mol/l) in 0.04 mol/l aqueous phosphate buffer at pH 7.4[17].

More often, chemically bonded non-polar phases (particle diameter 5-10 μm) were used for the reversed-phase ion-pair chromatography of sulphonic acids, because it is not necessary to coat the stationary phase and to pre-saturate the mobile phase. For example, various food, drug and cosmetic acid dyes containing sulphonic groups, such as F.D. & C. Yellow 5, Yellow 6, Red 2, Red 3, Red 40, Blue 1 and Blue 2, were separated in 6-30 min by chromatography on octadecylsilica columns using elution with various linear concentration gradients of methanol in aqueous solutions of tetrabutylammonium[18,19], tetraethylammonium[20] or tetramethylammonium[21] ions (phosphates, hydrogen sulphates, chlorides, acetates, 0.001-0.005 mol/l), usually buffered at pH ≃ 6-7.

Elution with a linear gradient of methanol in the mobile phase was used for the separation of phenol sulphates in algal extracts and urine by reversed-phase ion-pair chromatography on an octylsilica column[22]. A similar technique was also used for the separation of homologous linear alkylbenzenesulphonates in waste waters[23].

If mobile phases composed of water and organic solvents without further additives are employed in chromatography on alkyl bonded phases, sulphonic acids are usually eluted near (or even prior) to the column void volume, often as strongly distorted or split peaks with irreproducible shapes, owing to ion-exclusion effects. However, good separations of sulphonic acids on these columns are possible without the addition of ion-pairing agents. It is usually sufficient to add a salt (electrolyte) to the mobile phase containing water and an organic solvent to obtain retention of acids by virtue of the salting-out effect. Thus, for example, it was possible to separate a mixture of anthraquinone-1,5-,-2,6- and -1,8-disulphonic and -1- and -2-sulphonic acids on an octadecylsilica column in 10 min using stepwise elution with 0.133 mol/l sodium sulphate in 40% aqueous methanol in the first step, followed by elution with 60% aqueous methanol in the second step. Two-step elution with the first isocratic step (0.32 mol/l sodium sulphate in 12% aqueous methanol) followed by elution with a linear gradient of methanol (from 12 to 60%) in the second step was used to separate isomeric 1-aminonaphthalene-5-, -4-, -6-, -7- and -8-sulphonic acids in less than 15 min[24].

A mixture of ten naphthalenemono-, -di-, -tri- and -tetrasulphonic acids was separated in ca. 30 min on an octadecylsilica column by elution with a delayed linear concentration gradient of methanol in 0.4 mol/l aqueous sodium sulphate (Fig. 26.1)[24].

Using a linear gradient of methanol in 0.01 mol/l aqueous phosphate buffer, it was possible to separate eosine and azure dyes in 25 min by chromatography on an octadecylsilica column[25] and to resolve synthetic acid fast dyes used in alcoholic products by chromatography on an octylsilica column[26].

Dye and non-dye components of Brown Fk were separated by chromatography on a microparticulate chemically bonded aminopropyl phase using a linear gradient of acetonitrile in water for elution[27].

26.2. WATER-INSOLUBLE DYES

Water-insoluble disperse dyes in polyester textile samples (Navy D-26-133, Orange C B, Yellow D-3R, Scarlet B-R, Rubine C B, Red D-26, Red C-3B and Navy C-4B)

References on p. 423.

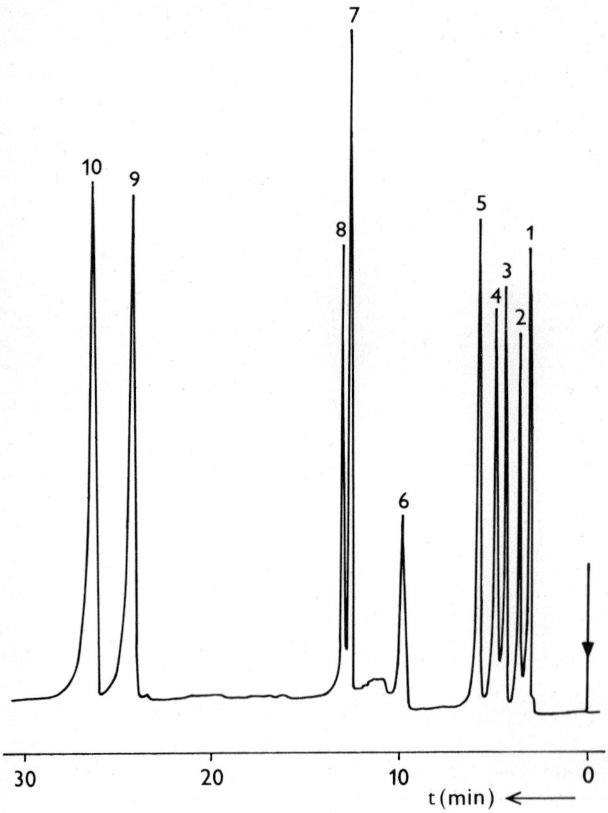

Fig. 26.1. Separation of naphthalenesulphonic acids by reversed-phase gradient elution salting-out chromatography. Column: C18 on LiChrosorb Si 100, 10 μm, 300 x 4.2 mm I.D. Gradient: two-step; first step, isocratic elution with solvent A, 5 ml; second step, linear gradient, 6.7% B/min. Solvents: A, 0.4 mol/l sodium sulphate in water; B, methanol-water (40:60). Flow-rate: 1 ml/min. Detection: UV at 254 nm, 0.5 a.u.f.s. Peaks: 1 = naphthalene-1,3,5,7-tetrasulphonic acid; 2 = naphthalene-1,3,6-trisulphonic acid; 3 = naphthalene-1,3,5-trisulphonic acid; 4 = naphthalene-1,3,7-trisulphonic acid; 5 = naphthalene-1,5-disulphonic acid; 6 = naphthalene-2,6-disulphonic acid; 7 = naphthalene-1,6-disulphonic acid; 8 = naphthalene-2,7-disulphonic acid; 9 = naphthalene-1-sulphonic acid; 10 = naphthalene-2-sulphonic acid. (Reproduced from ref. 24 with permission.)

were separated and determined by chromatography on a silica gel (10 μm) column using elution with a linear gradient of ethyl acetate in *n*-hexane[28].

A linear gradient of dichloromethane in *n*-hexane was used as the eluent for the separation of methyl-, amino- and methoxy-substituted N,N-dimethyl-*p*-aminoazobenzenes by chromatography on a silica gel (10 μm) column[29].

26.3. REFERENCES

1 P. Jandera and J. Churáček, *J. Chromatogr.*, 86 (1973) 423.
2 R.H. Stehl, *Anal. Chem.*, 42 (1970) 1802.
3 J.J. Kirkland, *Anal. Chem.*, 43 (1971) 37A.
4 D.M. Marmion, *J. Ass. Offic. Anal. Chem.*, 61 (1978) 668.
5 M. Singh, *J. Ass. Offic. Anal. Chem.*, 60 (1977) 173.
6 *LC Product Bulletin*, Du Pont, Wilmington, DE, August 1971.
7 J.E. Bailey and E.A. Cox, *J. Ass. Offic. Anal. Chem.*, 59 (1976) 5.
8 D.M. Marmion, *J. Ass. Offic. Anal. Chem.*, 59 (1976) 838.
9 S.J. Bell, *J. Ass. Offic. Anal. Chem.*, 59 (1976) 1294.
10 J.E. Bailey and E.A. Cox, *J. Ass. Offic. Anal. Chem.*, 58 (1975) 609.
11 D.M. Marmion, *J. Ass. Offic. Anal. Chem.*, 58 (1975) 719.
12 D.M. Marmion, *J. Ass. Offic. Anal. Chem.*, 60 (1977) 168.
13 M. Singh, *J. Ass. Offic. Anal. Chem.*, 60 (1977) 1105.
14 M. Singh, *J. Ass. Offic. Anal. Chem.*, 60 (1977) 1067.
15 C.J. Bailey, E.A. Cox and J.A. Springer, *J. Ass. Offic. Anal. Chem.*, 61 (1978) 1404.
16 M. Singh, *J. Ass. Offic. Anal. Chem.*, 58 (1975) 48.
17 K.G. Wahlund, *J. Chromatogr.*, 115 (1975) 411.
18 D. Popovich and D. Southern, in G. Charalambous (Editor), *Liquid Chromatographic Analysis of Food and Beverages*, Academic Press, London, 1979, p. 411.
19 *Food, Drug and Cosmetic dyes*, Waters Assoc., Milford, MA, January 1975.
20 *Spectra-Physics, Chromatogr. Rev.*, 3 No.1 (1977) 1.
21 R. Gloor and E.L. Johnson, *J. Chromatogr. Sci.*, 15 (1977) 413.
22 M.A. Ragan and M.D. Mackinnon, *J. Chromatogr.*, 178 (1979) 505.
23 W. Giger, E. Staub and L. Schaffner, *ACS Abstr.*, April 1979.
24 P. Jandera and J. Churáček, *J. Chromatogr.*, 197 (1980) 181.
25 *Eosine and Azure Dyes, Micromeritics Application Notes No. 66*, Micromeritics, Norcross, GA.
26 G.E. Martin, M. Tenenbaum and R.M. Dyer, *J. Ass. Offic. Anal. Chem.*, 61 (1978) 908.
27 A.D. Jones, D. Hoar and S.G. Sellings, *J. Chromatogr.*, 166 (1978) 619.
28 J.C. West, *J. Chromatogr.*, 208 (1981) 47.
29 *LC under Pressure, Application of LiChrosorb Si 60*, Merck, Darmstadt, G.F.R.

Chapter 27

PESTICIDES, ORGANOCHLORO COMPOUNDS

Pesticides with various structures are widely used in agriculture and, to a lesser extent, in the household and industry to destroy or suppress the growth of weeds (herbicides), insects (insecticides), fungi (fungicides), rodents (rodenticides) and others. According to their chemical nature, pesticides are divided into several sub-classes, such as chlorinated compounds, organophosphates, carbamates, phenoxy acids, *s*-triazines, substituted ureas, uracils, phenols and pyrethrins. New pesticide compounds are continually appearing on the market, which leads to increasing demands for suitable analytical methods capable not only of controlling the composition of commercial formulations, but also of allowing rapid, sensitive and selective determinations of trace amounts of pesticides and their residues and metabolites in fruits and vegetables, meat, milk, other food products, drinking water, soil, etc. Gas chromatography is still the most freqently used method for certain types of pesticides (organochlorine and organophosphorus compounds), but HPLC is becoming widely used because it is capable of meeting all (or most) of the above requirements.

Lawrence and Turton[1] presented a survey of HPLC applications for the analysis of a wide variety of pesticides, including several methods using the gradient elution technique.

UV detection (most frequently at 254 or 280 nm) is most commonly used for monitoring pesticides in effluents from analytical columns and this method is used, unless stated otherwise, in the applications cited below. However, for certain trace analyses of pesticides in environmental samples the sensitivity of this method appears to be insufficient and other detection systems have been tested.

27.1. SEPARATION OF MIXTURES OF PESTICIDES FROM DIFFERENT CLASSES

Complex mixtures of pesticides may be separated into groups before further analysis (GLC) by chromatography on a silica gel column, using stepwise elution with *n*-hexane, 60% benzene in *n*-hexane, benzene and 50% ethyl acetate in benzene[2]. This technique was used for the investigation of pesticides in samples of surface and drinking waters.

By using reversed-phase gradient elution chromatography, it was possible to separate and differentiate over 20 pesticides of different classes, included in the European Economic Community Directive on fruit and vegetables (phthalimide, dimethoate, atrazine, propoxur, carbaryl, 1-naphthol, paraoxon, thiram, captan, azinphos-methyl, chloroxuron, folpet, barban, malathion, fenitrothion, azinphos-ethyl, parathion, chlorfenson, methoxychlor, fenchlorphos, binapacryl and chlorbruside). An octadecylsilica column (Spherisorb ODS, 5 μm, 150 x 4.6 mm I.D.) was used and the elution was performed with a linear gradient of acetonitrile in water (25-75% in 50 min at 1 ml/min)[3].

27.2. CHLORINATED PESTICIDES, POLYCHLORINATED BIPHENYLS, DIOXINS

Gradient elution with a linear gradient of chlorobutane in trimethylpentane (2-100%) was used to separate *p,p'*-DDT and methoxychlor on a 250 x 2.1 mm I.D. Vydac silica gel column[4]. DDT and impurities were also separated on an octadecylsilica column using a linear gradient of acetonitrile in water and UV detection at 210 nm. Prior to reversed-phase gradient elution chromatography, samples were separated into fractions by steric exclusion chromatography on columns packed with microparticulate styrene-divinylbenzene resin[5].

Polychlorinated biphenyls are industrially used under the commercial name Aroclor. These preparations are complex mixtures of up to 210 compounds, including biphenyl and its mono- to deca-chlorinated derivatives. Reversed-phase gradient elution chromatography on superficially porous[6] and microparticulate totally porous octadecylsilica[7] and on silica with pyrolytically deposited carbon[8] was applied for the separation of some of the individual compounds in these products, using elution with exponential or linear gradients of methanol[6] or acetonitrile[7] in water or a ternary gradient of acetonitrile and ethanol in water[8]. By chromatography on a μBondapak C_{18} column (300 x 4 mm I.D.), it was possible to resolve over thirty peaks and to identify the components in various Aroclor products[7].

Non-phenolic impurities, including highly toxic dioxins (hexachlorodibenzo-*p*-dioxin and octachlorodibenzo-*p*-dioxin), in commercial pentachlorophenol used as a fungicide were separated and determined in ca. 80 min by chromatography on superficially porous octadecylsilica (Permaphase ODS) using elution with a linear gradient of methanol in water[9].

27.3. ORGANOPHOSPHATE PESTICIDES

Organophosphorus pesticides have been fractionated by classical liquid-solid column chromatography of polar adsorbents. Stepwise elution with light petroleum, benzene-light petroleum (5 : 1), chloroform and acetone was used for the fractionation of organophosphorus insecticides into groups[10]. By stepwise elution with diethyl ether-light petroleum (12 : 88) and diethyl ether-1,4-dioxane-light petroleum (50 : 4 : 946), it was possible to separate parathion, methyl-parathion, fenthion and malathion on a Florisil column[11].

Organophosphorus pesticides can be analysed by chromatography on silica gel columns. For example, fenthion and five of its metabolites were separated on silica gel columns using stepwise elution with an increasing concentration of acetone in benzene as the mobile phase (0, 1, 7.5, 10 and 100%). This method was applied in an investigation of fenthion metabolites in grass, maize and milk[12] and a similar approach was used for the analysis of disulfeton and its metabolites[13].

Mono- and diesters of phosphoric and phosphorothioic acids can be separated by anion-exchange chromatography. To achieve good separations on a Dowex 1-X8 (100-200 mesh) column (290 x 25 mm I.D.), it was necessary to perform the elution with a gradient of increasing concentrations of methanol and acetone in the mobile phase combined with a gradient of decreasing pH (from pH 2 to 1 mol/l hydrochloric acid). Using this method, O,O-dialkyl phosphorodithioate insecticides and metabolites were separated and determined in metabolic investigations of organophosphorus insecticides[14].

The above separations by classical liquid column chromatography were time consuming. In modern HPLC analysis of organophosphorus pesticides, the use of chemically bonded non-polar alkyl and aryl phases is more frequent than other systems. Octadecylsilica columns are usually employed and elution with linear gradients of acetonitrile in water is often applied. For example, the insecticides paraoxon, azinphos, fenchlorphos, chlorfenson, parathion and dimethoate in mixtures were separated using this method[3] and malathion was resolved from other pesticides[16].

Frequent applications of reversed-phase gradient elution chromatography under these conditions to the separation and determination of organophosphorus pesticides and their metabolites and residues in fruit and vegetables[17,18], soil surface dusts[17,18] and water samples have been reported[19,20]. In this way, azinphos and its metabolites[17], parathion and paraoxon and their residues[18], dursban[19] and abate[20] in environmental samples have been analysed. Abate was separated from parathion and semithion on a bonded phenyl phase (Bondapak Phenyl/Corasil) by elution with a linear gradient of acetonitrile in water[20].

Paraoxon and its degradation products were separated by chromatography on a medium-polarity column of Durapak Carbowax 400/Corasil by elution with a gradient of dioxane in *n*-hexane[18].

The organophosphorus pesticides fonofos and parathion and some carbamate and organochlorine pesticides were separated on superficially porous Vydac silica gel by elution with a linear gradient of chlorobutane in trimethylpentane[4].

References on p. 432.

27.4. CARBAMATE PESTICIDES

Anion-exchange chromatography on a column packed with Cellex AE (aminoethylcellulose) was used for the separation and determination of mobam (benzo[*b*]thien-4-yl methylcarbamate) and its metabolites in cow and goat milk and stepwise elution was performed with water, 0.5 mol/l formic acid, 0.1 mol/l ammonium formate (pH 4) and 1 mol/l ammonia in water[21]. Urinary metabolites of 3,4-dichlorobenzyl N-methylcarbamate were separated on column packed with DEAE-cellulose using a gradient (from 0.01 to 0.05 mol/l) in aqueous Tris-HCl buffer (pH 7.5)[22]. In these studies of carbamate pesticide metabolism, radiolabelled compounds were used and continuous radioactivity measurements were used for detection.

HPLC on silica gel columns can be used for the separation and determination of carbamate pesticides. Elution with a concave concentration gradient of isopropanol (1-10%) in *n*-heptane was employed for the separation of aldicarb from other pesticides (benomyl, bux) on a 250 x 2.1 mm I.D. column packed with 10 μm silica gel[23]. A μPorasil column and a gradient of methanol (1-15%) in chloroform served for the separation and determination of croneton in biological fluids[24]. Benomyl, carbofuran, aldicarb and their metabolites were analysed by chromatography on a silica gel column using linear or exponential gradients of acetone in *n*-hexane for elution[25].

Aldicarb, benomyl and bux were chromatographed on chemically bonded nitrile and amino phases using elution with a concave gradient of isopropanol in *n*-heptane (from 1 to 10%) and UV detection at 220 nm[23]. By elution with a linear gradient of dichloromethane in *n*-hexane (from 5 to 25%), it was possible to separate hopcide, isoprocarb, macbal, meobal, propoxur and tsumacide carbamate insecticides on a bonded amino phase (Micro Pak NH_2 column, 250 x 2.2 mm I.D.)[26,27].

Chemically bonded medium-polarity phases were also used for separations of carbamate pesticides in reversed-phase systems. Butylate, diallate, EPTC and pebulate were separated on an ethereal phase bonded on a superficially porous support (Permaphase ETH) by elution with a linear gradient of acetonitrile in water (from 10 to 30%)[23]. Separations of carbamate insecticides were compared on totally porous nitrile and octylsilica bonded phases, using elution with a linear concentration gradient of acetonitrile in water (from 12 to 70%) at 30°C. It was possible to separate completely sixteen carbamates and metabolites at the 1 ng level (aldicarb sulphoxide, aldicarb sulphone, oxamyl, methomyl, 3-hydroxycarbofuran, methiocarb sulphoxide, methiocarb sulphone, aldicarb, propoxur, carbofuran, bendiocarb, carbaryl, α-naphthol, landrin, methiocarb and bufencarb), with the exceptions of the pair aldicarb sulphone - oxamyl on the nitrile phase and the pair carbofuran - bendiocarb on the octylsilica phase[28].

Using elution with a linear or concave gradients of acetonitrile in water and columns packed with microparticulate octadecylsilica, it was possible to separate a number of carbamate pesticides, such as mixtures containing methomyl, aldicarb, isoban, baygon, carbofuran, mobam, carbaryl, landrin, propham, banol, mesurol, zectran, betanal, chloropropham, bux, eptam, pebulate, vernolate, eurex, captafol, barban, avadex, butylate[23] and mixtures of lannate, temik, baygon, carbofuran, sevin, propham, mesurol, zectran, chloropropham and barban[29]. On a superficially porous octadecylsilica column (Permaphase ODS), it was possible to separate sevin, landrin, metacin and furadan in 8 min by elution with a linear gradient of methanol in water[30]. Linear gradients of methanol in water were also applied to the separation of carbamate insecticides and metabolites (such as carbofuran and its five metabolites[31] and methomyl, carbofuran, carbaryl, methiocarb, 3-hydroxycarbofuran, propoxur and α-naphthol[32]) on microparticulate octadecylsilica columns. On these columns, aldicarb, carbaryl, carbofuran, propoxur, methiocarb and their metabolites in vegetable samples were also separated by elution with a concave gradient of dioxane in water[33].

3,4,4'-Trichlorocarbanilide and its metabolites in biological samples were separated and determined by chromatography on a microparticulate octadecylsilica column using a linear gradient of acetonitrile in 0.2% aqueous ammonium phosphate[34].

Chromatography on an octadecylsilica column with a linear gradient of acetonitrile in water was used for the analysis of agricultural fungicides based on carbamate - metal complexes (ferbam) and their decomposition products[35].

Various attempts have been made to increase the sensitivity of detection of carbamate pesticides with respect to the usual UV monitoring of the eluate. Polarographic, electron-capture, post-column reaction cholinesterase detectors and a post-column reaction detector utilizing the reaction of carbamates with dinitrophenylhydrazine were described[29]. The selectivity for nitrogen-containing pesticides can be increased by using a nitrogen-selective detector based on the combination of a moving-wire mass transport system and of a conductivity detector of the Hall type[25]. Moreover, a post-column reaction detector has been used in which the hydrolysis of carbamate in the effluent was performed after continuous addition of sodium hydroxide, followed by derivatization with *o*-phthalaldehyde, which allowed the continuous monitoring of fluorescence of the derivatives formed[28,32].

27.5. PHENOXY ACIDS

Phenoxycarboxylic acids can be conveniently separated by reversed-phase chromatography on chemically bonded octadecylsilica columns. Gradient elution may be applied to speed up the elution, such as in the separation of 2,4-dichlorophenoxyacetic acid and its phenolic degradation products on a μBondapak C_{18} column,

References on p. 432.

where elution with a linear concentration gradient of methanol in 0.01 mol/l phosphate buffer was used[36].

27.6. *s*-TRIAZINES

Ion-exchange chromatography on a Dowex 50W-X8 cation-exchange column was used for the separation of atrazine and its metabolites using a gradient of acetic acid in the mobile phase containing 2 mol/l ammonia in 30% methanol[37].

Gradient elution with a ternary gradient of dichloromethane and isopropanol in *n*-hexane was used to separate the triazine pesticides propazine, atrazine, simazine and cotoran in 14 min on a column packed with microparticulate silica gel[38].

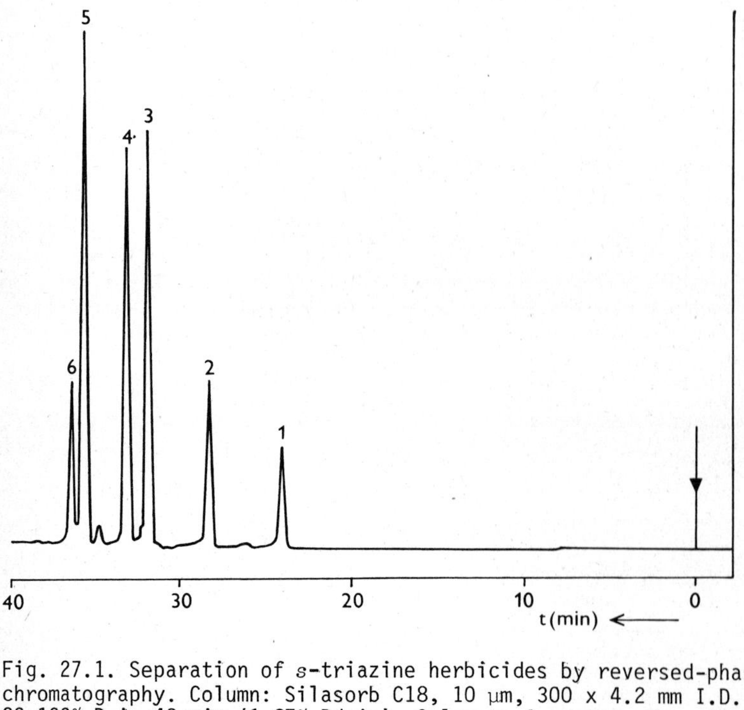

Fig. 27.1. Separation of *s*-triazine herbicides by reversed-phase gradient elution chromatography. Column: Silasorb C18, 10 μm, 300 x 4.2 mm I.D. Gradient: linear, 20-100% B in 48 min (1.67% B/min). Solvents A, water; B, methanol. Flow-rate: 1 ml/min. Detection: UV at 254 nm. Peaks: 1 = simazine; 2 = atrazine; 3 = metoprotrin; 4 = terbutylazine; 5 = prometrin; 6 = terbutrin.

s-Triazine herbicides can also be successfully separated using reversed-phase systems. Gradient elution with a concave gradient of acetonitrile in water was applied to the separation of some *s*-triazines on a μBondapak C_{18} column[39].

Gradients of methanol in water may also be used for this purpose (see Fig. 27.1, showing the separation of six *s*-triazine herbicides on an octadecylsilica column using elution with a linear gradient of methanol in water)[40].

27.7. SUBSTITUTED UREAS

Chromatography on a microparticulate silica gel column using elution with a linear gradient of dichloromethane and ethanol in *n*-hexane was applied for the separation of urea herbicides and chloroanilines as accompanying products. It was possible to separate a mixture containing 2,4,6-trichloroaniline, 2,5-dichloroaniline, 2,4,5-trichloroaniline, 2-chloroaniline, 3,4-dichloroaniline, 4-chloroaniline, metobromuron, methabenzthiazauron, diuron, monuron and fenuron in less than 20 min[41]. Linuron and seven related urea pesticides and degradation products were separated in ca. 30 min on a μPorasil column using elution with a linear gradient of propanol (containing 0.5% of ammonia and 1.5% of water) in *n*-hexane[42].

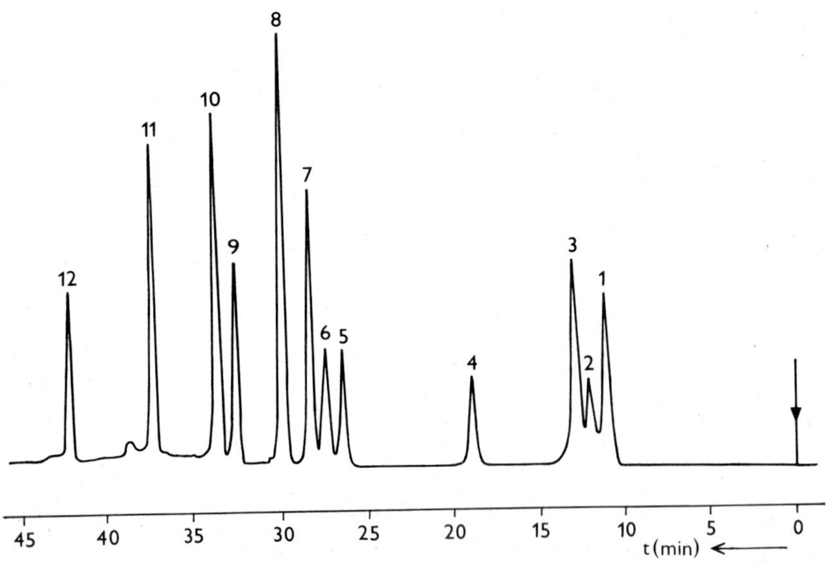

Fig. 27.2. Separation of urea herbicides by reversed-phase gradient elution chromatography. Column: Silasorb C18, 10 μm, 300 x 4.2 mm I.D. Gradient: linear, 20-100% B in 48 min (1.67% B/min). Solvents A, water; B, methanol. Flow-rate: 1 ml/min. Detection: UV at 254 nm. Peaks: 1 = hydroxymetoxuron; 2 = desfenuron; 3 = fenuron; 4 = metoxuron; 5 = isoproturon; 6 = fluometuron; 7 = chlorotoluron; 8 = diuron; 9 = linuron; 10 = chlorobromuron; 11 = neburon; 12 = N',N-bis(3-chloro-4-methylphenyl)urea.

References on p. 432.

The rapid separation of diuron and its two degradation products was possible in less than 6 min using a Micro Pak NH_2 column (bonded amino phase) and elution with a multi-linear convex gradient of isopropanol - methylene chloride (1 : 2) in isooctane[43].

Urea herbicides can be separated by reversed-phase gradient elution chromatography, as illustrated in Fig. 27.2, where the separation of twelve urea herbicides on an octadecylsilica column (Silasorb C18) using a linear gradient of methanol in water for elution is shown[40].

27.8. OTHER PESTICIDES

The reversed-phase separation of pyrethrine pesticides on a column of Permaphase ODS with the use of an exponential gradient of methanol in water (from 30 to 100%) for elution was reported[6].

A mixture of rotenoids (rotenolone, tephrosin, rotenone, deguelin, dehydrorotenone and dehydrodeguelin) was separated on a microparticulate octadecylsilica column by elution with a concave gradient of methanol in water with addition of 0.1% of phosphoric acid[44].

The fungicide triadimefon and its metabolic products (triadimenol, 4-chlorophenol, 4-chlorophenyl methylcarbamate and 1,2,4-triazole) were separated by reversed-phase chromatography on an octadecylsilica column using gradients of methanol or acetonitrile in water for elution[45].

The fungicide quinomethionate and its degradation products were chromatographed on a microparticulate silica gel column using elution with a linear gradient of ethanol in *n*-hexane (from 10 to 80%)[46].

Organic and inorganic arsenic compounds, used as herbicides, pesticides and wool preservatives (dimethylarsinate, monomethyl arsonate, *p*-aminophenyl arsonate and AsO_3^{3-} and AsO_4^{3-} ions), were separated in ca. 10 min by chromatography on a Dionex 500 x 3 mm I.D. anion-exchange column using elution with an exponential gradient of 0.003 mol/l sodium hydrogen carbonate - 0.0024 mol/l sodium carbonate solution as solvent B in 0.0025 mol/l aqueous sodium tetraborate as solvent A. Continuous detection was performed using an on-line arsine generator-atomic-absorption spectrometer flow-through detection system[15].

27.9. REFERENCES

1 J.F. Lawrence and D. Turton, *J. Chromatogr.*, 159 (1978) 207.
2 V. Leoni, *J. Chromatogr.*, 62 (1971) 63.
3 R.A. Hoodless, J.A. Sidwell, J.C. Skinner and R.D. Treble, *J. Chromatogr.*, 166 (1978) 279.
4 M.T. Jackson, *L. C. Application Briefs*, No. 117, Spectra Physics, San Jose, CA, 1974.

5 R.E. Majors and E.L. Johnson, *J. Chromatogr.*, 167 (1978) 17.
6 S.H. Byrne, J.A. Schmit and P.E. Johnson, *J. Chromatogr. Sci.*, 9 (1971) 592.
7 L.S. Kaminsky and M.J. Fasco, *J. Chromatogr.*, 155 (1978) 363.
8 T. Hanai and H.F. Walton, *Anal. Chem.*, 49 (1977) 1954.
9 C.D. Pfeiffer, *J. Chromatogr. Sci.*, 14 (1976) 386.
10 W. Sans, *J. Agr. Food Chem.*, 15 (1967) 192.
11 M. Duffy and R. Wong, *J. Agr. Food Chem.*, 15 (1967) 457.
12 M.C. Bowman and M. Beroza, *J. Agr. Food Chem.*, 16 (1968) 399.
13 M.C. Bowman, M. Beroza and C.R. Gentry, *J. Ass. Offic. Anal. Chem.*, 52 (1969) 157.
14 F.W. Plapp and J.E. Casida, *Anal. Chem.*, 30 (1958) 1622.
15 G.R. Ricci, L.S. Shepard, G. Colovos and N.E. Hester, *Anal. Chem.*, 53 (1981) 610.
16 R.E. Majors, *J. Chromatogr. Sci.*, 18 (1980) 571.
17 J. Kvalag, D.L. Elliott and F.A. Günther, *J. Ass. Offic. Anal. Chem.*, 60 (1977) 911.
18 J. Kvalag, D.L. Elliott, Y. Iwata and F.A. Gunther, *Bull. Environ. Contam. Toxicol.*, 17 (1977) 253.
19 W.A. Saner and J. Gilbert, *J. Liq. Chromatogr.*, 3 (1980) 1753.
20 A. Otsuki and T. Takaku, *Anal. Chem.*, 51 (1979) 833.
21 J.D. Robbins, J.E. Bakke and J.V. Feil, *J. Agr. Food Chem.*, 17 (1969) 236; 18 (1970) 130.
22 J.B. Knaak and L.J. Sullivan, *J. Agr. Food Chem.*, 16 (1968) 454.
23 C.M. Sparacino and J.W. Hines, *J. Chromatogr. Sci.*, 14 (1976) 549.
24 D.E. Nye, H.E. Hurst and H.W. Dorough, *J. Agr. Food Chem.*, 24 (1976) 371.
25 K.R. Hill and H.L. Crist, *J. Chromatogr. Sci.*, 17 (1979) 395.
26 Y. Ishi and T. Otake, *Bull. Agr. Chem. Insp. Std.*, 13 (1973) 32.
27 Y. Ishi, *VIA*, 9 (1975) 10, Varian, Palo Alto, CA.
28 R.T. Krause, *J. Chromatogr.*, 185 (1979) 615.
29 A. Moye, *J. Chromatogr. Sci.*, 13 (1975) 268.
30 L.F. Krzeminski, B.L. Cox and A.W. Neff, *Anal. Chem.*, 44 (1972) 126.
31 H. Sonobe, R.A. Carver and L.R. Kamps, *J. Chromatogr.*, 210 (1981) 356.
32 R.T. Krause, *J. Chromatogr. Sci.*, 16 (1978) 281.
33 H.A. Moye and P.A.S. John, *Ann. Meet. AOAC, Washington, DC, 1975.*
34 T.A. Taulli, J.T. Hill and G.W. Pounds, *J. Chromatogr. Sci.*, 15 (1977) 111.
35 E. Murrill, *Res./Develop.*, Aug. (1974) 30; *Application Highlight: Agricultural Fungicides*, Waters Assoc., Milford, MA, 1975.
36 A.D. Drinkwine, D.W. Bristol and J.R. Fleeker, *J. Chromatogr.*, 174 (1979) 264.
37 G.L. Lamoureux, R.H. Shimabukuro, H.R. Swanson and D.S. Frear, *J. Agr. Food Chem.*, 18 (1970) 81.
38 S. Reifsneider and J. Gotkis, *LC at Work*, 24, Varian, Palo Alto, CA.
39 D.J. Subach, *Chromatographia*, 14 (1981) 371.
40 P. Jandera, unpublished results.
41 J. Pribyl, *Chromatographia*, 10 (1977) 753.
42 G. Glad, T. Popoff and O. Theander, *J. Chromatogr. Sci.*, 16 (1978) 118.
43 G. Legendre and R.E. Majors, *LC at Work*, 25, Varian, Palo Alto, CA.
44 R.I. Freudenthal, D.C. Emmerling and R.L. Baron, *J. Chromatogr.*, 134 (1977) 207.
45 T. Clark, D.R. Clifford, A.H.B. Deads, P. Gendle and D.A.M. Watkins, *Pestic. Sci.*, 9 (1978) 497.
46 T. Clark and R.S.T. Leoffler, *Pestic. Sci.*, 11 (1980) 451.

Chapter 28

OTHER OXYGEN- AND NITROGEN-CONTAINING HETEROCYCLES OF BIOCHEMICAL IMPORTANCE

This chapter is concerned with some biologically important oxygen-containing heterocycles (flavones and similar compounds) and with organic pigments containing nitrogen heterocycles in their structures (porphyrins, chlorophylls, haemoglobins).

28.1. OXYGEN HETEROCYCLES

28.1.1. Flavonoids and coumarins

Flavones have been chromatographed on a silica gel (μPorasil) column using a convex gradient of dichloromethane - ethanol - acetic acid (97 : 3 : 0.2) in *n*-hexane for elution. Nine isoflavones (natural colorants) in soya beans were separated and determined by this method[1].

Isomeric glycoflavones (isoorientin-3'-O-glucoside and isoorientin-4'-O-glucoside) were separated in 15 min on a column packed with a microparticulate bonded amino phase using a linear gradient of acetonitrile in water for elution[2].

To mention a few earlier applications, polyamide was used as a column packing material for the separation of vitexin, homovitexin, saponaretin, homoorientin and orientin by elution with a linear gradient of ethanol concentration in water[3]. A number of flavonoids were separated on a column packed with Amberlite XAD-2 resin (200-400 mesh) by elution with a linear gradient of ethanol in water (from 20 to 90%). The flavonoids were eluted from the column in the order sugar esters, glycosides and aglycones. However, this early reversed-phase separation was lengthy, taking ca. 20 h[4].

In more recent HPLC applications of reversed-phase chromatography to separations of flavonoid compounds, columns packed with chemically bonded microparticulate octadecylsilica[5-8], octylsilica[9-11,38] or alkylphenylsilica[12] materials have been employed. A C-22 bonded phase was also tested for this purpose[8]. Linear[6-10,38], concave[5] and convex[12] gradients of methanol in water[5,6] and aqueous phosphate buffer[7], of methanol[9,10] and ethanol[12] in dilute aqueous acetic acid and ternary gradients of methanol and acetone in water[8], acetonitrile in phosphate buffer[38] and acetonitrile and tetrahydrofuran in water[11] were used for elution. These techniques were applied to the analysis of flavonoids in various plants[7,10,12,38] (such as of vitexin and isovitexin in *Passiflora*

References on p. 438.

incarnata used in pharmaceutical preparations[5]), soya beans (the isoflavonoids genistin, daidzin and their aglucones)[6], oats[9], orange juice (polymethoxylated flavones)[11], etc. UV detection at 254-365 nm or fluorimetric detection were employed in these separations.

Reversed-phase chromatography on microparticulate octadecylsilica columns was also used for the chromatography of coumarins (naturally occurring flavouring and fragrance materials). A concave gradient of methanol in aqueous acetic acid was used as the eluent for the separation of coumarin glucosides[13], and with a multi-linear gradient of tetrahydrofuran in 0.1% (v/v) aqueous acetic acid it was possible to separate coumarin and ten of its metabolites in rat livers (*o*-coumaric acid, isomeric hydroxycoumarins, *o*-hydroxyphenylpyruvic and *o*-hydroxyphenylacetic acids)[14].

Non-aqueous reversed-phase chromatography using gradients of *n*-hexane in tetrahydrofuran was applied to the separation and determination of neutral coumarins in roots of the citrus and poncirus[15].

Coumarins and related coumpounds in citrus oils 5-geranoxypsoralen, 5-geranoxy-7-methoxycoumarin, 5-geranoxy-8-methoxypsoralen, 5,7-dimethoxycoumarin and 5,8-dimethoxypsoralen) were determined by chromatography on a column packed with microparticulate silica gel using elution with a delayed linear gradient of chloroform in *n*-hexane and a combination of UV and fluorimetric detection[39].

28.1.2. Aflatoxins

Aflatoxins are strongly toxic and carcinogenic coumpounds, structurally related to flavonoids, which are produced by *Aspergillus flavus* fungi. Therefore, a sensitive and rapid method for the determination of these products in food and agricultural products would be very useful, and HPLC is well suited for this purpose.

The separation and determination of individual aflatoxins on a porous silica column (Corasil II) using elution with a linear gradient of tetrahydrofuran in isooctane has been reported[16].

Aflatoxins have also been analysed using reversed-phase chromatography on columns packed with microparticulate octacecylsilica. Linear gradients of methanol in water[17] or of acetonitrile in water - methanol (3 : 2)[18] served as the eluents and very sensitive detection was accomplished using continuous monitoring of the fluorescence in the eluate (λ_{ex} = 365 nm; λ_{em} = 400 nm).

28.2. NITROGEN HETEROCYCLES

28.2.1. Porphyrins and related compounds

Porphyrins are based on four pyrrole aza-heterocycles interconnected by means of methine groups. Their complexes with iron(II) ions together with a protein (globin) form the structural basis of blood haemoglobins. Photosynthetic pigments (chlorophylls) are complexes of porphyrins with magnesium.

Chromatography on columns packed with microparticulate silica gel has been used for the analysis of porphyrins. Gradient elution with an increasing concentration of ethyl acetate[19] or tetrahydrofuran[20] in *n*-hexane was used for the separation and determination of porphyrin methyl esters in biological fluids. The method can be applied to the investigation of porphyrin metabolism and for the clinical diagnosis of the type of porphyria. Free acid porphyrins in blood plasma were chromatographed using elution with a gradient of acetic acid in acetone[21]. A gradient of simultaneously increasing concentrations of toluene and chloroform in *n*-hexane was used in the chromatography of petroporphyrins in samples of oil and shale extracts[22].

The analysis of complex mixtures of petroporphyrins in geological samples is best performed in two steps, using chromatography on silica columns for preliminary purification and fractionation, followed by rechromatography of fractions of the eluate on octadecylsilica columns, after removing the solvents. In both steps, the gradient elution technique has been employed. A linear gradient of increasing concentrations of toluene and chloroform in *n*-hexane was used for elution from the silica gel column in the first step, and chromatography on octadecylsilica columns was performed with a linear ternary gradient of acetonitrile and chloroform in water[37].

Protoporphyrins and related compounds in biological samples were separated by reversed-phase chromatography on a column packed with octadecylsilica using elution with a linear gradient of methanol in aqueous acetic acid[23].

In these applications, absorption at 400 nm or fluorescence monitoring were used for detection.

Azopigments from bile (bilirubin formed by degradation of haemoglobins) and their glycosides in biological material were separated and determined by reversed-phase chromatography on a column packed with microparticulate Shimadzu PCH material, using elution with a linear gradient of acetonitrile in aqueous acetate buffer[24].

References on p. 438.

28.2.2. *Haemoglobins*

The protein part in the molecules of haemoglobins is most important for their chromatographic properties and therefore similar chromatographic conditions are used for the chromatography of haemoglobins as in the liquid column chromatography of proteins. Various haemoglobins, oxyhaemoglobin, carboxyhaemoglobin, most often in mixtures with albumin and other proteins, are separated by ion-exchange chromatography on DEAE-cellulose[25-27] or DEAE-Sephadex[28], on phosphocellulose[29] or on CM-cellulose[30,31] by elution with pH[25,26,29,30] or ionic strength[27,28,31] gradients in Tris-HCl[25-29] or phosphate[26,30] buffers.

Rapid separations of haemoglobins A_0, A_1 and A_2 were achieved by high-efficiency anion-exchange chromatography on a Syn Chropak AX 300 (10 μm) column, using elution with a gradient of increasing sodium acetate concentration in 0.02 mol/l Tris-HCl buffer[32].

Detection in the visible region at 410-415 nm is usually employed in the chromatography of haemoglobins.

28.2.3. *Chlorophylls*

Photosynthetic pigments (chlorophylls) in various plant extracts have been separated and determined by reversed-phase chromatography on octadecylsilica columns by elution with linear or multi-linear gradients of methanol in water[33] and ethanol[34] or ethyl acetate[35] in methanol. Columns packed with microparticulate polyethylene powder and a linear gradient of acetone in water were also applied for preparative separations of chlorophylls[36]. Chlorophylls are usually detected by monitoring the absorbance in the visible region at 440 nm.

28.3. REFERENCES

1 R.E. Carlson and D. Dolphin, *J. Chromatogr.*, 198 (1980) 193.
2 H. Becker, G. Wilking and K. Hostettmann, *J. Chromatogr.*, 136 (1977) 174.
3 R. Paris and S. Etchepare, *Ann. Pharm. Fr.*, 26 (1968) 51.
4 M. Hori, *Bull. Chem. Soc. Jap.*, 42 (1969) 2333.
5 V. Quercia, L. Turchetto, N. Pierini, V. Cuozzo and G. Percaccio, *J. Chromatogr.*, 161 (1978) 396.
6 P.A. Murphy, *J. Chromatogr.*, 211 (1981) 166.
7 C.T. Seitz and R.E. Wingard, *J. Agr. Food Chem.*, 26 (1978) 278.
8 D.A. Wellum and W. Kirby, *J. Chromatogr.*, 206 (1981) 400.
9 D. Strack, K. Fuisting and G. Popovici, *J. Chromatogr.*, 176 (1979) 270.
10 D. Strack and J. Krause, *J. Chromatogr.*, 156 (1978) 359.
11 R.L. Rouseff and S.V. Ting, *J. Chromatogr.*, 176 (1979) 75.
12 M. Vanhaelen and R. Vanhaelen-Fastre, *J. Chromatogr.*, 187 (1980) 255.
13 M. Williams, G. Hrazdina, M.M. Wilkinson, J.G. Sweeny and G.A. Iacobucci, *J. Chromatogr.*, 155 (1978) 389.
14 D.G. Walters, B.G. Lake and R.C. Cottrell, *J. Chromatogr.*, 196 (1980) 501.

15 H.E. Nordby and S. Nagy, *J. Chromatogr.*, 207 (1981) 21.
16 L.M. Seitz, *J. Chromatogr.*, 104 (1975) 81.
17 R. Knutti, C. Balsiger and K. Sutter, *Chromatographia*, 12 (1979) 349.
18 R.M. Beebe and D.M. Takahashi, *J. Agr. Food Chem.*, 28 (1980) 481.
19 S.A. Matlin, A.H. Jackson and N. Evans, *Methodol. Dev. Biochem.*, 5 (1976) 45.
20 R. Hörchner and T. Rietveld, *J. Chromatogr.*, 123 (1976) 414.
21 M.O. Longas and M.B. Poh-Fitzpatrick, *Anal. Biochem.*, 104 (1980) 268.
22 W.H. McFadden, D.C. Bradford, G. Eglinton, S.K. Hajibrahim and N. Nicolaides, *J. Chromatogr. Sci.*, 17 (1979) 518.
23 P. Culbreth, G. Walter and R. Carter, *Clin. Chem.*, 25 (1979) 605.
24 S. Onishi, S. Itoh, N. Kawade, K. Isobe and S. Sugiyama, *J. Chromatogr.*, 182 (1980) 105.
25 G.D. Efremov and T.H.J. Huisman, *J. Chromatogr.*, 89 (1974) 191.
26 B.F. Horton and A.I. Chernoff, *J. Chromatogr.*, 47 (1970) 493.
27 *Separation News*, 1 (1978) 1, Pharmacia, Uppsala, Sweden.
28 *Sephadex Ion Exchangers*, Pharmacia, Uppsala, Sweden, 1975.
29 B.F. Horton and A.I. Chernoff, *J. Chromatogr.*, 63 (1971) 414.
30 T.H.J. Huisman and R.N. Wrightstone, *J. Chromatogr.*, 92 (1974) 391.
31 W.A. Schroeder, L.A. Pace and T.H.J. Huisman, *J. Chromatogr.*, 118 (1976) 295.
32 K.M. Gooding, Kai-Chun Lu and F.E. Regnier, *J. Chromatogr.*, 164 (1979) 506.
33 D. Davies and E.S. Holdsworth, *J. Liq. Chromatogr.*, 3 (1980) 123.
34 T. Braumann and L.H. Grimme, *J. Chromatogr.*, 170 (1979) 264.
35 K. Eskins and H.J. Dutton, *Anal. Chem.*, 51 (1979) 1885.
36 H.-C. Chow, M.B. Caple and C.E. Strouse, *J. Chromatogr.*, 151 (1978) 357.
37 S.K. Hajibrahim, *J. Liq. Chromatogr.*, 4 (1981) 749.
38 R. Schuster, *Chromatographia*, 13 (1980) 379.
39 H.W. Latz and D.A. Ernes, *J. Chromatogr.*, 166 (1979) 189.

Chapter 29

ORGANOMETALLIC COMPOUNDS, METAL COORDINATION COMPLEXES AND INORGANIC IONS

29.1. ORGANOMETALLIC COMPOUNDS AND METAL COORDINATION COMPLEXES

In contrast to the frequently used ion-exchange separations of metal cations and inorganic anions, relatively few liquid chromatography applications to separations of organometallic and metal coordination compounds have been reported. A review of LC applications in this field has been published recently[29].

In addition to UV or visible detection of organometallic compounds, attempts to use atomic-absorption spectrometry (AAS) for detection purposes have been reported. Direct coupling of the outlet from the column to a conventional flame atomic-absorption spectrometer has several serious disadvantages, for example, a low detection sensitivity. Coupling of a liquid chromatograph with an electrothermal (graphite furnace) AAS detector has been described, which makes possible the determination of trace amounts of organometallic compounds at the nanogram level. Here, discontinuous automatic sampling from column effluent into the detector is employed and the chromatographic peak is recorded in the form of a set of "spikes", which necessitates numerical evaluation[7]. Direct on-line coupling has been accomplished between a liquid chromatograph and an atomic-absorption spectrometer equipped with an inductively coupled argon plasma emission source. Here, the plasma is generated by induction heating of a flowing gas (argon) by inducing a magnetic field around the top of an assembly of coaxial silica tubes which constitutes the burner[4].

The formation of complexes with chelating ligands was used for separations of certain metal ions. Nanogram amounts of mercury, nickel, copper and lead in the form of dialkyl dithiocarbamates, 1,2-diketobisthiobenzhydrazones, 1,2-diketobisthiosemicarbazones[1] and diacetylbisthiobenzhydrazones[2] were chromatographed on columns packed with silica gel (30 μm) using elution with gradients of benzene in *n*-heptane[1,2] or chloroform in *n*-hexane[1] and UV detection at 360 nm.

Geometrical isomers of octahedral cobalt(III) and chromium(III) chelates of various β-diketones were separated by chromatography on a column packed with silica gel using elution with a gradient of acetonitrile in dichloromethane and UV detection at 254 nm[30].

However, certain organometallic compounds are too strongly retained on silica gel and therefore chemically bonded phases yield better separations. For example,

References on p. 444.

six complexes of palladium(II) with N-methyl-N-phenylhydrazones of some aldehydes and ketones were resolved by chromatography on a column packed with LiChrosorb DIOL (10 μm), using elution with a linear gradient of dichloromethane in *n*-hexane. This method was used for controlling the synthesis and checking the purity of prepared complexes[3].

Reversed-phase chromatography on octadecylsilica[5-7] or octylsilica[4,31] columns is suitable for the separation and determination of various organometallics, which can be distinguished on the basis of differences in hydrophobicities (water solubilities) and molecular charges. For example, various iron and molybdenum carbonyl complexes were separated and determined using elution with gradients of methanol or ethanol in water[4,31]. Ion-pair reversed-phase chromatography with a gradient of tetrahydrofuran in water containing 0.005 mol/l *n*-heptanesulphonic acid and buffered to pH 3.5 was employed for the rapid separation of diester, monoester, monocarboxylate- and dicarboxylate derivatives of tris(2,2'-bipyridyl)-ruthenium(II)[5]. The ion-pair reversed-phase technique was also used for the chromatographic analysis of series of cobalt(III) bis(ethylenediamine) complexes with various amino acids (glycine, proline, valine, leucine and phenylalanine) in an investigation of cobalt(III)-induced synthesis of small peptides. A linear gradient of methanol in water containing *p*-toluenesulphonic acid was used for elution[6].

Gradients involving increasing concentrations of strong ligands were necessary to avoid extremely large retention volumes of alkyl- and arylmercury compounds. For this purpose, 2-mercaptoethanol dissolved in methanol was used as the stronger eluting solvent B and aqueous ammonium acetate as the starting solvent A[7].

29.2. METAL IONS

Various metal ions have been separated on ion-exchange resins. In separations on cation-exchange columns, differences in the selectivities of resins for various metal cations were utilized as the basis of the separation mechanism. For example, elution with a continuously increasing concentration of hydrochloric acid in the mobile phase was used for the separation of zinc from barium[8] and of manganese from other cations[9] on columns packed with Dowex 50W cation exchanger.

However, applications of complex-forming equilibria with various ligands added to the mobile phase are more frequent. For example, certain metal ions form anionic complexes with chloride ions. These complex-forming equilibria were utilized in the anion-exchange chromatography of metal chloride complex anions using elution with a stepwise decreasing concentration of hydrochloric acid in the mobile phase[10,11]. Continuous gradient elution was also employed for this purpose[12]. A similar principle was used in the chromatographic separation of nickel, manganese, cobalt, copper, iron, cadmium and zinc on columns packed with silica gel coated with tri-*n*-octylamine (which acts as a liquid anion exchanger), where the elution was performed with a gradually decreasing concentration of hydrochloric acid and an increasing concentration of nitric acid in the mobile phase[13].

Lanthanides and some other cations were chromatographed on cation-exchange columns with addition of complex-forming aliphatic hydroxycarboxylic acids to the mobile phase[11]. The separation was performed either by the stepwise elution technique or using continuous concentration gradients of citric[14] or α-hydroxyisobutyric acids[15-17] in a buffered aqueous mobile phase, and with pH gradients in aqueous solutions containing lactate[18] or α-hydroxyisobutyrate[19] anions.

In the earlier applications, which were rather lengthy, the metal concentrations were determined in fractions of the eluate, usually by spectrophotometric methods. Detailed information on these applications can be found elsewhere[11].

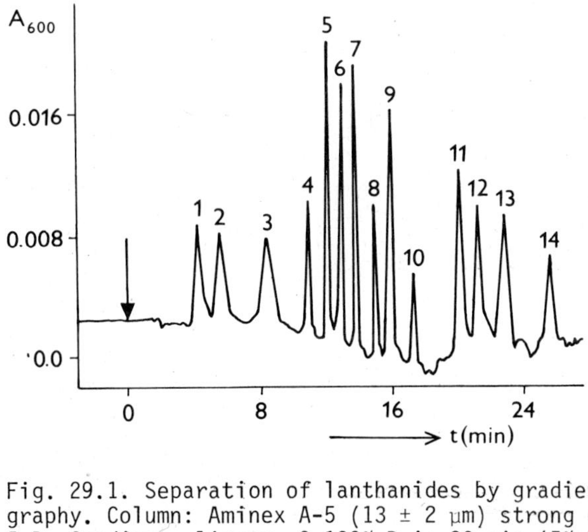

Fig. 29.1. Separation of lanthanides by gradient elution ion-exchange chromatography. Column: Aminex A-5 (13 ± 2 μm) strong cation-exchange resin, 100 x 4 mm I.D. Gradient: linear, 0-100% B in 20 min (5% B/min). Solvents: A, 0.17 mol/l α-hydroxyisobutyric acid in water; B, 1 mol/l α-hydroxyisobutyric acid in water; both at pH 4.6. Flow-rate: 0.8 ml/min. Detection: post-column reaction detector; reaction of lanthanides with Arsenazo I monitored continuously by measuring absorbance at 600 nm. Peaks: 1 = lutetium; 2 = ytterbium; 3 = thulium; 4 = erbium; 5 = holmium; 6 = dysprosium; 7 = terbium; 8 = gadolinium; 9 = europium; 10 = samarium; 11 = neodymium; 12 = praseodymium; 13 = cerium; 14 = lanthanum cations. (Reprinted with permission from ref. 15. Copyright 1979 American Chemical Society.)

The use of continuous monitoring of metal concentrations in the eluate based on post-column flow-through reaction detectors has obvious advantages over the earlier time-consuming methods. Post-column reactions with metallochromic indicators, such as Alizarin Red S, Arsenazo I or PAR |4(2-pyridylazo)resorcinol|

References on p. 444.

combined with detection in the visible region at 540-600 nm have been used for this purpose. The efficiency and speed of separation were increased by using cation-exchange resins with small particle diameters or chemically bonded cation exchangers on a microparticulate support, which allowed the separation to be accomplished in ca. 30 min[14,15]. An example of such a separation of a mixture of lanthanides is shown in Fig. 29.1[15]. This technique, in combination with direct on-column sample enrichment, was employed for the separation and determination of trace amounts of metal ions in river and drinking waters[14].

Complex-forming reagents may be chemically bonded on silica gel and then used as column packing materials for the liquid chromatography of metal ions. For example, the stability of metal oxinates (complexes with 8-hydroxyquinoline) depends on pH. Therefore, it was possible to separate a mixture containing manganese, cadmium, lead, zinc, nickel, cobalt and several lanthanides by chromatography on a column packed with chemically bonded oxine on silica gel using a concave gradient of nitric acid in water for elution[20].

29.3. INORGANIC ANIONS

Linear polycondensed phosphates in mixtures were separated by chromatography on anion-exchange columns using convex[21,22,25,26], linear[23] or multi-step[24] gradients of increasing concentration of ammonium acetate in water[25,26] and of sodium chloride in buffered aqueous solutions[21,23,24] or in aqueous solutions of ethylenediaminetetraacetic acid[22] as the mobile phase. Phosphate anions could be also separated by chromatography on a strong cation-exchange resin (18 μm) with an increasing gradient of water in dioxane as the eluent[27]. These methods were applied to the determination of phosphate anions in household detergents. In earlier applications, fractions of the eluate were collected and analysed for their phosphate content. More recently, an on-line post-column reaction detection system was described, making use of continuous hydrolysis of polyphosphates in the eluate to orthophosphate after addition of sodium hydroxide, followed by reaction with molybdate to give a yellow product, the concentration of which was monitored continuously using a UV or visible detector[22,24].

Polythionates in Wackenroder's solution were separated and determined by chromatography on an activated carbon column using elution with a concave gradient of increasing concentration of tetrahydrofuran in aqueous phosphate buffer and UV detection at 254 nm[28].

29.4. REFERENCES

1 P. Heizmann and K. Ballschmiter, *J. Chromatogr.*, 137 (1977) 153.
2 G. Schwedt, *Chromatographia*, 12 (1979) 613.

3 F. Gasparrini, D. Misiti, G. Natile and B. Galli, *J. Chromatogr.*, 161 (1978) 356.
4 C.H. Gast, J.C. Kraak, H. Poppe and F.J.M.J. Maessen, *J. Chromatogr.*, 185 (1979) 549.
5 S.J. Valenty and P.E. Behnken, *Anal. Chem.*, 50 (1978) 834.
6 D.A. Buckingham, C.R. Clark, R.F. Tasker and M.T.W. Hearn, *J. Liq. Chromatogr.*, 4 (1981) 689.
7 F.E. Brinckman, W.R. Blair, K.L. Jewett and W.P. Iverson, *J. Chromatogr. Sci.*, 15 (1977) 493.
8 R. Sitaram and S.M. Khopkar, *Chromatographia*, 5 (1972) 408.
9 S.B. Akki and S.M. Khopkar, *Chromatographia*, 3 (1970) 363.
10 K.A. Kraus and G.E. Moore, *J. Amer. Chem. Soc.*, 75 (1953) 1460.
11 O. Samuelson, *Ion Exchange Separations in Analytical Chemistry*, Wiley, New York, 1963.
12 H.G. Hicks, P.C. Stevenson and J.S. Schweiger, *J. Chromatogr. Sci.*, 16 (1978) 527.
13 B. Neef and H. Grosse-Ruyken, *J. Chromatogr.*, 79 (1973) 275.
14 R.M. Cassidy and S. Elchuk, *J. Chromatogr. Sci.*, 18 (1980) 217.
15 S. Elchuk and R.M. Cassidy, *Anal. Chem.*, 51 (1979) 1434.
16 C. Dubuquoy and G. Metzger, *Analusis*, 5 (1977) 314.
17 D.H. Sisson, V.A. Mode and D.O. Campbell, *J. Chromatogr.*, 66 (1972) 129.
18 Y. Takata and Y. Arikawa, *Bunseki Kagaku (Jap. Anal.)*, 24 (1975) 762.
19 Y.-M. Chen and N.-K. Sung, *J. Chin. Chem. Soc. (Taipei)*, 22 (1975) 113.
20 J.R. Jezorek and H. Freiser, *Anal. Chem.*, 51 (1979) 366.
21 S. Ohashi, N. Tsuji, Y. Ueno, M. Takeshita and M. Muto, *J. Chromatogr.*, 50 (1970) 349.
22 H. Yamaguchi, T. Nakamura, Y. Hirai and S. Ohashi, *J. Chromatogr.*, 172 (1979) 131.
23 T. Nakamura, T. Yano, T. Nunokawa and S. Ohashi, *J. Chromatogr.*, 161 (1978) 421.
24 R.S. Deelder, M.G.F. Kroll, A.J.B. Beeren and J.H.M. van den Berg, *J. Chromatogr.*, 149 (1978) 669.
25 K. Koguchi, H. Waki and S. Ohashi, *J. Chromatogr.*, 25 (1966) 398.
26 S. Ohashi and K. Koguchi, *J. Chromatogr.*, 27 (1967) 214.
27 K. Tanaka and T. Ishizuka, *J. Chromatogr.*, 190 (1980) 77.
28 J.N. Chapman and H.R. Beard, *Anal. Chem.*, 45 (1973) 2268.
29 B.R. Willeford and H. Veening, *J. Chromatogr.*, 251 (1982) 61.
30 P.C. Uden, I.E. Bigley and F.H. Walters, *Anal. Chim. Acta*, 100 (1978) 555.
31 C.H. Gast, F. Nooitgedacht and J.C. Kraak, *J. Organometal. Chem.*, 184 (1980) 221.

Chapter 30

TECHNICAL PRODUCTS AND INDUSTRIAL APPLICATIONS

This chapter is concerned with applications of gradient elution liquid column chromatography to the analysis of various technically important products. Some of these products involve mixtures of compounds belonging to various classes, which makes difficult a consistent classification of their separation according to the chemical structure of solutes. Other products could have been classified in this way; however, they are rather treated in this section because of their practical technical importance. Finally, some of the products are characterized only by their "fingerprint" chromatograms without further identification or determination of individual components.

30.1. FOSSIL FUELS AND RELATED PRODUCTS FROM THEIR PROCESSING

Concern about the efficient use of crude oil and other natural energy and petrochemical resources is obvious. Liquid chromatography is a valuable method for the characterization of oil and its refined products, coal liquefaction products, motor and lubricating oils and various additives to them and of asphalts. The gradient elution technique is often applied in the chromatography of these complex samples.

To characterize fossil fuels, their acid and base extracts could be prepared and subjected to chromatography on a column packed with microparticulate silica using elution with an increasing concentration gradient of dichloromethane containing ethanol and a lower carboxylic acid in *n*-heptane as the starting eluent. Fingerprint chromatograms are obtained, which can give a picture of the group separation of extracted compounds into various fractions (hydrocarbon; pyrrole/phenylaniline; aniline/thiazole; pyridine; phenol; carboxylic acid type). These chromatograms differ in the dependence on the alkyl chain length of the carboxylic acid used in the mobile phase (short alkyl chain acids work best with basic solutes, whereas acids with longer chains yield better results with acidic solutes), and therefore running the same sample in two or more gradient systems with different carboxylic acids may provide useful qualitative information on different chemical types of compounds in the sample[1].

Coal liquefaction products can be characterized using sequential analysis with gel permeation chromatography in the first step, which provides a picture of the molecular weight distribution and pre-separates the sample into fractions at the

References on p. 456.

same time. These fractions may be subjected to gas chromatographic analysis or to reversed-phase liquid chromatography on microparticulate octadecylsilica columns using elution with gradients of acetonitrile[2] or tetrahydrofuran[3,4] in water. The latter technique is especially useful for characterizing heavy fractions containing asphaltenes. This technique can be also applied to petroleum distillation residual fractions (petroleum pitches) and can be used to distinguish rapidly petroleum pitches from coal tars[4]. Non-aqueous gradient elution liquid chromatography on an octadecylsilica column using a linear gradient of dichloromethane in methanol[5] and chromatography on a column packed with a bonded nitrile phase using elution with a gradient of tetrahydrofuran in acetonitrile[6] have also been applied to the characterization of aromatics and asphaltenes in heavy and residual petroleum fractions. Fluorimetric detection may provide additional useful information to the chromatograms recorded using UV detection[2,5].

Pave bonding additives in asphalt were analysed quantitatively by liquid chromatography on an octadecylsilica column using a linear gradient of tetrahydrofuran in water for elution and UV detection[7].

Various detergent and antioxidant additives to motor lubricating oils were separated and determined by chromatography on an octadecylsilica column with a linear gradient of methanol[2] or tetrahydrofuran[3] in water as the eluent, using continuous UV detection. The identification of phosphorus compounds was improved by using a graphite furnace atomic-absorption spectrometric detector[8].

30.2. SYNTHETIC POLYMERS

Gel permeation (steric exclusion) chromatography is used for the characterization of the molecular weight distribution of polymers.

Mixtures of lower oligomeric components of polymers can also be separated on columns packed with both polar and non-polar packing materials, because of differences in the interactions of the individual oligomers with the stationary phase. As these mixtures often contain a large number of individual oligomers, the gradient elution technique is usually employed in these applications. In particular reversed-phase gradient elution chromatography usually provides separations of individual oligomers that are far superior to those obtained by gel permeation chromatography.

Polystyrene samples were subjected to chromatography on microparticulate silica gel columns with a concentration gradient of dichloromethane in *n*-hexane as the eluent. For better characterization and assignment of functional groups, infrared spectra of fractions collected in multiple repeated runs were measured[9].

Reversed-phase gradient elution liquid chromatography is very suitable for the separation of individual oligomers differing in the number of hycrocarbonaceous structural units. Using linear or slightly convex gradients of an organic solvent in water for elution, it is possible to achieve almost regular spacing of the peaks of the individual oligomers on the chromatogram and the oligomers are eluted in order of increasing molecular weight. In this way, low-molecular-weight polystyrene oligomers were separated on an octadecylsilica column using elution with a gradient of tetrahydrofuran in water[10] or in methanol[11] (see the example in Fig. 30.1).

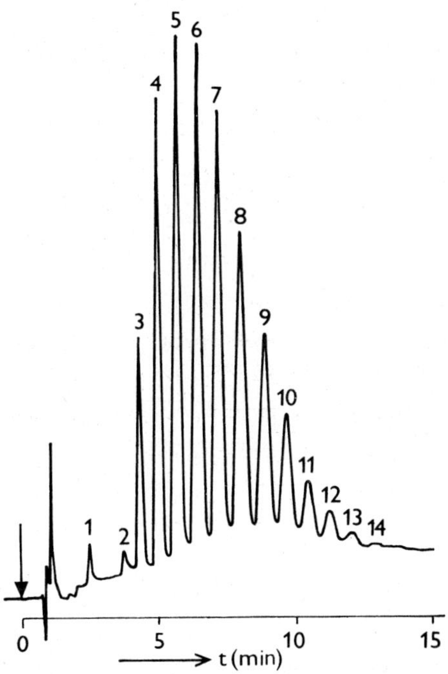

Fig. 30.1. Separation of polystyrene oligomers by reversed-phase gradient elution chromatography. Column: Partisil ODS-2. Gradient: convex, from 60 to 85% B in 35 min. Solvents A, water; B, tetrahydrofuran. Flow-rate: 2 ml/min. Detection: UV at 254 nm. Peaks: polystyrene oligomers, mono- to 14-mer, as indicated by the numbers. (Reproduced from ref. 10 with permission.)

This method has been used for separations of other oligomeric mixtures on columns packed with octadecylsilica materials. For example, oligomers of poly-2,2,4-trimethyl-1,2-dihydroquinoline were separated by elution with a convex gradient of tetrahydrofuran in water[11] and these gradients were also employed for the separation of linear oligomers of Novolak low-molecular-weight resins, prepared by condensation of formaldehyde and an excess of phenol in an acidic medium[12]. Oligomers from polyethylene extracts were separated using elution with a convex gradient of acetonitrile in water[21]. Mixtures of poly-(2,6-diphenyl-

References on p. 456.

p-phenylene oxide) oligomers formed by oxidative coupling of 2,6-diphenylphenol in the preparation of Tenax were resolved using a slightly convex gradient of dioxane in water and mixtures of poly(ethylene terephthalate) oligomers were completely separated by elution with a linear gradient of hexafluoroisopropanol in water[12].

Reversed-phase gradient elution liquid chromatography has frequently been applied to the analysis of epoxy resin formulations. These are widely used as thermosetting agents for various purposes, because of the excellent mechanical, electrical and chemical properties of cured epoxy resins. Epoxy resin formulations are complex mixtures of resins, curing agents, diluents, accelerators, etc. Chromatography on octadecylsilica columns using elution with linear gradients of tetrahydrofuran[13-15], acetonitrile[16-19] or dioxane[12] in water was used to provide fingerprint chromatograms for comparison of the quality of formulations[13-18], for the separation and determination of curing agents (amines) and accelerators (such as monuron and diuron) in these formulations[13,20] and for the resolution of individual oligomers[12,19]. In addition to complete separation of the main oligomers with the 2,3-epoxypropyloxy group as an end group, it was also possible to distinguish oligomeric by-products containing other end groups (2,3-dihydroxypropyloxy, 2-hydroxy-3-methoxypropyloxy or 2-hydroxy-3-*p-tert.*-butylphenoxypropyloxy)[19].

Distinct separations of individual oligomers up to 12- to 20-mers can be achieved in 1-1.5 h. However, certain special aspects of liquid chromatography of oligomers should be considered. Most important is the sample solubility. Usually, the samples are poorly soluble in the starting (polar) eluent used in reversed-phase liquid chromatography and the sample should be dissolved in a less polar solvent. This may, however, cause certain disturbancies in the separation of the earlier eluted sample components and result in poorer resolution of higher oligomers with the lowest solubility[12].

Chromatography on silica gel columns with linear or slightly concave gradients of tetrahydrofuran in *n*-octane or chloroform has been used to characterize epoxy resin formulations and to quantitate some additives (accelerators)[13].

Phthalate plasticizers are added to vinyl plastics to make them flexible. They may diffuse into various materials in contact with the plastics, such as foodstuffs, and therefore the maximum content of phthalates in plastics is limited and their analytical control is of interest. Rapid separations of various phthalate ester plasticizers (dimethyl, diethyl, diphenyl, di-*n*-butyl, di-*n*-octyl and diisodecyl) were achieved by reversed-phase chromatography on octadecylsilica columns, using a linear gradient of methanol in water for elution[22].

Because most of the oligomers and polymer additives absorb in the UV region, they are conveniently detected by common UV detectors.

30.3. ANTIOXIDANTS

Antioxidants are compounds added to various products, such as certain polymers (polyethylene), food products, pharmaceuticals and cosmetic preparations, that are susceptible to oxidation, with the aim of preventing the deterioration of the quality of such products owing to gradual oxidation by atmospheric oxygen. Most individual antioxidants are known under various commercial names, but the most commonly used compounds are phenolics, such as 2,6-di-*tert.*-butyl-4-methylphenol (BHT), 3-*tert.*-butyl-4-hydroxyanisole (BHA), propyl-, octyl- and dodecyl gallate and nordihydroguajaretic acid (NDGA). The maximum concentration of antioxidants in food products is usually regulated by law. HPLC is very suitable as a rapid and sensitive control method for the determination of antioxidants in various products.

Commercial antioxidants used as additives in polyethylene and in other products were separated on columns packed with pellicular[23] or microparticulate alumina[24] by elution with linear gradients of dioxane[23] or dichloromethane[24] in *n*-hexane. More recently, columns packed with microparticulate silica gel and elution with linear gradients of chloroform[25] or methylene chloride[26,27] in *n*-hexane or *n*-heptane have been used for this purpose.

Rapid separations of antioxidants in various food products were achieved by reversed-phase chromatography on octadecylsilica columns using elution with a linear gradient of methanol in water[28] and on octylsilica columns with a linear gradient of acetonitrile in 5% aqueous acetic acid[29] or acetonitrile-methanol (7 : 3) in water with addition of phosphoric acid[30].

Detection is performed by monitoring the UV absorption at 254 or 280 nm.

30.4. SURFACTANTS

Surfactants (surface-active agents) are organic compounds (or technical mixtures) that change the interfacial properties of liquids to which they are added. They are widely used as emulsifying, dispersing, wetting and foaming agents in the household, industry and agriculture, in textile, paper, leather and plastics manufacture, as additives in food and cosmetic products, etc. HPLC meets most requirements as an analytical method for surfactants and its use for this purpose has become increasingly popular[31].

Surfactant molecules are usually composed of a non-polar (hydrocarbonaceous) part and of a highly polar or ionized part involving one or more functional groups. Surfactants are usually classified into four classes: anionic, cationic, nonionic and amphoteric.

References on p. 456.

30.4.1. *Anionic surfactants*

The most frequently used anionic surfactants are long-chain sulphonates and sulphates, such as alkylbenzenesulphonates, α-olefinsulphonates and alkylsulphates. Similar chromatographic conditions may be used for separations of these compounds as for other sulphonic acids (Chapter 26). This means that reversed-phase chromatography on octadecyl- or octylsilica columns is especially suitable for anionic surfactants, by either the ion-pair or salting-out technique. Gradient elution may be helpful in improving the speed of separation of homologous sulphonates. Thus, linear gradients of methanol or acetonitrile in an aqueous solution of an ion-pairing reagent, such as tetrabutyl-, tetramethyl-, tetraethyl- or cetyltrimethylammonium salts (ca. 0.005-0.05 mol/l)[33] or in aqueous solutions of strong electrolytes (ca. 0.05-0.5 mol/l)[32] may be used for elution. Fig. 30.2 shows an example of a fingerprint chromatogram of a household detergent containing linear alkylbenzenesulphonates obtained by salting-out reversed-phase chromatography using a linear gradient of methanol in aqueous sodium sulphate[32].

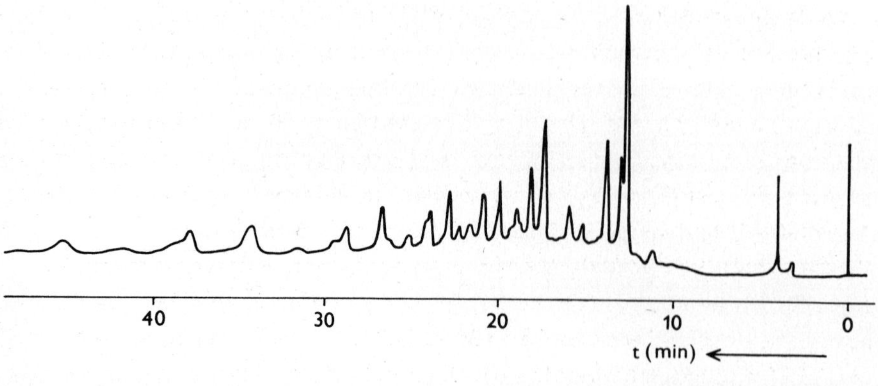

Fig. 30.2. Fingerprint chromatogram of a commercial household detergent containing alkylbenzenesulphonate surfactants. Column: C18 on LiChrosorb Si 100, 10 µm, 300 x 4.2 mm I.D. Gradient: linear, 0-100% B in 15 min (6.67% B/min). Solvents: A, 0.4 mol/l sodium sulphate in water; B, methanol - water (60 : 40). After the end of the gradient, elution was continued with pure B. Flow-rate: 1 ml/min. Detection: UV at 254 nm, 1.0 a.u.f.s.

30.4.2. *Nonionic surfactants*

Ethylene oxide adducts with fatty alcohols, fatty acids, alkylphenols, etc.; fatty acid glycerides, alkanolamides and fatty acid esters with hydroxy acids or sugars are the most frequently employed nonionic surfactants. Fatty acid glycerides may in principle be analysed using the same liquid chromatographic techniques as those described for lipids (Chapter 17). As in the chromatography of other lipids, detection presents problems with these compounds, which do not absorb in the region above 220 nm. However, the gradient elution technique is very useful in the liquid chromatography of commercial nonionic surfactants, which are usually complex mixtures. A mass-transport moving-wire flame-ionization detector can be employed, as for example in the chromatography of a food emulsifier based on lactylated monoglycerides on a microparticulate silica gel column using elution with a sequence of two concave gradients from *n*-heptane - diethyl ether (9 : 1) to *n*-heptane - diethyl ether - tetrahydrofuran (70 : 18 : 12) and then to *n*-heptane - diethyl ether - dioxane - isopropanol - water - acetic acid (20 : 30 : 40 : 10 : 1 : 0.5)[34]. Another possibility is to use UV detection in the region between 200 and 220 nm, as for example in the separation of partial glycerides and sorbitol fatty acid esters in food emulsifiers on a similar column, where elution with a gradient of 90% (v/v) aqueous ethanol in 1% solution of isopropanol in isooctane was employed[35]. Another application example of UV detection at 220 nm is the separation of glycerol- and polyglycerol esters in food emulsifiers on a Diol chemically bonded column packing material using gradient elution with an increasing concentration of isopropanol in *n*-hexane, which not only resolved glycerol mono-, 1,2-di-, 1,3-di- and triesters of fatty acids and also polyglycerol esters, but also separated some of the numerous polyglycerol ester positional isomers in ca. 1 h[36]. Separations of mono, di- and triesters of glycerol with fatty acids were also achieved on a column packed with a bonded aminocyano phase using elution with a linear gradient of simultaneously increasing concentrations of acetonitrile and chloroform in *n*-hexane and infrared detection at 5.72 μm (wavelength of absorption maximum of the carbonyl band)[37].

The separation of oligomeric ethylene oxide adducts in commercial nonionic surfactants could be achieved in either normal-phase or reversed-phase systems, often using the gradient elution technique. Thus, oligomers with 2-17 ethylene oxide units in mixtures of phenol and thiophenol condensates were separated on a column packed with silanized silica gel using elution with a linear gradient of acetone in dichloromethane[38]. Oligomeric ethoxylated alkylphenols were separated on columns packed with microparticulate silica gel using elution with a gradient of 90% aqueous ethanol in isooctane - isopropanol (4 : 6)[35] or with a linear gradient of aqueous acetic acid in ethyl acetate[39]. UV detection at 254 nm was employed in these applications, but this was not possible in the separation of oligomeric ethylene oxide adducts with alcohols, acids or amines. Here, a mass-transport moving-wire flame-ionization detector was used, such as in the chromatography of ethylene oxide adducts with C_{12} and C_{14} fatty alcohols and acids on microparticulate silica, where the elution was performed with a linear gradient of acetone in *n*-hexane[40] or with a gradient of simultaneously increasing concentrations of diethyl ether, dioxane, water and isopropanol in *n*-hexane[41].

References on p. 456.

To improve the possibilities of detection, ethylene oxide adducts in mixtures were converted into the UV-absorbing 3,5-dinitrobenzoyl esters, which could be separated by chromatography on microparticulate silica gel columns using gradient elution with an increasing concentration of aqueous ethanol in a solution of 1% isopropanol in isooctane[35], or on a column of microparticulate silica gel coated with Carbowax 20M by elution with a linear gradient of acetonitrile in dichloromethane - *n*-hexane (1 : 1), or on an octadecylsilica column with a linear gradient of tetrahydrofuran in water as the eluent[42]. Pre-column derivatization to benzoates was also used in the separation of free polyethylene glycols on a microparticulate octadecylsilica column using elution with a concave gradient of methanol in water[43] (free polyethylene glycols are often present in commercial nonionic surfactants). UV detection at 254 nm can conveniently be used in these separations.

Reversed-phase systems with chemically bonded alkyl phases generally have a low selectivity for the separation of oligomeric ethoxylated surfactants. This selectivity and even the order of elution of the individual oligomers depend on the type of the oligomeric series (type of ethylene oxide adducts), on the composition of the mobile phase, on the type of the bonded phase and on temperature, and it is therefore difficult to predict a priori. However, certain oligomeric series, such as ethoxylated non-substituted phenols, could be separated on a LiChrosorb RP-8 column using elution with a convex gradient of acetonitrile in water[44].

In the chromatography of ethoxylated alkyl phenols on octadecylsilica columns, it is not possible to separate oligomers using gradient elution with an increasing concentration of methanol in water, because the oligomers are eluted in one common peak. However, it is possible to separate ethoxylated alkylphenols according to the lengths of the alkyl chains and thus to distinguish, for example, ethoxylated octyl-, nonyl-, decyl- and dodecylphenols[57].

Successful separations of oligomeric ethylene oxide adducts with nonylphenol were achieved on a column packed with an amino phase bonded on a microparticulate silica gel support using elution with a gradient of increasing concentration of 90% aqueous isopropanol in tetrahydrofuran - *n*-hexane (1 : 4) as the starting eluent. In this way, resolution of oligomers with up to 30 ethylene oxide units is possible[12,41].

30.5. COSMETICS AND TOILETRIES

Liquid chromatography with gradient elution and UV detection can be used to characterize the quality of various plant oils and fragrances used in the production of perfumes and of other cosmetics. For example, fingerprint chromatograms of UV-absorbing compounds in rose[45] or lavender[46] oil obtained by chromatography on microparticulate silica gel columns using elution with concentration gradients of isopropanol in *n*-heptane are suitable for this purpose. Santale oil was characterized in a similar way by chromatograms obtained on an octadecylsilica column using elution with a linear gradient of methanol in water[47] and the quality of a fragrance concentrate was estimated from fingerprint chromatograms obtained on a similar column with a gradient of acetonitrile in water as the eluent[48].

A fingerprint chromatogram obtained on an octadecylsilica column by elution with a linear gradient of methanol in water was used in a similar way to characterize various shampoo products[49].

The alkyl esters of *p*-hydroxybenzoic acid (parabens) are widely used as antimicrobial preservatives in foods, beverages and cosmetics. The rapid separation of methyl-, ethyl-, *n*-propyl- and *n*-butyl-parabens in a skin cream formulation was achieved by reversed-phase chromatography on an octadecylsilica column using elution with a linear gradient of acetonitrile in water[50].

30.6. MISCELLANEOUS

Gradient elution liquid column chromatography can be used for the control of various other industrial products. To illustrate these possibilities, we will mention a few examples at this place.

Various explosives, such as nitroglycerine and pentaerythritol, were identified at trace levels by using chromatography on microparticulate silica or on bonded amino phases with a concave gradient of ethanol in isooctane as the eluent. Detection was performed with a nitrosyl-specific detector[51].

Flame retardants were analysed on pellicular octadecylsilica packing material using elution with a linear gradient of methanol in water[52].

Symmetrical and asymmetric dixanthogens resulting from the oxidation of alkyl xanthates were separated from each other and from xanthic esters and xanthates by chromatography on an octadecylsilica column with a linear gradient of methanol in water as the eluent. This method was employed to control xanthate salts used in the segregation of sulphide minerals by froth flotation[53].

Chromatography on an octadecylsilica column using elution with a linear gradient of tetrahydrofuran in water was employed for the analysis of filtered hydraulic fluids[54].

The purity of liquid crystals used in digital electronic displays was checked by reversed-phase chromatography on an octadecylsilica column using a linear gradient of isopropanol in water as the eluent[55].

HPLC on an octadecylsilica column with a linear gradient of acetonitrile in water separated vehicle components in ultraviolet curing inks[56].

References on p. 456.

30.7. REFERENCES

1 J.B. Green and R.J. Hoff, *J. Chromatogr.*, 209 (1981) 231.
2 M.J. Sepaniak and E.S. Yeung, *J. Chromatogr.*, 211 (1981) 95.
3 *Fossil Fuels Analysis by LC*, AN 154, Waters Assoc., Milford, MA, 1975.
4 *Distinguishing Coal Tar (Petroleum Pitches), Application Notes No. 47*, Micromeritics, Norcross, GA.
5 H. Elgass, *Dissertation*, Saarbrücken, 1978.
6 C. Reichert and E.L. Johnson, *LC at Work*, 57, Varian, Palo Alto, CA.
7 *Pave Bonding Additives in Asphalt, Application Notes No. 40*, Micromeritics, Norcross, GA.
8 P. Tittarelli and A. Mascherpa, *Anal. Chem.*, 53 (1981) 1466.
9 P.H. Sackett, R.W. Hannah and W. Slavin, *Chromatographia*, 11 (1978) 634.
10 *Polystyrene Oligomers, Application Notes No. 53*, Micromeritics, Norcross, GA.
11 R.P. Lattimer, D.J. Harmon and K.R. Welch, *Anal. Chem.*, 51 (1979) 1293.
12 F.P.B. van der Maeden, M.E.F. Biemond and P.C.G.M. Janssen, *J. Chromatogr.*, 149 (1978) 539.
13 G.L. Hagnauer and I. Setton, *J. Liq. Chromatogr.*, 1 (1978) 55.
14 *Oligomer Separations*, Waters Assoc., Milford, MA.
15 W.A. Dark, E.C. Conrad and L.W. Crossman, *J. Chromatogr.*, 91 (1974) 247.
16 *Epoxy Resins, Application Notes No. 55*, Micromeritics, Norcross, GA.
17 *Spectra Physics Chromatogr. Rev.*, 3, No.1 (1977) 3.
18 J.E. Twichell, J.Q. Walker and J.B. Maynard, *J. Chromatogr. Sci.*, 17 (1979) 259.
19 S. Shiono, I. Karino, A. Ishimura and J. Enomoto, *J. Chromatogr.*, 193 (1980) 243.
20 P.J. Dyness, R.M. Panos and C.L. Hamermesh, *J. Appl. Polym. Sci.*, 25 (1980) 1059.
21 *Series 3 Liquid Chromatograph*, Perkin-Elmer, Norwalk, CT. 1977.
22 *LC Analysis of Phthalate Plasticizers*, Waters Assoc., Milford, MA, 1973.
23 R.E. Majors, *Application Tips*, No. 89, Varian, Palo Alto, CA, 1971.
24 R.E. Majors, *Anal. Chem.*, 45 (1973) 755.
25 *Model LC-55 Detector for LC*, Perkin-Elmer, Norwalk, CT, 1974.
26 R.G. Lichtenthaler and E. Ranfelt, *J. Chromatogr.*, 149 (1978) 553.
27 J.F. Schabron and L.E. Fenska, *Anal. Chem.*, 52 (1980) 1411.
28 *Application Sheet*, No. 9, Applied Research Laboratories, Luton, Great Britain.
29 *Series 2 Liquid Chromatograph*, Perkin-Elmer, Norwalk, CT, 1980.
30 R. Schuster, *Application Note*, AN 232-5, Hewlett-Packard, Böblingen, 1978.
31 P. Jandera, in J.F. Lawrence (Editor), *Environmental Analysis by HPLC*, Humana Press, Clifton, NJ, 1984, p. 115.
32 P. Jandera and J. Churáček, *J. Chromatogr.*, 197 (1980) 181.
33 W. Giger, E. Staub and L. Schaffner, *ACS Abstr.*, April 1979.
34 K. Aitzetmüller, *J. Chromatogr. Sci.*, 13 (1975) 454.
35 J. Brüschweiler, *Mitt. Geb. Lebensmittelunters. Hyg.*, 68 (1977) 46.
36 N. Garti and A. Aserin, *J. Liq. Chromatogr.*, 4 (1981) 1173.
37 K. Payne-Wahl, G.F. Spencer, R.D. Plattner and R.O. Butterfield, *J. Chromatogr.*, 209 (1981) 61.
38 C. Calzolari, L. Favretto and B. Stancher, *J. Chromatogr.*, 47 (1970) 209.
39 C.F. Allen and L.I. Rice, *J. Chromatogr.*, 110 (1975) 151.
40 K. Nakamura and I. Matsumoto, *Nippon Kagaku Kaishi*, (1975) 1342.
41 H.J. Vonk, A.J. van Wely, L.G.J. van der Ven, A.J.J. de Breet, F.P.B. van der Maeden, M.E.F. Biemond, A. Venema and W.G.B. Huysmans, *Tr. Mezhdunar. Kongr. Poverekhn. Akt. Veshchestvam, 7th, Moscow*, Vol. 1, 1977, p. 435.
42 R.M. Cassidy, *J. Liq. Chromatogr.*, 1 (1978) 241.
43 R. Murphy, A.C. Selden, M. Fisher, E.A. Fagan and V.S. Chadwick, *J. Chromatogr.*, 211 (1981) 160.
44 W.R. Melander, A. Nahum and Cs. Horváth, *J. Chromatogr.*, 185 (1979) 129.
45 H. Engelhardt and H. Elgass, *J. Chromatogr.*, 112 (1975) 415.
46 *Packard 8200 Series*, Packard, Downers Grove, IL.

47 P. Kucera, *J. Chromatogr.*, 198 (1980) 93.
48 *Epoxy Resins, Application Notes No. 55,* Micromeritics, Norcross, GA.
49 *Rapid and Reliable Analysis of Cosmetics and Toiletries,* Waters Assoc., Milford, MA, 1977.
50 T. Alfredson and L. Tallman, *Liquid Chromatography at Work,* LC 115, Varian, Palo Alto, CA.
51 A.L. Lafleur and B.D. Morriseau, *Anal. Chem.*, 52 (1980) 1313.
52 R.E. Majors, *Application Tips,* No. 95, Varian, Palo Alto, CA, 1971.
53 J.G. Eckhardt, K. Stetzenbach, M.F. Burke and J.L. Moyers, *J. Chromatogr. Sci.,* 16 (1978) 510.
54 G.L. Hagnauer and B.M. Bowse, *J. Liq. Chromatogr.*, 3 (1980) 1.
55 *Liquid Crystals, Application Notes No. 62,* Micromeritics, Norcorss, GA.
56 *Ultra-violet Curing Inks, Application Notes No. 64,* Micromeritics, Norcross, GA.
57 A. Otsuki and H. Shiraishi, *Anal. Chem.*, 51 (1979) 2329.

Chapter 31

FOODS AND BEVERAGES

HPLC is becoming increasingly important as a rapid method for the control of agricultural and food products. These applications involve the determination of a wide variety of compounds, such as crabohydrates, amino acids, proteins, lipids, vitamins, food dyestuffs, emulsifiers and antioxidants and certain food contaminants (antibiotics, pesticides, aflatoxins)[24]. The applications of gradient elution liquid chromatography for the analysis of these compounds have been discussed in chapters on individual classes of organic compounds. Therefore, this chapter is concerned only with applications of gradient elution liquid chromatography that are specific for certain food products.

31.1. TANNINS AND CAFFEINE IN BEVERAGES

Tannins have been determined in tea extracts[1], ciders and wine[2] using reversed-phase chromatography on octadecylsilica[1] and octylsilica[2] columns using elution with a linear gradient of acetone in water containing 0.5% acetic acid[1] or with a linear gradient of methanol in water[2]. Caffeine in tea and coffee extracts, cola and similar beverages can be determined by reversed-phase chromatography on octadecylsilica columns using elution with a gradient of methanol in water (an example is shown in Fig. 31.1)[3].

31.2. CITRUS FRUIT AND BEVERAGES

Orange, grapefruit and other citrus juices, concentrates and beverages were characterized by fingerprint chromatograms obtained using chromatography on silica gel[4] or deactivated alumina columns[5] with gradients of methanol or methanol - dichloromethane in isooctane as the eluent. Citrus phenolics and phenolic acids in juices were determined by reversed-phase chromatography on octadecylsilica columns using elution with a linear gradient of methanol in 5% aqueous acetic acid[6] or with a linear gradient of acetonitrile[7] in 0.25 mol/l acetate buffer and with UV or fluorimetric detection. Glycosidic anthocyanine chalcones in grapefruit peel were analysed by chromatography on a microparticulate octadecylsilica column with a linear gradient of methanol in water containing 0.6% perchloric acid as the eluent, with UV detection at 280 nm[8].

References on p. 463.

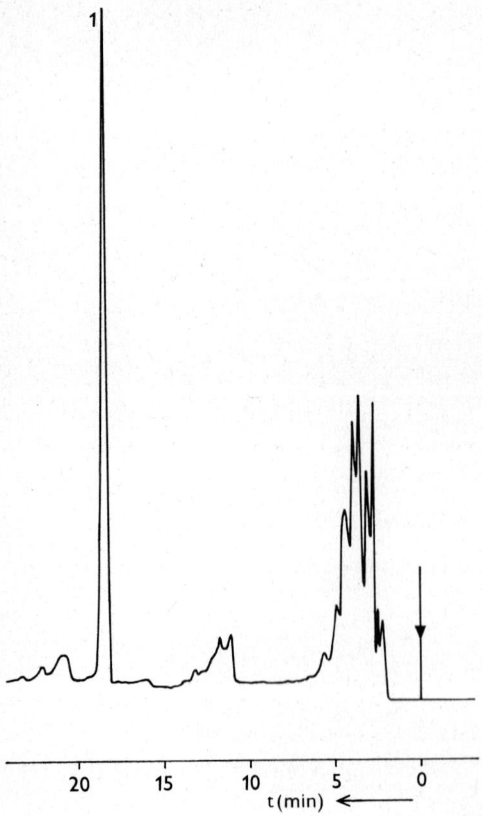

Fig. 31.1. Chromatogram of a coffee extract. Column: Silasorb C18, 10 μm, 300 x 4.2 mm I.D. Gradient: concave, 0-100% B in 30 min. Solvents: A, water; B, methanol. Flow-rate: 1 ml/min. Detection: UV at 254 nm, 1.0 a.u.f.s. Peak 1 = caffeine.

31.3. BEER AND HOPS

Reversed-phase chromatography with gradient elution has been employed for the analysis of beer and hops. Bitter hop compounds (humulone, cohumulone, adhumulone, isohumulone, isocohumulone, isoadhumulone, lupulone, colupulone, adlupulone, humulinic acids, etc.) could be separated and some of them identified and quantitated by chromatography on an octadecylsilica column using elution with a linear gradient of methanol in 1% aqueous sodium acetate[9] or with a linear gradient of acetonitrile in 0.1 mol/l aqueous phosphate buffer[10]. This method compares favourably with the gradient elution ion-exchange chromatography of hop compounds both in resolving power and in speed of separation. Bitter acids in beer were determined by chromatography on a microparticulate octadecylsilica column using elution with a linear gradient of methanol in 0.5% aqueous phosphoric acid[11]. UV detection at 280 or 254 nm was used in these applications.

Amino acids and vitamins in beer were analysed by chromatography on a column packed with a microparticulate bonded amino phase using elution with a linear gradient of increasing concentration of potassium dihydrogen phosphate and a simultaneously decreasing concentration of acetonitrile in water[12].

31.4. FISH PRODUCTS

Chromatography on an octadecylsilica column with a linear ternary gradient of increasing concentrations of acetonitrile and methanol in 0.02 mol/l aqueous acetic acid was used for the characterization of the quality of canned tuna[13]. Glyceride components of cod liver were analysed by non-aqueous reversed-phase chromatography on a microparticulate octadecylsilica column using a linear gradient of tetrahydrofuran in acetonitrile for elution and infrared monitoring of the eluate at 1740 cm^{-1} (absorption maximum of the carbonyl band)[14].

31.5. SPICES AND FOOD FLAVOURING INGREDIENTS

Several examples of gradient elution liquid chromatographic techniques used in the analysis of various flavouring compounds in spices and food ingredients are given in Table 31.1[15-20].

31.6. FOOD CONTAMINANTS AND SYNTHETIC ADDITIVES

Gradient elution liquid chromatography has been used for the analysis of various contaminants in food. For example, strigmatocystin, a mycotoxin occurring in moudly rice, was determined by reversed-phase liquid chromatography on an octadecylsilica column using elution with a linear gradient of methanol in water[21]. The naturally occurring toxicants myristicin and falcarinol and related aromatic ethers in carrots were determined by chromatography on a microparticulate chemically bonded nitrile phase using elution with a linear gradient of tetrahydrofuran in *n*-heptane[22].

Many food products contain preservatives to prevent their deterioration by microbes and fungi. The most frequently used compounds of this type are sorbic acid, benzoic acid, parabens such as *p*-hydroxybenzoic acid methyl, ethyl and *n*-propyl esters, biphenylol and biphenyl. These compounds were separated by chromatography on a microparticulate octylsilica column using elution with a linear gradient of acetonitrile in 0.005 mol/l aqueous ammonium acetate buffer (pH 4.4). This method can be employed for the identification and quantitation of preservatives in various food products[23].

References on p. 463.

TABLE 31.1

EXAMPLES OF APPLICATIONS OF GRADIENT ELUTION LIQUID CHROMATOGRAPHY IN THE ANALYSIS OF FLAVOURING COMPOUNDS IN SPICES AND FOOD INGREDIENTS

Compounds	Sample	Column	Solvent A	Solvent B	Gradient	Detection	Ref.
Benzaldehyde, eugenol, cinnamaldehyde, cinnamic acetate, cinnamic alcohol	Cinnamon oil	Corasil II, 37-50 μm, 900 x 2 mm	Cyclohexane	Ethyl acetate	Convex, 0-40% B in 30 min	UV, 260 nm	15
Fingerprint chromatogram	Dill oil	Micro Pak Si-10	*n*-Heptane	Dichloromethane-isopropanol (1:1)	Linear, 0-60% B, 2% B/min	UV, 250 nm	16
Capsaicin	Chilli powder	Partisil 10 ODS	Water	Tetrahydrofuran	Linear, 20-60% B, 1% B/min	UV, 254 nm	17
Fingerprint chromatogram	Liquorice	C18/Corasil, 37-50 μm, 500 x 3 mm	Water	Acetonitrile	Linear, 0-60% B, 1.5% B/min	UV, 254 nm	18
Glycyrrhizin	Liquorice roots	Permaphase AAX (pellicular anion exchanger)	Water	0.05 mol/l $NaClO_4$ in water	Linear, 0-100% B	UV, 254 nm	19
Fingerprint	Onion oil	C18/LiChrosorb Si 100, 10 μm, 300 x 4.2 mm	Water	Methanol	Linear, 20-100% B, 2.67% B/min	UV, 254 nm	20
Fingerprint chromatogram	Paprika powder	LiChrosorb Si 100, 10 μm, 300 x 4.2 mm	*n*-Heptane	Dichloromethane	Linear, 0-100% B, 1.67% B/min	UV, 254 nm	20

31.7. REFERENCES

1 *Model LC 750*, Applied Chromatography Systems, Luton, Great Britain.
2 A.G.H. Lea, *J. Chromatogr.*, 194 (1980) 62.
3 P. Jandera, unpublished results.
4 *Application Sheet E 7710*, Laboratory Data Control, Riviera Beach, FL.
5 E. Benk, H. Treiber and R. Bergmann, *Reichstoffe Arom. Körperpflegemittel*, 26 (1976) 216.
6 D. Popovich and D. Southern, in G. Charalambous (Editor), *Liquid Chromatographic Analysis of Food and Beverages*, Vol. 2, Academic Press, London, 1979, p. 411.
7 L.R. Rouseff, *Trends in Fluorescence*, 1 (1978) 10, Perkin-Elmer, Norwalk, CT.
8 N.W. Preston and C.F. Timberlake, *J. Chromatogr.*, 214 (1981) 222.
9 W.G. Schulze, P.L. Ting and H. Goldstein, in G. Charlambous (Editor), *Liquid Chromatographic Analysis of Food and Beverages*, Vol. 2, Academic Press, London, 1979, p. 441.
10 I. Molnár, *Chromatographia*, 12 (1979) 371.
11 C. Dewaele and M. Verzele, *J. Chromatogr.*, 197 (1980) 189.
12 R. Schuster, *Chromatographia*, 13 (1980) 379.
13 J.L. Mietz and E. Karmas, *J. Food Sci.*, 42 (1977) 155.
14 N.A. Parris, *J. Chromatogr.*, 149 (1978) 615.
15 M.S.F. Ross, *J. Chromatogr.*, 118 (1976) 273.
16 C.D. Carr, J. Smith and S. Abbott, *VIA*, 9 No. 2 (1975) 8, Varian, Palo Alto, CA.
17 *Capsaicin in Chili Powder, Application Notes No. 21*, Micromeritics, Norcross, GA.
18 T.H. Beasley, H.W. Ziegler and A.D. Bell, *J. Chromatogr.*, 175 (1979) 350.
19 Y. Akada and Y. Tanase, *Yakugaku Zasshi*, 98 (1976) 1035.
20 H. Elgass, *Dissertation*, Saarbrücken, 1978.
21 R. Schmidt, J. Mondani, E. Ziegenhagen and K. Dose, *J. Chromatogr.*, 207 (1981) 435.
22 L.W. Wulf, C.W. Nagel and A.L. Branen, *J. Chromatogr.*, 161 (1978) 271.
23 R. Schuster and K. Wessely, *Application Note*, AN 232-4, Hewlett-Packard, Böblingen, 1978.
24 G. Charalambous (Editor), *Liquid Chromatographic Analysis of Food and Beverages*, Vols. 1 and 2, Academic Press, New York, 1979.

Chapter 32

ENVIRONMENTAL APPLICATIONS

HPLC has become increasingly employed for the analytical control of environmental pollution levels. It appears very promising for the identification and determination of non-volatile or strongly polar compounds in air and in surface, waste and drinking waters. Polyaromatic hydrocarbons, phenolics, chlorinated aromatics, pesticides and their degradation products and certain surfactants are of primary concern among industrial, agricultural and household organic pollutants. As polluted samples are usually very complex mixtures of different compounds, the gradient elution technique is frequently used for separations of sample components that differ widely in their chromatographic properties.

32.1. AIR POLLUTION ANALYSIS

Polynuclear aromatic hydrocarbons in air[1], automobile exhausts[2,3] and cigarette smoke[4] samples can be collected on fibre-glass filters, extracted, the extracts concentrated and then analysed by reversed-phase chromatography on octadecylsilica columns using elution with a linear gradient of methanol[2], acetonitrile or tetrahydrofuran[3] in water. Chromatography on microparticulate silica gel columns with gradients of chloroform in *n*-hexane as the eluents[1] or separation on columns packed with a chemically bonded medium-polarity ethereal phase using elution with a gradient of *n*-propanol in water[4] have also been used for this purpose. UV detection may be employed, but the sensitivity and selectivity of detection is increased by using fluorimetric detectors.

Trace amounts of isocyanates are likely to occur in working atmospheres in large-scale polyurethane paint-spraying operations and a suitable method for the analysis of these toxic compounds is of interest. Airborne isocyanates may be collected and converted into stable fluorescent urea derivatives by reaction with 1-naphthalenemethylamine or N-methyl-1-naphthalenemethylamine. These derivatives are then analysed by reversed-phase chromatography on octadecylsilica columns using elution with a combination of concave and exponential gradients of ethanol in *n*-hexane[5,6] or with a linear gradient of methanol in chloroform - ethanol (99 : 1)[7]. Fluorimetric detection is the most sensitive, but UV detection at 254 nm can also be used.

References on p. 467.

Caffeine in samples of the New York City air was collected and determined by chromatography on an octadecylsilica column using a linear gradient of acetonitrile in water for elution, with UV detection at 254 nm[8].

32.2. WATER POLLUTION ANALYSIS

HPLC with the gradient elution technique is a powerful method for the analysis of water pollution by non-volatile organic compounds. Organic pollutants present in water in trace amounts can be concentrated by direct evaporation of water in vacuo[13] or extracted into a volatile organic solvent followed by concentration of the extracts by evaporation in vacuo or, better, subjected to sorption from large amounts of water on a short column or cartridge containing an appropriate sorbent, such as charcoal or octadecylsilica bonded material[14]. The sorption step is followed by extraction of the sorbed compounds into an appropriate organic solvent and the extracts are concentrated by evaporation.

Reversed-phase gradient elution chromatography itself offers an interesting possibility for coupling direct on-line sample enrichment with the separation, identification and quantitation of sample compounds. Because water has only a low eluting strength in reversed-phase chromatography, most organic compounds are strongly retained on octadecylsilica or octylsilica columns in pure water. This makes it possible to let pass large volumes of water samples (millilitres or even several hundred millilitres) directly through the analytical column, on the top of which most organic compounds from the sample become retained. Elution with a gradient of increasing concentration of an organic solvent in water as the mobile phase follows, which makes it possible to achieve the elution and separation of the organic compounds concentrated on the top of the column, usually without a significant loss of resolution[9-12]. A pre-column packed with the same material as the analytical column can be used for enrichment to increase the lifetime of the analytical column. Gradients of methanol, acetonitrile and tetrahydrofuran in water are used for this purpose.

Reversed-phase gradient elution chromatography on octadecylsilica or octylsilica columns, combined with direct on-line or various off-line sample enrichment techniques, was employed to provide an overall picture of surface, marine, drinking and waste water pollution by UV-absorbing organic compounds from fingerprint chromatograms[9-11,13,14]. This method was also used for the separation and quantitation of polynuclear aromatic hydrocarbons[12], phenolic and other pollutants in industrial sewages[12,15], chlorophenoxy acids[16] and organophosphorus, urea, triazine[17] and other[18] herbicides and their degradation products[19] in surface and drinking waters, nonionic surfactants of the alkoxylated alkylphenol type[20] and anionic surfactants (linear alkylbenzene sulphonates)[21] in surface and sewage waters.

In addition to the commonly used gradients of acetonitrile and methanol in water, gradients of these solvents in dilute aqueous solutions of acetic and other acids or in diluted aqueous solutions of buffers can be employed[13].

UV and fluorimetric detectors are generally used in these applications; however, specific detection techniques may be useful in certain analytical determinations, such as that of toxic N-nitroso compounds (N-nitrosoureas, -carbamates, -guanidines, etc.) formed in the environment by the degradation of widely used herbicides. Here, a nitrosamide-specific detector was developed, based on a post-column derivatization reaction of the nitroso compounds with Griess reagent. Sulphanilic acid and then N-1-naphthylenediamine dihydrochloride are continuously added to the column effluent and the coloration of the purple dye formed by reaction with nitroso compounds is continuously monitored by measuring the absorbance at 550 nm[19].

Polynuclear aromatic hydrocarbons in drinking waters have also been analysed by chromatography on a silica gel column using elution with a linear gradient of chloroform in *n*-hexane and on a column packed with chemically bonded nitrile phase, a linear gradient of methanol in *n*-hexane being used for elution[22].

Fatty acids in river waters were subjected to enrichment by extraction and evaporation and pre-separation on a Florisil column using elution with diethyl ether - methanol mixtures. The fraction of the eluate containing fatty acids was evaporated, the acids were converted into their phenacyl esters and then separated and determined by reversed-phase chromatography on two octadecylsilica columns in series using elution with a linear gradient of acetonitrile in water[23].

Organic compounds in dilute aqueous solutions, such as sewage effluents, have also been analysed by chromatography on a column packed with cation-exchange resin using elution with a gradient of acetate buffer[24].

The waste waters from paper mill plants contain relatively large amounts of soluble organics, mainly ligninsulphonic acids. These acids were analysed by anion-exchange chromatography after preliminary concentration on a charcoal column, elution with aqueous-ethanolic ammonia and evaporation of this eluate in vacuo. A column packed with Aminex A-27 anion-exchange resin and elution with a concentration gradient of ammonium acetate buffer were used for the separation of the acids, which were detected at 260 and 280 nm[25].

32.3. REFERENCES

1 W.C. Eisenberg, *J. Chromatogr. Sci.*, 4 (1978) 145.
2 A. Colmsjö and U. Stenberg, *Anal. Chem.*, 51 (1979) 145.
3 *Separation of Polynuclear Aromatics by LC*, AN 131, Waters Assoc., Milford, MA, 1973.
4 A.F. Haeberer, M.E. Snook and O.T. Chortyk, *Anal. Chim. Acta*, 80 (1975) 303.

5 K.L. Dunlap, R.L. Sandridge and J. Keller, *Anal. Chem.*, 48 (1976) 497.
6 S.P. Levine, J.H. Hoggatt, E. Chladek and G. Jungclaus, *Anal. Chem.*, 51 (1979) 1106.
7 L.H. Kormos, R.L. Sandridge and J. Keller, *Anal. Chem.*, 53 (1981) 1122.
8 M. Dong, D. Hoffmann, D.C. Locke and E. Ferrand, *Atmos. Environ.*, 11 (1977) 651.
9 W.E. May, S.N. Chesler, S.P. Cram, B.H. Gump, H.S. Hertz, D.P. Enagonio and S.M. Dyszel, *J. Chromatogr. Sci.*, 13 (1975) 535.
10 C.G. Creed, *Res. Dev.*, 27 (1976) 40.
11 A. Otsuki, *J. Chromatogr.*, 133 (1977) 402.
12 M. Schönmann, *Varian Instruments at Work*, LC 113, Varian, Palo Alto, CA, 1980.
13 B. Crathorne, C.D. Watts and M. Fielding, *J. Chromatogr.*, 185 (1979) 671.
14 W.A. Saner, J.R. Jadamec, R.W. Sager and T.J. Killeen, *Anal. Chem.*, 51 (1979) 2180.
15 *Industrial Effluent Monitoring, Application Notes No. 38*, Micromeritics, Norcross, GA.
16 *Half Hour Determination Method For Chlorphenoxy Acids and Esters Using LC*, J 22, Waters Assoc., Milford, MA, 1978.
17 *Screening Polar and Non-Polar Organics in a Single Analysis Using LC*, J 08, Waters Assoc., Milford, MA, 1978.
18 B. Crathorne and C.D. Watts, *J. Chromatogr.*, 169 (1979) 436.
19 G.M. Singer, S.S. Singer and D.G. Schmidt, *J. Chromatogr.*, 133 (1977) 59.
20 A. Otsuki and H. Shiraishi, *Anal. Chem.*, 51 (1979) 2329.
21 W. Giger, E. Staub and L. Schaffner, *ACS Abstr.*, April 1979.
22 A.D. Thruston, *J. Chromatogr. Sci.*, 6 (1978) 254.
23 D.A. Hullett and S.J. Eisenreich, *Anal. Chem.*, 51 (1979) 1953.
24 R.L. Jolley, S. Katz, J.E. Mrochek, W.W. Pitt and W.T. Rainey, *Chem. Technol.*, 5 (1975) 312.
25 A.W. Lis, R.K. McLaughlin, J. Trau, G.O. Davies and W.R. Anderson, *Tappi*, 59 (1976) 127.

Chapter 33

CLINICAL DIAGNOSTICS

Various compounds in physiological fluids can be analysed using HPLC, such as pharmaceuticals, drugs of abuse, vitamins, antibiotics and toxicants. Applications of gradient elution liquid chromatography to the separation and determination of such compounds and their metabolites have been mentioned in several earlier chapters (Chapters 17-25).

Recent medical research has indicated that hundreds of the molecular constituents of human body fluids may be related to bodily functions and to various inborne defects and other diseases. HPLC can be utilized for automated high-resolution analyses of UV-absorbing constituents in urine, blood serum and other physiological fluids. Originally, anion-exchange chromatography on columns packed with microparticulate anion-exchange resins (ca. 10 μm particle diameter) was proposed and used for monitoring compounds in physiological fluids. Gradient elution with linear, convex or multi-linear gradients of sodium acetate - acetic acid buffer concentration (from 0.015 to 6 mol/l) at a constant pH of 4.4 was employed to separate a variety of acidic, neutral and basic compounds in human urine. About 100-150 peaks were separated and some of them were identified[1-8]. The original time of analysis was more than 40 h with columns 1-3.2 m long, but it was later decreased to ca. 20-30 h by using a simultaneous temperature gradient (from 20 to 60°C)[9,10]. With shorter columns and higher flow-rates these separations have been accomplished recently in ca. 2 h at the expense of decreased resolution and peak capacity; however, the method modified in this way has become useful for routine clinical analyses[11].

Other gradient elution systems have been proposed, such as a gradient of increasing concentration of ammonium perchlorate in water (0-0.25 mol/l in 2 h)[12] or a gradient of acetonitrile in aqueous ammonium chloride[13]. UV detection at various wavelengths (200, 215, 225, 230, 254, 260 and 280 nm) and fluorimetric detection were used in these applications.

Chemically bonded weak anion exchangers with polyethylenimine, diethylamine or diethylaminoethyl functional groups on a controlled-porosity glass support were used for screening UV absorbing compounds in physiological materials (blood serum) using elution with linear or multi-linear gradients of sodium chloride in 0.05 mol/l Tris-HCl buffer[14].

Similar separations to those on anion-exchange resins, but considerably faster, have been achieved using reversed-phase gradient elution chromatography. For

References on p. 470.

example, on a standard analytical column packed with 5 μm octadecylsilica material, it was possible to resolve ca. 120 peaks of urinary acids and other compounds in urine samples in 30 min by elution with a gradient of increasing acetonitrile concentration in 0.1 mol/l aqueous phosphate[15,16]. An example of such a separation is shown on Fig. 4.16. Similar techniques have been used for the analysis of human blood serum[17-22] and other body fluids[19,22] (usually in ca. 30-45 min) on columns packed with microparticulate octadecylsilica material, using elution with gradients of methanol in aqueous potassium dihydrogen phosphate[17,18,22] or sodium acetate buffers[19,20] (pH in the range 2.5-5.6). UV detection at 254 or 280 nm is usually employed. Simultaneous UV and fluorimetric detection is helpful in the identification of sample compounds[16] and this may be further improved using more selective detectors, such as the amperometric type[22]. The use of microbore octadecylsilica columns (250-1000 mm long, 1 mm I.D.) in connection with gradient elution seems promising for body fluid analysis[21].

33.1. REFERENCES

1 S. Katz and C.A. Burtis, *J. Chromatogr.*, 40 (1969) 270.
2 C.D. Scott, J.E. Attrill and N.G. Anderson, *Proc. Soc. Exp. Biol. Med.*, 125 (1967) 181.
3 C.D. Scott, *Clin. Chem.*, 14 (1968) 521.
4 C.D. Scott, R.L. Jolley, W.W. Pitt and W.F. Johnson, *Amer. J. Clin. Pathol*, 53 (1970) 701.
5 C.D. Scott, F. Wayne and V.E. Walker, *Anal. Biochem.*, 32 (1969) 182.
6 C.A. Burtis and K.S. Warren, *Clin. Chem.*, 14 (1968) 290.
7 C.A. Burtis, W.C. Butts, W.T. Rainey and C.D. Scott, *Amer. J. Clin. Pathol.*, 53 (1970) 739.
8 S. Katz, *J. Chromatogr.*, 53 (1970) 415.
9 C.A. Burtis, *J. Chromatogr.*, 52 (1970) 97.
10 F. Geeraerts, L. Schimpfessel and R. Crokaert, *J. Chromatogr.*, 145 (1978) 63.
11 K. Seta, M. Washitake, T. Anmo, N. Takai and T. Okuyama, *J. Chromatogr.*, 181 (1980) 311.
12 K. Seta, M. Washitake, I. Tanaka, N. Takai and T. Okuyama, *J. Chromatogr.*, 221 (1980) 215.
13 H. Miyagi, J. Miura, Y. Takata and S. Ganno, *Clin. Chem.*, 25 (1979) 1617.
14 S. Ho Chang, K.M. Gooding and F.E. Regnier, *J. Chromatogr.*, 120 (1976) 321.
15 I. Molnár and Cs. Horváth, *J. Chromatogr.*, 143 (1977) 391.
16 I. Molnár, Cs. Horváth and P. Jatlow, *Chromatographia*, 11 (1978) 260.
17 A.M. Krstulovic, R.A. Hartwick, P.R. Brown and K. Lohse, *J. Chromatogr.*, 158 (1978) 365.
18 S.P. Assenza and P.R. Brown, *J. Chromatogr.*, 181 (1980) 169.
19 E.J. Knudzon, Y.C. Lau, H. Veening and D.A. Dayton, *Clin. Chem.*, 24 (1978) 686.
20 F.C. Senftleber, A.G. Halline and H. Veening, *Clin. Chem.*, 22 (1976) 1522.
21 R.P.W. Scott and P. Kucera, *J. Chromatogr.*, 185 (1979) 27.
22 A.M. Krstulovic, L. Bertani-Dziedzic, S.W. Dziedzic and S.E. Gitlow, *J. Chromatogr.*, 223 (1981) 305.

APPENDIXES

Appendix 1

RETENTION CHARACTERISTICS IN ISOCRATIC ELUTION CHROMATOGRAPHY IN DEPENDENCE OF THE CONCENTRATION OF THE STRONGER ELUTING AGENT IN THE MOBILE PHASE, c, FOR DIFFERENT $k' = f(c)$ FUNCTIONS

Function *	Eqn. No.
$k' = (a + bc)^{-m}$ function:	1.21
$V_R = V_m[1 + (a + bc)^{-m}]$	A1
$w = \frac{4V_m}{\sqrt{n}}[1 + (a + bc)^{-m}]$	A2
$\alpha = \frac{(a_1 + b_1c)^{m_1}}{(a_2 + b_2c)^{m_2}}$	A3
$R_s = \frac{\sqrt{n_2}}{4} \cdot \frac{(a_2 + b_2c)^{-m_2} - (a_1 + b_1c)^{-m_1}}{1 + (a_2 + b_2c)^{-m_2}}$	A4
$k' = k'_0c^{-m}$ function:	1.23
$V_R = V_m(1 + k'_0c^{-m})$	A5
$w = \frac{4V_m}{\sqrt{n}}(1 + k'_0c^{-m})$	A6
$\alpha = \frac{k'_{02}}{k'_{01}} \cdot \frac{c^{-m_2}}{c^{-m_1}} = \alpha_0c^{(m_1 - m_2)}$	A7
$R_s = \frac{\sqrt{n_2}}{4} \cdot \frac{k'_{02}c^{-m_2} - k'_{01}c^{-m_1}}{1 + k'_{02}c^{-m_2}}$	A8
$R_s = \frac{\sqrt{n_2}}{4}\left(1 - \frac{1}{\alpha_0}\right)\frac{k'_{02}}{k'_{02} + c^m}$ for $m_1 \approx m_2 = m$	A9
$k' = (a + bc)^{-1}$ function:	1.24
$V_R = V_m[1 + (a + bc)^{-1}]$	A10
$w = \frac{4V_m}{\sqrt{n}}[1 + (a + bc)^{-1}]$	A11
$\alpha = \frac{a_1 + b_1c}{a_2 + b_2c}$	A12
$R_s = \frac{\sqrt{n_2}}{4} \cdot \frac{(a_1 - a_2) + (b_1 - b_2)c}{(a_1 + b_1c)(1 + a_2 + b_2c)}$	A13

Appendix 1 (continued)

Function *	Eqn. No.
$k' = k'_0 10^{(-mc+pc^2)}$ function:	1.26
$V_R = V_m[1 + k'_0 10^{(-mc+pc^2)}]$	A14
$w = \frac{4V_m}{\sqrt{n}}[1 + k'_0 10^{(-mc+pc^2)}]$	A15
$\alpha = \frac{k'_{02}}{k'_{01}} \cdot \frac{10^{(-m_2c+p_2c^2)}}{10^{(-m_1c+p_1c^2)}} = \alpha_0 10^{[(m_1-m_2)c+(p_2-p_1)c^2]}$	A16
$R_s = \frac{\sqrt{n_2}}{4} \frac{k'_{02}10^{(-m_2c+p_2c^2)} - k'_{01}10^{(-m_1c+p_1c^2)}}{1 + k'_{02}10^{(-m_2c+p_2c^2)}}$	A17
$k' = k'_0 10^{-mc}$ function:	1.28
$V_R = V_m(1 + k'_0 10^{-mc})$	A18
$w = \frac{4V_m}{\sqrt{n}}(1 + k'_0 10^{-mc})$	A19
$\alpha = \frac{k'_{02}}{k'_{01}} \cdot \frac{10^{-m_2c}}{10^{-m_1c}} = \alpha_0 10^{(m_1-m_2)c}$	A20
$R_s = \frac{\sqrt{n_2}}{4} \cdot \frac{k'_{02}10^{-m_2c} - k'_{01}10^{-m_1c}}{1 + k'_{02}10^{-m_2c}}$	A21
$R_s = \frac{\sqrt{n_2}}{4} \cdot \frac{k'_{02} - k'_{01}}{10^{mc} + k'_{02}}$ for $m_1 \approx m_2 = m$	A22

* Retention volume V_R; bandwidth w; separation factor α; resolution R_s; eqns. 1.1, 1.5, 1.14, 1.13.

Appendix 2

SOLUTION OF THE BASIC EQUATION FOR GRADIENT ELUTION CHROMATOGRAPHY (EQN. 4.2) FOR DIFFERENT $k' = f(c)$ AND $c = \phi(V)$ FUNCTIONS FROM TABLES 2.1 AND 4.1

Reference numbers relate to Part II.

No. $c = \phi(V)$		$k' = f(c)$				
		$k' = (a + bc)^{-m}$	$k' = k'_0 c^{-m}$	$k' = (a + bc)^{-1}$	$k' = k'_0 10^{(-mc + pc^2)}$	$k' = k'_0 10^{-mc}$
1 $c + A + (DV + B)e^{-\kappa V}$	Algebraic solution	For $m = 1$ only; implicit	For $m = 1$ only; implicit	Implicit	No	No
	Refs.	56, 57	56, 57	56, 57		
	Eqn. no.	A23 ($m = 1$)	A48 ($m = 1$)	A23		
1a $c = A + Be^{-\kappa V}$	Algebraic solution	For m = integer only; implicit	For m = integer only; implicit	Implicit	No	By development into a polynomial; implicit
	Refs.	56, 57	23 ($m = 1, 2, 3$); 28 ($m = 1, 2, 3$); 27, 29 Numerical and graphical solutions	23		
	Eqn. no.	A24 ($m = 1$); A25 ($m = 2$); A26 ($m = 3$)	A49 ($m = 1$); A50 ($m = 2$); A51 ($m = 3$)	A24		A67
1b $c = Be^{-\kappa V}$	Algebraic solution	For m = integer only; implicit	Explicit for arbitrary m	Implicit	No	By development into a polynomial; implicit
	Refs.	56, 57	33, 66 26 ($m = 1$)	56, 57		67, numerical solution
	Eqn. no.	A27 ($m = 1$); A28 ($m = 2$); A29 ($m = 3$)	A52	A27		A68
2 $c = D - (A^{1/\kappa} + BV)^{\kappa}$	Algebraic solution	For m = integer only; implicit	For m = integer only; implicit	Implicit	No	For $1/\kappa$ = integer only. $\kappa = 1$, explicit; $\kappa \neq 1$, implicit
	Refs.	56, 57	56, 57	56, 57		8, 39, 50 ($\kappa = 1$)
	Eqn. no.	A30 ($m = 1$); A31 ($m = 2$); A32 ($m = 3$)	A53 ($m = 1$); A54 ($m = 2$)	A30		A69 ($\kappa = 1$); A70 ($\kappa = 1/2$); A71 ($\kappa = 1/3$)

	solution	only; implicit	arbitrary m			ger only. $\kappa = 1$, explicit; $\kappa \neq 1$. implicit
	Refs.	56, 57	8	56, 57		8, 39, 50
	Eqn. no.	A33 ($m = 1$); A34 ($m = 2$); A35 ($m = 3$)	A55	A33		A72 ($\kappa = 1$); A73 ($\kappa = 1/2$); A74 ($\kappa = 1/3$)
3a $c = A + BV$	Algebraic solution	Explicit for arbitrary m	Explicit for arbitrary m	Explicit	No *	Explicit for arbitrary m
	Refs.	39 ($m = -1$)	8 ($m =$ arbitrary)		39*, explicit solution involving "error function"	8, 17, 39, 67, numerical solution
	Eqn. no.	A39 ($m \neq -1$); A40 ($m = -1$)	A59 ($m \neq -1$); A60 ($m = -1$)	A39	*	A72 ($m \neq -1$); A77 ($m = -1$)
3b $c = BV^{\kappa}$	Algebraic solution	For $m =$ integer only; for $m = 1$ explicit; for $m \neq 1$ implicit	Explicit for arbitrary m	Explicit	No	For $1/\kappa =$ integer only; $\kappa = 1$: explicit; $\kappa \neq 1$: implicit
	Refs.	56, 57	8	56, 57		56, 57
	Eqn. no.	A41 ($m = 1$); A42 ($m = 2$); A43 ($m = 3$)	A61	A41		A78 ($\kappa = 1$); A79 ($\kappa = 1/2$); A80 ($\kappa = 1/3$)
3c $c = BV$	Algebraic solution	Explicit for arbitrary m	Explicit for arbitrary m	Explicit	No *	Explicit for arbitrary m
	Refs.	56, 57	23 ($m = 1$; $m = 2$; $m = 3$); 66 ($m = 2$); 8 ($m =$ arbitrary)		39 * (see 3a)	8, 17, 39

(Continued on p. 476)

Appendix 2 (continued)

No.	$c=\phi(V)$		$k'=f(c)$				
			$k'=(a+bc)^{-m}$	$k'=k'_0c^{-m}$	$k'=(a+bc)^{-1}$	$k'=k'_0 10^{(-mc+pc^2)}$	$k'=k'_0 10^{-mc}$
		Eqn. no.	A44 ($m \neq 1$); A46 ($m=-1$)	A62 ($m \neq -1$); A63 ($m=-1$)	A45		A78 ($m \neq -1$)
4	$c=A+BV^{\kappa}$	Algebraic solution	For m = integer only; implicit	For m = integer only; implicit	Implicit	No	For $1/\kappa$ = integer only; $\kappa=1$: explicit; $\kappa \neq 1$: implicit
		Refs.	56, 57	27, approximate solution	56, 57		56, 57
		Eqn. no.	A36 ($m=1$); A37 ($m=2$); A38 ($m=3$)	A56 ($m=1$); A57 ($m=2$); A58 ($m=3$)	A36		A72 ($\kappa=1$); A75 ($\kappa=1/2$); A76 ($\kappa=1/3$)
5	$c=\kappa \log(A^{1/\kappa}+BV)$	Algebraic solution	For m = integer only; implicit	For m = integer only; implicit	Implicit	No	Explicit for arbitrary m
		Refs.	56, 57	56, 57	56, 57		50
		Eqn. no.	A47 ($m=1$)	A64 ($m=1$); A65 ($m=2$); A66 ($m=3$)	A47		A81

* The solution leads to the integral $\int_{a_1}^{a_2(V'_g)} e^{-V^2}\, dV$.

Appendix 2 (continued)

EQUATIONS FOR RETENTION VOLUME IN GRADIENT ELUTION CHROMATOGRAPHY USING $k' = f(c)$ AND $c = \phi(V)$ FUNCTIONS FROM TABLES 2.1 AND 4.1

Function	Eqn. No.
Functions $k' = (a+bc)^{-m}$ $[k' = (a+bc)^{-1}]$:	
$m = 1$	
$V'_g = \frac{1}{\mathscr{H}} \ln\left[\frac{\mathscr{H} - (D/B)(\mathscr{H} V'_g + 1)}{(\mathscr{H}^2/bB)(V_m - aV'_g - bAV'_g) + \mathscr{H} - (D/B)}\right]$	A23
$m = 1$	
$V'_g = -\frac{1}{\mathscr{H}} \ln\left\{1 + \frac{\mathscr{H}[V_m - (a+bA)V'_g]}{bB}\right\}$	A24
$m = 2$	
$V'_g = -\frac{1}{2\mathscr{H}} \ln\left[1 - \frac{2\mathscr{H} V_m - 2\mathscr{H}(a+bA)^2 V'_g - 4bB(a+bA)(e^{-\mathscr{H} V'_g} - 1)}{b^2B^2}\right]$	A25
$m = 3$	
$V'_g = -\frac{1}{3\mathscr{H}} \ln\left[1 + \frac{3\mathscr{H} V_m - 3\mathscr{H}(a+bA)^3 V'_g + \frac{9}{2} b^2B^2(a+bA)(e^{-2\mathscr{H} V'_g} - 1) - 9bB(a+bA)^2(e^{-\mathscr{H} V'_g} - 1)}{b^3B^3}\right]$	A26
$m = 1$	
$V'_g = -\frac{1}{\mathscr{H}} \ln\left[1 - \frac{\mathscr{H}(V_m - aV'_g)}{bB}\right]$	A27
$m = 2$	
$V'_g = -\frac{1}{2\mathscr{H}} \ln\left[1 - \frac{2\mathscr{H} V_m + 2\mathscr{H} a^2 V'_g - 4abB(e^{-\mathscr{H} V'_g} - 1)}{b^2B^2}\right]$	A28
$m = 3$	
$V'_g = -\frac{1}{3\mathscr{H}} \ln\left[1 - \frac{3\mathscr{H} V_m + 3\mathscr{H} a^3 V'_g - \left(\frac{9}{2}\right) ab^2B^2(e^{-2\mathscr{H} V'_g} - 1) - 9a^2bB(e^{-\mathscr{H} V'_g} - 1)}{b^3B^3}\right]$	A29

(Continued on p. 478)

Appendix 2 (continued)

Function	Eqn. No.
$m = 1$	
$V'_g = \dfrac{A^{1/\mathscr{H}}}{B} - \dfrac{1}{B}\left[\dfrac{B}{b}(\mathscr{H}+1)V_m - \dfrac{(a+bD)B}{b}(\mathscr{H}+1)V'_g + A^{(\mathscr{H}+1)/\mathscr{H}}\right]^{1/(\mathscr{H}+1)}$	A30
$m = 2$	
$V'_g = \dfrac{A^{1/\mathscr{H}}}{B} - \dfrac{1}{B}\left[\dfrac{(B/b)(\mathscr{H}+1)V_m + 2(a+bD)A^{(\mathscr{H}+1)/\mathscr{H}} - b[(\mathscr{H}+1)/(2\mathscr{H}+1)]A^{(2\mathscr{H}+1)/\mathscr{H}} - (B/b)(a+bD)^2(\mathscr{H}+1)V'_g}{2(a+bD) - b[(\mathscr{H}+1)/(2\mathscr{H}+1)](A^{1/\mathscr{H}} - BV'_g)^{\mathscr{H}}}\right]^{1/(H+1)}$	A31
$m = 3$	
$V'_g = \dfrac{A^{1/\mathscr{H}}}{B} - \dfrac{1}{B}\left[\dfrac{(B/b)(\mathscr{H}+1)V_m + 3(a+bD)^2A^{[(\mathscr{H}+1)/\mathscr{H}]} - 3(a+bD)b[(\mathscr{H}+1)/(2\mathscr{H}+1)]A^{(2\mathscr{H}+1)/\mathscr{H}} + b^2[(\mathscr{H}+1)/(3\mathscr{H}+1)]A^{(3\mathscr{H}+1)/\mathscr{H}} - (B/b)(a+bD)^3(\mathscr{H}+1)V'_g}{3(a+bd)^2 - 3(a+bD)b[(\mathscr{H}+1)/(2\mathscr{H}+1)](A^{1/\mathscr{H}} - BV'_g)^{\mathscr{H}} + b^2[(\mathscr{H}+1)/(3\mathscr{H}+1)](A^{1/\mathscr{H}} - BV'_g)^{2\mathscr{H}}}\right]^{1/(\mathscr{H}+1)}$	A32
$m = 1$	
$V'_g = \dfrac{1}{B}[(B/b)(\mathscr{H}+1)(V_m - aV'_g) + A^{(\mathscr{H}+1)/\mathscr{H}}]^{1/(\mathscr{H}+1)} - \dfrac{A^{1/\mathscr{H}}}{b}$	A33
$m = 2$	
$V'_g = \dfrac{1}{B}\left[\dfrac{(B/b)(\mathscr{H}+1)V_m + 2aA^{(\mathscr{H}+1)/\mathscr{H}} + b[(\mathscr{H}+1)/(2\mathscr{H}+1)]A^{(2\mathscr{H}+1)/\mathscr{H}} - (Ba^2/b)(\mathscr{H}+1)V'_g}{2a + b(\mathscr{H}+1/2\mathscr{H}+1)(A^{1/\mathscr{H}} + BV'_g)^{\mathscr{H}}}\right]^{1/(\mathscr{H}+1)} - \dfrac{A^{1/\mathscr{H}}}{B}$	A34
$m = 3$	
$V'_g = \dfrac{1}{B}\left[\dfrac{(B/b)(\mathscr{H}+1)V_m + 3a^2A^{(\mathscr{H}+1)/\mathscr{H}} + 3ab[(\mathscr{H}+1)/(2\mathscr{H}+1)]A^{(2\mathscr{H}+1)/\mathscr{H}} + b^2[(\mathscr{H}+1)/(3\mathscr{H}+1)] - (Ba^3/b)(\mathscr{H}+1)V'_g}{3a^2 + 3ab[(\mathscr{H}+1)/(2\mathscr{H}+1)](A^{1/\mathscr{H}} + BV'_g)^{\mathscr{H}} + b^2[(\mathscr{H}+1)(3\mathscr{H}+1)](A^{1/\mathscr{H}} + BV'_g)^{2\mathscr{H}}}\right]^{1/(\mathscr{H}+1)} - \dfrac{A^{1/\mathscr{H}}}{B}$	A35
$m = 1$	
$V'_g = V_m\left[a + bA + \dfrac{bB}{\mathscr{H}+1}\cdot V_g'^{(\mathscr{H})}\right]^{-1}$	A36
$m = 2$	
$V'_g = V_m\left[(a+bA)^2 + \dfrac{2(a+bA)bB}{\mathscr{H}+1}\cdot V_g'^{(\mathscr{H})} + \dfrac{b^2B^2}{2\mathscr{H}+1}\cdot V_g'^{(2\mathscr{H})}\right]^{-1}$	A37

$m = 3$

$$V_g' = V_m\left[(a+bA)^3 + \frac{3(a+bA)^2bB}{\mathscr{H}+1}\cdot V_g'^{(\mathscr{H})} + \frac{3(a+bA)b^2B^2}{2\mathscr{H}+1}\cdot V_g'^{(2\mathscr{H})} + \frac{b^3B^3}{3\mathscr{H}+1}\cdot V_g'^{(3\mathscr{H})}\right]^{-1} \quad \text{A38}$$

$m \neq -1$

$$V_g' = \frac{1}{bB}\left[bB(m+1)V_m + (a+Ab)^{(m+1)}\right]^{1/(m+1)} - \frac{a+Ab}{bB} \quad \text{A39}$$

$m = -1$

$$V_g' = \frac{a+bA}{bB}(e^{bBV_m} - 1) \quad \text{A40}$$

$m = 1$

$$V_g' = V_m\left[a + \frac{bB}{\mathscr{H}+1}\cdot V_g'^{(\mathscr{H})}\right]^{-1} \quad \text{A41}$$

$m = 2$

$$V_g' = V_m\left[a^2 + \frac{2abB}{\mathscr{H}+1}\cdot V_g'^{(\mathscr{H})} + \frac{b^2B^2}{2\mathscr{H}+1} V_g'^{(2\mathscr{H})}\right]^{-1} \quad \text{A42}$$

$m = 3$

$$V_g' = V_m\left[a^3 + \frac{3a^2bB}{\mathscr{H}+1}\cdot V_g'^{(\mathscr{H})} + \frac{3ab^2B^2}{2\mathscr{H}+1}\cdot V_g'^{(2\mathscr{H})} + \frac{b^3B^3}{3\mathscr{H}+1} V_g'^{(3\mathscr{H})}\right]^{-1} \quad \text{A43}$$

$m \neq -1$

$$V_g' = \frac{1}{bB}[bB(m+1)V_m + a^{(m+1)}]^{1/(m+1)} - \frac{a}{bB} \quad \text{A44}$$

$m = 1$

$$V_g' = \frac{a}{bB}\left(\sqrt{1 + \frac{2bBV_m}{a^2}} - 1\right) \quad \text{A45}$$

$m = -1$

$$V_g' = \frac{a}{bB}(e^{bBV_m} - 1) \quad \text{A46}$$

$m = 1$

$$V_g' = \frac{1}{B}\exp\left\{\frac{\frac{2.31B}{b\mathscr{H}}[V_m - aV_g' + A^{1/\mathscr{H}}(\ln A^{1/\mathscr{H}} - 1)] + 1}{A^{1/\mathscr{H}} + BV_g'}\right\} - \frac{A^{1/\mathscr{H}}}{B} \quad \text{A47}$$

(Continued on p. 480)

Appendix 2 (continued)

Function	Eqn. No.
Function $k' = k'_0 c^{-m}$:	
$m = 1$	
$V'_g = -\frac{1}{\mathscr{H}} \ln\left[\frac{\mathscr{H} k'_0 V_m - \mathscr{H} A V'_g + B + D/\mathscr{H}}{B - D(V'_g + 1/\mathscr{H})}\right]$	A48
$m = 1$	
$V'_g = -\frac{1}{\mathscr{H}} \ln\left(\frac{B + \mathscr{H} k'_0 V_m - \mathscr{H} A V'_g}{B}\right)$	A49
$m = 2$	
$V'_g = -\frac{1}{2\mathscr{H}} \ln\left[\frac{B^2 - 2\mathscr{H} k'_0 V_m + 2\mathscr{H} A^2 V'_g + 4AB(e^{-\mathscr{H} V'_g} - 1)}{B^2}\right]$	A50
$m = 3$	
$V'_g = -\frac{1}{3\mathscr{H}} \ln\left[\frac{B^3 + 3\mathscr{H} k'_0 V_m - 3\mathscr{H} A^3 V'_g + (9/2)AB^2(e^{-2\mathscr{H} V'_g} - 1) - 9A^2 B(e^{-\mathscr{H} V'_g} - 1)}{B^3}\right]$	A51
$V'_g = -\frac{1}{\mathscr{H} m} \ln\left(1 - \frac{\mathscr{H} m k'_0 V_m}{B^m}\right)$	A52
$m = 1$	
$V'_g = \left(\frac{A^{1/\mathscr{H}}}{B} - \frac{1}{B}\right)[(\mathscr{H}+1)B(k'_0 V_m - DV'_g) + A^{(\mathscr{H}+1)/\mathscr{H}}]^{1/(\mathscr{H}+1)}$	A53
$m = 2$	
$V'_g = \frac{A^{1/\mathscr{H}}}{B} - \frac{1}{B}\left\{-(2\mathscr{H}+1)B(k'_0 V_m - D^2 V'_g) + \frac{2D(2\mathscr{H}+1)}{\mathscr{H}+1}[(A^{1/\mathscr{H}} - BV'_g)^{(\mathscr{H}+1)} - A^{(\mathscr{H}+1)/\mathscr{H}}] + A^{(2\mathscr{H}+1)/\mathscr{H}}\right\}^{1/(2\mathscr{H}+1)}$	A54
$V'_g = \frac{1}{B}[(\mathscr{H} m + 1)Bk'_0 V_m + A^{(\mathscr{H} m+1)/\mathscr{H}}]^{1/(\mathscr{H} m+1)} - \frac{A^{1/\mathscr{H}}}{B}$	A55
$m = 1$	
$V'_g = k'_0 V_m [A + \frac{B}{\mathscr{H}+1} \cdot V_g'^{(\mathscr{H})}]^{-1}$	A56

$m = 2$

$$V'_g = k'_0 V_m \left[A^2 + \frac{2AB}{\mathscr{H}+1} \cdot V'^{(\mathscr{H})}_g + \frac{B^2}{2\mathscr{H}+1} \cdot V'^{(2\mathscr{H})}_g \right]^{-1} \quad \text{A57}$$

$m = 3$

$$V'_g = k'_0 V_m \left[A^3 + \frac{3A^2B}{\mathscr{H}+1} \cdot V'^{(\mathscr{H})}_g + \frac{3AB^2}{2\mathscr{H}+1} \cdot V'^{(2\mathscr{H})}_g + \frac{B^3}{3\mathscr{H}+1} \cdot V'^{(3\mathscr{H})}_g \right]^{-1} \quad \text{A58}$$

$m \neq -1$

$$V'_g = \frac{1}{B}[(m+1)Bk'_0 V_m + A^{(m+1)}]^{1/(m+1)} - \frac{A}{B} \quad \text{A59}$$

$m = -1 \ (A \neq 0)$

$$V'_g = \frac{A}{B}(e^{Bk'_0 V_m} - 1) \quad \text{A60}$$

$$V'_g = \left[\frac{(\mathscr{H}m+1)k'_0 V_m}{B^m} \right]^{1/(\mathscr{H}m+1)} \quad \text{A61}$$

$m \neq -1$

$$V'_g = \left[\frac{(m+1)k'_0 V_m}{B^m} \right]^{1/(m+1)} \quad \text{A62}$$

$m = -1$

$$V'_g = e^{(Bk'_0 V_m + 1)} \quad \text{A63}$$

$m = 1$

$$V'_g = \frac{1}{B} \exp\left\{ \frac{2.31Bk'_0[V_m + A^{1/\mathscr{H}}(\ln A^{1/\mathscr{H}} - 1)]}{\mathscr{H}(A^{1/\mathscr{H}} + BV'_g)} + 1 \right\} - \frac{A^{1/\mathscr{H}}}{B} \quad \text{A64}$$

$m = 2$

$$V'_g = \frac{(2.31Bk'_0 V_m/\mathscr{H}^2) + A^{1/\mathscr{H}}[(\ln A^{1/\mathscr{H}})^2 - 2\ln A^{1/\mathscr{H}} + 2]}{B\left\{[\ln(A^{1/\mathscr{H}} + BV'_g)]^2 - 2\ln(A^{1/\mathscr{H}} + BV'_g) + 2\right\}} - \frac{A^{1/\mathscr{H}}}{B} \quad \text{A65}$$

$m = 3$

$$V'_g = \frac{(2.31Bk'_0 V_m/\mathscr{H}^3) + A^{1/\mathscr{H}}[(\ln A^{1/\mathscr{H}})^3 - 3(\ln A^{1/\mathscr{H}})^2 + 6\ln A^{1/\mathscr{H}} - 6]}{B\left\{[\ln(A^{1/\mathscr{H}} + BV'_g)]^3 - 3[\ln(A^{1/\mathscr{H}} + BV'_g)]^2 + 6\ln(A^{1/\mathscr{H}} + BV'_g) - 6\right\}} - \frac{A^{1/\mathscr{H}}}{B} \quad \text{A66}$$

(Continued on p. 482)

Appendix 2 (continued)

Function	Eqn. No.
Function $k' = k'_0 10^{-mc}$:	
$V'_g = k'_0 V_m 10^{-mA} + \frac{2.31mB}{\mathscr{H}}(1 - e^{-\mathscr{H}V_g}) - \frac{(2.31mB)^2}{4\mathscr{H}}(1 - e^{-2\mathscr{H}V_g}) + \frac{(2.31mB)^3}{18\mathscr{H}}(1 - e^{-3\mathscr{H}V_g}) \ldots$	A67
$V'_g = k'_0 V_m - \frac{2.31mB}{\mathscr{H}}(1 - e^{-\mathscr{H}V_g}) - \frac{(2.31mB)^2}{4\mathscr{H}}(1 - e^{-2\mathscr{H}V_g}) - \frac{(2.31mB)^3}{18\mathscr{H}}(1 - e^{-3\mathscr{H}V_g}) \ldots$	A68
$\mathscr{H} = 1$	
$V'_g = \frac{1}{mB} \log[1 - 2.31mBk'_0 V_m 10^{m(A-D)}]$	A69
$\mathscr{H} = 1/2$	
$V'_g = \frac{A^2}{B} - \frac{1}{B}\left\{\frac{1}{m} \log\left[\frac{(2.31mA+1)10^{-mA} + (1/2)(2.31m)^2 Bk'_0 V_m 10^{-mD}}{1 + 2.31m(A^2 - BV'_g)^{1/2}}\right]\right\}^2$	A70
$\mathscr{H} = 1/3$	
$V'_g = \frac{A^3}{B} - \frac{1}{B}\left\{\frac{1}{m} \log \frac{\{[2 + 2(2.31mA) + (2.31mA)^2] - (1/3)(2.31m)^3 Bk'_0 V_m 10^{-mD}\} 10^{-mA}}{2[2.31m(A^3 - BV'_g)^{1/3} + 1] + (2.31m)^2 (A^3 - BV'_g)^{2/3}}\right\}^3$	A71
$\mathscr{H} = 1$	
$V'_g = \frac{1}{mB} \log(2.31mBk'_0 V_m 10^{-mA} + 1)$	A72
$\mathscr{H} = 1/2$	
$V'_g = \frac{1}{B}\left\{\frac{1}{m} \log\left[\frac{(1/2)(2.31m)^2 Bk'_0 V_m + (2.31mA - 1)10^{mA}}{2.31m(A^2 + BV'_g)^{1/2} - 1}\right]\right\}^2 - \frac{A^2}{B}$	A73

$\mathcal{H} = 1/3$

$$V_g' = \frac{1}{B}\left(\frac{1}{m}\log\left\{\frac{-(1/3)(2.31m)^3 Bk_0'V_m + [(2.31m)^2A^2 - 2(2.31m)A + 2]10^{mA}}{(2.31m)^2(A^3 + BV_g')^{2/3} - 2[2.31m(A^3 + BV_g')^{1/3} - 1]}\right\}\right)^3 - \frac{A^3}{B} \quad \text{A74}$$

$\mathcal{H} = 1/2$

$$V_g' = \left\{\frac{1}{mB}\log\left[\frac{(1/2)(2.31mB)^2 k_0'V_m 10^{-mA} - 1}{2.31mB(V_g')^{1/2} - 1}\right]\right\}^2 \quad \text{A75}$$

$\mathcal{H} = 1/3$

$$V_g' = \left(\frac{1}{mB}\log\left\{\frac{(1/3)(2.31mB)^3 k_0'V_m 10^{-mA} + 2}{(2.31mB)^2(V_g')^{2/3} - 2[2.31mB(V_g')^{1/3} - 1]}\right\}\right)^3 \quad \text{A76}$$

$m = -1$

$$V_g' = \frac{A}{B}(e^{Bk_0'V_m} - 1) \quad \text{A77}$$

$\mathcal{H} = 1$

$$V_g' = \frac{1}{mB}\log(2.31mBk_0'V_m + 1) \quad \text{A78}$$

$\mathcal{H} = 1/2$

$$V_g' = \left\{\frac{1}{mB}\log\left[\frac{(1/2)(2.31mB)^2 k_0'V_m - 1}{2.31mB(V_g')^{1/2} - 1}\right]\right\}^2 \quad \text{A79}$$

$\mathcal{H} = 1/3$

$$V_g' = \left(\frac{1}{mB}\log\left\{\frac{(1/3)(2.31mB)^3 k_0'V_m + 2}{(2.31mB)^2(V_g')^{2/3} - 2[2.31mB(V_g')^{1/3} - 1]}\right\}\right)^3 \quad \text{A80}$$

$$V_g' = \frac{1}{B}[(\mathcal{H}m + 1)Bk_0'V_m + A^{(\mathcal{H}m+1)/\mathcal{H}}]^{1/(\mathcal{H}m+1)} - \frac{A^{1/\mathcal{H}}}{B} \quad \text{A81}$$

Appendix 3

BLOCK DIAGRAM OF THE PROGRAM FOR CALCULATION OF OPTIMIZED GRADIENT PROFILE IN GRADIENT ELUTION CHROMATOGRAPHY

(a) Main program; the optimization of the slope (B) and of the initial concentration of the more efficient eluting component in the mobile phase at the start of the gradient (A) at a given shape (curvature) of the gradient. The resolution of two appropriately chosen sample compounds (1,2) with adjacent bands should be kept at a required level, while the retention volume of another suitably chosen compound [$V'_{R(g)}$] should be minimal. The main program consists in searching the minimum of the function $V'_{R(g)} = f(A,B)$. An interval of practically possible values of A is defined; $A > 0$; $A < A_{max}$ (A_{max} is the concentration of the more efficient eluting agent in the mobile phase, at which te resolution required is just achieved under isocratic conditions). This interval is subsequently narrowed in the course of calculation as follows: the interval is divided into four equal parts by values $A_T - A_V$ and, for each of these values, a corresponding B is calculated that is necessary to achieve the required resolution ($B_T - B_V$) and the corresponding values of $V'_{R(g)T} - V'_{R(g)V}$ are evaluated and compared. The comparison of the values of $V'_{R(g)}$ corresponding to these values of A is used for a subsequent reduction of the interval of values of A (the interval is halved in each step) until the minimal value of $V'_{R(g)}$ is found with a pre-set precision of the parameters A and B (ε_A, ε_B). The value from the centre of this interval represents the required solution for A with the corresponding value of B. The approach is based on the assumption that there is only one minimum of $V'_{R(g)}$ in the interval of the values of A; it is universal and can be applied with different $k' = f(c)$ functions (k'_0, $m \equiv$ parameters of these functions) and gradient functions (parameters A, B, κ). An estimate of the gradient slope (B_1) should be introduced among the input constants at the start of calculation. (b) Subroutine B: calculation of B as a function of A and $V'_{R(g)}$ at a required level of resolution for compounds 1 and 2; it generally requires an iteration calculation. X_1 and X_2 are used for retention volumes of sample compounds 1 and 2, respectively.

(a)

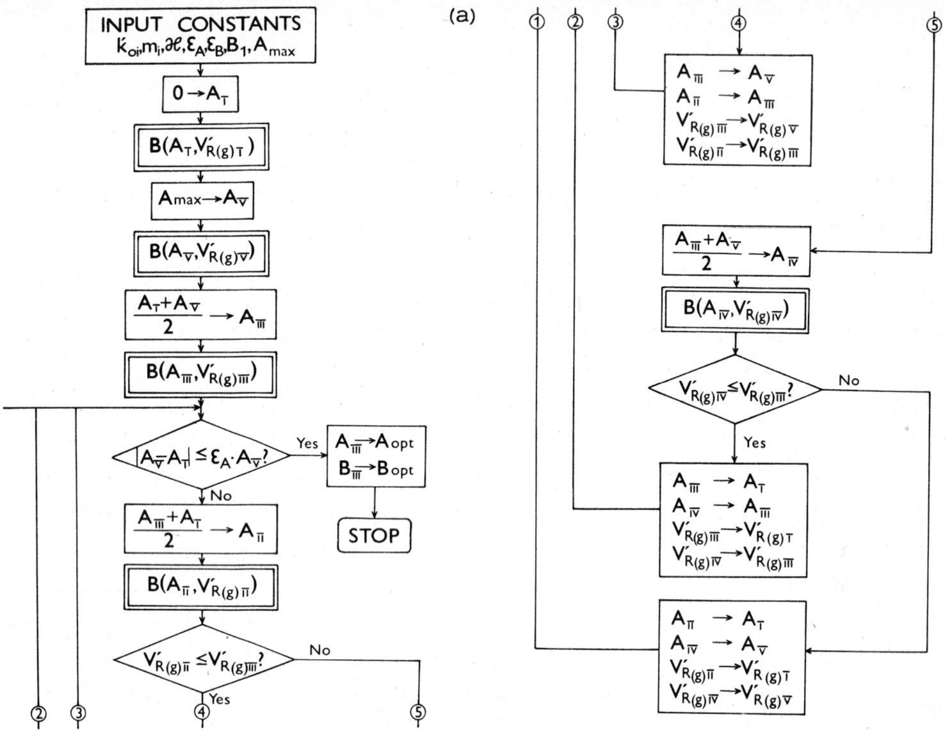

(b)

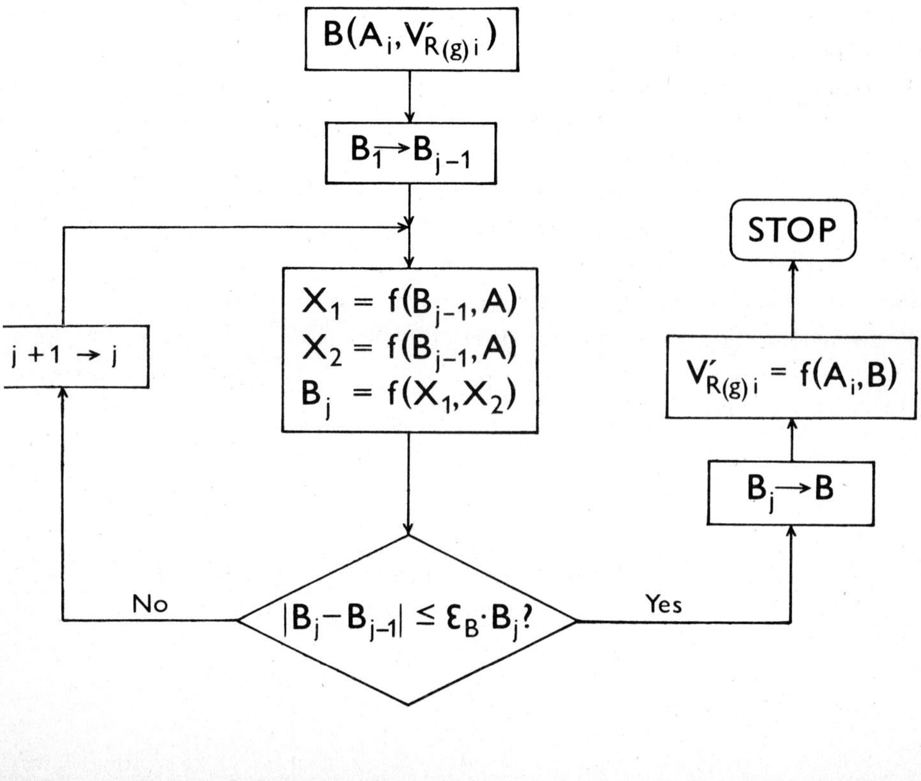
B(A_i, V'_R(g)i)
B_1 → B_j−1
X_1 = f(B_j−1, A)
X_2 = f(B_j−1, A)
B_j = f(X_1, X_2)
j + 1 → j
|B_j − B_j−1| ≤ ε_B · B_j?
No
Yes
B_j → B
V'_R(g)i = f(A_i, B)
STOP

Appendix 4

LIST OF SYMBOLS USED

a, b	The weaker (a) and the stronger (b) components of a two-component mobile phase.
a, b, m, p, k'_0	Different constants of the $k' = f(c)$ functions (Table 2.1).
$\bar{a}, \bar{b}$	Constants of the $\log V'_g = f(\log B)$ linear function.
c	Concentration of the stronger eluting agent in the binary mobile phase (molar fraction, volume fraction, molarity).
c_a, c_s	Concentration of the acid and of the salt, respectively, in the mobile phase in ion-exchange chromatography.
c_g	c at the end of the gradient.
$c_i[c_1, c_2, c_{(n-1)}, c_n]$	c in the step i [1, 2, $(n-1)$, n] in stepwise elution chromatography.
c_m, c_x	Constants in the equation for interaction indices.
c_{max}	c where maximum resolution for compounds 1 and 2 is achieved under isocratic conditions.
c_{mp}	Maximum proctically possible value of c in a mobile phase prepared from given two (or more) components.
c_0	c at the befinning of gradient elution; for two-chamber gradient devices c_0 is the initial concentration of solvent b in the mixing chamber close to the column.
c_t	Actual c at a given time t from the start of gradient elution.
c_x	Instantaneous concentration of the organic solvent x in the mobile phase at the outlet from the gradient mixer.
c_y	Instantaneous concentration of the organic solvent y in the mobile phase at the outlet from the gradient mixer.
c_A	Concentration of the pure component b in binary solvent A or instantaneous concentration of the mixed solvent A in the mobile phase at the outlet from the gradient mixer in ternary gradient elution making use of two reservoirs only.

c_{B}	Concentration of the pure component b in binary solvent B or instantaneous concentration of the mixed solvent B in the mobile phase at the outlet from the gradient mixer in ternary gradient elution making use of two reservoirs only.
$c_{\mathrm{A}x}$	Concentration of the organic solvent x in the mixed solvent (starting) A in reservoir 1 using ternary gradient elution in reversed-phase systems.
$c_{\mathrm{A}y}$	Concentration of the organic solvent y in the mixed solvent (starting) A in reservoir 1 using ternary gradient elution in reversed-phase systems.
$c_{\mathrm{B}x}$	Concentration of the organic solvent x in the mixed solvent B in reservoir 2 using ternary gradient elution in reversed-phase systems.
$c_{\mathrm{B}y}$	Concentration of the organic solvent y in the mixed solvent B in reservoir 2 using ternary gradient elution in reversed-phase systems.
c_{R}	Initial concentration of solvent b in the mixing chamber (or reservoir) more distant from the column for two-chamber gradient devices.
c_{RA}	Concentration of solvent b in an additional reservoir eventually connected to a two-chamber gradient device.
$c_{(\alpha=1)}$	c, where α = 1 for compounds 1 and 2 (isocratic conditions).
d	Inner diameter of the column.
d_p	Particle diameter of column packing material.
$f\ (P_{\mathrm{g}}/P_{\mathrm{c}})$	Increase factor for peak capacity in gradient elution chromatography in comparison with isocratic conditions.
h	Reduced height equivalent to a theoretical plate.
$i,\ q,\ s$	Charges of ions in ion-exchange chromatography.
k'	Capacity factor of sample solute.
k	Median capacity factor of sample solute during gradient elution.
k'_{a}	k' in pure solvent a (k' in the starting eluent of a linear solvent strength gradient).
k'_{b}	k' in pure solvent b.
k'_{ab}	k' in a binary mixture of solvents a and b.
k'_{f}	k'_i at L_i=L and t_i=t_{R}, i.e., instantaneous capacity factor at the time of elution of the peak maximum.
$k'_{\mathrm{g}} = V'_{\mathrm{g}}/V_{\mathrm{m}}$	Apparent capacity factor in gradient elution chromatography.

k'_i — Instantaneous capacity factor in gradient elution chromatography, corresponding to k' (isocratic) in the mobile phase of equal composition to the instantaneous composition of the mobile phase in contact with sample band maximum in the column at position L_i in time t_i; or k'_i $[k'_1, k'_2, k'_{(n-1)}, k'_n] = k'$ in step i $[1, 2\ (n-1)]$ in stepwise-elution chromatography.

$\bar{k}_o$ — Median value of k'_o in the ternary mobile phase under isocratic conditions.

$\bar{k}_{og}$ — Median value of k'_o in the course of gradient elution with a ternary solvent gradient.

k'_{ox}, k'_{oy}, m_x, m_y — Constants k'_o, m of the $k' = f(c)$ functions (eqn. 1.28) for the organic solvents x and y used in chromatography with ternary solvent gradients.

k'_A, k'_Z — k'_o for the first and the last compound, respectively, eluted in a gradient run.

$k'_{(A+B)}$ — k' in ternary mobile phases obtained by mixing of two solvent binaries, A and B, in the ratio $c_A : c_B$.

k'_I, k'_{II} — k' in mobile phases I and II, respectively.

n — Number of theoretical plates of the column.

n_b — Molecular area of the adsorbent surface occupied by a molecule of solvent b.

p', q' — Constants in eqn. 7.12.

p^* — Auxiliary factor for calculation of band compression factor (eqns. 4.39a-c)

p_s — Operating pressure.

Δp_s — Changes in p_s during the time period Δt.

$r = Li/L$ — Fraction of the column length migrated by the solute band maximum at time t_i.

r_i — Inner radius of the column.

t — Time.

t_c — Time from the start of the gradient elution necessary to achieve a given value of k'_c in the mobile phase at the inlet of the column (eqn. 4.4).

$\Delta t_{c/f}$ — Time difference necessary to change the composition of the mobile phase at the outlet of the column from the value corresponding to k'_c to that corresponding to k'_f (eqn. 7.9).

t_g — Retention time in gradient elution chromatography.

t_i — Time from the start of gradient elution before the elution of peak maximum, $t_i > 0$, $t_i < t_R$.

t_m	Retention time of an unretained compound.
t_L	Experimental analysis time.
t_M	Maximum acceptable analysis time.
t_R	Retention time.
t'_R	Corrected (net) retention time.
t_Z	Time delay of the gradient corresponding to V_Z.
u	Linear velocity of the mobile phase.
w	Bandwidth (in volume units) under isocratic conditions.
$w_{cor} = w_g \cdot G$	Bandwith in gradient elution chromatography corrected for band compression.
w_g	w in gradient elution chromatography.
w_{gi}	Contribution of the gradient step to bandwidth in gradient elution chromatography with an initial delay.
$w_i [w_n, w_{(n-1)}]$	w for a compound eluted in step i $[n, (n-1)]$ in stepwise elution chromatography.
x_b	Molar fraction of solvent b in the mobile phase.
y	Auxiliary quantity in eqn. 4.40a.
A, A', B, B', D, κ	Parameters of the gradient functions (Table 4.1) (D, A and A' characterize the initial concentration c_0 at the start of the gradient; B and B' represent gradient steepness, in unit concentration change of b per unit volume or unit time, respectively; κ is the gradient curvature parameter).
A, B	Solutions contained in reservoirs, from which the mobile phase gradient is formed (not necessarilly pure solvents a and b).
A, B, C	Terms in Van Deemter and Knox equations.
A_i, B	Weighting factors in eqns. 2.1 and 7.4.
A_s	Molecular area of the adsorbent surface occupied by a molecule of the solute.
A_x, A_y	Initial concentrations of solvents x and y, respectively, in elution with ternary gradients (constants of the linear gradient functions for each of the two solvents x and y, eqns. 6.5 and 6.6.
B_x, B_y	Slopes of the concentration change of the solvents x and y, respectively, in elution with ternary gradients (constants of the linear gradient functions for each of the solvents x and y, eqns. 6.5 and 6.6).
$B_{(1)}$	Gradient steepness B for $\alpha_g = 1$ (no resolution of the two sample compounds).

C	Capacitance of the flow-through capacitor used for measuring the flow-rate of the mobile phase.
COF	Chromatographic optimization function.
CRF	Chromatographic resolution function.
D_{m}	Diffusion coefficient of solute in the mobile phase.
$D_{\mathrm{H}_n\mathrm{B}}$	K_{o} of a nondissociated acid $\mathrm{H}_n\mathrm{B}$.
E	Compression criterion (length of the chromatogram in multiples of the relative part used for the separation of the first two sample compounds).
F_{m}	Volume flow-rate of the mobile phase.
G	Band compression factor.
H	Height equivalent to a theoretical plate.
I	Ionic strength of the mobile phase.
I_{i}	Interaction index.
I_{x}	I_i of the sample compound.
$I_{\mathrm{H_2O}}$	I_i of water.
I_{org}	I_i of the organic solvent in the mobile phase.
$K_{\mathrm{A}}^{\mathrm{B}}$ ($K_{\mathrm{L}}^{\mathrm{B}}$, $K_{\mathrm{M}}^{\mathrm{B}_m\mathrm{H}_b}$, $K_{\mathrm{A}}^{\mathrm{H}_i\mathrm{B}}$)	Selectivity constant in ion exchange.
K_i	Dissociation (activity) constant of a weak acid to the ith step.
$K_{\mathrm{M}_q\mathrm{B}}$	Complex formation constant for the complex $\mathrm{M}_q\mathrm{B}$.
L	Column length.
L_{A}, L_{B}	Widths of the orthogonal compartments (at the solvent level) used in two-compartment proportional volume gradient devices with a dividing flexible spacer foil.
L_i	Distance of the column through which the band maximum has migrated in time t_i.
N	Number of equal-volume reservoirs in the Varigrad system.
P_{c}	Peak capacity under isocratic conditions (eqn. 4.57).
P_{g}	Peak capacity in gradient elution chromatography (eqn. 4.58).
Q	Ion-exhange capacity of ion exchanger used for chromatography.
Q'	Effective capacity factor for gradient-elution chromatography according to Snyder and Saunders.
Q_1	Volume flow-rate of the solvent between the two mixing chambers in a two-chamber gradient device.
Q_2	Volume flow-rate of the solvent between the mixing chamber close to the column and the column for two-chamber device.

R	Universal gas constant.
R_s	Resolution of the two compounds 1 and 2 with adjacent bands.
R_g	R_s in gradient-elution chromatography.
$R_{g,o}$	Required optimum R_g.
R_i	Experimental R_s of the ith pair of sample compounds.
R_{id}	Desired R_s of the ith pair of sample compounds.
$R_{s_{min}}$, $R_{s_{max}}$	Limits of R_s on a given column using a binary mobile phase composed of given two (or more) components.
$S_{g/i}$	Ratio of the peak height in gradient elution chromatography to the peak height under isocratic conditions.
$\Delta S = \log k'_a - \log k'_b$	Difference between the logarithms of k' of solute in the two pure solvents, a and b used to prepare a binary gradient.
T	Absolute temperature (^oK).
T_c	Time of the duty cycle.
T_p	Time of the pump cycle.
V	Volume of the eluate.
V_c	Total (geometrical) inner volume of the void column.
V_C	Volume of the mixing chamber or of the flow-through capacitor.
ΔV_C	Change in V_C during the time period Δt.
V_{ei} (V_{e1}, V_{e2}, V_{en})	Volume of the mobile phase delivered on to the column in step i (1, 2, n) in stepwise elution chromatography.
V_g	V_R in gradient elution chromatography.
V'_g	V'_R in gradient elution chromatography.
V'_{gA}, V'_{gZ}	V'_g for the first and the last compound, respectively, eluted in a gradient run.
V'_{gi} [$V'_{g(n-1)}$, V'_{gn}]	Contribution of the step i [$(n-1)$, n] to the total net retention volume in stepwise elution chromatography and in chromatography with composed gradients.
V_G	Volume of the gradient $\equiv$ volume of the eluate from the start to the end of gradient elution.
V_m	Volume of the mobile phase in the column (column void volume).
V_{mi} (V_{m1}, V_{m2}, V_{mn})	Part of the column void volume migrated by a sample compound in step i (1, 2, n) in stepwise elution chromatography.
V_R	Retention volume.
V'_R	Corrected (net) retention volume.

$\Delta V'_g = V'_{g2} - V'_{g1}$	Difference in retention volumes between two sample compounds.
V_{R_A}, V_{R_Z}	V_R for the first and the last compound, respectively, eluted in an isocratic run.
V_s	Volume of the stationary phase in the column.
V_t	Total volume of solvents in all the N chambers of the Varigrad system.
V_z	Volume of the instrumentation causing gradient delay.
V_x	Molal volume of the sample compound.
V_1, V_2	Volume (total) of solvent reservoirs used to prepare the gradient in exponential or proportional gradient devices.
V_{01}, V_{02}	Volumes of the solutions in the reservoirs 1 and 2, respectively, at the start of the gradient elution.
α	Separation factor (retention ratio) for two compounds 1 and 2.
α_g	α in gradient elution chromatography.
$\alpha_0 = k'_{02}/k'_{01}$	α in the pure more polar (or, in the less polar) component of a binary mobile phase.
$\alpha_{0Z,A} = k'_{0Z}/k'_{0A}$	α_0 for the last eluted compound with respect to the first eluted compound.
$\alpha_{0B,A} = k'_{0B}/k'_{0A}$	α_0 for the second eluted compound with respect to the first eluted compound.
α'	Activity parameter of the adsorbent in adsorption chromatography.
β	Steepness of linear solvent strength gradient (eqn. 4.4).
δ	Solubility parameter.
δ_m	δ of the mobile phase.
δ_s	δ of the stationary phase.
δ_x	δ of the sample compound.
δ_a	δ of the pure solvent a.
δ_b	δ of the pure solvent b.
δ_{ab}	δ in a mixture of solvents a and b.
ε	Solvent strength parameter in adsorption chromatography.
ε_a	ε in the pure solvent a.
ε_b	ε in the pure solvent b.
ε_{ab}	ε in a mixture of solvents a and b.
ε_T	Total porosity of the packed column.
λ	Wavelength of the radiation used for detection.

ν	Reduced linear velocity of the mobile phase.
ρ_1	Initial density of the solution in the mixing chamber close to the column for two-chamber gradient device.
ρ_2	Initial density of the solution in the mixing chamber (or the reservoir) at a distance from the column for two-chamber gradient device.
σ	Band spreading in length units.
ϕ	Phase ratio in column [ratio of the volume of the stationary (V_s) and of the mobile (V_m) phases].
Φ	Instantaneous composition of the mobile phase in gradient elution chromatography expressed as % (v/v) $\cdot$ 10^{-2} of the mixed binary solution B in the mixture of two binary solvents A and B.

SUBJECT INDEX

A

C

D

E

F

H

I

N

Q

R

U

V

W